Fractional-Order Integral and Derivative Operators and Their Applications

Fractional-Order Integral and Derivative Operators and Their Applications

Editor

Hari Mohan Srivastava

MDPI • Basel • Beijing • Wuhan • Barcelona • Belgrade • Manchester • Tokyo • Cluj • Tianjin

Editor
Hari Mohan Srivastava
University of Victoria
Canada

Editorial Office
MDPI
St. Alban-Anlage 66
4052 Basel, Switzerland

This is a reprint of articles from the Special Issue published online in the open access journal *Mathematics* (ISSN 2227-7390) (available at: https://www.mdpi.com/journal/mathematics/special_issues/FOIDOTA).

For citation purposes, cite each article independently as indicated on the article page online and as indicated below:

LastName, A.A.; LastName, B.B.; LastName, C.C. Article Title. *Journal Name* **Year**, *Article Number*, Page Range.

ISBN 978-3-03936-650-7 (Hbk)
ISBN 978-3-03936-651-4 (PDF)

© 2020 by the authors. Articles in this book are Open Access and distributed under the Creative Commons Attribution (CC BY) license, which allows users to download, copy and build upon published articles, as long as the author and publisher are properly credited, which ensures maximum dissemination and a wider impact of our publications.

The book as a whole is distributed by MDPI under the terms and conditions of the Creative Commons license CC BY-NC-ND.

Contents

About the Editor vii

Hari Mohan Srivastava
Fractional-Order Integral and Derivative Operators and Their Applications
Reprinted from: *Mathematics* **2020**, *8*, 1016, doi:10.3390/math8061016 1

Yoritaka Iwata
Abstract Formulation of the Miura Transform
Reprinted from: *Mathematics* **2020**, *8*, 747, doi:10.3390/math8050747 5

Idris Ahmed, Poom Kumam, Kamal Shah, Piyachat Borisut, Kanokwan Sitthithakerngkiet and Musa Ahmed Demba
Stability Results for Implicit FractionalPantograph Differential Equations via ϕ-Hilfer Fractional Derivative with a NonlocalRiemann-Liouville Fractional Integral Condition
Reprinted from: *Mathematics* **2020**, *8*, 94, doi:10.3390/math8010094 13

Mohamed Jleli, Mokhtar Kirane and Bessem Samet
Absence of Global Solutions for a Fractional in Time and Space Shallow-Water System
Reprinted from: *Mathematics* **2019**, *7*, 1127, doi:10.3390/math7111127 35

Abdelghani Djeddi, Jalel Dib, Salem Abdelmalek and Ahmad Taher Azar
Fractional Order Unknown Inputs Fuzzy Observer for Takagi–Sugeno Systems with Unmeasurable Premise Variables
Reprinted from: *Mathematics* **2019**, *7*, 984, doi:10.3390/math7100984 47

Miraj Ul-Haq, Mohsan Raza, Muhammad Arif, Qaiser Khan, Huo Tang
q-Analogue of DifferentialSubordinations
Reprinted from: *Mathematics* **2019**, *7*, 724, doi:10.3390/math7080724 63

Hari M. Srivastava, Nazar Khan, Maslina Darus, Muhammad Tariq Rahim, Qazi Zahoor Ahmad * and Yousra Zeb
Properties of Spiral-Like Close-to-Convex Functions Associated with Conic Domains
Reprinted from: *Mathematics* **2019**, *7*, 706, doi:10.3390/math7080706 79

Carla M.A. Pinto, Ana R.M. Carvalho, Dumitru Baleanu and Hari M. Srivastava
Efficacy of the Post-Exposure Prophylaxis and of the HIV Latent Reservoir in HIV Infection
Reprinted from: *Mathematics* **2019**, *7*, 515, doi:10.3390/math7060515 91

Chengbiao Fu, Shu Gan, Xiping Yuan, Heigang Xiong and Anhong Tian
Impact of Fractional Calculus on Correlation Coefficient between Available Potassium and Spectrum Data in Ground Hyperspectral and Landsat 8 Image
Reprinted from: *Mathematics* **2019**, *7*, 488, doi:10.3390/math7060488 107

Hari M. Srivastava, Arran Fernandez, and Dumitru Baleanu
Some New Fractional-Calculus Connections between Mittag–Leffler Functions
Reprinted from: *Mathematics* **2019**, *7*, 485, doi:10.3390/math7060485 123

Mehmet Ali Özarslan and Ceren Ustaoğlu
Some Incomplete Hypergeometric Functions and Incomplete Riemann-LiouvilleFractional Integral Operators
Reprinted from: *Mathematics* **2019**, *7*, 483, doi:10.3390/math7050483 133

Ebrahim Analouei Adegani, Nak Eun Cho and Mostafa Jafari
Logarithmic Coefficients For Univalent Functions Defined by Subordination
Reprinted from: *Mathematics* 2019, 7, 408, doi:10.3390/math7050408 151

Hai-Yan Zhang, Rekha Srivastava and Huo Tang
Third-Order Hankel and Toeplitz Determinants for Starlike Functions Connected with the Sine Function
Reprinted from: *Mathematics* 2019, 7, 404, doi:10.3390/math7050404 163

Shuqin Zhang and Lei Hu
Unique Existence Result of Approximate Solution to Initial Value Problem for Fractional Differential Equation of Variable Order Involving the Derivative Arguments on the Half-Axis
Reprinted from: *Mathematics* 2019, 7, 286, doi:10.3390/math7030286 173

Saïd Abbas, Nassir Al Arifi, Mouffak Benchohra and Yong Zhou
Random Coupled Hilfer and Hadamard Fractional Differential Systems in Generalized Banach Spaces
Reprinted from: *Mathematics* 2019, 7, 285, doi:10.3390/math7030285 197

Onur Alp İlhan, Shakirbay G. Kasimov, Shonazar Q. Otaev and Haci Mehmet Baskonus
On the Solvability of a Mixed Problem for a High-Order Partial Differential Equation with Fractional Derivatives with Respect to Time, with Laplace Operators with Spatial Variables and Nonlocal Boundary Conditions in Sobolev Classes
Reprinted from: *Mathematics* 2019, 7, 235, doi:10.3390/math7030235 213

Harendra Singh, Rajesh K. Pandey and Hari Mohan Srivastava
Solving Non-Linear Fractional Variational Problems Using Jacobi Polynomials
Reprinted from: *Mathematics* 2019, 7, 224, doi:10.3390/math7030224 233

Lakshman Mahto, Syed Abbas, Mokhtar Hafayed and Hari M. Srivastava
Approximate Controllability of Sub-Diffusion Equation with Impulsive Condition
Reprinted from: *Mathematics* 2019, 7, 190, doi:10.3390/math7020190 257

Shuman Meng, and Yujun Cui
The Extremal Solution To Conformable Fractional Differential Equations Involving Integral Boundary Condition
Reprinted from: *Mathematics* 2019, 7, 186, doi:10.3390/math7020186 273

Hari M. Srivastava, Qazi Zahoor Ahmad, Nasir Khan, Nazar Khan and Bilal Khan
Hankel and Toeplitz Determinants for a Subclass of q-Starlike Functions Associated with a General Conic Domain
Reprinted from: *Mathematics* 2019, 7, 181, doi:10.3390/math7020181 283

Aydin Secer, Neslihan Ozdemir and Mustafa Bayram
A Hermite Polynomial Approach for Solving the SIR Model of Epidemics
Reprinted from: *Mathematics* 2018, 6, 305, doi:10.3390/math6120305 299

Mamoru Nunokawa, Janusz Sokół, Nak Eun Cho
On a Length Problem for Univalent Functions
Reprinted from: *Mathematics* 2018, 6, 266, doi:10.3390/math6110266 311

Aydin Secer and Selvi Altun
A New Operational Matrix of Fractional Derivatives to Solve Systems of Fractional Differential Equations via Legendre Wavelets
Reprinted from: *Mathematics* 2018, 6, 238, doi:10.3390/math6110238 319

About the Editor

Hari Mohan Srivastava has held the position of Professor Emeritus in the Department of Mathematics and Statistics at the University of Victoria in Canada since 2006, having joined the faculty there in 1969, first as an Associate Professor (1969–1974) and then as a Full Professor (1974–2006). He began his university-level teaching career right after having received his MSc degree in 1959 at 19 years of age, from the University of Allahabad in India. He compelted his PhD in 1965 while he was a full-time member of the teaching faculty at the Jai Narain Vyas University of Jodhpur in India. He holds numerous visiting research and honorary chair positions at many universities and research institutes in different parts of the world. Having received several DSc (honoris causa) degrees, as well as honorary memberships and honorary fellowships from many scientific academies and learned societies around the world, he is also actively associated editorially with numerous international scientific research journals. His current research interests include several areas of pure and applied mathematical sciences, such as real and complex analysis, fractional calculus and its applications, integral equations and transforms, higher transcendental functions and their applications, q-series and q-polynomials, analytic number theory, analytic and geometric inequalities, probability and statistics, and inventory modelling and optimization. He has published 33 books, monographs, and edited volumes, 36 book (and encyclopedia) chapters, 48 papers in international conference proceedings, and more than 1300 scientific research articles in peer-reviewed international journals, as well as the forewords, editorials and prefaces to many books and journals, etc. He is a Clarivate Analytics (Thomson-Reuters) (Web of Science) Highly Cited Researcher. For further details on his other professional achievements and scholarly accomplishments, as well as honors, awards, and distinctions, including lists of his most recent publications, such as journal articles, books, monographs and edited volumes, book chapters, encyclopedia chapters, papers in conference proceedings, forewords to books and journals, etc., please see: http://www.math.uvic.ca/~harimsri/.

Editorial

Fractional-Order Integral and Derivative Operators and Their Applications

Hari Mohan Srivastava [1,2,3]

1. Department of Mathematics and Statistics, University of Victoria, Victoria, BC V8W 3R4, Canada; harimsri@math.uvic.ca
2. Department of Medical Research, China Medical University Hospital, China Medical University, Taichung 40402, Taiwan
3. Department of Mathematics and Informatics, Azerbaijan University, 71 Jeyhun Hajibeyli Street, AZ1007 Baku, Azerbaijan

Received: 17 June 2020; Accepted: 17 June 2020; Published: 22 June 2020

The present volume contains the invited, accepted and published submissions (see [1–22]) to a Special Issue of the MDPI's journal, *Mathematics*, on the subject-area of "Fractional-Order Integral and Derivative Operators and Their Applications". Three successful predecessors of this volume happens to be the Special Issue of the MDPI's journal, *Mathematics*, on the subject-areas of "Recent Advances in Fractional Calculus and Its Applications", "Recent Developments in the Theory and Applications of Fractional Calculus" (see, for details, [23]) and "Operators of Fractional Calculus and Their Applications". In fact, encouraged by the noteworthy successes of this series of four Special Issues, as well as of (for example) two other Special Issues of *Axioms*, on the subject-areas of "Mathematical Analysis and Applications" and "Mathematical Analysis and Applications II", *Axioms* has already started the publication of a Topical Collection, entitled "Mathematical Analysis and Applications" (Collection Editor: H. M. Srivastava), with an open submission deadline. The interested reader should refer to and read the book format of several of these Special Issues (Guest Editor: H. M. Srivastava), which are cited below (see [23–26]).

In recent years, various families of fractional-order integral and derivative operators, such as those named after Riemann-Liouville, Weyl, Hadamard, Grunwald-Letnikov, Riesz, Erdelyi-Kober, Liouville-Caputo, and so on, have been found to be remarkably important and fruitful, due mainly to their demonstrated applications in numerous seemingly diverse and widespread areas of the mathematical, physical, chemical, engineering, and statistical sciences. Many of these fractional-order operators provide interesting, potentially useful tools for solving ordinary and partial differential equations, as well as integral, differintegral, and integro-differential equations; fractional-calculus analogues and extensions of each of these equations; and various other problems involving special functions of mathematical physics and applied mathematics, as well as their extensions and generalizations in one or more variables.

In this Special Issue, we invited and welcomed review, expository, and original research articles dealing with the recent advances in the theory of fractional-order integral and derivative operators and their multidisciplinary applications.

The suggested topics of interest for the call of papers for this Special Issue included, but by no means limited to, the following keywords:

- Operators of fractional calculus and their applications;
- Chaos and fractional dynamics;
- Fractional-order ODEs and PDEs;
- Fractional-order differintegral equations;
- Fractional-order integro-differential equations;
- Fractional-order integrals and fractional-order derivatives associated with special functions of mathematical physics and applied mathematics;

- Identities and inequalities involving fractional-order integrals and fractional-order derivatives;
- Dynamical systems based upon fractional calculus.

Here, in this Editorial, we choose first to briefly describe the status of the Special Issue as follows:

Papers included in this volume deal extensively with various theoretical as well applied topics of fractional calculus and its applications of current research interests. Some of the notable contributions in this volume happen to have successfully addressed such topics of fractional calculus and related mathematical analysis as (for example) operational matrix of fractional-order derivatives for solving systems of fractional differential equations via Legendre wavelets, Hermite polynomial approach for solving the SIR model of epidemics, the extremal solution to conformable fractional differential equations involving integral boundary condition, approximate controllability of sub-diffusion equation with impulsive condition, incomplete hypergeometric functions and incomplete Riemann-Liouville fractional integral operators, random coupled Hilfer and Hadamard fractional differential systems in generalized Banach spaces, uniqueness and existence of approximate solution to initial value problem for fractional differential equation of variable order involving the derivative arguments on the half-axis, solvability of a mixed problem for a high-order partial differential equation with fractional derivatives with respect to time, with Laplace operators with spatial variables and nonlocal boundary conditions in Sobolev classes, fractional-calculus connections between Mittag–Leffler functions, impact of fractional calculus on correlation coefficient between available potassium and spectrum data in ground hyperspectral and Landsat 8 image, efficacy of the post-exposure prophylaxis and of the HIV latent reservoir in HIV infection, fractional-order unknown inputs fuzzy observer for Takagi–Sugeno systems with unmeasurable premise variables, stability results for implicit fractional pantograph differential equations via φ-Hilfer fractional derivative with a nonlocal Riemann-Liouville fractional integral condition, and so on. In connection with such works as (for example) [4,18], and indeed also many papers included in the published volumes [23–26], the recent survey-cum-expository review articles [27,28] will be potentially useful in order to motivate further researches and developments involving a wide variety of operators of basic (or q-) calculus and fractional q-calculus and their widespread applications in Geometric Function Theory of Complex Analysis.

I take this opportunity to thank all of the participating authors, and the referees and the peer-reviewers, for their *invaluable* contributions toward the remarkable success of each of the above-mentioned Special Issues. I do also greatly appreciate the editorial and managerial help and assistance provided efficiently and generously by Ms. Grace Wang and Ms. Cynthia Chen, and also many of their colleagues and associates in the Editorial Office of *Mathematics*.

Funding: This research received no external funding.

Conflicts of Interest: The author declares no conflict of interest.

References

1. Secer, A.; Altun, S. A New Operational Matrix of Fractional Derivatives to Solve Systems of Fractional Differential Equations via Legendre Wavelets. *Mathematics* **2018**, *6*, 238. [CrossRef]
2. Nunokawa, M.; Sokół, J.; Cho, N.E. On a Length Problem for Univalent Functions. *Mathematics* **2018**, *6*, 266. [CrossRef]
3. Secer, A.; Ozdemir, N.; Bayram, M. A Hermite Polynomial Approach for Solving the SIR Model of Epidemics. *Mathematics* **2018**, *6*, 305. [CrossRef]
4. Srivastava, H.M.; Ahmad, Q.Z.; Khan, N.; Khan, N.; Khan, B. Hankel and Toeplitz Determinants for a Subclass of q-Starlike Functions Associated with a General Conic Domain. *Mathematics* **2019**, *7*, 181. [CrossRef]
5. Meng, S.; Cui, Y. The Extremal Solution To Conformable Fractional Differential Equations Involving Integral Boundary Condition. *Mathematics* **2019**, *7*, 186. [CrossRef]
6. Mahto, L.; Abbas, S.; Hafayed, M.; Srivastava, H.M. Approximate Controllability of Sub-Diffusion Equation with Impulsive Condition. *Mathematics* **2019**, *7*, 190. [CrossRef]

7. Singh, H.; Pandey, R.K.; Srivastava, H.M. Solving Non-Linear Fractional Variational Problems Using Jacobi Polynomials. *Mathematics* **2019**, *7*, 224. [CrossRef]
8. İlhan, O.A.; Kasimov, S.G.; Otaev, S.Q.; Baskonus, H.M. On the Solvability of a Mixed Problem for a High-Order Partial Differential Equation with Fractional Derivatives with Respect to Time, with Laplace Operators with Spatial Variables and Nonlocal Boundary Conditions in Sobolev Classes. *Mathematics* **2019**, *7*, 235. [CrossRef]
9. Abbas, S.; Arifi, N.A.; Benchohra, M.; Zhou, Y. Random Coupled Hilfer and Hadamard Fractional Differential Systems in Generalized Banach Spaces. *Mathematics* **2019**, *7*, 285. [CrossRef]
10. Zhang, S.; Hu, L. Unique Existence Result of Approximate Solution to Initial Value Problem for Fractional Differential Equation of Variable Order Involving the Derivative Arguments on the Half-Axis. *Mathematics* **2019**, *7*, 286. [CrossRef]
11. Zhang, H.-Y.; Srivastava, R.; Tang, H. Third-Order Hankel and Toeplitz Determinants for Starlike Functions Connected with the Sine Function. *Mathematics* **2019**, *7*, 404. [CrossRef]
12. Adegani, E.A.; Cho, N.E.; Jafari, M. Logarithmic Coefficients for Univalent Functions Defined by Subordination. *Mathematics* **2019**, *7*, 408. [CrossRef]
13. Özarslan, M.A.; Ustaoğlu, C. Some Incomplete Hypergeometric Functions and Incomplete Riemann-Liouville Fractional Integral Operators. *Mathematics* **2019**, *7*, 483. [CrossRef]
14. Srivastava, H.M.; Fernandez, A.; Baleanu, D. Some New Fractional-Calculus Connections between Mittag–Leffler Functions. *Mathematics* **2019**, *7*, 485. [CrossRef]
15. Fu, C.; Gan, S.; Yuan, X.; Xiong, H.; Tian, A. Impact of Fractional Calculus on Correlation Coefficient between Available Potassium and Spectrum Data in Ground Hyperspectral and Landsat 8 Image. *Mathematics* **2019**, *7*, 488. [CrossRef]
16. Pinto, C.M.A.; Carvalho, A.R.M.; Baleanu, D.; Srivastava, H.M. Efficacy of the Post-Exposure Prophylaxis and of the HIV Latent Reservoir in HIV Infection. *Mathematics* **2019**, *7*, 515. [CrossRef]
17. Srivastava, H.M.; Khan, N.; Darus, M.; Rahim, M.T.; Ahmad, Q.Z.; Zeb, Y. Properties of Spiral-Like Close-to-Convex Functions Associated with Conic Domains. *Mathematics* **2019**, *7*, 706. [CrossRef]
18. Ul-Haq, M.; Raza, M.; Arif, M.; Khan, Q.; Tang, H. q-Analogue of Differential Subordinations. *Mathematics* **2019**, *7*, 724. [CrossRef]
19. Djeddi, A.; Dib, D.; Azar, A.T.; Abdelmalek, S. Fractional Order Unknown Inputs Fuzzy Observer for Takagi–Sugeno Systems with Unmeasurable Premise Variables. *Mathematics* **2019**, *7*, 984. [CrossRef]
20. Jleli, M.; Kirane, M.; Samet, B. Absence of Global Solutions for a Fractional in Time and Space Shallow-Water System. *Mathematics* **2019**, *7*, 1127. [CrossRef]
21. Ahmed, I.; Kumam, P.; Shah, K.; Borisut, P.; Sitthithakerngkiet, K.; Ahmed Demba, M. Stability Results for Implicit Fractional Pantograph Differential Equations via φ-Hilfer Fractional Derivative with a Nonlocal Riemann-Liouville Fractional Integral Condition. *Mathematics* **2020**, *8*, 94. [CrossRef]
22. Iwata, Y. Abstract Formulation of the Miura Transform. *Mathematics* **2020**, *8*, 747. [CrossRef]
23. Srivastava, H.M. *Mathematical Analysis and Applications*; Printed Edition of the Special Issue "Mathematical Analysis and Applications"; Published in *Axioms*; MDPI Publishers: Basel, Switzerland, 2019; p. viii, 209p; ISBN: 978-3-03897-400-0 (Pbk); ISBN: 978-3-03897-401-7 (PDF).
24. Srivastava, H.M. *Integral Transforms and Operational Calculus*; Printed Edition of the Special Issue "Integral Transforms and Operational Calculus"; Published in *Symmetry*; MDPI Publishers: Basel, Switzerland, 2019; ISBN: 978-3-03921-618-5 (Pbk); ISBN: 978-3-03921-619-2 (PDF).
25. Srivastava, H.M. *Operators of Fractional Calculus and Their Applications*; Printed Edition of the Special Issue "Operators of Fractional Calculus and Their Applications"; Published in *Mathematics*; MDPI Publishers: Basel, Switzerland, 2019; p. viii, 125p; ISBN: 978-3-03897-340-9 (Pbk); ISBN: 978-3-03897-341-6 (PDF).

26. Srivastava, H.M. *Mathematical Analysis and Applications II*; Printed Edition of the Special Issue "Mathematical Analysis and Applications II"; Published in *Axioms*; MDPI Publishers: Basel, Switzerland, 2020; p. viii, 215p; ISBN 978-3-03928-384-2 (Pbk); ISBN 978-3-03928-385-9 (PDF).
27. Srivastava, H.M. Operators of Basic (or q-) Calculus and Fractional q-Calculus and Their Applications in Geometric Function Theory of Complex Analysis. *Iran. J. Sci. Technol. Trans. A Sci.* **2020**, *44*, 327–344. [CrossRef]
28. Srivastava, H.M. Fractional-Order Derivatives and Integrals: Introductory Overview and Recent Developments. *Kyungpook Math. J.* **2020**, *60*, 73–116, doi:10.5666/KMJ.2020.60.1.73.

© 2020 by the author. Licensee MDPI, Basel, Switzerland. This article is an open access article distributed under the terms and conditions of the Creative Commons Attribution (CC BY) license (http://creativecommons.org/licenses/by/4.0/).

Article
Abstract Formulation of the Miura Transform †

Yoritaka Iwata

Faculty of Chemistry, Materials and Bioengineering, Kansai University, Osaka 564-8680, Japan;
iwata_phys@08.alumni.u-tokyo.ac.jp
† This article has been presented in ICRAAM 2020 Conference, Kuala Lumper, Malaysia, 4–6 February 2020.

Received: 10 March 2020; Accepted: 29 April 2020; Published: 8 May 2020

Abstract: Miura transform is known as the transformation between Korweg de-Vries equation and modified Korweg de-Vries equation. Its formal similarity to the Cole-Hopf transform has been noticed. This fact sheds light on the logarithmic type transformations as an origin of a certain kind of nonlinearity in the soliton equations. In this article, based on the logarithmic representation of operators in infinite-dimensional Banach spaces, a structure common to both Miura and Cole-Hopf transforms is discussed. In conclusion, the Miura transform is generalized as the transform in abstract Banach spaces, and it is applied to the higher order abstract evolution equations.

Keywords: Miura transform; soliton equations; logarithm

1. Introduction

The Korteweg-de-Vries equation (KdV equation, for short) and the modified Korweg de-Vries equation (mKdV equation, for short) are known as nonlinear equations holding the soliton solutions. Let u and v be the solutions of the KdV equation and mKdV equation, respectively. Let functions u and v be the general solutions that satisfy

$$[\text{KdV}] \quad \partial_t u - 6u \partial_x u + \partial_x^3 u = 0,$$

$$[\text{mKdV}] \quad \partial_t v - 6v^2 \partial_x v + \partial_x^3 v = 0$$

without identifying the details such as the initial and boundary conditions of the mixed problem. For a recent result associated with the well-posedness of the KdV equations, the existence and uniqueness of the solution of semilinear KdV equations in non-parabolic domain is obtained in [1] by using the parabolic regularization method, the Faedo-Galerkin method, and the approximation of a non-parabolic domain by a sequence of regularizable subdomains. Meanwhile, interesting studies on the family of KdV-type equations have been recently carried out in [2]. Let a set of all the real numbers be denoted by $\mathbb{R}$. Although u and v are functions of $t \in \mathbb{R}$ and $x \in \mathbb{R}$, they are not apparently shown if there is no confusion. The Miura transform [3] $\mathcal{M} : u \to v$ reads

$$u = \partial_x v + v^2, \tag{1}$$

which is formally the same as the Riccati's differential equation of a variable x if u is assumed to be a known function. In this article, the Miura transform is generalized as the transform in the abstract spaces. The essence of several nonlinear transforms are pined downed within the theory of abstract equations defined in a general Banach spaces. In conclusion, the structure of the general solutions of second order abstract evolution equations are presented in association with the Miura transform.

2. Operator Logarithm as Nonlinear Transform

2.1. Nonlinear Transform Associated with the Riccati's Equation

Following the method for solving the Riccati's equation, the logarithmic type transform appears as

$$v = \psi^{-1} \partial_x \psi, \qquad (2)$$

which corresponds to $v = \partial_x \log \psi$ if $\log \psi$ and its derivative are well-defined. This is formally the same as the Cole-Hopf transform [4–6]. By applying this transform to the Miura transform, the Miura transform is written by

$$\begin{aligned} u &= \partial_x v + v^2 \\ &= \partial_x(\psi^{-1}\partial_x\psi) + (\psi^{-1}\partial_x\psi)^2 \\ &= -\psi^{-2}(\partial_x\psi)^2 + \psi^{-1}(\partial_x^2\psi) + \psi^{-1}(\partial_x\psi)\psi^{-1}(\partial_x\psi). \end{aligned} \qquad (3)$$

If ψ and $\partial_x \psi$ commute,

$$\begin{aligned} u &= -(\partial_x\psi)^2\psi^{-2} + (\partial_x^2\psi)\psi^{-1} + (\partial_x\psi)^2\psi^{-2} = (\partial_x^2\psi)\psi^{-1} \\ \Leftrightarrow\ \partial_x^2\psi &= u\psi. \end{aligned} \qquad (4)$$

This is the second order evolution equation in which u plays a role of infinitesimal generator, and the evolution direction is fixed to x. It is remarkable that, after the combination with the Cole-Hopf transform, the Miura transform $\mathcal{M} : u \to \psi$ is a transform between nonlinear KdV equation and linear equation. In other words, it provides the transform between the evolution operator and its infinitesimal generator. In the following the obtained transform from u to ψ is called the combined Miura transform.

2.2. Miura Transform and Cole-Hopf Transform

It is worth differentiating Equation (2) for clarifying the identity of the Miura transform. Under the commutation assumption, the formal calculation without taking the differentiability into account leads to

$$\partial_x^2(\log \psi) := \partial_x(\psi^{-1}\partial_x\psi) = (\partial_x^2\psi)\psi^{-1} - (\psi^{-1}\partial_x\psi)^2, \qquad (5)$$

where the first term of the right hand side corresponds to the combined Miura transform, and the second term of the right hand side is the square of the Cole-Hopf transform. It simply means that $\partial_x^2(\log \psi)$ being defined by the right hand side of (5) can be defined by the combined Miura transform and Cole-Hopf transform simultaneously. As is well known in the theory of integrable systems, $\partial_x^2(\log \psi)$ corresponds to one typical type of Hirota's methods [7], thus a typical type of linear to nonlinear transformation. This type is known to be associated with the Bäcklund transform and KP theory (for a textbook, see [8]). That is, the second order derivative can be represented by the two transforms.

2.3. Logarithmic Representation of Infinitesimal Generators

Let X be a Banach space, $B(X)$ is a set of bounded linear operators on X, and Y be a dense Banach subspace of X. The Cauchy problem for the first order abstract evolution equation of hyperbolic type [9,10] is defined by

$$\begin{aligned} du(t)/dt - A(t)u(t) &= f(t), \qquad t \in [0,T] \\ u(0) &= u_0 \end{aligned} \qquad (6)$$

in X, where $A(t) : Y \to X$ is assumed to be the infinitesimal generator of the evolution operator $U(t,s) \in B(X)$ satisfying the strong continuity and the semigroup property:

$$U(t,s) = U(t,r)U(r,s)$$

for $0 \le s \le r \le t < T$. $U(t,s)$ is a two-parameter C_0-semigroup of operator (for definition, see [11–13]) that is a generalization of one-parameter C_0-semigroup and therefore an abstract generalization of the exponential function of operator. If $A(t)$ is confirmed to be an infinitesimal generator, then the solution $u(t)$ is represented by $u(t) = U(t,s)u_s$ with $u_s \in X$ for a certain $0 \le s \le T$ (cf. Hille-Yosida Theorem; for example, see [11–13]). Then, for a certain complex number κ, the alternative bounded infinitesimal generator $a(t,s) = \mathrm{Log}(U(t,s) + \kappa I)$ to $A(t)$ is well defined [14,15], where Log denotes the principal branch of logarithm.

Lemma 1 (Logarithmic representation of infinitesimal generators [14]). *Let t and s satisfy $0 \le t, s \le T$, and Y be a dense subspace of X. Let $a(t,s) \in B(X)$ be defined by $a(t,s) = \mathrm{Log}U(t,s)$. If $A(t)$ and $U(t,s)$ commute, infinitesimal generators $\{A(t)\}_{0 \le t \le T}$ are represented by means of the logarithm function; there exists a certain complex number $\kappa \ne 0$ such that*

$$A(t)\, u = (I - \kappa e^{-a(t,s)})^{-1}\, \partial_t a(t,s)\, u, \tag{7}$$

where u is an element of a dense subspace Y of X, and the logarithm of operator is defined by the Riesz-Dunford integral.

Proof. Only formal discussion is given here (for the detail, see [14,16]). Since $a(t)$ is defined by $a(t,s) = \mathrm{Log}(U(t,s) + \kappa I)$, $\partial_t a(t,s) = (U(t,s) + \kappa I)^{-1}\partial_t U(t,s)$,

$$(U(t,s) + \kappa I)\partial_t a(t,s) = (U(t,s) + \kappa I)(U(t,s) + \kappa I)^{-1}\partial_t U(t,s).$$

Under the commutation relation between $U(t,s)$ and $A(t)$,

$$\begin{aligned}
A(t)\, u &:= \partial_t \mathrm{Log}U(t,s)u \\
&= U(t,s)^{-1}\partial_t U(t,s)u \\
&= U(t,s)^{-1}(U(t,s) + \kappa I)\partial_t a(t,s)\, u \\
&= (I + \kappa(e^{a(t,s)} - \kappa I)^{-1})\partial_t a(t,s)\, u \\
&= (e^{a(t,s)} - \kappa I + \kappa I)(e^{a(t,s)} - \kappa I)^{-1}\partial_t a(t,s)\, u \\
&= (I - \kappa e^{-a(t,s)})^{-1}\partial_t a(t,s)\, u,
\end{aligned}$$

where u is an element in Y. □

The commutation assumption is trivially satisfied if $A(t)$ is independent of t. Equation (7) is the logarithmic representation of infinitesimal generator $A(t)$. This representation is the generalization of the Cole-Hopf transform [4,5].

3. Main Result

3.1. Generalization of Miura Transform

Let $\mathcal{X}$ be a Banach space, and $\mathcal{Y}$ be a dense Banach subspace of $\mathcal{X}$. By focusing on establishing the definition of infinitesimal generator, the discussion is limited to the autonomous case. The Cauchy problem for the second order abstract evolution equation of hyperbolic type (for example, see [17]) is defined by

$$d^2 u(t)/dt^2 - \mathcal{A}(t) u(t) = 0, \qquad t \in [0, T]$$

$$u(0) = u_0 \tag{8}$$

in $\mathcal{X}$ and $\mathcal{A}(t) : \mathcal{Y} \to \mathcal{X}$. Let the Cauchy problem be solvable; i.e., it admits the well-defined evolution operator $\mathcal{U}(t,s) \in B(\mathcal{X})$ satisfying the strong continuity and the semigroup property:

$$\mathcal{U}(t,s) = \mathcal{U}(t,r) \mathcal{U}(r,s)$$

for $0 \le s \le r \le t < T$. The solution is represented by $u(t) = \mathcal{U}(t,s) u_s$ with $u_s \in \mathcal{X}$ for a certain $0 \le s \le T$. For the second order equation, $\mathcal{U}(t,s)$ is not equal to the abstraction of $\exp(\int \mathcal{A}(t) dt)$, so that $\mathcal{A}(t)$ is not the infinitesimal generator of $\mathcal{U}(t,s)$, and $\mathcal{A}(t)$ is the infinitesimal generator of $\exp(\int \mathcal{A}(t) dt)$ instead. In the following, the combined Miura transform is shown to be equivalent to the logarithmic representation for the infinitesimal generator of the second order abstract evolution equations.

The master equation of (8) is also written as a system of equations:

$$\begin{cases} du(t)/dt - v(t) = 0, \\ dv(t)/dt - \mathcal{A}(t) u(t) = 0, \end{cases} \tag{9}$$

where, by focusing on the representation of $v(t)$, $v(t)$ is formally represented by $v(t) = \partial_t \mathcal{U}(t,s) v_s$ for a certain $v_s \in D(\partial_t \mathcal{U}(t,s))$ that is compatible with the original $u(t) = \mathcal{U}(t,s) u_s$ for a certain $u_s \in D(\mathcal{U}(t,s)) = \mathcal{X}$.

Lemma 2 (Logarithmic representation of the derivative). *Let κ be a certain complex number. For the evolution operator of Equation (8), let $\mathcal{U}(t,s)$ be included in the C^1 class in terms of variables t and s, and the first order derivative $\mathcal{V}(t,s) := \partial_t \mathcal{U}(t,s)$ be further assumed to be bounded on $\mathcal{X}$ and strongly continuous for $0 \le t, s \le T$. Then, for $\mathcal{V}(t,s)$,*

$$\partial_t \mathrm{Log} \mathcal{V}(t,s) := (I - \kappa e^{-\hat{\alpha}(t,s)})^{-1} \partial_t \hat{\alpha}(t,s) \tag{10}$$

is well defined if $\mathcal{V}(t,s)$ and $\partial_t \mathcal{V}(t,s)$ commute, where $\hat{\alpha}(t,s) : \partial_t \mathcal{U}(t,s) \to \mathcal{X}$ is an operator defined by $\hat{\alpha}(t,s) = \mathrm{Log}(\mathcal{V}(t,s) + \kappa I)$.

Proof. The statement follows from Lemma 1. □

On the other hand, another logarithmic representation

$$\partial_t \mathrm{Log} \mathcal{U}(t,s) := (I - \kappa e^{-\alpha(t,s)})^{-1} \partial_t \alpha(t,s) \tag{11}$$

is trivially well-defined by the assumption. According to the representations (10) and (11), the abstract version of the Miura transform is obtained as the product of two logarithmic representations.

Theorem 1 (Abstract formulation of the Miura transform). *Let t and s satisfy $0 \le t, s \le T$, κ be a certain complex number, and $\mathcal{Y}$ and $\mathcal{Y}'$ be a dense subspace of $\mathcal{X}$. The operator $\partial_t \alpha(t,s)$ is assumed to be a closed operator from $\mathcal{Y}$ to $\mathcal{X}$, and $\partial_t \hat{\alpha}(t,s)$ is assumed to be a closed operator from $\mathcal{Y}'$ to $\mathcal{X}$. If $\mathcal{U}(t,s)$, $\partial_t \mathcal{U}(t,s)$, and $\partial_t^2 \mathcal{U}(t,s)$ commute with each other within a properly given domain space, the operators $\{\mathcal{A}(t)\}_{0 \le t \le T}$ are represented by means of the logarithm function; there exists a certain complex number $\kappa \ne 0$ such that*

$$\mathcal{A}(t) u = (I - \kappa e^{-\hat{\alpha}(t,s)})^{-1} \partial_t \hat{\alpha}(t,s) \, (I - \kappa e^{-\alpha(t,s)})^{-1} \partial_t \alpha(t,s) \, u, \tag{12}$$

for an element u of a dense subspace $\{ u \in \mathcal{Y}; \; (I - \kappa e^{-\alpha(t,s)})^{-1} \partial_t \alpha(t,s) u \subset D(\partial_t \hat{\alpha}(t,s)) \}$ of $\mathcal{X}$, and

$$\mathcal{A}(t)\, \hat{u} = (I - \kappa e^{-\alpha(t,s)})^{-1} \partial_t \alpha(t,s)(I - \kappa e^{-\hat{\alpha}(t,s)})^{-1} \partial_t \hat{\alpha}(t,s)\, \hat{u}, \tag{13}$$

for an element $\hat{u}$ of a dense subspace $\{ \hat{u} \in \mathcal{Y}'; \; (I - \kappa e^{-\hat{\alpha}(t,s)})^{-1} \partial_t \hat{\alpha}(t,s) \hat{u} \subset D(\partial_t \alpha(t,s)) \}$ of $\mathcal{X}$.

Proof. The autonomous Equation (8), which corresponds to the abstract form of the combined Miura transform (4), is written by

$$\partial_t^2 \mathcal{U}(t,s) u = \mathcal{A}(t) \mathcal{U}(t,s) u$$

for any $u \in \mathcal{X}$, so that it follows that $\partial_t^2 \mathcal{U}(t,s) = \mathcal{A} \mathcal{U}(t,s)$ is valid as an operator equation. Under the assumption of commutative property between $\mathcal{A}$ and $\mathcal{U}(t,s)$, the operator $\mathcal{A}$ is represented by

$$\mathcal{A}(t) = \mathcal{U}(t,s)^{-1} \partial_t^2 \mathcal{U}(t,s)$$
$$= \mathcal{U}(t,s)^{-1} \partial_t \mathcal{U}(t,s)\, (\partial_t \mathcal{U}(t,s))^{-1} \partial_t^2 \mathcal{U}(t,s),$$

where the former part $\mathcal{U}(t,s)^{-1} \partial_t \mathcal{U}(t,s)$ and the latter part $(\partial_t \mathcal{U}(t,s))^{-1} \partial_t^2 \mathcal{U}(t,s)$ of the right hand side correspond to the logarithmic representation $(I - \kappa e^{-\alpha(t,s)})^{-1} \partial_t \alpha(t,s)$ and $(I - \kappa e^{-\hat{\alpha}(t,s)})^{-1} \partial_t \hat{\alpha}(t,s)$ respectively. In this equation $\mathcal{U}(t,s)$, $\partial_t \mathcal{U}(t,s)$, and $\partial_t^2 \mathcal{U}(t,s)$ are assumed to commute with each other, so that the logarithmic representation of $\mathcal{A}(t)$ follows. □

In particular, for the commutation between two generally-unbounded operators, the intermediate domain space can be different depending on the order of operators. Here is a reason why two different orders of representations (12) and (13) are obtained.

3.2. Second Order Abstract Evolution Equations

Corollary 1 (Logarithmic representation of infinitesimal generator). *Let operators $\partial_t \mathcal{U}(t,s)$ and $\partial_t \mathcal{V}(t,s) = \partial_t^2 \mathcal{U}(t,s)$ satisfy the sectorial property. If either Equation (12) or Equation (13) is well-defined, then their square roots being represented by either*

$$\pm \mathcal{A}(t)^{1/2} = \pm \left\{ (I - \kappa e^{-\hat{\alpha}(t,s)})^{-1}\, \partial_t \hat{\alpha}(t,s) (I - \kappa e^{-\alpha(t,s)})^{-1} \partial_t \alpha(t,s) \right\}^{1/2} \tag{14}$$

or

$$\pm \mathcal{A}(t)^{1/2} = \pm \left\{ (I - \kappa e^{-\alpha(t,s)})^{-1} \partial_t \alpha(t,s)\, (I - \kappa e^{-\hat{\alpha}(t,s)})^{-1}\, \partial_t \hat{\alpha}(t,s) \right\}^{1/2} \tag{15}$$

are the infinitesimal generators of Equation (8) in the sense that the solution of $d^2 u(t)/dt^2 = \mathcal{A}(t) u(t)$ is represented by

$$u(t) = \mathcal{U}(t,s) u_0 = \exp\left[+\mathcal{A}(t)^{1/2}\right] u_+ + \exp\left[-\mathcal{A}(t)^{1/2}\right] u_-, \tag{16}$$

where u_+ and u_- are the elements of $\mathcal{X}$. The representation (14) is valid if (12) is true, and the representation (15) is valid if (13) is true.

Proof. Under the commutation relation, the autonomous Equation (8) is also formally factorized as

$$\left(\partial_t \mathcal{U}(t,s)^{1/2} + \mathcal{A}^{1/2} \mathcal{U}(t,s)^{1/2} \right) \left(\partial_t \mathcal{U}(t,s)^{1/2} - \mathcal{A}^{1/2} \mathcal{U}(t,s)^{1/2} \right) u = 0$$

for any $u \in \mathcal{X}$. It leads to the decomposition such that

$$\begin{cases} \partial_t \mathcal{U}(t,s)^{1/2} u_+ + \mathcal{A}^{1/2} \mathcal{U}(t,s)^{1/2} u_+ = 0, \\ \partial_t \mathcal{U}(t,s)^{1/2} u_- - \mathcal{A}^{1/2} \mathcal{U}(t,s)^{1/2} u_- = 0. \end{cases}$$

The representation shown in Equation (16) is understood.

It is necessary to confirm the possibility of defining the fractional power of $\mathcal{A}$. The possibility of defining square root of operator is justified if it is possible to define the exponential of

$$\text{Log}\left[\mathcal{A}(t)^{1/2}\right] := \tfrac{1}{2}\left\{\text{Log}\left[\partial_t \text{Log}\mathcal{U}(t,s)\right] + \text{Log}\left[\partial_t \text{Log}\mathcal{V}(t,s)\right]\right\},$$

where note that the logarithms of the right hand side are the formal form, and it does not matter whether they are well defined or not. According to the commutation assumption between $\partial_t \mathcal{U}(t,s)$ and $\partial_t \mathcal{V}(t,s) = \partial_t^2 \mathcal{U}(t,s)$, the exponential of each logarithm of the right hand side are independently well-defined. According to the commutation relation between $\mathcal{U}(t,s)$, $\partial_t \mathcal{U}(t,s)$ and $\partial_t \mathcal{V}(t,s) = \partial_t^2 \mathcal{U}(t,s)$, the exponential function of right hand side is equal to

$$[\partial_t \text{Log}\mathcal{U}(t,s)]^{1/2}\ [\partial_t \text{Log}\mathcal{V}(t,s)]^{1/2} = [\partial_t \text{Log}\mathcal{V}(t,s)]^{1/2}\ [\partial_t \text{Log}\mathcal{U}(t,s)]^{1/2},$$

and each square root is well-defined by the sectorial assumption (for the sectorial property, see Section 2.10 of Chapter 5 in [11]), where the logarithms of $\mathcal{U}(t,s)$ and $\mathcal{V}(t,s)$ leading to the definition of fractional powers of $\mathcal{U}(t,s)$ and $\mathcal{V}(t,s)$ are valid as seen in Equations (10) and (11). Consequently, the logarithmic representations of infinitesimal generators $\pm\mathcal{A}(t)^{1/2}$ are true. □

4. Conclusions

In this article, the Miura transform is generalized in the following sense:

- it is not only the transform between the KdV and mKdV equations;
- the spatial dimension of the equation is not necessarily equal to 1;
- the differential in Equation (12) is not necessarily for the spatial variable x;

where, in terms of applying to theory of higher order abstract evolution equations, the variable is taken as t in this article. For the preceding work dealing with the general choice of the evolution direction, see [16,18]. Consequently, the generalized Miura transform is obtained as the product of two logarithmic representations of operators in a general Banach space framework and they are applied to clarify the structure of the general solutions of second order abstract evolution equations defined in finite and infinite dimensional Banach spaces. Since the linear operator $\mathcal{A}$ is a generalized concept of matrices, the presented result potentially includes any matrix situations.

Funding: This work was partially supported by JSPS KAKENHI Grant No. 17K05440.

Acknowledgments: The author is grateful to Emeritus Hiroki Tanabe for valuable comments. The referee's comment on the Reference [6] is acknowledged for giving an idea for initiating this research.

Conflicts of Interest: The author declares no conflict of interest.

References

1. Benia, Y.; Scapellato, A. Existence of solution to Korteweg-de Vries equation in a non-parabolic domain. *Nonlinear Anal.* **2020**, *195*, 111758. [CrossRef]
2. Ruggieri, M.; Speciale, M.P. Quasi self-adjoint coupled KdV-like equations. *AIP Conf. Proc.* **2013**, *1558*, 1220–1223.
3. Miura, R. Korteweg-de Vries Equation and Generalizations. I. A Remarkable Explicit Nonlinear Transformation. *J. Math. Phys.* **1968**, *9*, 1202–1204. [CrossRef]
4. Cole, D. On a quasi-linear parabolic equation occurring in aerodynamics. *Q. Appl. Math.* **1951**, *9*, 225–236. [CrossRef]
5. Hopf, E. The partial differential equation $u_t + uu_x = \mu u_{xx}$. *Commun. Pure Appl. Math.* **1950**, *3*, 201–230. [CrossRef]
6. Iwata, Y. Abstract formulation of the Cole-Hopf transform. *Methods Funct. Anal. Topol.* **2019**, *25*, 142–151.
7. Hirota, R. Exact solution of the Korteweg-de Vries equation for multiple collisions of solitons. *Phys. Rev. Lett.* **1971**, *27*, 1192–1194. [CrossRef]

8. Jackson, E.A. *Perspectives of Nonlinear Dynamics Vols. I and II*; Cambridge University Press: Cambridge, UK, 1991.
9. Kato, T. Linear evolution equation of "hyperbolic" type. *J. Fac. Sci. Univ. Tokyo* **1970**, *17*, 241–258. [CrossRef]
10. Kato, T. Linear evolution equation of "hyperbolic" type II. *J. Math. Soc. Jpn.* **1973**, *25*, 648–666. [CrossRef]
11. Kato, T. *Perturbation Theory for Linear Operators*; Springer: Berlin/Heidelberg, Germany, 1966.
12. Tanabe, H. *Equations of Evolution*; Pitman: London, UK, 1979.
13. Yosida, K. *Functional Analysis*; Springer: Berlin/Heidelberg, Germany, 1965.
14. Iwata, Y. Infinitesimal generators of invertible evolution families. *Methods Funct. Anal. Topol.* **2017**, *23*, 26–36. [CrossRef]
15. Iwata, Y. Alternative infinitesimal generator of invertible evolution families. *J. Appl. Math. Phys.* **2017**, *5*, 822–830. [CrossRef]
16. Iwata, Y. Theory of $B(X)$-module. *arXiv* **2019**, arXiv:1907.08767.
17. Brezis, H. *Functional Analysis, Sobolev Spaces and Partial Differential Equations*; Springer: Berlin/Heidelberg, Germany, 2010.
18. Iwata, Y. Relativistic formulation of abstract evolution equations. *AIP Conf. Proc.* **2019**, *2075*, 100007.

© 2020 by the authors. Licensee MDPI, Basel, Switzerland. This article is an open access article distributed under the terms and conditions of the Creative Commons Attribution (CC BY) license (http://creativecommons.org/licenses/by/4.0/).

Article

Stability Results for Implicit Fractional Pantograph Differential Equations via ϕ-Hilfer Fractional Derivative with a Nonlocal Riemann-Liouville Fractional Integral Condition

Idris Ahmed [1,2,3], Poom Kumam [2,4,]*, Kamal Shah [5,6], Piyachat Borisut [1,2], Kanokwan Sitthithakerngkiet [7] and Musa Ahmed Demba [1,2,8]

- [1] KMUTTFixed Point Research Laboratory, Room SCL 802 Fixed Point Laboratory, Science Laboratory Building, Department of Mathematics, Faculty of Science, King Mongkut's University of Technology Thonburi (KMUTT), 126 Pracha-Uthit Road, Bang Mod, Thrung Khru, Bangkok 10140, Thailand; idris.ahamed@mail.kmutt.ac.th (I.A.); piyachat.borisut@mail.kmutt.ac.th (P.B.); musa.demba@mail.kmutt.ac.th (M.A.D.)
- [2] Center of Excellence in Theoretical and Computational Science (TaCS-CoE), Science Laboratory Building, King Mongkut's University of Technology Thonburi (KMUTT), 126 Pracha-Uthit Road, Bang Mod, Thrung Khru, Bangkok 10140, Thailand
- [3] Department of Mathematics and Computer Science, Sule Lamido University, Kafin-Hausa, Jigawa State P.M.B 048, Nigeria
- [4] Department of Medical Research, China Medical University Hospital, China Medical University, Taichung 40402, Taiwan
- [5] Department of Mathematics, University of Malakand, Chakadara Dir(L), Khyber Pakhtunkhwa 18800, Pakistan; kamal@uom.edu.pk
- [6] Department of Mathematics and Basic Sciences, Prince Sultan University, Riyadh 11586, Saudi Arabia
- [7] Intelligent and Nonlinear Dynamic Innovations Research Center, Department of Mathematics, Faculty of Applied Science, King Mongkut's University of Technology North Bangkok (KMUTNB), Wongsawang, Bangsue, Bangkok 10800, Thailand; kanokwan.s@sci.kmutnb.ac.th
- [8] Department of Mathematics, Faculty of Computing and Mathematical Sciences, Kano University of Science and Technology, Wudil, Kano State P.M.B 3244, Nigeria
- * Correspondence: poom.kumam@mail.kmutt.ac.th

Received: 27 November 2019; Accepted: 18 December 2019; Published: 7 January 2020

Abstract: This paper presents a class of implicit pantograph fractional differential equation with more general Riemann-Liouville fractional integral condition. A certain class of generalized fractional derivative is used to set the problem. The existence and uniqueness of the problem is obtained using Schaefer's and Banach fixed point theorems. In addition, the Ulam-Hyers and generalized Ulam-Hyers stability of the problem are established. Finally, some examples are given to illustrative the results.

Keywords: Hilfer fractional derivative; Ulam stability; pantograph differential equation; nonlocal integral condition

MSC: 26A33; 34A34; 34D20; 34A12

1. Introduction

The fractional-order differential equation is the oldest theory in the field of science and engineering. This theory has been used over the years, as the outcomes were found to be important in the field of economics, control theory and material sciences see [1–4]. Because of the nonlocal property of fractional-order differential equation, researchers are allowed to select the most appropriate operator and use it in order to get a better description of the complex phenomena in the real world. The generalization of classical calculus are the fractional calculus. Nevertheless, there are various definitions of fractional integrals and derivatives of arbitrary order with different types of operator. Recently, Furati et al. [5] proposed a Hilfer fractional derivatives which interpolates with Riemann-Liouville and Caputo fractional derivatives. These fractional operator provide an extra degree of freedom when choosing the initial condition. Furthermore, models based on this operator provide an excellent results compared with the integer-order derivatives, for example, we refer the interesting reader to see [6–18].

Qualitative analysis of fractional differential equations plays a vital role in the field of fractional differential equations. However, many researchers studied the existence and uniqueness of solution of differential equation with different types of fractional integral and derivatives. More recently, motivated by classical Riemann-Liouville, Caputo fractional derivative, Hilfer-fractional derivative, ψ-Riemann-Liouville integral and ψ-Caputo fractional derivatives, Sousa and Oliveira [19] initiated an interesting fractional differential operator called ψ-Hilfer fractional derivatives, that is a fractional derivative of a function with respect to another function ψ. These fractional derivatives generalized the aforementioned fractional derivatives and integrals. The main advantages of these operator is the freedom of choice of the function ψ and its merge and acquire the properties of the aforementioned fractional operators. Results based on these setting can be found in [18–34]. The Ulam-Hyers stability point of view, is the vital and special type of stability that attracts many researchers in the field of mathematical analysis. Moreover, the Ulam-Hyers and Ulam-Hyers-Rassias stability of linear, implicit and nonlinear fractional differential equations were examined in [17,35–49].

Pantograph differential equations are a special class of delay differential equation arising in deterministic situations and are of the form:

$$\begin{cases} g'(s) = kg(s) + lg(\lambda s), & s \in [0,b],\ b > 0,\ 0 < \lambda < 1, \\ g(0) = g_0. \end{cases} \quad (1)$$

The pantograph is a device used in electric trains to collects electric current from the overload lines. This equation was modeled by Ockendon and Tayler [50]. Pantograph equation play a vital role in physics, pure and applied mathematics, such as control systems, electrodynamics, probability, number theory, and quantum mechanics. Motivated by their importance, a lot of researchers generalized these equation in to various forms and introduced the solvability aspect of such problems both theoretically and numerically, (for more details see [16,51–57] and references therein). However, very few works have been proposed with respect to pantograph fractional differential equations.

In [48], the authors considered an implicit fractional differential equations with nonlocal condition described by:

$$\begin{cases} D_{0+}^{\alpha,\beta} w(\tau) = f(\tau, w(\tau), D_{0+}^{\alpha,\beta} w(\tau)), & \tau \in I = [0,T], \\ I_{0+}^{1-\gamma} w(0) = \sum_{i=1}^{m} c_i w(\eta_i), & \alpha \leq \gamma = \alpha + \beta - \alpha\beta, \eta_i \in [0,T], \end{cases} \quad (2)$$

where $D_{0+}^{\alpha,\beta}(\cdot)$ is the Hilfer fractional derivative of order $(0 < \alpha < 1)$ and type $0 \leq \beta \leq 1$. The existence and uniqueness results were obtained by applying Schaefer's fixed point theorem and Banach's contraction principle. Moreover, the authors discussed the stability analysis via Gronwall's lemma. Sousa and Oliveira [47] discussed the existence, uniqueness and Ulam-Hyers-Rassias stability for a class of φ-Hilfer fractional differential equations described by:

$$\begin{cases} {}^H\mathfrak{D}_{a+}^{\alpha,\beta;\varphi} g(t) = f(t, g(t), {}^H\mathfrak{D}_{a+}^{\alpha,\beta;\varphi} g(t)), & t \in \mathcal{J} = [a,T], \\ \mathfrak{J}_{a+}^{1-\gamma;\varphi} g(a) = g_a, & \alpha \leq \gamma = \alpha + \beta - \alpha\beta,\ T > a, \end{cases} \quad (3)$$

where $^H\mathfrak{D}_{0^+}^{\alpha,\beta;\varphi}(\cdot)$ is the φ-Hilfer fractional derivative of order $(0 < \alpha \le 1)$ and operator $(0 \le \beta \le 1)$, $\mathfrak{I}_{0^+}^{1-\gamma;\varphi}(\cdot)$, is the Riemann-Liouville fractional integral of order $1 - \gamma$, with respect to the function φ, $f : [a, T] \times \mathbb{R}^2 \to \mathbb{R}$ is a continuous function. Recently Harikrishman et. al [58] established existence and uniqueness of nonlocal initial value problem for fractional pantograph differential equation involving ψ-Hilfer fractional derivative of the form:

$$\begin{cases} ^HD_{a^+}^{\alpha,\beta;\psi}v(s) = f(s, v(s), v(\lambda s)), & s \in (a, b], \quad s > a, \quad 0 < \lambda < 1, \\ \mathrm{I}_{a^+}^{1-\gamma;\psi}v(a) = \sum_{j=1}^{k} c_j v(\tau_j), & \tau_j \in (a, b], \quad \gamma = \alpha + \beta - \alpha\beta, \end{cases} \quad (4)$$

where $^HD_{a^+}^{\alpha,\beta;\psi}(\cdot)$ is the ψ-Hilfer fractional derivative of order $0 < \alpha < 1$ and type $0 \le \beta \le 1$, $\mathrm{I}_{a^+}^{1-\gamma;\psi}(\cdot)$, is the Riemann-Liouville fractional integral of order $1 - \gamma$, with respect to the continuous function ψ such that $\psi'(\cdot) > 0$, $f \in C(t \in (a, b], \mathbb{R}^2, \mathbb{R})$.

Motivated by the papers [21,47,48] and some familiar results on fractional pantograph differential equations [16,52,55,58]. We discuss the existence and uniqueness of the solution of the implicit pantograph fractional differential equations involving ϕ-Hilfer fractional derivatives. Furthermore, the Ulam-Hyers and generalized Ulam-Hyers stability are also discussed. The implicit pantograph fractional differential equations involving ϕ-Hilfer fractional derivatives is of the form

$$\begin{cases} ^H\mathcal{D}_{0^+}^{r,p;\phi}z(t) = f(t, z(t), z(\gamma t), ^H\mathcal{D}_{0^+}^{r,p;\phi}z(\gamma t)), & t \in \mathcal{J} = (0, T], \ 0 < \gamma < 1, \\ \mathcal{I}_{0^+}^{1-q;\phi}z(0^+) = \sum_{i=1}^{m} b_i \mathcal{I}_{0^+}^{\rho;\phi}z(\xi_i), & r \le q = r + p - rp, \end{cases} \quad (5)$$

where $^H\mathcal{D}_{0^+}^{r,p;\phi}(\cdot)$ is the generalized ϕ-Hilfer fractional derivatives of order $(0 < r < 1)$ and type $(0 \le p \le 1)$, $\mathcal{I}_{0^+}^{1-q;\phi}(\cdot)$ and $\mathcal{I}_{0^+}^{\rho;\phi}(\cdot)$ are ϕ-Riemann-Lioville fractional integral of order $1 - q$ and $\rho > 0$ respectively with respect to the continuous function ϕ such that $\phi'(\cdot) \ne 0$, $f : (0, T] \times \mathbb{R}^3 \to \mathbb{R}$ is a given continuous function, $T > 0$, $b_i \in \mathbb{R}$ and $\xi_i \in \mathcal{J}$ satisfying $0 < \xi_1 \le \xi_2 \le \cdots \le \xi_m < T$ for $i = 1, 2, \cdots, m$.

As far as we know, to the best of our understanding, results of Ulam-Hyers and generalized Ulam-Hyers stability with respect to the pantograph differential equation are very few and in fact most authors discuss existence and uniqueness, while we study existence, uniqueness and stability analysis for a class of implicit pantograph fractional differential equations with ϕ-Hilfer derivatives and nonlocal Riemann-Liouville fractional integral condition.

This paper contributes to the growth of qualitative analysis of fractional differential equation in particular pantograph fractional differential equation when ϕ-Hilfer fractional derivatives involved and the nonlocal initial condition proposed in this paper generalized the following initial conditions:

- If $\rho \to 0$, the initial condition reduces to multi-point nonlocal condition.
- If $\rho \to 1$, the initial condition coincide with the nonlocal integral condition.
- In physical problems, the nonlocal condition yields an excellent results compared with the initial condition $z(0) = z_0$ [59,60].

In addition, we notice that the function $f(s, v(s), v(\lambda s))$, $s \in (a, b]$, $0 < \lambda < 1$, defined in Equation (4) is not well-define for some choices of λ.

Therefore, the paper is organized as follows: In Section 2, it recalls some basic and fundamental definitions and lemmas. In Section 3, we prove existence and uniqueness of the proposed problem (5). Ulam-Hyers and generalized Ulam-Hyers stability for the proposed problem were discussed in Section 4. While in Section 5, two examples were given to illustrate the applicability of our results. Lastly, the conclusion part of the paper is given in Section 6.

2. Preliminaries

This section will recall some useful prerequisites facts, definitions and some fundamental lemmas with respect to fractional differential equations.

Throughout the paper, we denote $C[\mathcal{J}, \mathbb{R}]$ the Banach space of all continuous functions from $\mathcal{J}$ into $\mathbb{R}$ with the norm defined by [1]

$$\|f\| = \sup_{t \in \mathcal{J}}\{|f(t)|\}.$$

The weighted space $\mathcal{C}_{q,\phi}[\mathcal{J}, \mathbb{R}]$ of continuous function f on the interval $[a, T]$ is defined by

$$\mathcal{C}_{q,\phi}[\mathcal{J}, \mathbb{R}] = \{f(t) : (a, T] : (\phi(t) - \phi(0))^q f(t) \in C[\mathcal{J}, \mathbb{R}]\},$$

with the norm

$$\|f\|_{\mathcal{C}_{q,\phi}[\mathcal{J},\mathbb{R}]} = \|(\phi(t) - \phi(0))^q f(t)\| = \max |(\phi(t) - \phi(0))^q f(t) : t \in \mathcal{J}|.$$

Moreover, for each $n \in \mathbb{N}$ and $0 \le q < 1$ with $q = r + p - rp$

$$\mathcal{C}_{q,\phi}^n[\mathcal{J}, \mathbb{R}] = \{f^n \in \mathcal{C}_{q;\phi}[\mathcal{J}, \mathbb{R}]\}$$

$$\mathcal{C}_{q;\phi}^{r,p}[\mathcal{J}, \mathbb{R}] = \{f \in \mathcal{C}_{q;\phi}[\mathcal{J}, \mathbb{R}] : \mathcal{D}_{0^+}^{r,p;\phi} \in \mathcal{C}_{q;\phi}[\mathcal{J}, \mathbb{R}]\}.$$

Indeed, for $n = 0$, we have

$$\mathcal{C}_{q;\phi}^0[\mathcal{J}, \mathbb{R}] = \mathcal{C}_{q;\phi}[\mathcal{J}, \mathbb{R}],$$

with the norm

$$\|f\|_{\mathcal{C}_{q;\phi}^n[\mathcal{J},\mathbb{R}]} = \sum_{k=0}^{n-1} \|f^k\|_{C[\mathcal{J},\mathbb{R}]} + \|f^n\|_{\mathcal{C}_{q;\phi}^n[\mathcal{J},\mathbb{R}]}.$$

Furthermore, we present the following space $\mathcal{C}_{1-q;\phi}^{r,p}[\mathcal{J}, \mathbb{R}]$ and $\mathcal{C}_{1-q;\phi}^q[\mathcal{J}, \mathbb{R}]$ defined as:

$$\mathcal{C}_{1-q;\phi}^{r,p}[\mathcal{J}, \mathbb{R}] = \{f \in \mathcal{C}_{1-q;\phi}[\mathcal{J}, \mathbb{R}], \mathcal{D}_{0^+}^{r,p;\phi} \in \mathcal{C}_{1-q;\phi}[\mathcal{J}, \mathbb{R}]\}$$

and

$$\mathcal{C}_{1-q;\phi}^q[\mathcal{J}, \mathbb{R}] = \{f \in \mathcal{C}_{1-q;\phi}[\mathcal{J}, \mathbb{R}], \mathcal{D}_{0^+}^{q;\phi} \in \mathcal{C}_{1-q;\phi}[\mathcal{J}, \mathbb{R}]\}.$$

Clearly, $\mathcal{C}_{1-q;\phi}^q[\mathcal{J}, \mathbb{R}] \subset \mathcal{C}_{1-q;\phi}^{r,p}[\mathcal{J}, \mathbb{R}]$.

Definition 1 ([1]). *Let $(0, b]$ be a finite or infinite interval on the half-axis $\mathbb{R}^+$, and $\phi(\xi) \ge 0$ be monotone function on $(a, b]$ whose $\phi'(\xi)$ is continuous on $(0, b)$. The ϕ-Hilfer Riemann-Liouville fractional integral of order $r \in \mathbb{R}^+$ of function w is defined by*

$$(\mathcal{I}_{0^+}^{r,\phi} w)(\xi) = \frac{1}{\Gamma(r)} \int_{0^+}^{\xi} \phi'(s)(\phi(\xi) - \phi(s)) w(s) ds, \quad \xi > 0, \tag{6}$$

where $\Gamma(\cdot)$ represent the Gamma function.

Definition 2 ([5]). *Let $n - 1 < r < n, 0 \le p \le 1$. The left-sided Hilfer fractional derivative of order r and parameter p of function w is defined by*

$$\mathcal{D}_{0^+}^{r,p} w(\xi) = \left(\mathcal{I}_{0^+}^{p(n-r)} \mathcal{D}^n \mathcal{I}_{0^+}^{(1-p)(n-r)} w \right)(\xi), \tag{7}$$

where $\mathcal{D}^n = \left(\frac{d}{d\xi} \right)^n$.

The following Definition generalized Euqation (7).

Definition 3 ([19]). *Let $f, \phi \in C^n[\mathcal{J}, \mathbb{R}]$ be two functions such that $\phi(\xi) \geq 0$ and $\phi'(t) \neq 0$ for all $\xi \in [\mathcal{J}, \mathbb{R}]$ and $n - 1 < r < n$ with $n \in \mathbb{N}$. The left-side ϕ-Hilfer fractional derivative of a function w of order r and type $(0 \leq p \leq 1)$ is defined by*

$$^H\mathcal{D}_{0^+}^{r,p;\phi} w(\xi) = \mathcal{I}_{0^+}^{p(n-r);\phi} \left(\frac{1}{\phi'(\xi)} \frac{d}{d\xi} \right)^n \mathcal{I}_{0^+}^{(1-p)(n-r);\phi} w(\xi). \tag{8}$$

The following lemma shows the semigroup properties of ϕ-Hilfer fractional integral and derivative.

Lemma 1 ([5]). *Let $r \geq 0, 0 \leq p < 1$ and $w \in L^1[\mathcal{J}, \mathbb{R}]$. Then*

$$\mathcal{I}_{0^+}^{r;\phi} \mathcal{I}_{0^+}^{p;\phi} w(\xi) = \mathcal{I}_{0^+}^{r+p;\phi} w(\xi),$$

a.e $\xi \in \mathcal{J}$.
In particular, if $w \in C_{q,\phi}[\mathcal{J}, \mathbb{R}]$ and $w \in C[\mathcal{J}, \mathbb{R}]$, then

$$\mathcal{I}_{0^+}^{r;\phi} \mathcal{I}_{0^+}^{p;\phi} w(\xi) = \mathcal{I}_{0^+}^{r+p;\phi} w(\xi),$$

for all $\xi \in (0, T]$ and

$$^H\mathcal{D}_{0^+}^{r;\phi} \mathcal{I}_{0^+}^{r;\phi} w(\xi) = w(\xi),$$

for all $\xi \in \mathcal{J}$.

The composition of the ϕ-Hilfer fractional integral and derivative operator is given by the following lemmas.

Lemma 2 ([21]). *Let $r \geq 0, 0 \leq p < 1$ and $q = r + p - rp$. If $w(\xi) \in C_{1-q}^q[\mathcal{J}, \mathbb{R}]$, then*

$$\mathcal{I}_{0^+}^{q;\phi} {}^H\mathcal{D}_{0^+}^{q;\phi} w(\xi) = \mathcal{I}_{0^+}^{r;\phi} {}^H\mathcal{D}_{0^+}^{r,p;\phi} w(\xi)$$

and

$$^H\mathcal{D}_{0^+}^{q;\phi} \mathcal{I}_{0^+}^{r;\phi} w(\xi) = {}^H\mathcal{D}_{0^+}^{p(1-r);\phi} w(\xi).$$

Lemma 3 ([6,19]). *If $w \in C^n[\mathcal{J}, \mathbb{R}]$ and let $n - 1 < r < n, 0 \leq p \leq 1$ and $q = r + p - rp$. Then*

$$\mathcal{I}_{0^+}^{r;\phi} {}^H\mathcal{D}_{0^+}^{r,p;\phi} w(\xi) = w(\xi) - \sum_{k=1}^n \frac{(\phi(\xi) - \phi(0))^{q-k}}{\Gamma(q-k+1)} w_\phi^{[n-k]} \mathcal{I}_{0^+}^{(1-p)(n-r);\phi} w(0),$$

for all $\xi \in \mathcal{J}$. Moreover, if $0 < r < 1$, we have

$$\mathcal{I}_{0^+}^{r;\phi} {}^H\mathcal{D}_{0^+}^{r,p;\phi} w(\xi) = w(\xi) - \frac{(\phi(\xi) - \phi(0))^{q-1}}{\Gamma(q)} \mathcal{I}_{0^+}^{(1-p)(1-r);\phi} w(0).$$

In addition, if $w \in C_{1-q;\phi}[\mathcal{J}, \mathbb{R}]$ and $\mathcal{I}_{0^+}^{1-q;\phi} w \in C_{1-q;\phi}^1[\mathcal{J}, \mathbb{R}]$, then

$$\mathcal{I}_{0^+}^{q;\phi} {}^H\mathcal{D}_{0^+}^{q;\phi} w(\xi) = w(\xi) - \frac{(\phi(\xi) - \phi(0))^{q-1}}{\Gamma(q)} \mathcal{I}_{0^+}^{(1-q);\phi} w(0),$$

for all $0 < q < 1$ and $t \in \mathcal{J}$.

Lemma 4 ([6]). *Let $r > 0, 0 \leq q < 1$ and $w \in C_{q;\phi}[\mathcal{J}, \mathbb{R}]$. If $r > q$, then $\mathcal{I}_{0^+}^{r;\phi} w \in C[\mathcal{J}, \mathbb{R}]$ and*

$$\mathcal{I}_{0^+}^{r;\phi} w(0) = \lim_{\xi \to 0} \mathcal{I}_{0^+}^{r;\phi} w(\xi) = 0.$$

Lemma 5 ([21]). *Let $r > 0, 0 \leq p \leq 1$ and $q = r + p - rp$. If $w \in C_{1-q;\phi}^q[\mathcal{J}, \mathbb{R}]$, then*

$$\mathcal{I}_{0^+}^{q;\phi} \mathcal{D}_{0^+}^{q;\phi} w(\xi) = \mathcal{I}_{0^+}^{r;\phi} \mathcal{D}_{0^+}^{r,p;\phi} w(\xi)$$

and

$$^H\mathcal{D}_{0^+}^{q;\phi} \mathcal{I}_{0^+}^{r;\phi} w(\xi) = \mathcal{D}_{0^+}^{q(1-r);\phi} w(\xi).$$

Lemma 6. *Let $f \in L^1(\mathcal{J})$ such that $\mathcal{D}_{0^+}^{p(1-r);\phi} w \in L^1(\mathcal{J})$ exists, then*

$$\mathcal{D}_{0^+}^{r,p;\phi} \mathcal{I}_{0^+}^{r;\phi} w(\xi) = \mathcal{I}_{0^+}^{p(1-r);\phi} \mathcal{D}_{0^+}^{p(1-r);\phi} w(\xi).$$

Next, we take into account some important properties of ϕ-fractional derivative and integral operator as follows:

Proposition 1 ([1]). *Let $\xi > 0, r \geq 0$ and $s > 0$. Then, ϕ-fractional integral and derivative of a power function are given by*

$$^H\mathcal{D}_{0^+}^{r;\phi} (\phi(\xi) - \phi(0))^{s-1} = \frac{\Gamma(s)}{\Gamma(s-r)} (\phi(\xi) - \phi(0))^{r+s-1}$$

and

$$\mathcal{I}_{0^+}^{r;\phi} (\phi(\xi) - \phi(0))^{s-1} = \frac{\Gamma(s)}{\Gamma(s+r)} (\phi(\xi) - \phi(0))^{r+s-1}.$$

Furthermore, if $0 < r < 1$, then

$$^H\mathcal{D}_{0^+}^{r;\phi} (\phi(\xi) - \phi(0))^{r-1} = 0.$$

Theorem 1 ([19]). *If $w \in C^1[\mathcal{J}, \mathbb{R}], 0 < r < 1$ and $0 \leq p \leq 1$. Then we have the followings:*
(i) $^H\mathcal{D}_{0^+}^{r,p;\phi} \mathcal{I}_{0^+}^{r;\phi} w(\xi) = w(\xi)$.
(ii) $\mathcal{I}_{0^+}^{r;\phi} {}^H\mathcal{D}_{0^+}^{r,p;\phi} w(\xi) = w(\xi) - \frac{(\phi(\xi)-\phi(0))^{q-1}}{\Gamma(q)} \mathcal{I}_{0^+}^{(1-p)(1-r);\phi} w(\xi)$.

Lemma 7 ([6]). *Let $h : \mathcal{J} \times \mathbb{R} \to \mathbb{R}$ such that for any $z \in C_{1-q;\phi}[\mathcal{J}, \mathbb{R}], h \in C_{1-q;\phi}[\mathcal{J}, \mathbb{R}]$. A function $z \in C_{1-q;\phi}^q[\mathcal{J}, \mathbb{R}]$ is a solution of the fractional initial value problem:*

$$\begin{cases} ^H\mathcal{D}_{0^+}^{r,p;\phi} z(t) = h(t), & 0 < r \leq 1, \quad 0 \leq p \leq 1, \\ \mathcal{I}_{0^+}^{1-q;\phi} z(0^+) = z_0 \in \mathbb{R}, & q = r + p - rp, \end{cases}$$

if and only if z satisfies the following integral equation,

$$z(t) = \frac{z_0}{\Gamma(q)} (\phi(t) - \phi(0))^{q-1} + \frac{1}{\Gamma(r)} \int_0^t \phi'(s)(\phi(t) - \phi(s))^{r-1} h(s) ds.$$

3. Main Results

In this section, we first adopt some techniques from Lemma 7 in order to establish an important mixed-type integral equation of problem (5). Thus, we need the following auxiliary lemma.

Lemma 8. Let $0 < r < 1$, $0 \leq p \leq 1$ and $q = r + p - rp$. Suppose $f : \mathcal{J} \times \mathbb{R}^3 \to \mathbb{R}$ is a function such that $f \in \mathcal{C}_{1-q;\phi}[\mathcal{J}, \mathbb{R}]$ for any $z \in \mathcal{C}_{1-q;\phi}[\mathcal{J}, \mathbb{R}]$. If $z \in \mathcal{C}^q_{1-q;\phi}[\mathcal{J}, \mathbb{R}]$ then z satisfies the problem (5) if and only if z satisfies the mixed-type integral equation:

$$z(t) = \frac{\delta \Gamma(\rho + q)}{\Gamma(q)\Gamma(\rho + r)}(\phi(t) - \phi(0))^{q-1} \sum_{i=1}^{m} b_i \int_{0^+}^{\xi_i} \phi'(s)(\phi(\xi_i) - \phi(s))^{\rho+r-1} T_z(s)ds \qquad (9)$$
$$+ \frac{1}{\Gamma(r)} \int_{0^+}^{t} \phi'(s)(\phi(t) - \phi(s))^{r-1} T_z(s) ds,$$

where

$$\delta = \frac{1}{\Gamma(\rho + q) - \sum_{i=1}^{m} b_i(\phi(\xi_i) - \phi(0))^{\rho+q-1}}, \qquad (10)$$

such that $\Gamma(\rho + q) \neq \sum_{i=1}^{m} b_i(\phi(\xi_i) - \phi(0))^{\rho+q-1}$.

For simplicity, we take

$$T_z(t) = {}^H D_{0^+}^{r,p;\phi} z(t) = f(t, z(t), z(\gamma t), T_z(t)). \qquad (11)$$

Proof. Suppose $z \in \mathcal{C}^q_{1-q;\phi}[\mathcal{J}, \mathbb{R}]$ is a solution to the problem (5), then, we show that z is also a solution of (5). Indeed, from Lemma 7, we have

$$z(t) = \frac{(\phi(t) - \phi(0))^{q-1}}{\Gamma(q)} \mathcal{I}_{0^+}^{1-q;\phi} z(0) + \frac{1}{\Gamma(r)} \int_{0^+}^{t} \phi'(s)(\phi(t) - \phi(s))^{r-1} T_z(s) ds. \qquad (12)$$

Now, if we substitute $t = \xi_i$ and multiply both sides by b_i in Equation (12), we obtain

$$b_i z(\xi_i) = \frac{(\phi(\xi_i) - \phi(0))^{q-1}}{\Gamma(q)} b_i \mathcal{I}_{0^+}^{1-q;\phi} z(0) + \frac{b_i}{\Gamma(r)} \int_{0^+}^{\xi_i} \phi'(s)(\phi(\xi_i) - \phi(s))^{r-1} T_z(s) ds. \qquad (13)$$

Next, by applying $\mathcal{I}_0^{\rho;\phi}$ to both sides of Equation (13) and using Lemma 1 and Proposition 1, we get

$$\mathcal{I}_{0^+}^{\rho;\phi} b_i z(\xi_i) = \frac{(\phi(\xi_i) - \phi(0))^{\rho+q-1}}{\Gamma(\rho + q)} b_i \mathcal{I}_0^{1-q;\phi} z(0) \qquad (14)$$
$$+ \frac{b_i}{\Gamma(\rho + r)} \int_{0^+}^{\xi_i} \phi'(s)(\phi(\xi_i) - \phi(s))^{\rho+r-1} T_z(s) ds.$$

This implies that

$$\sum_{i=1}^{m} \mathcal{I}_{0^+}^{\rho;\phi} b_i z(\xi_i) = \frac{1}{\Gamma(\rho + q)} \left(\sum_{i=1}^{m} b_i(\phi(\xi_i) - \phi(0))^{\rho+q-1} \right) \mathcal{I}_{0^+}^{1-q;\phi} z(0) \qquad (15)$$
$$+ \frac{1}{\Gamma(\rho + r)} \sum_{i=1}^{m} b_i \int_{0^+}^{\xi_i} \phi'(s)(\phi(\xi_i) - \phi(s))^{\rho+r-1} T_z(s) ds.$$

Inserting the initial condition: $\mathcal{I}_{0^+}^{1-q;\phi}z(0^+) = \sum_{i=1}^{m}\mathcal{I}_{0^+}^{\rho;\phi}b_i z(\xi_i)$ in Equation (15) we have

$$\mathcal{I}_{0^+}^{1-q;\phi}z(0) = \frac{1}{\Gamma(\rho+q)}\left(\sum_{i=1}^{m}b_i(\phi(\xi_i)-\phi(0))^{\rho+q-1}\right)\mathcal{I}_{0^+}^{1-q;\phi}z(0) \\ + \frac{1}{\Gamma(\rho+r)}\sum_{i=1}^{m}b_i\int_{0^+}^{\xi_i}\phi'(s)(\phi(\xi_i)-\phi(s))^{\rho+r-1}T_z(s)ds, \quad (16)$$

which implies that

$$\frac{1}{\Gamma(\rho+r)}\sum_{i=1}^{m}b_i\int_{0^+}^{\xi_i}\phi'(s)(\phi(\xi_i)-\phi(s))^{\rho+r-1}T_z(s)ds \\ = \left(1 - \frac{1}{\Gamma(\rho+q)}\sum_{i=1}^{m}b_i(\phi(\xi_i)-\phi(0))^{\rho+q-1}\right)\mathcal{I}_{0^+}^{1-q;\phi}z(0) \quad (17) \\ = \frac{1}{\delta\Gamma(\rho+q)}\mathcal{I}_{0^+}^{1-q;\phi}z(0).$$

Thus,

$$\mathcal{I}_{0^+}^{1-q;\phi}z(0) = \frac{\delta\Gamma(\rho+q)}{\Gamma(\rho+r)}\sum_{i=1}^{m}b_i\int_{0^+}^{\xi_i}\phi'(s)(\phi(\xi_i)-\phi(s))^{\rho+r-1}T_z(s)ds. \quad (18)$$

Hence, the result follows by putting Equation (18) in Equation (12). This implies that $z(t)$ satisfies Equation (9).

Conversely, suppose that $z \in C_{1-q;\phi}^{q}$ satisfies the mixed-type integral Equation (9), then, we show that z satisfies Equation (5). Applying $\mathcal{D}_{0^+}^{q;\phi}$ to both sides of Equation (9) and using Lemma 2 and Proposition 1, we get

$$\mathcal{D}_{0^+}^{q;\phi}z(t) = \mathcal{D}_{0^+}^{q;\phi}\left(\frac{\delta\Gamma(\rho+q)}{\Gamma(q)\Gamma(r)}(\phi(t)-\phi(0))^{q-1}\sum_{i=1}^{m}b_i\int_{0^+}^{\xi_i}\phi'(s)(\phi(\xi_i)-\phi(s))^{\rho+r-1}T_z(s)ds\right) \\ + \mathcal{D}_{0^+}^{q;\phi}\left(\frac{1}{\Gamma(r)}\int_{0^+}^{t}\phi'(s)(\phi(t)-\phi(s))^{r-1}T_z(s)ds\right) \quad (19) \\ = \mathcal{D}_{0^+}^{p(1-r);\phi}f(t,z(t),z(\gamma t),\mathcal{D}_{0^+}^{r,p;\phi}z(\gamma t)).$$

Since $\mathcal{D}_{0^+}^{r,p;\phi}z \in \mathcal{C}_{1-q;\phi}[\mathcal{J},\mathbb{R}]$, then by definition of $\mathcal{C}_{1-q;\phi}^{q}[\mathcal{J},\mathbb{R}]$ and make use of Equation (19), we have

$$\mathcal{D}_{0^+}^{p(1-r);\phi}f = D\mathcal{I}_{0^+}^{1-p(1-r);\phi}f \in \mathcal{C}_{1-q;\phi}[\mathcal{J},\mathbb{R}].$$

For every $f \in \mathcal{C}_{1-q;\phi}[\mathcal{J},\mathbb{R}]$ and Lemma 3, we can see that $\mathcal{I}_{0^+}^{1-p(1-r);\phi}f \in \mathcal{C}_{1-q;\phi}[\mathcal{J},\mathbb{R}]$, which implies that $\mathcal{I}_{0^+}^{1-p(1-r);\phi}f \in \mathcal{C}_{1-q;\phi}^{1}[\mathcal{J},\mathbb{R}]$ from the definition of $\mathcal{C}_{q;\phi}^{n}[\mathcal{J},\mathbb{R}]$. Applying $\mathcal{I}_{0^+}^{p(1-r);\phi}$ on both sides of Equation (19) and using Lemma 3, we have

$$\mathcal{I}_{0^+}^{p(1-r);\phi}\mathcal{D}_{0^+}^{q;\phi}z(t) = \mathcal{I}_{0^+}^{p(1-r);\phi}\mathcal{D}_{0^+}^{p(1-r);\phi}T_z(t) \\ = T_z(t) - \frac{\left(\mathcal{I}_{0^+}^{1-p(1-r);\phi}T_z\right)(0^+)}{\Gamma(p(1-r))}(\phi(t)-\phi(0))^{p(r-1)-1} \quad (20) \\ = T_z(t) = f(t,z(t),z(\gamma t),\mathcal{D}_{0^+}^{r,p;\phi}z(\gamma t)).$$

Finally, we show that if $z \in C^q_{1-q}[\mathcal{J}, \mathbb{R}]$ satisfies Equation (9), it also satisfies the initial condition. Thus, by applying $\mathcal{I}^{1-q;\phi}_{0^+}$ to both sides of Equation (9) and using Lemma 1 and Proposition 1, we obtain

$$\begin{aligned}
\mathcal{I}^{1-q;\phi}_{0^+}&z(t) \\
&= \mathcal{I}^{1-q;\phi}_{0^+}\left(\frac{\delta\Gamma(\rho+q)}{\Gamma(q)\Gamma(\rho+r)}(\phi(t)-\phi(0))^{q-1}\sum_{i=1}^{m}b_i\int_{0^+}^{\xi_i}\phi'(s)(\phi(\xi_i)-\phi(s))^{\rho+r-1}T_z(s)ds\right) \\
&+ \mathcal{I}^{1-q;\phi}_{0^+}\left(\frac{1}{\Gamma(r)}\int_{0^+}^{t}\phi'(s)(\phi(t)-\phi(s))^{r-1}T_z(s)ds\right) \\
&= \frac{\delta\Gamma(\rho+q)}{\Gamma(\rho+r)}\sum_{i=1}^{m}b_i\int_{0^+}^{\xi_i}\phi'(s)(\phi(\xi_i)-\phi(s))^{\rho+r-1}T_z(s)ds + \mathcal{I}^{1-p(1-r);\phi}_{0^+}T_z(t).
\end{aligned} \quad (21)$$

Using Lemma 4 and the fact that $1-q < 1-p(1-r)$, then taking limit as $t \to 0$ in Equation (21) yields

$$\mathcal{I}^{1-q;\phi}_{0^+}z(0^+) = \frac{\delta\Gamma(\rho+q)}{\Gamma(\rho+r)}\sum_{i=1}^{m}b_i\int_{0^+}^{\xi_i}\phi'(s)(\phi(\xi_i)-\phi(s))^{\rho+r-1}T_z(s)ds. \quad (22)$$

Now, substituting $t = \xi_i$ and multiplying through by b_i in Equation (9), we get

$$\begin{aligned}
b_i z(\xi_i) &= \frac{\delta\Gamma(\rho+q)}{\Gamma(q)\Gamma(\rho+r)}b_i(\phi(\xi_i)-\phi(0))^{q-1}\sum_{i=1}^{m}b_i\int_{0^+}^{\xi_i}\phi'(s)(\phi(\xi_i)-\phi(s))^{\rho+r-1}T_z(s)ds \\
&+ \mathcal{I}^{1-q;\phi}_{0^+}\frac{b_i}{\Gamma(r)}\int_{0^+}^{\xi_i}\phi'(s)(\phi(\xi_i)-\phi(s))^{r-1}T_z(s)ds.
\end{aligned} \quad (23)$$

Applying $\mathcal{I}^{\rho;\phi}_{0^+}$ to both sides of Equation (23), we obtain

$$\begin{aligned}
\mathcal{I}^{\rho;\phi}_{0^+}b_i z(\xi_i) &= \frac{\delta b_i(\phi(\xi_i)-\phi(0))^{\rho+q-1}}{\Gamma(\rho+r)}\sum_{i=1}^{m}b_i\int_{0^+}^{\xi_i}\phi'(s)(\phi(\xi_i)-\phi(s))^{\rho+r-1}T_z(s)ds \\
&+ \frac{b_i}{\Gamma(\rho+r)}\int_{0^+}^{\xi_i}\phi'(s)(\phi(\xi_i)-\phi(s))^{\rho+r-1}T_z(s)ds,
\end{aligned} \quad (24)$$

which implies

$$\begin{aligned}
\sum_{i=1}^{m}&b_i\mathcal{I}^{\rho;\phi}_{0^+}z(\xi_i) \\
&= \frac{\delta}{\Gamma(\rho+r)}\sum_{i=1}^{m}b_i\int_{0^+}^{\xi_i}\phi'(s)(\phi(\xi_i)-\phi(s))^{\rho+r-1}T_z(s)ds\sum_{i=1}^{m}b_i(\phi(\xi_i)-\phi(0))^{\rho+q-1} \\
&+ \frac{1}{\Gamma(\rho+r)}\sum_{i=1}^{m}b_i\int_{0^+}^{\xi_i}\phi'(s)(\phi(\xi_i)-\phi(s))^{\rho+r-1}T_z(s)ds \\
&= \frac{1}{\Gamma(\rho+r)}\sum_{i=1}^{m}b_i\int_{0^+}^{\xi_i}\phi'(s)(\phi(\xi_i)-\phi(s))^{\rho+r-1}T_z(s)ds\times \\
&\left(1+\delta\sum_{i=1}^{m}b_i(\phi(\xi_i)-\phi(0))^{\rho+q-1}\right)
\end{aligned} \quad (25)$$

and

$$\mathcal{I}^{1-q;\phi}_{0^+}z(0^+) = \frac{\delta\Gamma(\rho+q)}{\Gamma(\rho+r)}\sum_{i=1}^{m}b_i\int_{0^+}^{\xi_i}\phi'(s)(\phi(\xi_i)-\phi(s))^{\rho+r-1}T_z(s)ds. \quad (26)$$

Therefore, in view of Equations (22) and (26), we have

$$\mathcal{I}_{0^+}^{1-q;\phi} z(0^+) = \sum_{i=1}^{m} b_i \mathcal{I}_{0^+}^{\rho;\phi} z(\xi_i). \tag{27}$$

□

3.1. Existence Result Via Schaefer'S Fixed Point Theorem

This subsection will provide the proof of the existence results of Equation (5) using Schaefer's fixed point theorem.

Theorem 2 ([61]). *Let $\mathcal{A} : \mathcal{X} \to \mathcal{X}$ be a completely continuous operator. Suppose that the set $\mathcal{E}(\mathcal{A}) = \{p \in \mathcal{X} : p = \varrho \mathcal{A} p, \text{ for some } \varrho \in [0,1]\}$ is bounded, then $\mathcal{A}$ has a fixed point.*

Thus we need the following assumptions:
(A_1) Let $f : \mathcal{J} \times \mathbb{R}^3 \to \mathbb{R}$ be a function such that $f \in \mathcal{C}_{1-q;\phi}[\mathcal{J}, \mathbb{R}]$ for any $z \in \mathcal{C}_{1-q;\phi}[\mathcal{J}, \mathbb{R}]$.
(A_2) There exist $k, l, m, n \in \mathcal{C}_{1-q;\phi}[\mathcal{J}, \mathbb{R}]$ with $k^* = \sup_{t \in \mathcal{J}} |k(t)| < 1$ such that

$$|f(t, u, v, w)| \leq k(t) + l(t)|x| + m(t)|y| + n(t)|z|, \quad t \in \mathcal{J}, \quad u, v, w \in \mathbb{R}.$$

Theorem 3. *Let $0 < r < 1$, $0 \leq p \leq 1$ and $q = r + p - rp$. Suppose that the assumptions (A_1) and (A_2) are satisfied. Then there exist at least one solution of the problem (5) in the space $\mathcal{C}_{1-q;\phi}^{r,p}[\mathcal{J}, \mathbb{R}]$.*

Proof. Define the operator $F : \mathcal{C}_{1-q;\phi}[\mathcal{J}, \mathbb{R}] \to \mathcal{C}_{1-q;\phi}[\mathcal{J}, \mathbb{R}]$ by

$$\begin{aligned}(Fz)(t) =& \frac{\delta \Gamma(\rho+q)}{\Gamma(q)\Gamma(\rho+r)}(\phi(t) - \phi(0))^{q-1} \sum_{i=1}^{m} b_i \int_{0^+}^{\xi_i} \phi'(s)(\phi(\xi_i) - \phi(s))^{\rho+r-1} T_z(s) ds \\ &+ \frac{1}{\Gamma(r)} \int_{0^+}^{t} \phi'(s)(\phi(t) - \phi(s))^{r-1} T_z(s) ds,\end{aligned} \tag{28}$$

then, clearly the operator F is well-defined. The proof is given in the following steps: Step 1: the operator F is continuous. Let z_n be a sequence such that $z_n \to z$ in $\mathcal{C}_{1-q;\phi}[\mathcal{J}, \mathbb{R}]$. Then for each $t \in \mathcal{J}$, we have

$$\begin{aligned}&|((Fz_n)(t) - (Fz)(t))(\phi(t) - \phi(0))^{1-q}| \\ &\leq \frac{|\delta|\Gamma(\rho+q)}{\Gamma(q)\Gamma(\rho+r)} \sum_{i=1}^{m} b_i \int_{0^+}^{\xi_i} \phi'(s)(\phi(\xi_i) - \phi(s))^{\rho+r-1}|T_{z_n}(s) - T_z(s)| ds \\ &\quad + \frac{1}{\Gamma(r)}(\phi(t) - \phi(0))^{1-q} \int_{0^+}^{t} \phi'(s)(\phi(t) - \phi(s))^{r-1}|T_{z_n}(s) - T_z(s)| ds \\ &\leq \frac{|\delta|\Gamma(\rho+q)}{\Gamma(q)\Gamma(\rho+r)} B(q, \rho+r) \sum_{i=1}^{m} b_i(\phi(\xi_i) - \phi(s))^{\rho+r+q-1} \|T_{z_n}(\cdot) - T_z(\cdot)\|_{\mathcal{C}_{1-q;\phi}} \\ &\quad + \frac{B(q,r)}{\Gamma(r)}(\phi(T) - \phi(0))^r \|T_{z_n}(\cdot) - T_z(\cdot)\|_{\mathcal{C}_{1-q;\phi}} \\ &\leq \left[\frac{|\delta|\Gamma(\rho+q)}{\Gamma(q)\Gamma(\rho+r)} B(q, \rho+r) \sum_{i=1}^{m} b_i(\phi(\xi_i) - \phi(s))^{\rho+r+q-1}\right. \\ &\quad \left. + \frac{B(q,r)}{\Gamma(r)}(\phi(T) - \phi(0))^r\right] \|T_{z_n}(\cdot) - T_z(\cdot)\|_{\mathcal{C}_{1-q;\phi}}.\end{aligned} \tag{29}$$

Since f is continuous, this implies that T_z is also continuous. Therefore, we have

$$\|T_{z_n} - T_z\|_{\mathcal{C}_{1-q;\phi}} \to 0, \quad \text{as} \quad n \to \infty.$$

Step 2: F maps bounded sets into bounded sets in $\mathcal{C}_{1-q;\phi}[\mathcal{J}, \mathbb{R}]$.

Indeed, it suffices to show that for any $\kappa > 0$, there exist a $\mu > 0$ such that for any $z \in \mathbf{B}_\kappa = \{z \in \mathcal{C}_{1-q;\phi}[\mathcal{J}, \mathbb{R}] : \|z\| \leq \kappa\}$, thus we have $\|F(z)\|_{\mathcal{C}_{1-q;\phi}} \leq \mu$.

For simplicity, we put

$$E_1 = \frac{|\delta|\Gamma(\rho+q)}{\Gamma(q)\Gamma(\rho+r)} \sum_{i=1}^{m} b_i \int_{0^+}^{\xi_i} \phi'(s)(\phi(\xi_i) - \phi(s))^{\rho+r-1} |T_z(s)| ds \tag{30}$$

and

$$E_2 = \frac{1}{\Gamma(r)} (\phi(t) - \phi(0))^{1-q} \int_{0^+}^{t} \phi'(s)(\phi(t) - \phi(s))^{r-1} |T_z(s)| ds. \tag{31}$$

It follows from assumption (A_2) that

$$\begin{aligned}|T_z(t)| &= |f(t, z(t), z(\gamma t), T_z(t))| \\ &\leq k(t) + l(t)|z| + m(t)|z| + n(t)|T_z(t)| \\ &\leq \frac{k^* + (l^* + m^*)|z(t)|}{1 - n^*}.\end{aligned} \tag{32}$$

Thus, in view of Equations (30)–(32), we get

$$\begin{aligned}E_1 &\leq \frac{|\delta|\Gamma(\rho+q)}{\Gamma(q)(1-n^*)} \sum_{i=1}^{m} b_i \left(\frac{k^*}{\Gamma(\rho+r+1)} (\phi(\xi_i) - \phi(0))^{\rho+r} \right. \\ &\quad \left. + (l^* + m^*) \frac{(\phi(\xi_i) - \phi(0))^{\rho+r+q-1}}{\Gamma(\rho+r)} B(q, \rho+r) \|z\|_{\mathcal{C}_{1-q;\phi}} \right) \\ E_2 &\leq \frac{1}{(1-n^*)} \left(\frac{k^*}{\Gamma(r+1)} (\phi(T) - \phi(0))^{\rho+r-q+1} \right. \\ &\quad \left. + \frac{(l^* + m^*) B(q, r)}{\Gamma(r)} (\phi(T) - \phi(0))^r \|z\|_{\mathcal{C}_{1-q;\phi}} \right).\end{aligned}$$

This implies that,

$$\begin{aligned}&|(Fz)(t)((\phi(t)-\phi(0))^{q-1})| \\ &\leq \frac{k^*}{(1-n^*)} \left[\frac{|\delta|\Gamma(\rho+q)}{\Gamma(q)\Gamma(\rho+r+1)} \sum_{i=1}^{m} b_i (\phi(\xi_i) - \phi(0))^{\rho+r} \right. \\ &\quad \left. + \frac{k^*}{\Gamma(r+1)} (\phi(T) - \phi(0))^{\rho+r-q+1} \right] \\ &\quad + \frac{(l^* + m^*)}{(1-n^*)} \left[\frac{|\delta|\Gamma(\rho+q)}{\Gamma(q)\Gamma(\rho+r)} B(q, \rho+r) \sum_{i=1}^{m} b_i (\phi(\xi_i) - \phi(0))^{\rho+r+q-1} \right. \\ &\quad \left. + \frac{B(q,r)}{\Gamma(r)} (\phi(T) - \phi(0))^r \right] \|z\|_{\mathcal{C}_{1-q;\phi}} \\ &= \mu.\end{aligned} \tag{33}$$

Step 3: F maps bounded sets into equicontinuous set of $\mathcal{C}_{1-q;\phi}[\mathcal{J},\mathbb{R}]$. Let $t_1, t_2 \in \mathcal{J}$ such that $t_1 \geq t_2$ and B_κ be a bounded set of $\mathcal{C}_{1-q;\phi}[\mathcal{J},\mathbb{R}]$ as defined in Step 2. Let $z \in B_\kappa$, then

$$|((\phi(t_1) - \phi(a))^{q-1})(Fz)(t_1) - ((\phi(t_2) - \phi(0))^{q-1})(Fz)(t_2)|$$
$$\leq \left| \frac{1}{\Gamma(r)}(\phi(t_1) - \phi(0))^{1-q} \int_{0+}^{t_1} \phi'(s)(\phi(t_1) - \phi(s))^{r-1} T_z(s) ds \right.$$
$$\left. - \frac{1}{\Gamma(r)}(\phi(t_2) - \phi(0))^{1-q} \int_{0+}^{t_2} \phi'(s)(\phi(t_2) - \phi(s))^{r-1} T_z(s) ds \right|$$
$$\leq \frac{1}{\Gamma(r)} \left| \int_{0+}^{t_1} \phi'(s) \left[((\phi(t_1) - \phi(0))^{q-1})(\phi(t_1) - \phi(s))^{r-1} \right. \right. \tag{34}$$
$$\left. \left. - ((\phi(t_2) - \phi(0))^{q-1})(\phi(t_2) - \phi(s))^{r-1} T_z(s) ds \right] \right|$$
$$+ \left| \frac{(\phi(t_2) - \phi(0))^{q-1}}{\Gamma(r)} \int_{t_1}^{t_2} \phi'(s)(\phi(t_1) - \phi(0))^{r-1} T_z(s) ds \right|$$
$$\to 0, \quad \text{as} \quad t_1 \to t_2.$$

Thus, steps 1–3, together with the Arzela–Ascoli theorem, show that the operator F is completely continuous.

Step 4: a priori bounds.

It is enough to show that the set $\chi = \{z \in \mathcal{C}_{1-q;\phi}[\mathcal{J},\mathbb{R}] : z = \sigma(Fz), 0 < \sigma < 1\}$ is bounded. Now, let $z \in \chi$, $z = \sigma(Fz)$ for some $0 < \sigma < 1$. Thus for each $t \in \mathcal{J}$, we obtain

$$z(t) = \sigma \left[\frac{\delta \Gamma(\rho+q)}{\Gamma(q)\Gamma(\rho+r)}(\phi(t) - \phi(0))^{q-1} \sum_{i=1}^{m} b_i \int_{0+}^{\xi_i} \phi'(s)(\phi(\xi_i) - \phi(s))^{\rho+r-1} T_z(s) ds \right.$$
$$\left. + \frac{1}{\Gamma(r)} \int_{0+}^{t} \phi'(s)(\phi(t) - \phi(s))^{r-1} T_z(s) ds \right].$$

It follows from assumption (A_2), that for every $t \in \mathcal{J}$,

$$|z(t)(\phi(t) - \phi(0))^{1-q}| \leq |(Fz)(t)(\phi(t) - \phi(0))^{1-q}|$$
$$\leq \frac{k^*}{(1-n^*)} \left[\frac{|\delta| \Gamma(\rho+q)}{\Gamma(q)\Gamma(\rho+r+1)} \sum_{i=1}^{m} b_i (\phi(\xi_i) - \phi(0))^{\rho+r} \right.$$
$$\left. + \frac{k^*}{\Gamma(r+1)}(\phi(T) - \phi(0))^{\rho+r-q+1} \right]$$
$$+ \frac{(l^* + m^*)}{(1-n^*)} \left[\frac{|\delta| \Gamma(\rho+q)}{\Gamma(q)\Gamma(\rho+r)} \mathcal{B}(q, \rho+r) \sum_{i=1}^{m} b_i (\phi(\xi_i) - \phi(0))^{\rho+r+q-1} \right. \tag{35}$$
$$\left. + \frac{\mathcal{B}(q,r)}{\Gamma(r)}(\phi(T) - \phi(0))^r \right] \|z\|_{\mathcal{C}_{1-q;\phi}}$$
$$< \infty.$$

This shows that the set χ is bounded. Hence, by the Schaefer's fixed point theorem, problem (5) has at least one solution. □

3.2. Existence Result Via Banach Contraction Principle

Now, we prove the uniqueness of problem (5) by means of Banach contraction principle. Therefore, the following hypotheses are needed.

(A_3) There exist constants $K, L > 0$ such that

$$|f(t, u, v, w) - f(t, \bar{u}, \bar{v}, \bar{w})| \leq K(|u - \bar{u}| + |v - \bar{v}|) + L|w - \bar{w}|$$

for any $u, v, w, \bar{u}, \bar{v}, \bar{w} \in \mathbb{R}$ and $t \in \mathcal{J}$.

(A_4) Suppose that
$$\left(\frac{2K}{1-L}\right)\Omega < 1,$$

where
$$\Omega = \frac{|\delta|\Gamma(\rho+q)}{\Gamma(q)\Gamma(\rho+r)}\mathcal{B}(q,\rho+r)\sum_{i=1}^{m}b_i(\phi(\xi_i)-\phi(0))^{\rho+r+q-1} + \frac{\mathcal{B}(q,r)}{\Gamma(r)}(\phi(T)-\phi(0))^r. \tag{36}$$

Theorem 4. *Let $0 < r < 1$, $0 \le p \le 1$ and $q = r + p - rp$. Suppose that the hypotheses (A_1), (A_3) and (A_4) are satisfied. Then, problem (5) has a unique solution in the space $\mathcal{C}_{1-q;\phi}^{r,p}[\mathcal{J},\mathbb{R}]$.*

Proof. Define the operator $F : \mathcal{C}_{1-q;\phi}[\mathcal{J},\mathbb{R}] \to \mathcal{C}_{1-q;\phi}[\mathcal{J},\mathbb{R}]$ by

$$(Fz)(t) = \frac{\delta\Gamma(\rho+q)}{\Gamma(q)\Gamma(\rho+r)}(\phi(t)-\phi(0))^{q-1}\sum_{i=1}^{m}b_i\int_{0^+}^{\xi_i}\phi'(s)(\phi(\xi_i)-\phi(s))^{\rho+r-1}T_z(s)ds$$
$$+ \frac{1}{\Gamma(r)}\int_{0^+}^{t}\phi'(s)(\phi(t)-\phi(s))^{r-1}T_z(s)ds, \tag{37}$$

then, clearly the operator F is well-defined. Let $z_1, z_2 \in \mathcal{C}_{1-q;\phi}^{r,p}[\mathcal{J},\mathbb{R}]$ and $t \in \mathcal{J}$, then, we have

$$|((Fz_1)(t) - (Fz_2)(t))(\phi(t)-\phi(0))^{1-q}|$$
$$\le \frac{|\delta|\Gamma(\rho+q)}{\Gamma(q)\Gamma(\rho+r)}\sum_{i=1}^{m}b_i\int_{0^+}^{\xi_i}\phi'(s)(\phi(\xi_i)-\phi(s))^{\rho+r-1}|T_{z_1}(s) - T_{z_2}(s)|ds \tag{38}$$
$$+ \frac{1}{\Gamma(r)}(\phi(t)-\phi(0))^{1-q}\int_{0^+}^{t}\phi'(s)(\phi(t)-\phi(s))^{r-1}|T_{z_1}(s) - T_{z_2}(s)|ds$$

and

$$|T_{z_1}(t) - T_{z_2}(t)| = |f(t, z_1(t), z_1(\gamma t)), T_{z_1}(t) - f(t, z_2(t), z_2(\gamma t), T_{z_2}(t))|$$
$$\le K(|z_1(t) - z_2(t)| + |z_1(\gamma t) - z_2(\gamma t)|) + L|(T_{z_1})(t) - (T_{z_2})(t)| \tag{39}$$
$$\le \left(\frac{2K}{1-L}\right)|z_1(t) - z_2(t)|.$$

Thus, by substituting Equation (39) in Equation (38), we obtain

$$|((Fz_1)(t) - (Fz_2)(t))(\phi(t)-\phi(0))^{1-q}|$$
$$\le \frac{|\delta|\Gamma(\rho+q)}{\Gamma(q)\Gamma(\rho+r)}\sum_{i=1}^{m}b_i\left(\frac{2K}{1-L}\int_{0^+}^{\xi_i}\phi'(s)(\phi(\xi_i)-\phi(s))^{\rho+r-1}ds\right)\|z_1(t) - z_2(t)\|_{\mathcal{C}_{1-q;\phi}}$$
$$+ \frac{1}{\Gamma(r)}(\phi(t)-\phi(0))^{1-q}\frac{2K}{(1-L)}\left(\int_{0^+}^{t}\phi'(s)(\phi(t)-\phi(s))^{r-1}ds\right)\|z_1(t) - z_2(t)\|_{\mathcal{C}_{1-q;\phi}} \tag{40}$$
$$\le \frac{2K}{(1-L)}\left(\frac{|\delta|\Gamma(\rho+q)}{\Gamma(q)\Gamma(\rho+r)}\mathcal{B}(q,\rho+r)\sum_{i=1}^{m}b_i(\phi(\xi_i)-\phi(0))^{\rho+r+q-1}\right.$$
$$\left.+ \frac{\mathcal{B}(q,r)}{\Gamma(r)}(\phi(T)-\phi(0))^r\right)\|z_1(t) - z_2(t)\|_{\mathcal{C}_{1-q;\phi}}.$$

Also,

$$\|(Fz_1) - (Fz_2)\|_{\mathcal{C}_{1-q;\phi}} \leq \frac{2K}{(1-L)} \left(\frac{|\delta|\Gamma(\rho+q)}{\Gamma(q)\Gamma(\rho+r)} \mathcal{B}(q, \rho+r) \sum_{i=1}^{m} b_i(\phi(\xi_i) - \phi(0))^{\rho+r+q-1} \right.$$
$$\left. + \frac{\mathcal{B}(q,r)}{\Gamma(r)} (\phi(T) - \phi(0))^r \right) \|z_1(t) - z_2(t)\|_{\mathcal{C}_{1-q;\phi}}. \tag{41}$$

It follows from hypotheses (A_4) that F is a contraction map. Therefore, by Banach contraction principle, we can conclude that problem (5) has a unique solution. □

4. Ulam-Hyers Stabilty

Two types of Ulam stability for (5) are discussed in this section, namely Ulam-Hyers and generalized Ulam-Hyers stability.

Definition 4. *Problem (5) is said to be Ulam-Hyers stable if there exists $\omega \in \mathbb{R}_+ \setminus \{0\}$, such that for each $\epsilon > 0$ and solution $x \in \mathcal{C}^1_{1-q;\phi}[\mathcal{J}, \mathbb{R}]$ of the inequality*

$$|{}^H\mathcal{D}^{r,p;\phi}_{0+} x(t) - f(t, x(t), x(\gamma t), {}^H\mathcal{D}^{r,p;\phi}_{0+} x(\gamma t))| \leq \epsilon, \quad t \in \mathcal{J}, \tag{42}$$

there exists a solution $z \in \mathcal{C}^1_{1-q;\phi}[\mathcal{J}, \mathbb{R}]$ of equation (5), such that

$$|x(t) - z(t)| \leq \omega \epsilon, \quad t \in \mathcal{J}.$$

Definition 5. *Problem (5) is said to be generalized Ulam-Hyers stable if there exist $\Phi \in \mathcal{C}(\mathbb{R}_+, \mathbb{R}_+)$, $\Phi_f(0) = 0$, such that for each solution $x \in \mathcal{C}^1_{1-q;\phi}[\mathcal{J}, \mathbb{R}]$ of the (42), there exists a solution $z \in \mathcal{C}^1_{1-q;\phi}[\mathcal{J}, \mathbb{R}]$ of Equation (5), such that*

$$|x(t) - z(t)| \leq \Phi_f \epsilon, \quad t \in \mathcal{J}.$$

Remark 1. *A function $x \in \mathcal{C}_{1-q;\phi}[\mathcal{J}, \mathbb{R}]$ is a solution of the inequality (42), if and only if there exist a function $g \in \mathcal{C}_{1-q;\phi}[\mathcal{J}, \mathbb{R}]$ such that:*

(i) $|g(t)| \leq \epsilon, \quad t \in \mathcal{J}.$

(ii) ${}^H\mathcal{D}^{r,p;\phi}_{0+} x(t) = f(t, x(t), x(\gamma t), {}^H\mathcal{D}^{r,p;\phi}_{0+} x(\gamma t)) + g(t), \quad t \in \mathcal{J}.$

Lemma 9. *Let $0 < r < 1, 0 \leq p \leq 1$, if a function $x \in \mathcal{C}_{1-q;\phi}[\mathcal{J}, \mathbb{R}]$ is a solution of the inequality (42), then x is a solution of the following integral inequality*

$$\left| x(t) - A_x - \frac{1}{\Gamma(r)} \int_{0+}^{t} \phi'(s)(\phi(t) - \phi(s))^{r-1} T_x ds \right| \leq \Omega \epsilon. \tag{43}$$

Proof. Clearly it follow from Remark 1 that

$${}^H\mathcal{D}^{r,p;\phi}_{0+} x(t) = f(t, x(t), x(\gamma t), {}^H\mathcal{D}^{r,p;\phi}_{0+} x(\gamma t)) + g(t)$$
$$= T_x(t) + g(t),$$

and

$$x(t) = \frac{\delta\Gamma(\rho+q)}{\Gamma(q)\Gamma(\rho+r)}(\phi(t)-\phi(0))^{q-1}\sum_{i=1}^{m} b_i \left(\int_{0^+}^{\xi_i} \phi'(s)(\phi(\xi_i)-\phi(s))^{\rho+r-1} T_x(s)ds \right.$$
$$\left. + \int_{0^+}^{\xi_i} \phi'(s)(\phi(\xi_i)-\phi(s))^{\rho+r-1} g(s)ds \right) + \frac{1}{\Gamma(r)}\int_{0^+}^{t} \phi'(s)(\phi(t)-\phi(s))^{r-1} T_x(s)ds \quad (44)$$
$$+ \frac{1}{\Gamma(r)}\int_{0^+}^{t} \phi'(s)(\phi(t)-\phi(s))^{r-1} g(s)ds.$$

Hence

$$\left| x(t) - A_x - \frac{1}{\Gamma(r)} \int_{0^+}^{t} \phi'(s)(\phi(t)-\phi(s))^{r-1} T_x ds \right|$$
$$= \left| \frac{\delta\Gamma(\rho+q)}{\Gamma(q)\Gamma(\rho+r)}(\phi(t)-\phi(0))^{q-1}\sum_{i=1}^{m} b_i \int_{0^+}^{\xi_i} \phi'(s)(\phi(\xi_i)-\phi(s))^{\rho+r-1} g(s)ds \right.$$
$$\left. + \frac{1}{\Gamma(r)}\int_{0^+}^{t} \phi'(s)(\phi(t)-\phi(s))^{r-1} g(s)ds \right| \quad (45)$$
$$\leq \frac{|\delta|\Gamma(\rho+q)}{\Gamma(q)\Gamma(\rho+r)}(\phi(t)-\phi(0))^{q-1}\sum_{i=1}^{m} |b_i| \int_{0^+}^{\xi_i} \phi'(s)(\phi(\xi_i)-\phi(s))^{\rho+r-1} |g(s)|ds$$
$$+ \frac{1}{\Gamma(r)}\int_{0^+}^{t} \phi'(s)(\phi(t)-\phi(s))^{r-1} |g(s)|ds$$
$$\leq \Omega\epsilon.$$

□

Theorem 5. *Suppose that the hypotheses (A_1), (A_3) and (A_4) are satisfied. Then problem (5) is both Ulam-Hyers and generalized Ulam-Hyers stable on $\mathcal{J}$.*

Proof. Let $\epsilon > 0$ and $x \in \mathcal{C}_{1-q;\phi}[\mathcal{J},\mathbb{R}]$ be a function which satisfies the inequality (42) and let $z \in \mathcal{C}_{1-q;\phi}[\mathcal{J},\mathbb{R}]$ be a unique solution of the following implicit fractional pantograph differential equation

$$^H\mathcal{D}_{0^+}^{r,p;\phi} z(t) = f(t,z(t),z(\gamma t), {}^H\mathcal{D}_{0^+}^{r,p;\phi} z(\gamma t))| \leq \epsilon, \quad t \in \mathcal{J}, \quad 0 < r < 1, 0 \leq p \leq 1,$$
$$\mathcal{I}_{0^+}^{1-q;\phi} z(0^+) = \mathcal{I}_{0^+}^{1-q;\phi} z(0^+) = \sum_{i=1}^{m} b_i \mathcal{I}_{0^+}^{\rho;\phi} z(\xi_i), \quad \xi_i \in (0,T], \quad q = r+p-rp.$$

Using Lemma 9, we have

$$z(t) = A_z - \frac{1}{\Gamma(r)} \int_{0^+}^{t} \phi'(s)(\phi(t)-\phi(s))^{r-1} T_z(s)ds,$$

where

$$A_z = \frac{\delta\Gamma(\rho+q)}{\Gamma(q)\Gamma(\rho+r)}(\phi(t)-\phi(a))^{q-1}\sum_{i=1}^{m} b_i \int_{0^+}^{\xi_i} \phi'(s)(\phi(\xi_i)-\phi(s))^{\rho+r-1} T_z(s)ds.$$

Clearly, if $z(\xi_i) = x(\xi_i)$ and $\mathcal{I}_{0+}^{1-q;\phi}z(0^+) = \mathcal{I}_{0+}^{1-q;\phi}z(0^+)$, we get $A_z = A_x$ and that

$$|A_z - A_x|$$
$$= \frac{|\delta|\Gamma(\rho+q)}{\Gamma(q)\Gamma(\rho+r)}(\phi(t) - \phi(0))^{q-1}\sum_{i=1}^{m}b_i\int_{0+}^{\xi_i}\phi'(s)(\phi(\xi_i) - \phi(s))^{\rho+r-1}|T_z(s) - T_x(s)|ds$$
$$\leq \frac{|\delta|\Gamma(\rho+q)}{\Gamma(q)\Gamma(\rho+r)}(\phi(t) - \phi(0))^{q-1}\left(\frac{2K}{1-L}\right)\sum_{i=1}^{m}b_i\mathcal{I}_{0+}^{\rho+r;\phi}|z(s) - x(s)|(\xi_i)$$
$$= 0.$$

Now for any $t \in \mathcal{J}$ and Lemma 9, we have

$$|x(t) - z(t)| = \left|x(t) - A_x - \frac{1}{\Gamma(r)}\int_{0+}^{t}\phi'(s)(\phi(t) - \phi(s))^{r-1}T_x(s)ds\right|$$
$$+ \frac{1}{\Gamma(r)}\int_{0+}^{t}\phi'(s)(\phi(t) - \phi(s))^{r-1}|T_x(s) - T_z(s)|ds$$
$$\leq \left|x(t) - A_x - \frac{1}{\Gamma(r)}\int_{0+}^{t}\phi'(s)(\phi(t) - \phi(s))^{r-1}T_x(s)ds\right|$$
$$+ \left(\frac{2K}{1-L}\right)\frac{1}{\Gamma(r)}\int_{0+}^{t}\phi'(s)(\phi(t) - \phi(s))^{r-1}|x(s) - z(s)|ds$$
$$\leq \Omega\epsilon + \left(\frac{2K}{1-L}\right)\frac{\mathcal{B}(r,q)(\phi(T) - \phi(0))^r}{\Gamma(r)}|x(t) - z(t)|ds.$$

Thus,
$$|x(t) - z(t)| \leq \omega\epsilon,$$

where
$$\omega = \frac{\Omega(1-L)\Gamma(r)}{(1-L)\Gamma(r) - 2K(\phi(T) - \phi(0))^r\mathcal{B}(r,q)}.$$

Therefore, problem (5) is Ulam-Hyers stable. Moreover, if we set $\Phi_f(\epsilon) = \omega\epsilon$ such that $\Phi_f(0) = 0$, then problem (5) is generalized Ulam-Hyers stable. □

5. Examples

Example 1. *Consider the implicit fractional pantograph differential equation which involves ϕ-Hilfer fractional derivative of the following form:*

$$\begin{cases} {}^H\mathcal{D}_{0+}^{\frac{2}{3},\frac{1}{2};t}z(t) = \dfrac{1}{3(5^{2t}+5)[1+|z(t)|+|z(\frac{1}{2}t)|+|{}^H\mathcal{D}_{1+}^{\frac{2}{3},\frac{1}{2};t}z(\frac{1}{2}t)|]}, & t \in \mathcal{J} = (0,2], \\ \mathcal{I}_{0+}^{1-\frac{5}{6};t}z(0) = 3\mathcal{I}_{0+}^{\frac{1}{2};t}z(\frac{3}{2}), & \frac{2}{3} \leq \frac{5}{6} = \frac{2}{3} + (\frac{1}{2}) - (\frac{2}{3})(\frac{1}{2}). \end{cases} \quad (46)$$

By comparing (5) with (46), we have:
$r = \frac{2}{3}$, $p = \gamma = \rho = \frac{1}{2}$, $q = \frac{5}{6}$, $T = 2$ and $\phi(\cdot) = t$. Also from the initial condition we can easily see that $b_1 = 3$ since $m = 1$, $\xi_1 = \frac{3}{2} \in \mathcal{J}$ and $f: \mathcal{J} \times \mathbb{R}^3 \to \mathbb{R}$ is a function defined by

$$f(t, u, v, w) = \frac{1}{3(5^{2t}+5)(1+|u|+|v|+|w|)}, t \in \mathcal{J}, \quad u, v, w \in \mathbb{R}_+.$$

Obviously, f is continuous and for all $u, v, w, \bar{u}, \bar{v}, \bar{w} \in \mathbb{R}_+$ and $t \in \mathcal{J}$, we have
$|f(t, u, v, w) - f(t, \bar{u}, \bar{v}, \bar{w})| \leq \frac{1}{90} (|u - \bar{u}| + |v - \bar{v}| + |w - \bar{w}|)$. Thus, it follows that conditions (A_1) and (A_3) are true with $K = L = \frac{1}{90}$. Therefore, by simple calculation, we get $|\delta| \approx 0.3935$ and

$$\left(\frac{2K}{1-L} \right) \Omega \approx 0.0642 < 1.$$

Since, all the assumptions of Theorem 4 are satisfied. Then problem (5) has a unique solution on $\mathcal{J}$. However, we can also find out that $\Omega \approx 2.8551 > 0$ and $\omega = 2.9321 > 0$. Hence, by Theorem 5, problem (5) is both Ulam-Hyers and also generalized Ulam-Hyers stable.

Example 2. *Consider the implicit fractional pantograph differential equation which involves ϕ-Hilfer fractional derivative of form:*

$$\begin{cases} {}^H\mathcal{D}_{0^+}^{\frac{1}{3},\frac{2}{3};\sqrt{t}} z(t) = \dfrac{2 + |z(t)| + |z(\frac{3}{2}t)| + \left|{}^H\mathcal{D}_{0^+}^{\frac{1}{3},\frac{2}{3};\sqrt{t}} z(\frac{3}{2}t)\right|}{95 e^{2t} \cos 2t \left(1 + |z(t)| + |z(\frac{3}{2}t)| + \left|{}^H\mathcal{D}_{0^+}^{\frac{1}{3},\frac{2}{3};\sqrt{t}} z(\frac{3}{2}t)\right| \right)}, & t \in \mathcal{J} = (0,1], \\ \mathcal{I}_{0^+}^{1-\frac{7}{9};\sqrt{t}} z(0) = z(\frac{1}{2}) + 3z(\frac{4}{5}), & \frac{2}{3} < \frac{7}{9} = \frac{1}{3} + (\frac{2}{3}) - (\frac{1}{3})(\frac{2}{3}). \end{cases} \quad (47)$$

By comparing Equation (47) with Equation (5), we obtain that:
$r = \frac{1}{3}, p = \frac{2}{3}, q = \frac{7}{9}, \rho = 0, \gamma = \frac{3}{2}, T = 1$ and $\phi(\cdot) = \sqrt{t}$. Also we can easily see that $b_1 = 1, b_2 = 3$ since $m = 2, \xi_1 = \frac{1}{2}, \xi_2 = \frac{4}{5} \in \mathcal{J}$ and $f : \mathcal{J} \times \mathbb{R}^3 \to \mathbb{R}$ is a function defined by

$$f(t, u, v, w) = \frac{2 + |u| + |v| + |w|}{95 e^{2t} \cos 2t \, (1 + |u| + |v| + |w|)}, \quad t \in \mathcal{J}, \quad u, v, w \in \mathbb{R}_+.$$

Thus, f is continuous and we can see that, for all $u, v, w, \bar{u}, \bar{v}, \bar{w} \in \mathbb{R}_+$ and $t \in \mathcal{J}$,
$|f(t, u, v, w) - f(t, \bar{u}, \bar{v}, \bar{w})| \leq \frac{1}{95} (|u - \bar{u}| + |v - \bar{v}| + |w - \bar{w}|)$.
So assumptions (A_1) and (A_3) are fulfilled with $K = L = \frac{1}{95}$. Furthermore,

$$|f(t, u, v, w)| \leq \frac{1}{95 e^{2t} \cos 2t} (2 + |u| + |v| + |w|), \quad t \in \mathcal{J}.$$

The above implies that (A_2) is true with $k(t) = \frac{2}{95 e^{2t} \cos 2t}$, $l(t) = m(t) = n(t) = \frac{1}{95 e^{2t} \cos 2t}$ and $k^* = \frac{2}{95}$, $l^* = m^* = n^* = \frac{1}{95}$. Therefore, all the hypotheses of Theorem 4 are satisfied, which means that problem (5) has at least one solution on $\mathcal{J}$. Moreover, by using the same procedure as in example 5.2, we obtain, that $|\delta| \approx 1.1025, \Omega \approx 3.6662 > 0$ and

$$\left(\frac{2K}{1-L} \right) \Omega \approx 0.0782 < 1.$$

Thus, all the hypotheses of Theorem 4 holds. Hence, problem (5) has a unique solution on $\mathcal{J}$.

Example 3. *Consider the implicit fractional pantograph differential equation which involves ϕ-Hilfer fractional derivative of the following form:*

$$\begin{cases} {}^H\mathcal{D}_{0^+}^{\frac{1}{2},\frac{1}{3};t} z(t) = \dfrac{1}{4^{t+3}[1 + |z(t)| + |z(\frac{1}{6}t)| + |{}^H\mathcal{D}_{0^+}^{\frac{1}{2},\frac{1}{3};t} z(\frac{1}{6}t)|]}, & t \in \mathcal{J} = (0,3], \\ \mathcal{I}_{0^+}^{1-q;t} z(0) = \sqrt{2}\mathcal{I}_{0^+}^{\frac{5}{3};t} z(2) + \sqrt{5}\mathcal{I}_{0^+}^{\frac{2}{5};t} z(\frac{5}{2}), & q = \frac{1}{2} + (\frac{1}{3}) - (\frac{1}{2})(\frac{1}{3}). \end{cases} \quad (48)$$

By comparing Equation (5) with Equation (48), we get the followings values:
$r = \frac{1}{2}$, $p = \frac{1}{3}$ $\gamma = \frac{1}{6}$ $\rho = \frac{2}{5}$, $q = \frac{2}{3}$, $T = 3$ and $\phi(\cdot) = t$. Also from the initial condition we can easily see that $b_1 = \sqrt{2}$ $b_1 = \sqrt{5}$ since $m = 2$, $\xi_1 = 2$ $\xi_2 = \frac{5}{2}$ and $f : \mathcal{J} \times \mathbb{R}^3 \to \mathbb{R}$ is a function defined by

$$f(t, u, v, w) = \frac{1}{4^{t+3}(1 + |u| + |v| + |w|)}, t \in \mathcal{J}, \quad u, v, w \in \mathbb{R}_+.$$

Thus, f is continuous and for all $u, v, w, \bar{u}, \bar{v}, \bar{w} \in \mathbb{R}_+$ and $t \in \mathcal{J}$, yields
$|f(t, u, v, w) - f(t, \bar{u}, \bar{v}, \bar{w})| \leq \frac{1}{64}(|u - \bar{u}| + |v - \bar{v}| + |w - \bar{w}|)$. Hence, it follows that conditions (A_1) and (A_3) are true with $K = L = \frac{1}{90}$. Therefore, by substitution these values, we get $|\delta| \approx 0.3456$, $\Omega \approx 7.4535 > 0$ and

$$\left(\frac{2K}{1-L}\right) \Omega \approx 0.2366 < 1,$$

which implies that, all the assumptions of Theorem 4 are satisfied. Thus, problem (5) has a unique solution on $\mathcal{J}$.

6. Conclusions

In our study, Firstly, we established the equivalence between problem (5) and the Volterra integral equation. Secondly, Banach and Schaefer's fixed point theorems were used to establish the existence and uniqueness solutions for implicit fractional pantograph differential equation which involves ϕ-Hilfer fractional derivatives. Based on ϕ-Hilfer fractional derivatives, we found that the stability of Ulam-Hyers and generalized Ulam-Hyers allowed on the implicit fractional pantograph differential equation, supplemented with a nonlocal Riemann-Liouville condition. In addition, examples were given to illustrate our main results. Moreover, it worthy to mention the following remarks:

- If $\rho \to 0$ and $\phi(t) = t$, we obtain the results of [48] and [52]. Furthermore, if $\rho \to 0$ we obtain the Ulam-Hyers and generalized Ulam-Hyers stability for the implicit fractional pantograph differential equations with ϕ-Hilfer fractional derivatives [52,58] and if $q = 0$ we obtain [51].
- If $\rho \to 1$, the nonlocal Riemann-Liouville integral condition reduces to a nonlocal integral condition which plays an important role in computational fluid dynamics, ill-posed problems and mathematical models [62].
- If $\rho \to 0$, the initial condition reduces to multi-point nonlocal condition.
- If $t \in [a, b]$ as defined in paper [58], the function $f(t, x(t), x(\lambda t))$ is not well-defined for some choice of $0 < \lambda < 1$. Thus, our results modify and improve the above cited remarks and can be considered as the development of the qualitative analysis of fractional differential equations. The study of Ulam-Hyers stability in the frame of ϕ-Hilfer fractional derivative with a generalized nonlocal boundary condition proposed in this paper and other coupled system will be presented in the near future.

Author Contributions: The authors contributed equally in writing this article. All authors have read and agreed to the published version of the manuscript.

Funding: Petchra Pra Jom Klao Doctoral Scholarship for Ph.D. program of King Mongkut's University of Technology Thonburi (KMUTT). The Center of Excellence in Theoretical and Computational Science (TaCS-CoE), KMUTT. Center of Excellence in Theoretical and Computational Science (TaCS-CoE), KMUTT. King Mongkut's University of Technology North Bangkok, Contract no. KMUTNB-63-KNOW-033.

Acknowledgments: The authors acknowledge the financial support provided by the Center of Excellence in Theoretical and Computational Science (TaCS-CoE), KMUTT. The first author was supported by the "Petchra Pra Jom Klao Ph.D. Research Scholarship from King Mongkut's University of Technology Thonburi" (Grant No. 13/2561). Moreover, this research work was financially supported by King Mongkut's University of Technology Thonburi through the KMUTT 55th Anniversary Commemorative Fund.

Conflicts of Interest: The authors declare no conflict of interest. The funders had no role in the design of the study; in the collection, analyses, or interpretation of data; in the writing of the manuscript, or in the decision to publish the results.

References

1. Kilbas, A.A.; Srivastava, H.M.; Trujillo, J.J. *Theory and Applications of fractional derivatial Equations*; Elsevier Science Limited: Amsterdam, The Netherlands, 2006; Volume 204.
2. Samko, S.G.; Kilbas, A.A.; Marichev, O.I.; *Fractional Integrals and Derivatives*; Gordon and Breach Science Publishers: Yverdon Yverdon-les-Bains, Switzerland, 1993; Volume 1993.
3. Podlubny, I. *Fractional Differential Equations: An Introduction to Fractional Derivatives, Fractional Differential Equations, to Methods of Their Solution and Some of Their Applications*; Elsevier: Amsterdam, The Netherlands, 1998.
4. Mainardi, F. *Fractional Calculus and Waves in Linear Viscoelasticity: An Introduction to Mathematical Models*; World Scientific: Singapore, 2010.
5. Furati, K.; Kassim, M.; Tata, N. Existence and uniqueness for a problem involving Hilfer fractional derivative. *Comput. Math. Appl.* **2012**, *64*, 1616–1626. [CrossRef]
6. Oliveira, D.S.; de Oliveira, E.C. Hilfer–Katugampola fractional derivatives. *Comput. Appl. Math.* **2018**, *37*, 3672–3690. [CrossRef]
7. Osler, T.J. The fractional derivative of a composite function. *SIAM J. Math. Anal.* **1970**, *1*, 288–293. [CrossRef]
8. Gambo, Y.Y.; Jarad, F.; Baleanu, D.; Abdeljawad, T. On Caputo modification of the Hadamard fractional derivatives. *Adv. Differ. Equ.* **2014**, *2014*, 10. [CrossRef]
9. Jarad, F.; Abdeljawad, T.; Baleanu, D. Caputo-type modification of the Hadamard fractional derivatives. *Adv. Differ. Equ.* **2012**, *2012*, 142. [CrossRef]
10. Jarad, F.; Abdeljawad, T.; Baleanu, D. On the generalized fractional derivatives and their Caputo modification. *J. Nonlinear Sci. Appl.* **2017**, *10*, 2607–2619. [CrossRef]
11. Marin, M.; Baleanu, D.; Vlase, S. Effect of microtemperatures for micropolar thermoelastic bodies. *Struct. Eng. Mech.* **2017**, *61*, 381–387. [CrossRef]
12. Gladkov, S.; Bogdanova, S. On the question of the magnetic susceptibility of fractal ferromagnetic wires. *Russ. Phys. J.* **2014**, *57*, 469–473. [CrossRef]
13. Hilfer, R. Fractional calculus and regular variation in thermodynamics. In *Applications of Fractional Calculus in Physics*; World Scientific: Singapore, 2000; pp. 429–463.
14. Hilfer, R. Fractional time evolution. In *Applications of Fractional Calculus in Physics*; World Scientific: Singapore, 2000; pp. 87–130.
15. Gerolymatou, E.; Vardoulakis, I.; Hilfer, R. Modelling infiltration by means of a nonlinear fractional diffusion model. *J. Phys. Appl. Phys.* **2006**, *39*, 4104. [CrossRef]
16. Vivek, D.; Kanagarajan, K.; Sivasundaram, S. Dynamics and stability of pantograph equations via Hilfer fractional derivative. *Nonlinear Stud.* **2016**, *23*, 685–698.
17. Vivek, D.; Kanagarajan, K.; Harikrishnan, S. Analytic study on nonlocal initial value problems for pantograph equations with Hilfer-Hadamard fractional derivative. *Int. J. Math. Its Appl.* **2018**, *55*, 7.
18. Abdo, M.S.; Panchal, S.K.; Bhairat, S.P. Existence of solution for Hilfer fractional differential equations with boundary value conditions. *arXiv* **2019**, arXiv:1909.13680.
19. Sousa, J.V.d.C.; de Oliveira, E.C. On the ψ-Hilfer fractional derivative. *Commun. Nonlinear Sci. Numer. Simul.* **2018**, *60*, 72–91. [CrossRef]
20. Ravichandran, C.; Logeswari, K.; Jarad, F. New results on existence in the framework of Atangana–Baleanu derivative for fractional integro-differential equations. *Chaos Solitons Fractals* **2019**, *125*, 194–200. [CrossRef]
21. Abdo, M.S.; Panchal, S.K. Fractional integro-differential equations involving ψ-Hilfer fractional derivative. *Adv. Appl. Math. Mech.* **2019**, *11*, 1–22.
22. Almeida, R. A Caputo fractional derivative of a function with respect to another function. *Commun. Nonlinear Sci. Numer. Simul.* **2017**, *44*, 460–481. [CrossRef]
23. Ahmad, B.; Nieto, J.J. Existence results for nonlinear boundary value problems of fractional integrodifferential equations with integral boundary conditions. *Bound. Value Probl.* **2009**, *2009*, 708576. [CrossRef]
24. Ahmad, B.; Sivasundaram, S. Existence of solutions for impulsive integral boundary value problems of fractional order. *Nonlinear Anal. Hybrid Syst.* **2010**, *4*, 134–141. [CrossRef]
25. Wang, G.; Ghanmi, A.; Horrigue, S.; Madian, S. Existence Result and Uniqueness for Some Fractional Problem. *Mathematics* **2019**, *7*, 516. [CrossRef]
26. Ali, K.B.; Ghanmi, A.; Kefi, K. Existence of solutions for fractional differential equations with Dirichlet boundary conditions. *Electron. J. Differ. Equ.* **2016**, *2016*, 1–11.

27. Nieto, J.; Ouahab, A.; Venktesh, V. Implicit fractional differential equations via the Liouville–Caputo derivative. *Mathematics* **2015**, *3*, 398–411. [CrossRef]
28. Zhang, W.; Liu, W.; Xue, T. Existence and uniqueness results for the coupled systems of implicit fractional differential equations with periodic boundary conditions. *Adv. Differ. Equ.* **2018**, *2018*, 413. [CrossRef]
29. Staněk, S. Existence results for implicit fractional differential equations with nonlocal boundary conditions. *Mem. Differ. Equations Math. Phys.* **2017**, *72*, 119–130.
30. Srivastava, H. Some families of Mittag-Leffler type functions and associated operators of fractional calculus (Survey). *Turk. World Math. Soc. J. Pure Appl. Math.* **2016**, *7*, 123–145.
31. Benchohra, M.; Bouriah, S.; Nieto, J.J. Existence of periodic solutions for nonlinear implicit Hadamard's fractional differential equations. *Revista de la Real Academia de Ciencias Exactas Físicas y Naturales. Serie A Matemáticas* **2018**, *112*, 25–35. [CrossRef]
32. Srivastava, H.; El-Sayed, A.; Gaafar, F. A class of nonlinear boundary value problems for an arbitrary fractional-order differential equation with the Riemann-Stieltjes functional integral and infinite-point boundary conditions. *Symmetry* **2018**, *10*, 508. [CrossRef]
33. Borisut, P.; Kumam, P.; Ahmed, I.; Sitthithakerngkiet, K. Nonlinear Caputo Fractional Derivative with Nonlocal Riemann-Liouville Fractional Integral Condition Via Fixed Point Theorems. *Symmetry* **2019**, *11*, 829. [CrossRef]
34. Živorad Tomovski.; Hilfer, R.; Srivastava, H.M. Fractional and operational calculus with generalized fractional derivative operators and Mittag–Leffler type functions. *Integral Transform. Spec. Funct.* **2010**, *21*, 797–814.
35. Kharade, J.P.; Kucche, K.D. On the Impulsive Implicit ψ-Hilfer Fractional Differential Equations with Delay. *arXiv* **2019**, arXiv:1908.07793.
36. Hyers, D.H. On the stability of the linear functional equation. *Proc. Natl. Acad. Sci. USA* **1941**, *27*, 222. [CrossRef]
37. Ulam, S.M. *Problems in Modern Mathematics*; Courier Corporation: North Chelmsford, MA, USA, 2004.
38. Ulam, S.M. *A Collection of Mathematical Problems*; Interscience Publishers: Geneva, Switzerland, 1960; Volume 8.
39. Aoki, T. On the stability of the linear transformation in Banach spaces. *J. Math. Soc. Jpn.* **1950**, *2*, 64–66. [CrossRef]
40. Rassias, T.M. On the stability of the linear mapping in Banach spaces. *Proc. Am. Math. Soc.* **1978**, *72*, 297–300. [CrossRef]
41. Abbas, S.; Benchohra, M.; Lagreg, J.; Alsaedi, A.; Zhou, Y. Existence and Ulam stability for fractional differential equations of Hilfer-Hadamard type. *Adv. Differ. Equ.* **2017**, *2017*, 180. [CrossRef]
42. Wang, J.; Lin, Z. Ulam's type stability of Hadamard type fractional integral equations. *Filomat* **2014**, *28*, 1323–1331. [CrossRef]
43. de Oliveira, E.C.; Sousa, J.V.d.C. Ulam-Hyers-Rassias stability for a class of fractional integro-differential equations. *Results Math.* **2018**, *73*, 111. [CrossRef]
44. Sousa, J.V.d.C.; Oliveira, D.d.S.; Capelas de Oliveira, E. On the existence and stability for noninstantaneous impulsive fractional integrodifferential equation. *Math. Methods Appl. Sci.* **2019**, *42*, 1249–1261. [CrossRef]
45. Liu, K.; Wang, J.; O'Regan, D. Ulam-Hyers–Mittag-Leffler stability for ψ-Hilfer fractional-order delay differential equations. *Adv. Differ. Equ.* **2019**, *2019*, 50. [CrossRef]
46. Liu, K.; Fečkan, M.; O'Regan, D.; Wang, J. Hyers-Ulam Stability and Existence of Solutions for Differential Equations with Caputo-Fabrizio Fractional Derivative. *Mathematics* **2019**, *7*, 333. [CrossRef]
47. Sousa, J.V.d.C.; de Oliveira, E.C. On the Ulam-Hyers-Rassias stability for nonlinear fractional differential equations using the ψ-Hilfer operator. *J. Fixed Point Theory Appl.* **2018**, *20*, 96. [CrossRef]
48. Vivek, D.; Kanagarajan, K.; Elsayed, E. Some existence and stability results for Hilfer-fractional implicit differential equations with nonlocal conditions. *Mediterr. J. Math.* **2018**, *15*, 15. [CrossRef]
49. Ahmed, H.M.; El-Borai, M.M.; El-Owaidy, H.M.; Ghanem, A.S. Impulsive Hilfer fractional differential equations. *Adv. Differ. Equations* **2018**, *2018*, 226. [CrossRef]
50. Ockendon, J.R.; Tayler, A.B. The dynamics of a current collection system for an electric locomotive. *Proc. R. Soc. London. Math. Phys. Sci.* **1971**, *322*, 447–468. [CrossRef]
51. Balachandran, K.; Kiruthika, S.; Trujillo, J. Existence of solutions of nonlinear fractional pantograph equations. *Acta Math. Sci.* **2013**, *33*, 712–720. [CrossRef]

52. Vivek, D.; Kanagarajan, K.; Sivasundaram, S. Theory and analysis of nonlinear neutral pantograph equations via Hilfer fractional derivative. *Nonlinear Stud.* **2017**, *24*, 699–712.
53. Anguraj, A.; Vinodkumar, A.; Malar, K. Existence and stability results for random impulsive fractional pantograph equations. *Filomat* **2016**, *30*, 3839–3854. [CrossRef]
54. Bhalekar, S.; Patade, J. Series Solution of the Pantograph Equation and Its Properties. *Fractal Fract.* **2017**, *1*, 16. [CrossRef]
55. Shah, K.; Vivek, D.; Kanagarajan, K. Dynamics and Stability of ψ-fractional Pantograph Equations with Boundary Conditions. *Boletim da Sociedade Paranaense de Matemática* **2018**, *22*, 1–13.
56. Elsayed, E.M.; Harikrishnan, S.; Kanagarajan, K. Analysis of nonlinear neutral pantograph differential equations with Hilfer fractional derivative. *MathLAB* **2018**, *1*, 231–240.
57. Harikrishnan, S.; Ibrahim, R.; Kanagarajan, K. Establishing the existence of Hilfer fractional pantograph equations with impulses. *Fundam. J. Math. Appl.* **2018**, *1*, 36–42. [CrossRef]
58. Harikrishnan, S.; Elsayed, E.; Kanagarajan, K. Existence and uniqueness results for fractional pantograph equations involving ψ-Hilfer fractional derivative. *Dyn. Contin. Discret. Impuls. Syst.* **2018**, *25*, 319–328.
59. Ahmad, B.; Sivasundaram, S. Some existence results for fractional integro-differential equations with nonlinear conditions. *Commun. Appl. Anal.* **2008**, *12*, 107.
60. Ntouyas, S. Nonlocal initial and boundary value problems: a survey. In *Handbook of Differential Equations: Ordinary Differential Equations*; Elsevier: Amsterdam, The Netherlands, 2006; Volume 2, pp. 461–557.
61. Yong, Z.; Jinrong, W.; Lu, Z. *Basic Theory of Fractional Differential Equations*; World Scientific: Singapore, 2016.
62. Ciegis, R.; Bugajev, A. Numerical approximation of one model of bacterial self-organization. *Nonlinear Anal. Model. Control.* **2012**, *17*, 253–270. [CrossRef]

© 2020 by the authors. Licensee MDPI, Basel, Switzerland. This article is an open access article distributed under the terms and conditions of the Creative Commons Attribution (CC BY) license (http://creativecommons.org/licenses/by/4.0/).

Article

Absence of Global Solutions for a Fractional in Time and Space Shallow-Water System

Mohamed Jleli [1], Mokhtar Kirane [2,3,*] and Bessem Samet [1]

[1] Department of Mathematics, College of Science King Saud University, P.O. Box 2455, Riyadh 11451, Saudi Arabia; jleli@ksu.edu.sa (M.J.); bsamet@ksu.edu.sa (B.S.)
[2] LaSIE, Pôle Sciences et Technologies, Université de La Rochelle, Avenue Michel Crépeau, 17031 La Rochelle, France
[3] Peoples' Friendship University of Russia (RUDN University), 6 Miklukho-Maklaya Street, 117198 Moscow, Russia
* Correspondence: mokhtar.kirane@univ-lr.fr

Received: 12 October 2019; Accepted: 15 November 2019; Published: 18 November 2019

Abstract: An initial boundary value problem for a fractional in time and space shallow-water system involving ψ-Caputo fractional derivatives of different orders is considered. Using the test function method, sufficient criteria for the absence of global in time solutions of the system are obtained.

Keywords: global solutions; fractional in time and space shallow-water system; ψ-Caputo fractional derivative

MSC: 35B44; 26A33

---

1. Introduction

We consider the fractional in time and space shallow-water system

$$\begin{cases} \partial_{0|t}^{\alpha,\psi}\eta + \partial_{0|x}^{\beta,\psi}(\eta u) = 0, & t>0, 0<x<L, \\ \frac{1}{2}\left[\frac{1}{\psi'(t)}\partial_t(\eta u) + \partial_{0|t}^{\alpha,\psi}(\eta u)\right] + \partial_{0|x}^{\beta,\psi}(\eta u^2) + \partial_{0|x}^{\beta,\psi}(\eta^2) = 0, & t>0, 0<x<L \end{cases} \quad (1)$$

with

$$(u(0,\cdot),\eta(0,\cdot)) = (u_0,\eta_0) \quad (2)$$

and

$$\eta(\cdot,0) = \eta(\cdot,L) \equiv 0; \quad (3)$$

here $\eta = \eta(t,x)$, $u = u(t,x)$, $L > 0$, $0 < \alpha, \beta < 1$, $\psi \in C^1([0,\infty))$, $\lim_{x\to\infty}\psi(x) = +\infty$, $\psi'(x) > 0$, $x \geq 0$, $\partial_{0|t}^{\alpha,\psi}$ is the ψ-Caputo derivative in time of fractional order α and $\partial_{0|x}^{\beta,\psi}$ is the ψ-Caputo derivative in space of fractional order β. Using the test function method [1], we get sufficient criteria for which problem (1)–(2)–(3) has no global solutions in time.

The considered problem is a fractional version of the shallow-water system

$$\begin{cases} \partial_t \eta + \partial_x(\eta u) = 0, & t>0, 0<x<L, \\ \partial_t(\eta u) + \partial_x(\eta u^2) + \partial_x(\eta^2) = 0, & t>0, 0<x<L, \end{cases} \quad (4)$$

which models the motion of an incompressible fluid in a gravitational field when the fluid height above the channel bottom is small with respect to the characteristic flow length. Here u is the velocity of the fluid particle and η is the height of the fluid above the horizontal flat bottom [2–4].

In [2], Korpusov and Yushkov derived sufficient criteria for the non-existence of global in time solutions of problem (4) under different types of boundary conditions. In particular, under the boundary conditions (3), they proved that if for some $0 < T_0 < \infty$, the problem admits a solution $(u, \eta) \in C^1([0, T_0] \times [0, L]) \times C^1([0, T_0] \times [0, L])$, and

$$\int_0^L x \eta_0(x) u_0(x)\, dx > 0,$$

then there exist no solutions on intervals larger than $[0, T_\infty]$, where

$$T_\infty = \frac{L^2 \int_0^L \eta_0(x)\, dx}{\int_0^L x \eta_0(x) u_0(x)\, dx}.$$

It was shown in many published works that the theory of fractional calculus provides useful tools for modeling various phenomena from physics (see e.g., [5–8]). Specifically, it was found that fractional order models of many real-world phenomena are more adequate than the classical integer order models. This fact motivated researchers to take an interest in the study of fractional in time and/or space evolution equations. In particular, the study of analytic and numerical solutions of fractional shallow-water equations was investigated by many authors (see e.g., [5,9–12]). For the study of existence and non-existence of global solutions for fractional in time and/or space evolution equations, we refer to [13–15] and references therein.

Motivated by the above contributions, the study of the absence of global in time solutions for problem (1)–(2)–(3) is investigated in this work. In the considered problem, we use ψ-Caputo fractional derivative (in time and space) [16], which depends of a function $\psi \in C^1([0, \infty))$. In the special case $\psi(t) = t$, the considered fractional operator reduces to Caputo fractional derivative. Let us mention that in this paper we are concerned essentially with the mathematical study of problem (1)–(2)–(3). For the physical interpretation of this model, we are not able to check if it is more adequate than the standard model (4)–(2)–(3). For a such study, some physical experiments and numerical simulations are needed; this is not the goal of this paper. Nevertheless, let us notice that Tao in [17] proposed a possible scenario for obtaining blowing-up solutions of the Navier–Stokes system; he showed that it is possible for a body of fluid to form a sort of computer, which can build a self—replicating fluid robot that keeps transferring its energy to smaller and smaller copies of itself until the fluid "blows up." He tried to devise a system that would incorporate a *delay* at each step—a sort of timer that would push the energy cleanly from one size scale to the next at just the right moment (according to Erica Klarreich, A Fluid New Path in Grand Math Challenge, Quantamagazine, 24 February 2014). From here, one can speculate any form of delay in time or space for fluid dynamical systems.

In Section 2, we provide some preliminary results that will be needed afterwards. A key lemma is established in Section 3. In the next section, we present and establish our principal results. Specifically, we first establish a mass conservation law for problem (1)–(2)–(3). Next, we obtain sufficient criteria for which the considered problem has no global in time solutions.

2. Preliminaries

Let $c_1, c_2 \in \mathbb{R}$, $c_1 < c_2$, $\mathcal{R} \in L^1(c_1, c_2)$ and $\mu > 0$. The Riemann-Liouville fractional integrals of order μ of $\mathcal{R}$ are given by (see e.g., [18])

$$I_{c_1}^\mu \mathcal{R}(x) = [\Gamma(\mu)]^{-1} \int_{c_1}^x (x - \sigma)^{\mu-1} \mathcal{R}(\sigma)\, d\sigma$$

and

$$I_{c_2}^\mu \mathcal{R}(x) = [\Gamma(\mu)]^{-1} \int_x^{c_2} (\sigma - x)^{\mu-1} \mathcal{R}(\sigma)\, d\sigma,$$

for a.e. $x \in [c_1, c_2]$, where Γ denotes the gamma function.

Let ψ be a C^1 function in $[0, \infty)$ satisfying

$$\lim_{x \to \infty} \psi(x) = +\infty \quad \text{and} \quad \psi'(x) > 0, \quad x \geq 0.$$

Please note that under the above conditions, the function $\psi: [0, \infty) \to [\psi(0), \infty)$ is bijective. Let $\tau > 0$ and $\mathcal{R} \in L^1((0, \tau), \psi'(\sigma) d\sigma)$, i.e.,

$$\int_0^\tau |\mathcal{R}(\sigma)| \psi'(\sigma) d\sigma < \infty.$$

The ψ-fractional integrals of order μ of $\mathcal{R}$ are given by (see [16])

$$I_0^{\mu,\psi} \mathcal{R}(x) = [\Gamma(\mu)]^{-1} \int_0^x (\psi(x) - \psi(\sigma))^{\mu-1} \psi'(\sigma) \mathcal{R}(\sigma) d\sigma$$

and

$$I_\tau^{\mu,\psi} \mathcal{R}(x) = [\Gamma(\mu)]^{-1} \int_x^\tau (\psi(\sigma) - \psi(x))^{\mu-1} \psi'(\sigma) \mathcal{R}(\sigma) d\sigma, \tag{5}$$

for a.e. $x \in [0, \tau]$.

If $\mathcal{R} \in C([0, \tau])$, then $I_0^{\mu,\psi} \mathcal{R}, I_\tau^{\mu,\psi} \mathcal{R} \in C([0, \tau])$ and $I_0^{\mu,\psi} \mathcal{R}(0) = I_\tau^{\mu,\psi} \mathcal{R}(\tau) = 0$.

Lemma 1. *For $\mathcal{R} \in L^1((0, \tau), \psi'(\sigma) d\sigma)$, it holds*

$$\left(I_0^{\mu,\psi} \mathcal{R} \right)(x) = \left(I_{\psi(0)}^\mu \mathcal{R} \circ \psi^{-1} \right)(\psi(x)), \quad \text{a.e. } x \in [0, \tau] \tag{6}$$

and

$$\left(I_\tau^{\mu,\psi} \mathcal{R} \right)(x) = \left(I_{\psi(\tau)}^\mu \mathcal{R} \circ \psi^{-1} \right)(\psi(x)), \quad \text{a.e. } x \in [0, \tau], \tag{7}$$

where $\circ$ stands for the composition of mappings.

Proof. For a.e. $x \in [0, \tau]$, one has

$$\begin{aligned}
\left(I_0^{\mu,\psi} \mathcal{R} \right)(x) &= [\Gamma(\mu)]^{-1} \int_0^x (\psi(x) - \psi(\sigma))^{\mu-1} \psi'(\sigma) \mathcal{R}(\sigma) d\sigma \\
&= [\Gamma(\mu)]^{-1} \int_{\psi(0)}^{\psi(x)} (\psi(x) - t)^{\mu-1} \mathcal{R}\left(\psi^{-1}(t)\right) dt \\
&= \left(I_{\psi(0)}^\mu \mathcal{R} \circ \psi^{-1} \right)(\psi(x)),
\end{aligned}$$

which proves (6). Proceeding as above, one obtains (7). $\square$

Lemma 2 (see e.g., [18]). *Let $(\mathcal{R}, \mathcal{S}) \in L^1(a, b) \times C([a, b])$. Then*

$$\int_a^b \mathcal{R}(t) \left(I_a^\mu \mathcal{S} \right)(t) dt = \int_a^b \mathcal{S}(t) \left(I_b^\mu \mathcal{R} \right)(t) dt.$$

Lemma 3. *Let $(\mathcal{R}, \mathcal{S}) \in L^1((0, \tau), \psi'(\sigma) d\sigma) \times C([0, \tau])$. Then*

$$\int_0^\tau \mathcal{R}(\sigma) \left(I_0^{\mu,\psi} \mathcal{S} \right)(\sigma) \psi'(\sigma) d\sigma = \int_0^\tau \mathcal{S}(\sigma) \left(I_\tau^{\mu,\psi} \mathcal{R} \right)(\sigma) \psi'(\sigma) d\sigma.$$

Proof. Using (6), one obtains

$$\int_0^\tau \mathcal{R}(\sigma) \left(I_0^{\mu,\psi}\mathcal{S}\right)(\sigma)\psi'(\sigma)\,d\sigma = \int_0^\tau \mathcal{R}(\sigma) \left(I_{\psi(0)}^{\mu}\mathcal{S} \circ \psi^{-1}\right)(\psi(\sigma))\psi'(\sigma)\,d\sigma$$

$$= \int_{\psi(0)}^{\psi(\tau)} \mathcal{R}(\psi^{-1}(t)) \left(I_{\psi(0)}^{\mu}\mathcal{S} \circ \psi^{-1}\right)(t)\,dt.$$

Next, using Lemma 2, one deduces that

$$\int_0^\tau \mathcal{R}(\sigma) \left(I_0^{\mu,\psi}\mathcal{S}\right)(\sigma)\psi'(\sigma)\,d\sigma = \int_{\psi(0)}^{\psi(\tau)} \mathcal{S} \circ \psi^{-1}(t) \left(I_{\psi(\tau)}^{\mu}\mathcal{R} \circ \psi^{-1}\right)(t)\,dt.$$

Hence, by (7), the desired result follows. □

Let $\mathcal{R} \in C^1([0,\tau])$ and $0 < \theta < 1$. The ψ-Caputo fractional derivative of order θ of $\mathcal{R}$ is given by (see [16])

$$\left(\partial_{0|x}^{\theta,\psi}\mathcal{R}\right)(x) = \left(I_0^{1-\theta,\psi}\frac{\mathcal{R}'}{\psi'}\right)(x), \quad 0 \le x \le \tau, \tag{8}$$

i.e.,

$$\left(\partial_{0|x}^{\theta,\psi}\mathcal{R}\right)(x) = [\Gamma(1-\theta)]^{-1} \int_0^x (\psi(x) - \psi(\sigma))^{-\theta}\mathcal{R}'(\sigma)\,d\sigma.$$

Lemma 4 (see [16]). *Let $\mathcal{R} \in C^1([0,\tau])$ and $0 < \theta < 1$. One has*

$$I_0^{\theta,\psi}\left(\partial_{0|x}^{\theta,\psi}\mathcal{R}\right)(x) = \mathcal{R}(x) - \mathcal{R}(0), \quad 0 \le x \le \tau.$$

3. A Key Lemma

The following lemma will be useful for proving our principal result.

Lemma 5. *Let $0 < \theta < 1$ and $a > 0$. Suppose that for some $0 < T_0 < \infty$, $J \in C^1([0,T_0])$ is a function satisfying $J(0) > 0$ and*

$$\frac{1}{\psi'(t)}J'(t) + \left(\partial_{0|t}^{\theta,\psi}J\right)(t) \ge aJ^2(t), \quad 0 < t < T_0. \tag{9}$$

Let

$$T_\infty := \sup\left\{\tau > 0 : J \in C^1([0,\tau)) \text{ satisfies (9) for all } 0 < t < \tau\right\}.$$

Then

$$T_0 \le T_\infty \le \psi^{-1}\left(\psi(0) + M(a,\theta)\right) < \infty, \tag{10}$$

where

$$M(a,\theta) = \sup\{X > 0 : f(X) \le 0\} < \infty, \quad f(X) = \mathcal{B}X^{2-\theta} + \mathcal{C}X - \mathcal{D}X^{2-2\theta} - 1 \tag{11}$$

and

$$\mathcal{B} = \frac{a}{\Gamma(4-\theta)}J(0), \quad \mathcal{C} = \frac{a}{2}J(0), \quad \mathcal{D} = \frac{1}{(3-2\theta)\Gamma(3-\theta)^2}.$$

Proof. First, since $f(0) = -1 < 0$ and f is continuous, $\{X > 0 : f(X) \le 0\} \ne \emptyset$. Furthermore, since $\mathcal{B} > 0$ (because $J(0) > 0$), on has $\lim_{X \to +\infty} f(X) = +\infty$. Hence, one deduces that $0 < M(a,\theta) < \infty$.

Next, let $\tau > 0$ be such that $J \in C^1([0,\tau))$ satisfies (9) for all $0 < t < \tau$. Then, for all $0 < T < \tau$, $J \in C^1([0,T])$ satisfies (9) for all $0 < t < T$. Fix $0 < T < \tau$ and introduce the function

$$\varphi(t) = \kappa(0)^{-2}\kappa(t)^2\psi'(t) := \mathcal{Z}(t)\psi'(t), \tag{12}$$

for all $0 \leq t \leq T$, where
$$\kappa(t) = \psi(T) - \psi(t).$$

Using (9), one obtains
$$a \int_0^T J^2(t)\varphi(t)\, dt \leq \int_0^T \left(\partial_{0|t}^{\theta,\psi} J\right)(t)\varphi(t)\, dt + \int_0^T J'(t) \mathcal{Z}(t)\, dt. \tag{13}$$

On the other hand, using (8) and Lemma 3, one has
$$\int_0^T \left(\partial_{0|t}^{\theta,\psi} J\right)(t)\varphi(t)\, dt = \int_0^T \left(I_0^{1-\theta,\psi} \frac{J'}{\psi'}\right)(t)\varphi(t)\, dt$$
$$= \int_0^T J'(t)\left(I_T^{1-\theta,\psi} \mathcal{Z}\right)(t)\, dt.$$

Integrating by parts, it holds
$$\int_0^T \left(\partial_{0|t}^{\theta,\psi} J\right)(t)\varphi(t)\, dt = J(T)\left(I_T^{1-\theta,\psi} \mathcal{Z}\right)(T) - J(0)\left(I_T^{1-\theta,\psi} \mathcal{Z}\right)(0) - \int_0^T J(t)\left(I_T^{1-\theta,\psi} \mathcal{Z}\right)'(t)\, dt. \tag{14}$$

Using (5), an elementary calculation gives us that
$$\left(I_T^{1-\theta,\psi} \mathcal{Z}\right)(t) = \frac{2}{\Gamma(4-\theta)} \kappa(0)^{-2} \kappa(t)^{3-\theta} \tag{15}$$

and
$$\left(I_T^{1-\theta,\psi} \mathcal{Z}\right)'(t) = -\frac{2}{\Gamma(3-\theta)} \kappa(0)^{-2} \kappa(t)^{2-\theta} \psi'(t), \tag{16}$$

for all $0 \leq t \leq T$. Using (14) and (15), one deduces that
$$\int_0^T \left(\partial_{0|t}^{\theta,\psi} J\right)(t)\varphi(t)\, dt = -\frac{2}{\Gamma(4-\theta)} \kappa(0)^{1-\theta} J(0) - \int_0^T J(t)\left(I_T^{1-\theta,\psi} \mathcal{Z}\right)'(t)\, dt. \tag{17}$$

Again, integrating by parts, it holds
$$\int_0^T J'(t) \mathcal{Z}(t)\, dt = -J(0) - \int_0^T J(t) \mathcal{Z}'(t)\, dt. \tag{18}$$

Hence, it follows from (13), (17) and (18) that
$$\mathcal{A} J(0) + a \int_0^T J^2(t)\varphi(t)\, dt \leq \int_0^T |J(t)| \left|\left(I_T^{1-\theta,\psi} \mathcal{Z}\right)'(t)\right| dt + \int_0^T |J(t)||\mathcal{Z}'(t)|\, dt, \tag{19}$$

where
$$\mathcal{A} = \frac{2}{\Gamma(4-\theta)} \kappa(0)^{1-\theta} + 1.$$

On the other hand, by Young's inequality with parameter $\frac{a}{2} > 0$, one has
$$\int_0^T |J(t)| \left|\left(I_T^{1-\theta,\psi} \mathcal{Z}\right)'(t)\right| dt = \int_0^T \sqrt{a\varphi(t)} |J(t)| \frac{\left|\left(I_T^{1-\theta,\psi} \mathcal{Z}\right)'(t)\right|}{\sqrt{a\varphi(t)}}\, dt$$
$$\leq \frac{a}{2} \int_0^T J^2(t)\varphi(t)\, dt + \frac{1}{2a} \int_0^T \frac{\left|\left(I_T^{1-\theta,\psi} \mathcal{Z}\right)'(t)\right|^2}{\varphi(t)}\, dt. \tag{20}$$

Similarly, one gets

$$\int_0^T |J(t)||\mathcal{Z}'(t)|\,dt \leq \frac{a}{2}\int_0^T J^2(t)\varphi(t)\,dt + \frac{1}{2a}\int_0^T \frac{|\mathcal{Z}'(t)|^2}{\varphi(t)}\,dt. \tag{21}$$

Combining (19)–(21), it comes that

$$2aAJ(0) \leq \int_0^T \frac{\left|\left(I_T^{1-\theta,\psi}\mathcal{Z}\right)'(t)\right|^2}{\varphi(t)}\,dt + \int_0^T \frac{|\mathcal{Z}'(t)|^2}{\varphi(t)}\,dt. \tag{22}$$

Furthermore, using (12) and (16), one obtains

$$\frac{\left|\left(I_T^{1-\theta,\psi}\mathcal{Z}\right)'(t)\right|^2}{\varphi(t)} = \left[\frac{2}{\Gamma(3-\theta)}\right]^2 \kappa(0)^{-2}\kappa(t)^{2-2\theta}\psi'(t),$$

for all $0 < t < T$, which yields

$$\int_0^T \frac{\left|\left(I_T^{1-\theta,\psi}\mathcal{Z}\right)'(t)\right|^2}{\varphi(t)}\,dt = \frac{1}{(3-2\theta)}\left[\frac{2}{\Gamma(3-\theta)}\right]^2 \kappa(0)^{1-2\theta}. \tag{23}$$

Similar calculations yield

$$\int_0^T \frac{|\mathcal{Z}'(t)|^2}{\varphi(t)}\,dt = 4\kappa(0)^{-1}. \tag{24}$$

It follows from (22)–(24) that

$$2aAJ(0) \leq \frac{1}{(3-2\theta)}\left[\frac{2}{\Gamma(3-\theta)}\right]^2 \kappa(0)^{1-2\theta} + 4\kappa(0)^{-1},$$

which yields

$$f(\kappa(0)) \leq 0.$$

Therefore, one deduces that

$$\kappa(0) \leq M(a,\theta).$$

Hence, it holds

$$T \leq \psi^{-1}\left(\psi(0) + M(a,\theta)\right), \quad \text{for all } 0 < T < \tau,$$

which implies that

$$\tau \leq \psi^{-1}\left(\psi(0) + M(a,\theta)\right),$$

and (10) follows. □

Remark 1. *Taking $\psi(t) = t$ and the limit as $\theta \to 1^-$, (9) reduces to*

$$J'(t) \geq \frac{a}{2}J^2(t), \quad 0 < t < T_0.$$

Hence, under the assumptions of Lemma 5, passing to the limit as $\theta \to 1^-$ in (10), it holds

$$T_0 \leq T_\infty \leq \frac{2}{aJ(0)},$$

which is the same estimate as in ([19], Corollary 1).

4. Non-Existence of Global in Time Solutions for Problem (1)–(2)–(3)

We assume that

(i) $0 < \alpha, \beta < 1$, $L > 0$.
(ii) $\psi \in C^1([0, \infty))$, $\lim_{x \to \infty} \psi(x) = +\infty$, $\psi'(x) > 0$, $x \geq 0$.
(iii) $\eta_0, u_0 \in C^1([0, L])$.

We first establish the following mass conservation law.

Proposition 1. *Suppose that for some $0 < T_0 < \infty$, $(\eta, u) \in C^1([0, T_0] \times [0, L]) \times C^1([0, T_0] \times [0, L])$, $\eta \geq 0$, is a solution of problem (1)–(2)–(3). Then*

$$\int_0^L \mathcal{K}_L(x)^{\beta-1} \eta(t, x) \psi'(x)\, dx = \int_0^L \mathcal{K}_L(x)^{\beta-1} \eta_0(x) \psi'(x)\, dx := m_0, \quad 0 \leq t \leq T_0, \tag{25}$$

where

$$\mathcal{K}_L(x) = \psi(L) - \psi(x). \tag{26}$$

Proof. From the first equation in (1), one has

$$-\partial_{0|t}^{\alpha,\psi} \eta(t, x) = \partial_{0|x}^{\beta,\psi} (\eta u)(t, x), \quad (t, x) \in (0, T_0] \times (0, L),$$

whereupon

$$-\partial_{0|t}^{\alpha,\psi} \left(I_0^{\beta,\psi} \eta(t, \cdot) \right)(L) = \left(I_0^{\beta,\psi} \partial_{0|x}^{\beta,\psi} (\eta u)(t, \cdot) \right)(L).$$

Using Lemma 4 and the boundary conditions (3), one obtains

$$\left(I_0^{\beta,\psi} \partial_{0|x}^{\beta,\psi} (\eta u)(t, \cdot) \right)(L) = \eta(t, L) u(t, L) - \eta(t, 0) u(t, 0) = 0.$$

Hence, it holds

$$\partial_{0|t}^{\alpha,\psi} \left(I_0^{\beta,\psi} \eta(t, \cdot) \right)(L) = 0,$$

i.e.,

$$\partial_{0|t}^{\alpha,\psi} \int_0^L \mathcal{K}_L(x)^{\beta-1} \psi'(x) \eta(t, x)\, dx = 0,$$

which implies that

$$I_0^{\alpha,\psi} \partial_{0|t}^{\alpha,\psi} \int_0^L \mathcal{K}_L(x)^{\beta-1} \psi'(x) \eta(t, x)\, dx = 0.$$

Again, using Lemma 4, one deduces that

$$\int_0^L \mathcal{K}_L(x)^{\beta-1} \psi'(x) \eta(t, x)\, dx - \int_0^L \mathcal{K}_L(x)^{\beta-1} \psi'(x) \eta(0, x)\, dx = 0,$$

which yields (25). □

Our principal result is the following.

Theorem 1. *Suppose that for some $0 < T_0 < \infty$, $(\eta, u) \in C^1(Q) \times C^1(Q)$, $Q = [0, T_0] \times [0, L]$, $\eta \geq 0$, is a solution of problem (1)–(2)–(3). Let*

$$T_{\max} := \sup \left\{ \tau > 0 : (\eta, u) \in C^1([0, \tau] \times [0, L]) \times C^1([0, \tau] \times [0, L]) \text{ is a solution of } (1) - (2) - (3) \right\}.$$

If

$$J(0) := \int_0^L \eta_0(x) u_0(x) \mathcal{K}_L(x)^{\frac{\beta}{2}-1} \psi'(x)\, dx > 0, \tag{27}$$

where $\mathcal{K}_L$ is given by (26), then

$$T_0 \leq T_{max} \leq \psi^{-1}(\psi(0) + M(a,\alpha)) < \infty, \quad (28)$$

where $M(a,\alpha)$ is given by (11) (with $\theta = \alpha$),

$$a = 2\frac{\Gamma\left(1+\frac{\beta}{2}\right)}{\Gamma\left(1-\frac{\beta}{2}\right)} \frac{\mathcal{K}_L(0)^{-\frac{\beta}{2}}}{m_0}$$

and m_0 is given by (25).

Proof. We introduce the function

$$\varphi(x) = \mathcal{K}_L(x)^{\frac{\beta}{2}-1}\psi'(x), \quad 0 \leq x < L. \quad (29)$$

Multiplying the second equation in (1) by $\varphi(x)$ and integrating over $(0,L)$, one obtains

$$\frac{1}{2}\int_0^L \varphi(x)\frac{1}{\psi'(t)}\partial_t(\eta u)(t,x)\,dx + \frac{1}{2}\int_0^L \varphi(x)\partial_{0|t}^{\alpha,\psi}(\eta u)(t,x)\,dx + \int_0^L \varphi(x)\partial_{0|x}^{\beta,\psi}(\eta u^2)(t,x)\,dx$$
$$+ \int_0^L \varphi(x)\partial_{0|x}^{\beta,\psi}(\eta^2)(t,x)\,dx = 0, \quad 0 < t < T_0,$$

which yields

$$\frac{1}{2\psi'(t)}J'(t) + \frac{1}{2}\left(\partial_{0|t}^{\alpha,\psi}J\right)(t) = -\int_0^L \varphi(x)\partial_{0|x}^{\beta,\psi}(\eta u^2)(t,x)\,dx - \int_0^L \varphi(x)\partial_{0|x}^{\beta,\psi}(\eta^2)(t,x)\,dx, \quad 0 < t < T_0, \quad (30)$$

where

$$J(t) = \int_0^L \varphi(x)(\eta u)(t,x)\,dx, \quad 0 \leq t \leq T_0.$$

On the other hand, using (8), one has

$$\int_0^L \varphi(x)\partial_{0|x}^{\beta,\psi}(\eta u^2)(t,x)\,dx = \int_0^L \left(I_0^{1-\beta,\psi}\frac{\partial_x(\eta u^2)(t,\cdot)}{\psi'}\right)(x)\frac{\varphi(x)}{\psi'(x)}\psi'(x)\,dx.$$

Hence, by Lemma 3, one obtains

$$\int_0^L \varphi(x)\partial_{0|x}^{\beta,\psi}(\eta u^2)(t,x)\,dx = \int_0^L \frac{\partial_x(\eta u^2)(t,x)}{\psi'(x)}\left(I_L^{1-\beta,\psi}\frac{\varphi}{\psi'}\right)(x)\psi'(x)\,dx$$
$$= \int_0^L \partial_x(\eta u^2)(t,x)\left(I_L^{1-\beta,\psi}\frac{\varphi}{\psi'}\right)(x)\,dx.$$

Next, using an integration by parts and the boundary conditions (3), one deduces that

$$\int_0^L \varphi(x)\partial_{0|x}^{\beta,\psi}(\eta u^2)(t,x)\,dx = -\int_0^L \eta(t,x)u^2(t,x)\partial_x\left(I_L^{1-\beta,\psi}\frac{\varphi}{\psi'}\right)(x)\,dx. \quad (31)$$

Similarly, one has

$$\int_0^L \varphi(x)\partial_{0|x}^{\beta,\psi}(\eta^2)(t,x)\,dx = -\int_0^L \eta^2(t,x)\partial_x\left(I_L^{1-\beta,\psi}\frac{\varphi}{\psi'}\right)(x)\,dx. \quad (32)$$

It follows from (30)–(32) that

$$\frac{1}{2\psi'(t)}J'(t) + \frac{1}{2}\left(\partial_{0|t}^{\alpha,\psi}J\right)(t) = \int_0^L (\eta u^2)(t,x)\partial_x\left(I_L^{1-\beta,\psi}\frac{\varphi}{\psi'}\right)(x)\,dx + \int_0^L \eta^2(t,x)\partial_x\left(I_L^{1-\beta,\psi}\frac{\varphi}{\psi'}\right)(x)\,dx. \quad (33)$$

Next, using (29), for $x \in (0, L)$, an elementary calculation gives us that

$$\left(I_L^{1-\beta,\psi} \frac{\varphi}{\psi'} \right)(x) = \frac{\Gamma\left(\frac{\beta}{2}\right)}{\Gamma\left(1-\frac{\beta}{2}\right)} \mathcal{K}_L(x)^{-\frac{\beta}{2}}.$$

Hence, it holds

$$\partial_x \left(I_L^{1-\beta,\psi} \frac{\varphi}{\psi'} \right)(x) = \frac{\Gamma\left(1+\frac{\beta}{2}\right)}{\Gamma\left(1-\frac{\beta}{2}\right)} \mathcal{K}_L(x)^{-\frac{\beta}{2}-1} \psi'(x) > 0, \quad 0 < x < L. \tag{34}$$

It follows from (33) and (34) that

$$\frac{1}{2\psi'(t)} J'(t) + \frac{1}{2} \left(\partial_{0|t}^{\alpha,\psi} J \right)(t) \geq \int_0^L (\eta u^2)(t,x) \partial_x \left(I_L^{1-\beta,\psi} \frac{\varphi}{\psi'} \right)(x) \, dx. \tag{35}$$

On one hand, by Hölder's inequality, one has

$$J^2(t)$$

$$\leq \left(\int_0^L \eta(t,x) |u(t,x)| \varphi(x) \, dx \right)^2$$

$$= \left(\int_0^L \sqrt{\eta(t,x)} |u(t,x)| \sqrt{\partial_x \left(I_L^{1-\beta,\psi} \frac{\varphi}{\psi'} \right)(x)} \sqrt{\frac{\eta(t,x)}{\partial_x \left(I_L^{1-\beta,\psi} \frac{\varphi}{\psi'} \right)(x)}} \varphi(x) \, dx \right)^2 \tag{36}$$

$$\leq \left(\int_0^L (\eta u^2)(t,x) \partial_x \left(I_L^{1-\beta,\psi} \frac{\varphi}{\psi'} \right)(x) \, dx \right) \left(\int_0^L \frac{\eta(t,x)}{\partial_x \left(I_L^{1-\beta,\psi} \frac{\varphi}{\psi'} \right)(x)} \varphi^2(x) \, dx \right).$$

On the other hand, using (29) and (34), one obtains

$$\int_0^L \frac{\eta(t,x)}{\partial_x \left(I_L^{1-\beta,\psi} \frac{\varphi}{\psi'} \right)(x)} \varphi^2(x) \, dx$$

$$= \frac{\Gamma\left(1-\frac{\beta}{2}\right)}{\Gamma\left(1+\frac{\beta}{2}\right)} \int_0^L \mathcal{K}_L(x)^{\beta-1} \psi'(x) \eta(t,x) \mathcal{K}_L(x)^{\frac{\beta}{2}} \, dx$$

$$\leq \frac{\Gamma\left(1-\frac{\beta}{2}\right)}{\Gamma\left(1+\frac{\beta}{2}\right)} \mathcal{K}_L(0)^{\frac{\beta}{2}} \int_0^L \mathcal{K}_L(x)^{\beta-1} \psi'(x) \eta(t,x) \, dx.$$

Furthermore, using the mass conservation law (25), one deduces that

$$\int_0^L \frac{\eta(t,x)}{\partial_x \left(I_L^{1-\beta,\psi} \frac{\varphi}{\psi'} \right)(x)} \varphi^2(x) \, dx \leq \frac{\Gamma\left(1-\frac{\beta}{2}\right)}{\Gamma\left(1+\frac{\beta}{2}\right)} \mathcal{K}_L(0)^{\frac{\beta}{2}} m_0. \tag{37}$$

Next, (36) and (37) yield

$$\int_0^L (\eta u^2)(t,x) \partial_x \left(I_L^{1-\beta,\psi} \frac{\varphi}{\psi'} \right)(x) \, dx \geq \frac{\Gamma\left(1+\frac{\beta}{2}\right)}{\Gamma\left(1-\frac{\beta}{2}\right)} \frac{\mathcal{K}_L(0)^{-\frac{\beta}{2}}}{m_0} J^2(t), \quad 0 < t < T. \tag{38}$$

It follows from (35) and (38) that

$$\frac{1}{\psi'(t)}J'(t) + \left(\partial_{0|t}^{\alpha,\psi}J\right)(t) \geq 2\frac{\Gamma\left(1+\frac{\beta}{2}\right)}{\Gamma\left(1-\frac{\beta}{2}\right)}\frac{\mathcal{K}_L(0)^{-\frac{\beta}{2}}}{m_0}J^2(t), \quad 0 < t < T_0.$$

Hence, using (27) and Lemma 5, the estimate (28) follows. □

Example 1. *Consider the system*

$$\begin{cases} {}^C D_{0|t}^{\alpha}\eta + {}^C D_{0|x}^{\beta}(\eta u) &= 0, \quad t > 0, 0 < x < L, \\ \frac{1}{2}\left[\partial_t(\eta u) + {}^C D_{0|t}^{\alpha}(\eta u)\right] + {}^C D_{0|x}^{\beta}(\eta u^2) + {}^C D_{0|x}^{\beta}(\eta^2) &= 0, \quad t > 0, 0 < x < L \end{cases} \quad (39)$$

under the initial and boundary conditions (2) and (3). Here ${}^C D_{0|t}^{\alpha}$ is the Caputo derivative in time of fractional order $0 < \alpha < 1$ and ${}^C D_{0|x}^{\beta}$ is the Caputo derivative in space of fractional order $0 < \beta < 1$. System (39) is a special case of (1) with $\psi(s) = s$. Hence, by Theorem 1, one deduces that if $(\eta, u) \in C^1([0, T_0] \times [0, L]) \times C^1([0, T_0] \times [0, L])$ is a solution of problem (39)–(2)–(3) for some $0 < T_0 < \infty$, and

$$J(0) := \int_0^L \eta_0(x)u_0(x)(L-x)^{\frac{\beta}{2}-1}\,dx > 0,$$

then

$$T_0 \leq T_{max} \leq M(a, \alpha) < \infty,$$

where $M(a, \alpha)$ is given by (11) (with $\theta = \alpha$) and

$$a = 2\frac{\Gamma\left(1+\frac{\beta}{2}\right)}{\Gamma\left(1-\frac{\beta}{2}\right)}L^{-\frac{\beta}{2}}\left(\int_0^L (L-x)^{\beta-1}\eta_0(x)\,dx\right)^{-1}.$$

5. Conclusions

A fractional in time and space shallow-water system is investigated in this paper. The considered fractional derivative depends of a function $\psi \in C^1([0,\infty))$, and generalizes Caputo fractional derivative, which corresponds to the case $\psi(t) = t$. Using the test function method, it is shown that under certain conditions imposed on the initial data, the system admits no global in time solutions. Furthermore, an upper bound of the lifespan is obtained.

Author Contributions: Investigation, M.J., M.K. and B.S. M.J., M.K. and B.S. contributed equally to this work.

Funding: M. Jleli is supported by Researchers Supporting Project number (RSP-2019/57), King Saud University, Riyadh, Saudi Arabia. M. Kirane is supported by "RUDN University program 5-100".

Conflicts of Interest: The authors declare no conflict of interest.

References

1. Mitidieri, E.; Pokhozhaev, S.I. A priori estimates and blow-up of solutions to nonlinear partial differential equations and inequalities. *Proc. Steklov Inst. Math.* **2011**, *234*, 1–362.
2. Korpusov, M.O.; Yushkov, E.V. Solution blowup for systems of shallow-water equations. *Theor. Math. Phys.* **2013**, *177*, 1505–1514. [CrossRef]
3. Landau, L.D.; Lifshitz, E.M. *Fluid Mechanics*; Nauka: Moscow, Russia, 1986. (In Russian)
4. Rozhdestvenskij, B.L.; Yanenko, N.N. Systems of Quasilinear Equations and their Applications to Gas Dynamics Series. In *Translations of Mathematical Monographs, American Mathematical Society*, 2nd ed.; AMS: Providence, RI, USA, 1983; p. 676.

5. Arshad, S.; Sohail, A.; Maqbool, K. Nonlinear shallow water waves: A fractional order approach. *Alex. Eng. J.* **2016**, *55*, 525–532. [CrossRef]
6. Atanackovic, T.; Pilipovic, S.; Stankovic, B.; Zorica, D. *Fractional Calculus with Applications in Mechanics: Vibrations and Diffusion Processes*; Wiley-ISTE: London, UK, 2014.
7. Hilfer, R. (Ed.) *Applications of Fractional Calculus in Physics*; World Scientific: Singapore, 1999.
8. Uchaikin, V. *Fractional Derivatives for Physicists and Engineers*; Springer: Berlin, Germany, 2013.
9. Khater, M.M.; Kumar, D. New exact solutions for the time fractional coupled Boussinesq-Burger equation and approximate long water wave equation in shallow water. *J. Ocean Eng. Sci.* **2017**, *2*, 223–228. [CrossRef]
10. Kumar, S. A numerical study for solution of time fractional nonlinear shallow water equation in oceans. *Z. Naturforsch. A* **2013**, *68a*, 1–7. [CrossRef]
11. Kumar, D.; Darvishi, M.T.; Joardar, A.K. Modified Kudryashov method and its application to the fractional version of the variety of Boussinesq-like equations in shallow water. *Opt. Quantum Electron.* **2018**, *50*, 1–17. [CrossRef]
12. Sahoo, S.; Ray, S.S. New double-periodic solutions of fractional Drinfeld-Sokolov-Wilson equation in shallow water waves. *Nonlinear Dyn.* **2017**, *88*, 1869–1882. [CrossRef]
13. Bai, Z.; Chen, Y.; Lian, H.; Sun, S. On the existence of blow up solutions for a class of fractional differential equations. *Fract. Calc. Appl. Anal.* **2014**, *17*, 1175–1187. [CrossRef]
14. Hu, J.; Xin, J.; Lu, H. The global solution for a class of systems of fractional nonlinear Schrödinger equations with periodic boundary condition. *Comput. Math. Appl.* **2011**, *62*, 1510–1521. [CrossRef]
15. Kirane, M.; Malik, S.A. The profile of blowing-up solutions to a nonlinear system of fractional differential equations. *Nonlinear Anal. Theory Methods Appl.* **2010**, *73*, 3723–3736. [CrossRef]
16. Almeida, R. A Caputo fractional derivative of a function with respect to another function. *Commun. Nonlinear Sci. Numer. Simul.* **2017**, *44*, 460–481. [CrossRef]
17. Tao, T. Finite time blowup for an averaged three-dimensional Navier-Stokes equation. *J. Am. Math. Soc.* **2016**, *29*, 601–674. [CrossRef]
18. Samko, S.G.; Kilbas, A.A.; Marichev, O.I. *Fractional Integrals and Derivatives: Theory and Applications*; Gordon and Breach: Yverdon, Switzerland, 1993.
19. Panin, A.A. Local solvability and blowup of the solution of the Rosenau-Bürgers equation with different boundary conditions. *Theor. Math. Phys.* **2013**, *177*, 1361–1376. [CrossRef]

© 2019 by the authors. Licensee MDPI, Basel, Switzerland. This article is an open access article distributed under the terms and conditions of the Creative Commons Attribution (CC BY) license (http://creativecommons.org/licenses/by/4.0/).

Article

Fractional Order Unknown Inputs Fuzzy Observer for Takagi–Sugeno Systems with Unmeasurable Premise Variables

Abdelghani Djeddi [1,†], Djalel Dib [1,†], Ahmad Taher Azar [2,3,*,†] and Salem Abdelmalek [4,†]

1. Department of Electrical Engineering, Larbi Tebessi University, Tebessa 12002, Algeria; abdelghani.djeddi@univ-tebessa.dz (A.D.); dibdjalel@gmail.com (D.D.)
2. College of Engineering, Robotics and Internet-of-Things Lab (RIOTU), Prince Sultan University, Riyadh 12435, Saudi Arabia
3. Faculty of Computers and Artificial Intelligence, Benha University, Benha 13511, Egypt
4. Department of Mathematics, Larbi Tebessi University, Tebessa 12002, Algeria; salem.abdelmalek@univ-tebessa.dz
* Correspondence: aazar@psu.edu.sa or ahmad.azar@fci.bu.edu.eg
† These authors contributed equally to this work.

Received: 21 August 2019; Accepted: 11 October 2019; Published: 16 October 2019

Abstract: This paper presents a new procedure for designing a fractional order unknown input observer (FOUIO) for nonlinear systems represented by a fractional-order Takagi–Sugeno (FOTS) model with unmeasurable premise variables (UPV). Most of the current research on fractional order systems considers models using measurable premise variables (MPV) and therefore cannot be utilized when premise variables are not measurable. The concept of the proposed is to model the FOTS with UPV into an uncertain FOTS model by presenting the estimated state in the model. First, the fractional-order extension of Lyapunov theory is used to investigate the convergence conditions of the FOUIO, and the linear matrix inequalities (LMIs) provide the stability condition. Secondly, performances of the proposed FOUIO are improved by the reduction of bounded external disturbances. Finally, an example is provided to clarify the proposed method. The obtained results show that a good convergence of the outputs and the state estimation errors were observed using the new proposed FOUIO.

Keywords: fractional order unknown input fuzzy observer; fractional order Takagi–Sugeno models; L_2 optimization; linear matrix inequalities; unmeasurable premise variables

1. Introduction

Recently, the interest in fractional derivatives and integral applications, as well as in theoretical and practical works, has grown immensely, see for example [1–6]. The main aspects, concept and several applications of fractional calculus are outlined, for example, in [7–14]. This is essentially due to the fact that various physical systems are well described by a fractional order state equation [15–17].

Growing applications have attracted interest in studying the state estimation of fractional differential equations in a linear case [18–21] and in a nonlinear case [22–25]. It is well known that the study of the problem of stabilization of the fractional order system is particularly important for the synthesis of the observer [26–36].

Takagi–Sugeno (TS) fuzzy models have also attracted attention in recent years. The main feature of this class of nonlinear models is to represent the local dynamics of each fuzzy implication (rule) by linear system models. It has been effectively employed in the implementation of nonlinear systems [37–41]. Takagi–Sugeno models have been broadly utilized to represent nonlinear integer-order systems. However, the fuzzy Takagi–Sugeno scheme remains very efficient for nonlinear fractional order systems (FOS) [42–44]. Therefore, the use of fractional-order Takagi–Sugeno (FOTS) models to represent nonlinear FOS will be introduced in this paper. Several approaches confirm that the validity functions of Takagi–Sugeno representation rely on measurable premise variables, whereas various applications, like diagnosis, consider that those variables rely on the input and state variables of the system that are usually immeasurable [45–48].

Takagi–Sugeno uses premise variables for computing weighting functions. Premise variables can be known (inputs or outputs of the system), or unknown variables taken as the state of the system to be estimated. State variables are usually unmeasurable, but they can be measured by the introduction of sensors, with an additional cost, but the right choice is to estimate the state variables in order to avoid the effects of sensor and shareholder faults that may have appeared on the inputs or outputs of the system considered. This justifies the research works on the state estimation of systems [48,49]. In order to use the state of the system as premise variables, then the states must be estimated, hence the need to synthesize an adequate observer able to estimate the state of the system despite the presence of unknown inputs and disturbances. Hence, it is motivating to deem the common state of unmeasurable variables such as system states. The problem appears especially in the structure of the state of the TS observer.

To implement a fuzzy observer for TS systems with unmeasurable premise variables (UPV), several methods have been evolved, comprising those which take account of analytical advances of an estimation error [50–52], and those which use the error description by a TS model with uncertainty or unorganized disruption [49]. The present work presents the Takagi–Sugeno unknown input fractional order observers design for FOTS models with UPV.

The main objective of the current paper is to found new stability and stabilization conditions using FOTS systems with UPV in the continuous case, to implement observers for nonlinear systems. The case where the weighting functions rely on premise variables depending on unmeasurable system states is considered. First, the representation of FOTS systems with UPV and their observers will be considered, which are given under the linear matrix inequalitie (LMI) formulation. Then, an analysis of the stability of the state estimation error studied by using the minimization of the L_2 norm of the transfer from bounded unknown exogenous disturbances to the state estimation error will be established. An application example is designed to demonstrate the performance of the suggested approach.

This paper is organized in the following way. The next section provides some background on the fractional calculus. The FOTS model is presented in Section 3. The main results of the paper, namely the synthesis of the fractional fuzzy observer based unmeasurable premise variables, are presented in Section 4. A new proposed method for unknown input estimation of the fractional order Takagi–Sugeno unknown input observer is given in Section 5. A numerical example is given in Section 6 to demonstrate the efficiency and validity of the proposed approach. Finally, the paper ends with concluding remarks and future perspectives in Section 7.

2. A Brief Introduction to Fractional Calculus

The fractional differo-integral operators are symbolized by $_aD_t^\alpha f(t)$, where a and t, are the bounds of the operation and $\alpha \in \mathbb{R}$ is a generalization of the standard integration and differentiation to an arbitrary order, which can be rational, irrational or even complex. The basic continuous differo-integral operator is given by the following:

$$_aD_t^\alpha := \begin{cases} \frac{d^\alpha}{dt^\alpha}, & \text{for } \alpha > 0, \\ 1, & \text{for } \alpha = 0, \\ \int_a^t (d\tau)^\alpha, & \text{for } \alpha < 0. \end{cases} \quad (1)$$

In the literature, different definitions can be found concerning fractional order systems. The best most commonly used definitions of fractional order derivatives are:

The Riemann–Liouville (RL) definition [53]:

$$_aD_t^\alpha f(t) = \frac{1}{\Gamma(m-\alpha)} \left(\frac{d}{dt}\right)^m \int_a^t \frac{f(\tau)}{(t-\tau)^{1-(m-\alpha)}} d\tau. \tag{2}$$

The Grunwald–Letnikov (GL) definition [53]:

$$_aD_t^\alpha f(t) = \lim_{h \to 0} \frac{1}{h^\alpha} \sum_{k=0}^{(t-a)/h} (-1)^k \binom{\alpha}{k} f(t-kh). \tag{3}$$

The Caputo definition of the fractional differ-integral operator for the function $f(t)$ is adopted in this paper as the Caputo definition allows the initial values of classical integer-order derivatives with clear physical interpretation to be used as follows [51,53]:

$$_aD_t^\alpha f(t) = \frac{1}{\Gamma(m-\alpha)} \int_a^t \frac{f^m(\tau)}{(t-\tau)^{1-(m-\alpha)}} d\tau. \tag{4}$$

In these expressions $m-1 < \alpha < m$, and $\Gamma(.)$ is the well-known Eulers gamma function:

$$\Gamma(x) = \int_0^\infty e^{-t} t^{(x-1)} dt, \quad x > 0. \tag{5}$$

3. Fractional Order Takagi–Sugeno Model

Consider the following nonlinear system given by [54]:

$$\begin{cases} _aD_t^\alpha x(t) = f(x,u), \\ y(t) = g(x,u), \end{cases} \tag{6}$$

where $x(t) \in \mathbb{R}^n$ and α is the fractional order derivative. f and g are nonlinear functions.

Using the well-known transformation by nonlinear sector, the following TS fuzzy system is given [49]:

$$\begin{cases} _aD_t^\alpha x(t) = \sum_{i=0}^M h_i(\xi(t))[A_i x(t) + B_i u(t)], \\ y(t) = \sum_{i=0}^M h_i(\xi(t))[C_i x(t) + D_i u(t)], \end{cases} \tag{7}$$

where $u(t) \in \mathbb{R}^m$ is the input vector, $x(t) \in \mathbb{R}^n$ is the state vector, and $y(t) \in \mathbb{R}^p$ represents the output vector. $A_i \in \mathbb{R}^{n \times n}$, $B_i \in \mathbb{R}^{n \times n_u}$, $C_i \in \mathbb{R}^{n \times n_y}$ and $D_i \in \mathbb{R}^{n_y \times n_u}$ are known matrices. $h_i(\xi(t))$ are the weighting functions relying on the premise variables $\xi(t)$ which can be measurable (input or output of the system) or unmeasurable variables (state of the system). It can also be an external signal. These functions confirm the following so-called convex sum property:

$$\begin{cases} \sum_{i=0}^M h_i(\xi(t)) = 1, \\ 0 \leq h_i(\xi(t)) \leq 1, \quad \forall i \in \{1,2,...,M\}. \end{cases} \tag{8}$$

In this paper, the target is to model a fractional order fuzzy TS observer with UPV. Thus, the work is dedicated to the problem of state estimation for nonlinear fractional order systems characterized by continuous time TS models, with unknown input $\bar{u}(t)$.

Under the hypothesis $C_1 = C_2 = \cdots = C$ and $D_i = 0$, the FOTS model in the presence of unknown inputs and a measurable premise variable can be defined as follows:

$$\begin{cases} _aD_t^\alpha x(t) = \sum_{i=0}^M h_i(x(t))[A_i x(t) + B_i u(t) + E_i \bar{u}(t)], \\ y(t) = Cx(t) + E\bar{u}(t), \end{cases} \tag{9}$$

where $\bar{u}(t) \in \mathbb{R}^q$ ($q < p$) is the input vector, and $E_i \in \mathbb{R}^{n \times n_{\bar{u}}}$ and $E \in \mathbb{R}^{n_y \times n_{\bar{u}}}$ are known matrices. This can be rewritten as:

$$\begin{cases} {}_aD_t^\alpha x(t) = \sum_{i=0}^{M} h_i(\hat{x}(t))\left[A_i x(t) + B_i u(t) + E_i \bar{u}(t) + (h_i(x(t)) - h_i(\hat{x}(t)))(A_i x(t) + B_i u(t) + E_i \bar{u}(t))\right] \\ y(t) = Cx(t) + E\bar{u}(t). \end{cases} \quad (10)$$

After rewriting the model (9), we obtain:

$$\begin{cases} {}_aD_t^\alpha x(t) = \sum_{i=0}^{M} h_i(\hat{x}(t))\left[A_i x(t) + B_i u(t) + E_i \bar{u}(t) + \omega(t)\right], \\ y(t) = Cx(t) + E\bar{u}(t). \end{cases} \quad (11)$$

This form corresponds to a perturbed FOTS model with measurable premise variables (estimated state of the system), where:

$$\omega(t) = \sum_{i=1}^{M}(h_i(x(t)) - \hat{x}(t)))\left[A_i x(t) + B_i u(t) + E_i \bar{u}(t)\right]. \quad (12)$$

This term is considered a global bounded and asymptotically vanishing perturbation.

4. Fractional Order Takagi–Sugeno Unknown Input Observer

The proposed fractional order Takagi–Sugeno fuzzy unknown input observer is given by the following equations:

$$\begin{cases} {}_aD_t^\alpha x(t) = \sum_{i=0}^{M} h_i(\hat{x}(t))\left[N_i z(t) + G_i u(t) + L_i y(t)\right], \\ \hat{x}(t) = z(t) + Hy(t). \end{cases} \quad (13)$$

The state and the output estimation can be defined as:

$$\begin{aligned} \tilde{x}(t) &= x(t) - \hat{x}(t), \\ &= x(t) - z(t) + HCx(t) + HE\bar{u}(t), \\ &= Px(t) - z(t) + HE\bar{u}(t), \end{aligned} \quad (14)$$

where

$$P = I + HC. \quad (15)$$

Hence, the dynamics of the state estimation error is

$$\begin{aligned} {}_{t_0}D_t^\alpha \tilde{x}(t) &= P_{t_0}D_t^\alpha x(t) - {}_{t_0}D_t^\alpha z(t) + HE_{t_0}D_t^\alpha \bar{u}(t) \\ &= \sum_{i=1}^{M} h_i(\hat{x}(t))\left[PA_i x(t) + PB_i u(t) + PE_i \bar{u}(t) \right. \\ &\quad \left. + P\omega(t) - N_i z(t) - G_i u(t) - L_i y(t)\right] + HE_{t_0}D_t^\alpha \bar{u}(t), \end{aligned} \quad (16)$$

replacing $y(t)$ and $z(t)$ by their respective expressions given by (11) and (13), the state error is given as follows:

$$\begin{aligned} {}_{t_0}D_t^\alpha \tilde{x}(t) &= \sum_{i=1}^{M} h_i(\hat{x}(t))\left[(PA_i - N_i - K_i C)x(t) + (PB_i - G_i)u(t) + (PE_i - K_i E)\bar{u}(t) + P\omega(t) \right. \\ &\quad \left. + N_i e(t)\right] + HE_{t_0}D_t^\alpha \bar{u}(t), \end{aligned} \quad (17)$$

with $K_i = N_i H + L_i$.

If the next conditions are satisfied:

$$HE = 0, \quad (18)$$

$$N_i = PA_i - K_iC, \qquad (19)$$

$$PB_i = G_i, \qquad (20)$$

$$PE_i = K_iE \qquad (21)$$

and

$$L_i = K_i - N_iH. \qquad (22)$$

Then, the dynamics of the state estimation error become:

$$_{t_0}D_t^\alpha \tilde{x}(t) = \sum_{i=1}^{M} h_i(\hat{x}(t))[N_i\tilde{x}(t) + P\omega(t)], \qquad (23)$$

thus showing that the dynamics of the state estimation error is disturbed by $\omega(t)$.

To synthesize the matrices of the observer (13), two techniques are proposed.

4.1. First Approach

It is assumed that the term $\omega(t)$ defined in (12) satisfies the following Lipschitz condition:

$$|\omega(t)| \leq \delta |\tilde{x}(t)|, \qquad (24)$$

where δ is a positive constant.

Lemma 1 (see [55]). *Let M and N be matrices of the appropriate sizes, then the following property holds:*

$$M^T N + N^T M \leq \eta M^T M + \eta^{-1} N^T N, \qquad \eta > 0. \qquad (25)$$

Theorem 1. *A fractional order unknown input observer (13) for system (11) exists if there exists a positive definite matrix X, matrices M_i, S, positive scalars η and δ satisfying the following conditions for all $i = 1, \cdots, M$:*

$$\begin{bmatrix} \Theta_i & (X + SC) \\ (X + SC)^T & -\lambda I \end{bmatrix} < 0, \qquad (26)$$

$$SE = 0, \qquad (27)$$

$$(X + SC)E_i = M_iE, \qquad (28)$$

where

$$\Theta_i = A_i^T\left(X + C^T S\right) + (X + SC)A_i - C^T M_i^T - M_i C + \eta \delta^2 I. \qquad (29)$$

Then, the fractional order observer (13) is completely defined by:

$$H = X^{-1}S, \qquad (30)$$

$$K_i = X^{-1}M_i, \qquad (31)$$

$$N_i = (I + HC)A_i - K_iC, \qquad (32)$$

$$L_i = K_i - N_iH, \qquad (33)$$

and

$$G_i = (I + HC)B_i. \qquad (34)$$

Proof of Theorem 1. In order to found the existence conditions of the fractional order observer in Theorem 1, Lemma 1 can be introduced:

Considering the following quadratic Lyapunov function:

$$V(t) = \tilde{x}(t)^T X \tilde{x}(t), \qquad X = X^T > 0, \tag{35}$$

its derivative with regard to time is given by:

$$_{t_0}D_t^\alpha V(t) \leq {}_{t_0}D_t^\alpha \tilde{x}(t)^T X \tilde{x}(t)(t) + \tilde{x}(t)^T X {}_{t_0}D_t^\alpha \tilde{x}(t). \tag{36}$$

By substituting (24), the dynamic of the quadratic Lyapunov function becomes:

$$_{t_0}D_t^\alpha V(t) \leq \sum_{i=1}^{M} \mu_i(\hat{x}(t)) \left[\tilde{x}(t)^T \left(N_i^T X + X N_i\right) \tilde{x}(t) + \tilde{x}(t)^T X P \omega(t) + \omega(t)^T P^T X \tilde{x}(t)\right], \tag{37}$$

and when using Lemma 1 and (25), this allows for the following:

$$\tilde{x}^T X P \omega + \omega^T P^T X \tilde{x} \leq \eta \omega^T \omega + \eta^{-1} \tilde{x}^T X P P^T X \tilde{x}$$
$$\leq \eta \gamma^2 \tilde{x}^T \tilde{x} + \eta^{-1} \tilde{x}^T X P P^T X \tilde{x}. \tag{38}$$

Substituting (38) in the fractional derivative of the Lyapunov function (36) yields:

$$_{t_0}D_t^\alpha V = \sum_{i=1}^{M} \mu_i(\hat{x}) \tilde{x}^T \left(N_i^T X + X N_i + \eta \gamma^2 I + \eta^{-1} X P P^T X\right) \tilde{x}. \tag{39}$$

Since the activation functions satisfy condition (8), the fractional derivative of the Lyapunov function is negative if:

$$N_i^T X + X N_i + \eta \gamma^2 I + \eta^{-1} X P P^T X < 0. \tag{40}$$

According to (19), Equation (40) becomes:

$$(PA_i - K_i C)^T X + X (PA_i - K_i C) + \eta \delta^2 I + \eta^{-1} X P P^T X < 0. \tag{41}$$

It is noted unfortunately that the matrix inequality (41) gives a disadvantage since it is nonlinear with respect to the variables K_i, X and η (more precisely bilinear). A numerical procedure of resolution by linearization is obtained in the following section.

In order to convert these conditions into an LMI formulation, the following change of variables is considered:

$$M_i = X K_i \tag{42}$$

and by using the Schur complement [16], the linear matrix inequality is obtained:

$$\begin{bmatrix} A_i^T P^T X + X P A_i - C^T M_i^T + \eta \delta^2 I & XP \\ P^T X & -\eta I \end{bmatrix} < 0. \tag{43}$$

To satisfy condition (18), the equality can be solved:

$$XHE = 0. \tag{44}$$

Using the change of variable $S = XH$, linear matrix equality is obtained:

$$SE = 0. \tag{45}$$

The conditions (21) must be satisfied simultaneously, and using the change of variable (42) gives:

$$(X + SC) E_i = M_i E. \tag{46}$$

Since $P = I + HC$, replacing P in (43), the matrix inequality of Theorem 1 can be obtained. The conditions (26)–(28) of Theorem 1 are thus demonstrated. □

4.2. Second Approach

In the case where hypothesis (24) is not satisfied, meaning that the information on its bounded δ is not available, the method established in the previous section cannot be applied.

In this section, another method based on the use of the L_2 approach is proposed.

Theorem 2. *A fractional order unknown input observer (13) for system (11) exists if there exists a positive definite matrix X, matrices M_i, S and positive scalars $\bar{\delta}$ satisfying the following conditions for all $i = 1, \cdots, M$:*

$$\begin{bmatrix} \Theta_i & X + SC \\ (X + SC)^T & -\bar{\gamma} I \end{bmatrix} < 0, \qquad (47)$$

$$SE = 0 \qquad (48)$$

and

$$(X + SC)\, E_i = M_i E, \qquad (49)$$

where

$$\Theta_i = A_i^T \left(X + C^T S \right) + (X + SC)\, A_i - C^T M_i^T - M_i C + I. \qquad (50)$$

Then, the fractional order UI observer (13) is completely defined by (30)–(34).

Proof of Theorem 2. To prove Theorem 2, the real bounded Lemma 1 [56] is used.
The dynamics of the fractional state estimation error are given by:

$$_{t_0}D_t^\alpha \tilde{x}(t) = \sum_{i=1}^{M} h_i\left(\hat{x}(t)\right) \left[N_i \tilde{x}(t) + P\omega(t) \right]. \qquad (51)$$

Consider the following Lyapunov quadratic function:

$$V(t) = \tilde{x}(t)^T X \tilde{x}(t),\ X = X^T > 0. \qquad (52)$$

Its derivative with regard to time is specified by:

$$_{t_0}D_t^\alpha V(t) \leq\, _{t_0}D_t^\alpha \tilde{x}(t)^T X \tilde{x}(t) + \tilde{x}(t)^T X\, _{t_0}D_t^\alpha \tilde{x}(t). \qquad (53)$$

By substituting (23), the dynamic of the quadratic Lyapunov function is obtained:

$$_{t_0}D_t^\alpha V(t) \leq \sum_{i=1}^{M} \mu_i\left(\hat{x}(t)\right) \left[\tilde{x}(t)^T \left(N_i^T X + X N_i \right) \tilde{x}(t) + \tilde{x}(t)^T X P \omega(t) + \omega(t)^T P^T X \tilde{x}(t) \right]. \qquad (54)$$

In order to mitigate the impact of $\omega(t)$ on the state estimation error, the L_2 [52] will be used. It can guarantee:

$$\frac{\|\tilde{x}(t)\|_2}{\|\omega(t)\|_2} < \delta, \qquad \delta > 0. \qquad (55)$$

The system of the state estimation error is stable and the gain L_2 noted δ of the transfer from $\omega(t)$ to $\tilde{x}(t)$ is bounded, if:

$$_{t_0}D_t^\alpha V\left(\tilde{x}(t)\right) + \tilde{x}(t)^T \tilde{x}(t) - \delta^2 \omega(t)^T \omega(t) < 0. \qquad (56)$$

By substituting $_{t_0}D_t^\alpha V(\tilde{x}(t))$, the inequality (56) becomes:

$$\sum_{i=1}^{M} \mu_i(\hat{x}(t)) \left[\tilde{x}(t)^T \left(N_i^T X + X N_i\right) \tilde{x}(t) + \tilde{x}(t)^T X P \omega(t) + \omega(t)^T P^T X \tilde{x}(t) \right. \\ \left. + \tilde{x}(t)^T \tilde{x}(t) - \delta^2 \omega(t)^T \omega(t)\right]. \tag{57}$$

The estimation error converges to zero and the gain L_2 of the transfer from $\omega(t)$ to $\tilde{x}(t)$ is bounded by δ if the following inequality is verified:

$$\sum_{i=1}^{M} \mu_i(\hat{x}(t)) \begin{bmatrix} N_i^T X + X N_i + I & X P \\ P^T X & -\delta^2 I \end{bmatrix} < 0. \tag{58}$$

The convex sum property of the activation functions makes it possible to write the following sufficient condition:

$$\begin{bmatrix} N_i^T X + X N_i + I & X P \\ P^T X & -\delta^2 I \end{bmatrix} < 0. \tag{59}$$

Using the expression (19) of N_i and the changes of variables $M_i = X K_i$ and $\bar{\delta} = \delta^2$ (48) becomes:

$$\begin{bmatrix} \Theta_i & X P \\ P^T X & -\bar{\delta} I \end{bmatrix} < 0, \forall i = 1, ..., M \tag{60}$$

where

$$\Theta_i = A_i^T P^T X + X P A_i - C^T M_i^T - M_i C + I.$$

To satisfy condition (18), the equality can be solved:

$$X H E = 0. \tag{61}$$

Using the change of variable $S = X H$, the linear matrix equality is obtained:

$$S E = 0. \tag{62}$$

The conditions (21) must be satisfied simultaneously, and by using the change of variable (42) gives:

$$(X + S C) E_i = M_i E. \tag{63}$$

Since $P = I + H C$, replacing P in (60), the matrix inequality of Theorem 2 is obtained. The conditions (47)–(49) of Theorem 2 are thus demonstrated. □

5. Unknown Inputs Estimation

In system (11), the unknown input $\bar{u}(t)$ appears with the influence matrix:

$$\Phi(t) = \begin{bmatrix} \sum_{i=1}^{M} h_i(\hat{x}(t)) E_i \\ E \end{bmatrix}. \tag{64}$$

For the estimation of the unknown input, it is necessary that the rank of the matrix $\Phi(t)$ is verified at each time t for the following condition:

$$\text{rank}(\Phi(t)) = q, \tag{65}$$

where q is the dimension of $\bar{u}(t)$. If this condition is satisfied, $\Phi(t)$ is full-rank column and its pseudo-inverse left $\Phi^{-1}(t)$ exists:

$$\Phi^-(t) = \left(\Phi^T(t)\Phi(t)\right)^{-1}\Phi^T(t). \tag{66}$$

The unknown input can then be calculated according to the state estimated as follows:

$$\hat{\bar{u}}(t) = \Phi^- \begin{bmatrix} {}_{t_0}D_t^\alpha \hat{x}(t) - \sum_{i=1}^M h_i\left(\hat{x}(t)\right)[A_i\hat{x}(t) + B_iu(t)] \\ y(t) - C\hat{x}(t) \end{bmatrix}, \tag{67}$$

under condition (65) the asymptotic convergence from $\hat{x}(t)$ to $x(t)$ results in the asymptotic convergence of $\hat{\bar{u}}(t)$ to $\bar{u}(t)$.

6. Example and Comparisons

To validate the advantages of the proposed fractional order unknown input observer, the system (68) represented by the FOTS model with UPV is considered with $\alpha = 0.8$. The state estimation is carried out by means of two fuzzy unknown input observers, the first with integer order and the second one with fractional order. The unknown inputs considered may be noise, faults or modeling uncertainties.

Example and Simulation Results

Consider the FOTS model (8), which is defined as follows:

$$\begin{cases} {}_{t_0}D_t^\alpha \hat{x}(t) = \sum_{i=1}^M h_i\left(x(t)\right)[A_ix(t) + B_iu(t) + F_i\bar{u}(t)], \\ y(t) = Cx(t) + G\bar{u}(t), \end{cases} \tag{68}$$

where: $A_1 = \begin{bmatrix} -2 & 1 & 1 \\ 1 & -3 & 0 \\ 2 & 1 & -4 \end{bmatrix}$, $A_2 = \begin{bmatrix} -3 & 2 & -2 \\ 5 & -3 & 0 \\ 0.5 & 0.5 & -4 \end{bmatrix}$, $B_1 = \begin{bmatrix} 1 \\ 0.3 \\ 0.5 \end{bmatrix}$, $B_2 = \begin{bmatrix} 0.5 \\ 1 \\ 0.25 \end{bmatrix}$,

$C = \begin{bmatrix} 1 & 1 & 1 \\ 1 & 0 & 1 \end{bmatrix}$, $F_1 = \begin{bmatrix} 0.5 \\ -1 \\ 0.25 \end{bmatrix}$, $F_2 = \begin{bmatrix} -1 \\ 0.52 \\ 1 \end{bmatrix}$, $G = \begin{bmatrix} 0.9 \\ 0.9 \end{bmatrix}$.

The activation functions are chosen in the form:

$$\begin{cases} h_1(x) = \frac{1-\tanh(x_1)}{2}, \\ h_2(x) = 1 - h_1(x) = \frac{1+\tanh(x_1)}{2}. \end{cases} \tag{69}$$

Two cases are considered for simulation, the first one in the absence of unknown inputs (unknown inputs are null), and the second one in the presence of unknown inputs. The unknown input considered is accompanied by an additive noise. To have a treatment close to reality, the initial values of the system are chosen non-null, but the initial values of the two unknown input observers are chosen equal to zero.

The outputs and the states of the FOTS system with their estimations given by the fuzzy integer and fractional order unknown input observers, and the unknown inputs and their estimates will be compared and analyzed.

Case 1: Absence of unknown input.

At first, the case of the absence of unknown inputs (unknown inputs are null) will be evaluated. Figure 1 shows the two outputs of the considered FOTS system (ys_1, ys_2), the outputs estimated by the FOUIO (yo_1, yo_2) and the fractional order unknown input observer (FOUIO) (yfo_1, yfo_2) in the absence of unknown inputs. Figure 2 shows the outputs estimation error (a and b) in the absence of unknown inputs ($ys_1 - yo_1, ys_2 - yo_2$) and ($ys_1 - yfo_1, ys_2 - yfo_2$). The two Figures 1 and 2 show that the FOUIO gives better output estimation for the considered system. The decreased quality of the output estimation at the moment $t = 0$ is because of the choice of the initial values.

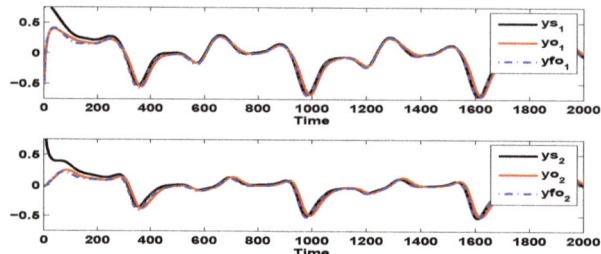

Figure 1. Outputs of the fractional-order Takagi–Sugeno (FOTS) system and its estimation by FOUIO and fractional order unknown input observer (FOUIO) in a free of fault case.

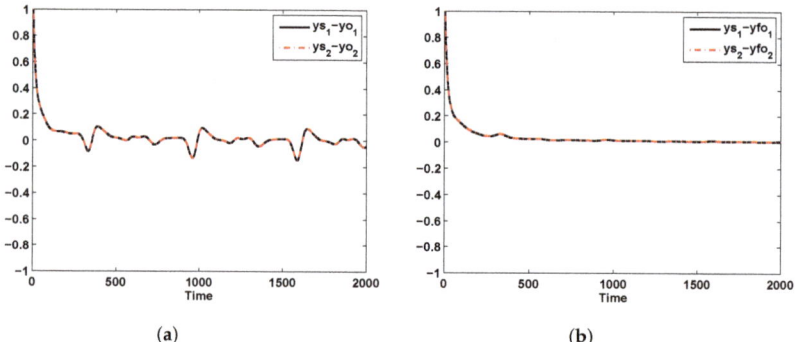

Figure 2. Output estimation Error in a free of fault case. (**a**) Error output estimation of the FOTS system by FUIO in a free of fault case. (**b**) Error output estimation of the FOTS system by FOUIO in a free of fault case.

Figure 3 presents the states of the FOTS system (xs_1, xs_2, xs_3) and their estimations given by the FUIO (xo_1, xo_2, xo_3) and the FOUIO (xfo_1, xfo_2, xfo_3) in the absence of unknown inputs. Figure 4 shows the state estimation errors (a and b) in the absence of unknown inputs ($xs_1 - xo_1, xs_2 - xo_2, xs_3 - xo_3$) and ($xs_1 - xfo_1, xs_2 - xfo_2, xs_3 - xfo_3$). The two Figures 3 and 4 show that the fuzzy observer with unknown inputs gives a better state estimation of the FOTS system. The decreased quality of the state estimation at the moment $t = 0$ is because of the choice of the initial values.

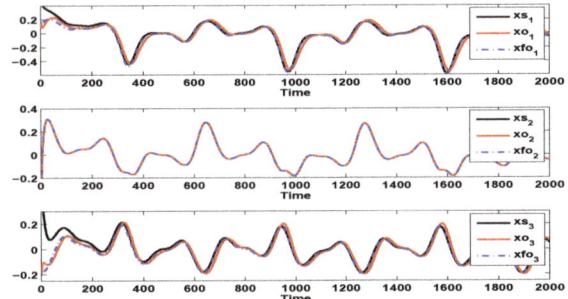

Figure 3. State of the FOTS system and its estimation by FUIO and FOUIO in a free of fault case.

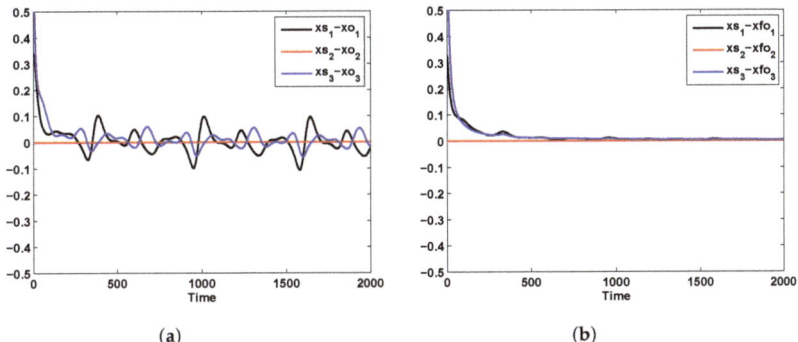

Figure 4. State estimation error in free of fault case. (**a**) State estimation error between the FOTS system and FUIO in a free of fault case. (**b**) State estimation error between the FOTS system and FOUIO in a free of fault case.

Now, the case of the presence of unknown inputs will be evaluated.

Case 2: Presence of unknown input and measurement noise simultaneously.

Figure 5 shows the outputs of the FOTS system (ys_1, ys_2), and their estimations given by the FUIO (yo_1, yo_2) and the fuzzy FOUIO (yfo_1, yfo_2) in the presence of unknown inputs. Figure 6 shows the outputs estimation error (a and b) in the presence of unknown inputs $(ys_1 - yo_1, ys_2 - yo_2)$ and $(ys_1 - yfo_1, ys_2 - yfo_2)$. The two Figures 5 and 6 show that the FOUIO gives a better output estimation for the FOTS system. The decreased quality of the outputs estimation at the moment $t = 0$ is because of the choice of the initial values.

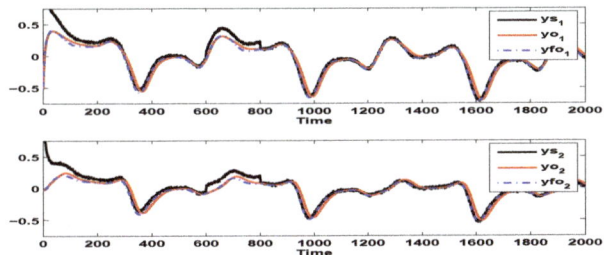

Figure 5. Outputs for the FOTS system and its estimation by FUIO and FOUIO in a faulty case.

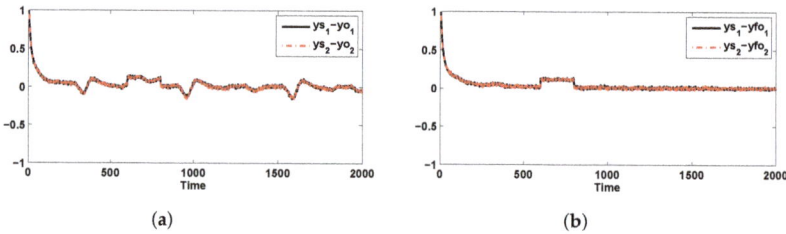

(a) (b)

Figure 6. Output for error estimation faulty case. (**a**) Outputs for error estimation between the FOTS system and FUIO in a faulty case. (**b**) Outputs error estimation between the FOTS system and FOUIO in a faulty case.

Figure 7 presents the states of the FOTS system (xs_1, xs_2, xs_3) and their estimations given by the FUIO (xo_1, xo_2, xo_3) and the FOUIO (xfo_1, xfo_2, xfo_3) in the presence of unknown inputs. Figure 8 shows the state estimation errors (a and b) in the presence of unknown inputs $(xs_1 - xo_1, xs_2 - xo_2, xs_3 - xo_3)$ and $(xs_1 - xfo_1, xs_2 - xfo_2, xs_3 - xfo_3)$. The two Figures 7 and 8 show that the FOUIO gives a better state estimation for the FOTS system. The decreased quality of the state estimation at the moment $t = 0$ is because of the choice of the initial values.

Analyzing the convergence conditions of the proposed FOUIO, if the condition (23) on the term $\omega(t)$ is not satisfied or the value of the constant δ is very important (impossibility of finding a solution with Theorem 1, Theorem 2) offers the possibility of designing the observer with unknown input.

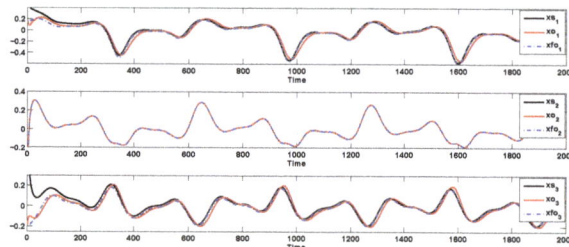

Figure 7. State estimation for the FOTS system and its estimations by FUIO and FOUIO in a faulty case.

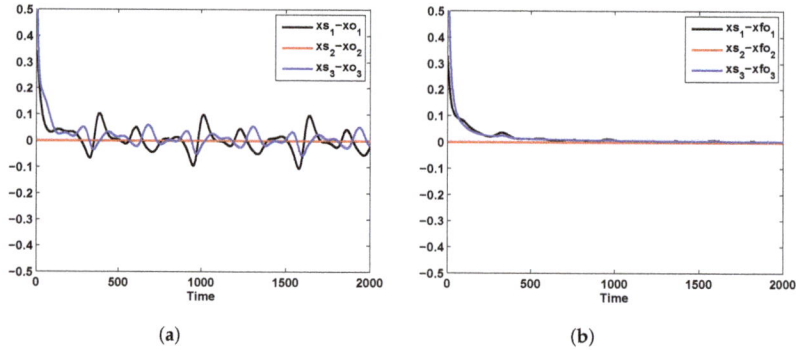

(a) (b)

Figure 8. Output estimation error in faulty case. (**a**) State estimation error between the FOTS system and FUIO in a faulty case. (**b**) State estimation error between the FOTS system and FOUIO in a faulty case.

Figure 9 shows the considered unknown input with normal noise (ubar), their estimations given by the FUIO (ubar FUIO), FOUIO (ubar FOUIO) and the unknown input without noise (ubar without noise).

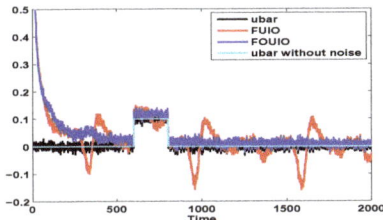

Figure 9. Unknown input and their estimations.

Figure 10 shows the unknown input estimation errors (ubar-ubar FUIO, ubar-ubar FOUIO).

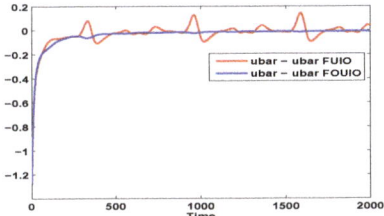

Figure 10. Unknown input errors estimation.

The two Figures 9 and 10 show that the FOUIO gives a better unknown input estimation of the FOTS system, but it cannot be decoupled from the noise. The decreased quality of the unknown input estimation at the moment $t = 0$ is because of the choice of the initial values.

In the presence of adding random measurement noises bounded by 0.01, the unknown input estimated based on the proposed observer is noisy. Indeed, the presence of measurement noise, at high frequency, decreases the quality of reconstruction of the unknown input.

7. Conclusions

In this paper, a new approach is proposed for designing a fractional order Takagi–Sugeno unknown input observer for a nonlinear system described by FOTS models with UPV. The first step is to rewrite the FOTS system in the form of a disturbed equivalent FOTS and with measurable premise variables. After that, two cases are considered; the first one uses the hypothesis that the perturbation, which appears after rewriting the FOTS model, verifies a Lipschitz condition, while the second one does not use this hypothesis. In this second case, another method is developed and based on an L_2 approach. The convergence conditions of the proposed observers are given in the form of linear matrix inequalities (LMI) that can be easily solved with conventional digital tools.

The obtained results show that a good convergence of the outputs and the state estimation errors is observed using the new proposed FOUIO. The state of the system can be estimated even in the presence of an unknown input varying rapidly since it is totally decoupled from the state. An improvement in the dynamics of the proposed observer is possible by placing the poles.

In future work, it would be interesting to study the decoupling of the noise and the estimation of the unknown inputs using the augmented systems.

Author Contributions: Conceptualization, A.D. and D.D.; methodology, D.D. and A.T.A.; software, A.D., D.D. and S.A.; validation, A.D. and A.T.A.; formal analysis, A.T.A.; investigation, S.A.; resources, S.A.; writing–original draft preparation, A.D. and S.A.; writing–review and editing, A.D., D.D. and A.A.; visualization, D.D. and S.A.; funding acquisition, A.T.A.

Funding: This research was funded by Prince Sultan University, Riyadh, Saudi Arabia.

Acknowledgments: The authors would like to thank Prince Sultan University, Riyadh, Saudi Arabia for supporting and funding this work. Special acknowledgment to Robotics and Internet-of-Things Lab (RIOTU) at Prince Sultan University, Riyadh, SA. Also, the authors wish to acknowledge the editor and anonymous reviewers for their insightful comments, which have improved the quality of this publication.

Conflicts of Interest: The authors declare no conflict of interest.

Abbreviations

The following abbreviations are used in this manuscript:

FOTS	Fractional Order Takagi–Sugeno
FOS	Fractional Order Systems
FUIO	Fuzzy Unknown Input Observer
FOUIO	Fractional Order Unknown Input Observer
LMI	Linear Matrix Inequalities
UPV	Unmeasurable Premise Variables
MPV	Measurable Premise Variables

References

1. Machado, J.T.; Kiryakova, V.; Mainardi, F. Recent history of fractional calculus. *Commun. Nonlinear Sci. Numer. Simul.* **2011**, *16*, 1140–1153. [CrossRef]
2. Oldham, K.B.; Spanier, J. *The Fractional Calculus, Theory and Applications of Differentiation and Integration of Arbitrary Order*; Academic Press: New York, NY, USA, 1974.
3. Miller, K.S.; Ross, B. *The Fractional Calculus, an Introduction to the Fractional Calculus and Fractional Deferential Equations*; John Wiley & Sons Inc.: New York, NY, USA, 1993.
4. George, A.; Argyros, I.K. *Intelligent Numerical Methods: Applications to Fractional Calculus*; Springer International Publishing: Cham, Switzerland, 2016; Volume 624.
5. Li, W.; Zhao, H.M. Rational function approximation for fractional order differential and integral operators. *Acta Autom. Sin.* **2011**, *37*, 999–1005.
6. Xe, D.Y.; Zhao, C.N.; Chen, Y.Q. Robust control for fractional order four-wing hyperchaotic system using LMI. In Proceedings of the IEEE Conference on Mechatronics and Automation, Luoyang, China, 25–28 June 2006; pp. 1043–1048.
7. Li, C.L.; Su, K.L.; Zhang, J.; Wei, D.Q. Rational function approximation for fractional order differential and integral operators. *Opt. Int. J. Light Electron Opt.* **2013**, *124*, 5807–5810. [CrossRef]
8. Yuan, J.; Shi, B.; Ji, W.Q. Adaptive sliding mode control of a novel class of fractional chaotic systems. *Adv. Math. Phys.* **2013**, *2013*, 6709. [CrossRef]
9. Li, W.; Peng, C.; Wang, Y. Frequency domain subspace identification of commensurate fractional order input time delay systems. *Int. J. Control. Autom. Syst.* **2011**, *9*, 310–316.
10. Vinagre, B.M.; Podlubny, I.; Dorcak, L.; Feliu, V. On fractional PID controllers: A frequency domain approach. In Proceedings of the IFAC Workshop on Digital Control: Past, Present and Future of PID Control, Terrasa, Spain, 5–7 April 2000; pp. 53–58.
11. Aldair, A.A.; Wang, W.J. Design of fractional order controller based on evolutionary algorithm for a full vehicle nonlinear active suspension systems. *Int. J. Control. Autom.* **2010**, *3*, 33–46.
12. Ostalczyk, P. Discrete Fractional Calculus: Applications in Control and Image Processing. In *Series in Computer Vision*; World Scientific Publishing Co.: Singapore, 2016; Volume 4.
13. Mozyrska, D.; Wyrwas, M. The Z-transform method and delta type fractional difference operators. *Discret. Dyn. Nat. Soc.* **2015**, *2–3*, 1–12. [CrossRef]
14. Das, S. *Functional Fractional Calculus for System Identification and Controls*; Springer: Berlin/Heidelberg, Germany, 2008.

15. Ibrir, S. Robust state estimation with q-integral observers. In Proceedings of the American Control Conference, Boston, MA, USA, 30 June–2 July 2004; pp. 3466–3471.
16. Farges, C.; Moze, M.; Sabatier, J. Pseudo-state feedback stabilization of commensurate fractional order systems. *Automatica* **2010**, *46*, 1730–1734. [CrossRef]
17. Stanisławski, R.; Rydel, M.; Latawiec, K.J. Modeling of discrete-time fractional-order state space systems using the balanced truncation method. *J. Frankl. Inst.* **2017**, *354*, 3008–3020. [CrossRef]
18. Doye, I.; Darouach, M.; Voos, H.; Zasadzinsk, M. Design of unknown input fractional-order observer for fractional-order systems. *Int. J. Appl. Math. Comput. Sci.* **2013**, *23*, 491–500. [CrossRef]
19. Wei, Y.H.; Sun, Z.Y.; Hu, Y.S.; Wang, Y. On fractional order adaptive observer. *Int. J. Autom. Comput.* **2015**, *12*, 664–670. [CrossRef]
20. Sabatier, J.; Farges, C.; Merveillaut, M.; Feneteau, L. On Observability and Pseudo State Estimation of Fractional Order Systems. *Eur. J. Control.* **2012**, *18*, 260–271. [CrossRef]
21. Safarinejadian, B.; Asad, M.; Sha Sadeghi, M. Simultaneous state estimation and parameter identification in linear fractional order systems using colored measurement noise. *Int. J. Control.* **2016**, *89*, 2277–2296. [CrossRef]
22. Li, F.; Wu, R.; Liang, S. Observer-based state estimation for non-linear fractional systems. *Int. J. Dyn. Syst. Differ. Equ.* **2015**, *5*, 322–335. [CrossRef]
23. Fuli, Z.; Hui, L.; Shouming, Z. State estimation based on fractional order sliding mode observer method for a class of uncertain fractional-order nonlinear systems. *Signal Process.* **2016**, *127*, 168–184.
24. Kong, S.; Saif, M.; Liu, B. Observer design for a class of nonlinear fractional-order systems with unknown input. *J. Frankl. Inst.* **2017**, *354*, 5503–5518. [CrossRef]
25. Djeghali, N.; Djennoune, S.; Bettayeb, M.; Ghanes, M.; Barbot, J.P. Observation and sliding mode observer for nonlinear fractional-order system with unknown input. *ISA Trans.* **2016**, *63*, 1–10. [CrossRef]
26. Ding, B.; Sun, H.; Yang, P. Further studies on LMI based relaxed stabilization conditions for nonlinear systems in Takagi–Sugeno's form. *Automatica* **2006**, *42*, 503–508. [CrossRef]
27. Kruszewski, A.; Wang, R.; Guerra, T.M. Nonquadratic stabilization conditions for a class of uncertain nonlinear discrete time TS fuzzy models: A new approach. *IEEE Trans. Autom. Control.* **2008**, *53*, 606–611. [CrossRef]
28. N'Doye, I.; Darouach, M.; Zasadzinski, M.; Radhy, N.E. Robust stabilization of uncertain descriptor fractional-order systems. *Automatica* **2013**, *49*, 1907–1913. [CrossRef]
29. Lu, J.G.; Chen, Y.Q. Robust stability and stabilization of fractional-order interval systems with the fractional order α: The case $0 < \alpha < 1$. *IEEE Trans. Autom. Control.* **2010**, *55*, 152–158.
30. Trigeassou, J.; Maamri, N.; Sabatier, J.; Oustaloup, A. A Lyapunov approach to the stability of fractional differential equations. *Signal Process.* **2011**, *91*, 437–445. [CrossRef]
31. Yu, W.; Li, Y.; Wen, G.; Yu, X.; Cao, J. Observer design for tracking consensus in second-order multi-agent systems: Fractional order less than two. *IEEE Trans. Autom. Control.* **2017**, *62*, 894–900. [CrossRef]
32. Park, J.H.; Park, T.S.; Kim, S.H. Approximation-Free Output-Feedback Non-Backstepping Controller for Uncertain SISO Nonautonomous Nonlinear Pure-Feedback Systems. *Mathematics* **2019**, *7*, 456. [CrossRef]
33. Faieghi, M.; Mashhadi, S.K.M.; Baleanu, D. Sampled-data nonlinear observer design for chaos synchronization: A Lyapunov-based approach. *Commun. Nonlinear Sci. Numer. Simul.* **2014**, *19*, 2444–2453. [CrossRef]
34. Zhang, X.; Ding, F.; Xu, L.; Alsaedi, A.; Tasawar, H. A Hierarchical Approach for Joint Parameter and State Estimation of a Bilinear System with Autoregressive Noise. *Mathematics* **2019**, *7*, 356. [CrossRef]
35. Ibrir, S.; Bettayeb, M. New sufficient conditions for observer-based control of fractional-order uncertain systems. *Automatica* **2015**, *59*, 216–223. [CrossRef]
36. Song, C.; Fei, S.; Cao, J.; Huang, C. Robust Synchronization of Fractional-Order Uncertain Chaotic Systems Based on Output Feedback Sliding Mode Control. *Mathematics* **2019**, *7*, 599. [CrossRef]
37. Liu, S.; Dong, X.; Zhang, Y. A New State of Charge Estimation Method for Lithium-Ion Battery Based on the Fractional Order Model. *IEEE Access* **2019**, *7*, 122949–122954. [CrossRef]
38. Shi, S.L.; Li, J.X.; Fang, Y.M. Fractional-disturbance-observer-based Sliding Mode Control for Fractional Order System with Matched and Mismatched Disturbances. *Int. J. Control. Autom. Syst.* **2019**, *17*, 1184–1190. [CrossRef]

39. Trinh, H.; Huong, D.C.; Nahavandi, S. Observer design for positive fractional-order interconnected time-delay systems. *Trans. Inst. Meas. Control. Publ.* **2018**, *41*, 378–391. [CrossRef]
40. Dabiri, A.; Butcher, E. Optimal observer-based feedback control for linear fractional-order systems with periodic coefficients. *J. Vib. Control.* **2019**, *25*, 1379–1392. [CrossRef]
41. Kong, S.; Saif, M.; Cui, G. Estimation and Fault Diagnosis of Lithium-Ion Batteries: A Fractional-Order System Approach. *Math. Probl. Eng.* **2018**, 8705363, 1–12. [CrossRef]
42. Yang, B.; Yu, T.; Shu, H.; Zhu, D.; Sang, Y.; Jiang, L. Passivity-based fractional-order sliding-mode control design and implementation of grid-connected photovoltaic systems. *J. Renew. Sustain. Energy* **2018**, *10*, 43701. [CrossRef]
43. Chadli, M.; Karimi, H.R. Robust observer design for unknown inputs Takagi–Sugeno models. *IEEE Trans. Fuzzy Syst.* **2013**, *21*, 158–164. [CrossRef]
44. Liu, S.; Li, X.; Wang, H.; Yan, J. Adaptive fault estimation for T-S fuzzy systems with unmeasurable premise variables. *Adv. Differ. Equ.* **2018**, *105*. [CrossRef]
45. Takagi, T.; Sugeno, M. Fuzzy identification of systems and its applications to modeling and control. *IEEE Trans. Syst. Man Cybern. Part B Cybern.* **1985**, *SMC-15*, 116–132. [CrossRef]
46. Krokavec, D.; Filasova, A. On observer design methods for a class of Takagi Sugeno fuzzy systems. In Proceedings of the Third International Conference on Advanced Information Technologies & Applications, Dubai, UAE, 7–8 November 2014; pp. 279–290.
47. Djeddi, A.; Harkat, M.F.; Soufi, Y. A New Approach for State Estimation of Uncertain Multiple model with Unknown Inputs. Application to Sensor Fault Diagnosis. *Mediterr. J. Meas. Control.* **2016**, *12*, 537–545.
48. Chadli, M.; Guerra, T.M. LMI solution for robust static output feedback control of Takagi–Sugeno fuzzy models. *IEEE Trans. Fuzzy Syst.* **2012**, *20*, 1060–1065. [CrossRef]
49. Oukacine, S.; Djamah, T.; Djennoune, S.; Mansouri, R.; Bettayeb, M. Multi-model identification of a fractional nonlinear system. *IFAC Proc.* **2013**, *46*, 48–53. [CrossRef]
50. Junmin, L.; Yuting, L. Robust Stability and Stabilization of Fractional Order Systems Based on Uncertain Takagi–Sugeno Fuzzy Model With the Fractional Order $1 \leq v < 2$. *ASME J. Comput. Nonlinear Dyn.* **2013**, *8*, 41005.
51. Gao, Z.; Liao, X. Observer-based fuzzy control for nonlinear fractional-order systems via fuzzy T-S models: The $1 < \alpha < 2$ case. In Proceedings of the 19th World Congress, The International Federation of Automatic Control, Cape Town, South Africa, 24–29 August 2014.
52. Ichalal, D.; Marx, B.; Mammar, S.; Maquin, D.; Ragot, J. How to cope with unmeasurable premise variables in Takagi–Sugeno observer design: Dynamic extension approach. *Eng. Appl. Artif. Intell.* **2018**, *67*, 430–435. [CrossRef]
53. Shantanu, D. *Functional Fractional Calculus*, 2nd ed.; Springer: Berlin/Heidelberg, Germany, 2011.
54. Petras, I. *Fractional-Order Nonlinear Systems Modeling, Analysis and Simulation*; Higher Education Press: Beijing, China; Springer: Berlin/Heidelberg, Germany, 2011.
55. Akhenak, A.; Chadli, M.; Ragot, J.; Maquin, D. Estimation of state and unknown inputs of a nonlinear system represented by a multiple model. *IFAC Proc. Vol.* **2004**, *37*, 385–390. [CrossRef]
56. Boyd, S.; El Ghaoui, L.; Feron, E.; Balakrishnan, V. *Linear Matrix Inequalities in System and Control Theory*; SIAM: Philadelphia, PA, USA, 1994.

© 2019 by the authors. Licensee MDPI, Basel, Switzerland. This article is an open access article distributed under the terms and conditions of the Creative Commons Attribution (CC BY) license (http://creativecommons.org/licenses/by/4.0/).

Article

q-Analogue of Differential Subordinations

Miraj Ul-Haq [2], Mohsan Raza [3], Muhammad Arif [2], Qaiser Khan [2] and Huo Tang [1,*]

1. School of Mathematics and Statistics, Chifeng University, Chifeng 024000, China
2. Department of Mathematics, Abdul Wali Khan University Mardan, Mardan 23200, Pakistan
3. Department of Mathematics, Government College University Faisalabad, Faisalabad 38000, Pakistan
* Correspondence: thth2009@163.com

Received: 24 June 2019; Accepted: 8 August 2019; Published: 9 August 2019

Abstract: In this article, we study differential subordnations in q-analogue. Some properties of analytic functions in q-analogue associated with cardioid domain and limacon domain are considered. In particular, we determine conditions on α such that $1 + \alpha \frac{z \partial_q h(z)}{(h(z))^n}$ ($n = 0, 1, 2, 3$) are subordinated by Janowski functions and $h(z) \prec 1 + \frac{4}{3}z + \frac{2}{3}z^2$. We also consider the same implications such that $h(z) \prec 1 + \sqrt{2}z + \frac{1}{2}z^2$. We apply these results on analytic functions to find sufficient conditions for q-starlikeness related with cardioid and limacon.

Keywords: q-calculus; differential subordination; Janowski function; cardioid domain; limacon domain

MSC: 30C45; 30C50

1. Introduction

We recall here some basic notions from the literature of Geometric Function Theory which are essential for clarity and understandings of the upcoming work. We start with the symbol $\mathcal{A}$ which represents the family of analytic functions in $\mathcal{D} = \{z : |z| < 1\}$ and any function f in $\mathcal{A}$ satisfies the conditions $f(0) = 0$ and $f'(0) - 1 = 0$. That is, if f in $\mathcal{A}$, then it has the Taylor series expansion as:

$$f(z) = z + \sum_{n=2}^{\infty} a_n z^n, \quad z \in \mathcal{D}. \tag{1}$$

Also let $\mathcal{S}$ denote a subclass of $\mathcal{A}$ which contains univalent functions in $\mathcal{D}$. The notion of subordinations between analytic functions is represented by $f \prec g$ and is defined as; a function f is subordinated by function g, if we can find an analytic function w with the properties $w(0) = 0$ and $|w(z)| < |z|$ such that $f(z) = g(w(z))$. Further, if g is univalent in $\mathcal{D}$, then we have:

$$f(z) \prec g(z) \Leftrightarrow f(0) = g(0) \ \& \ f(\mathcal{D}) \subset g(\mathcal{D}). \tag{2}$$

Ma and Minda [1] studied the function φ which is analytic and normalized by $\varphi(0) = 1$ and $\varphi'(0) > 0$ with $Re\{\varphi(z)\} > 0$ in $\mathcal{D}$. The function φ maps $\mathcal{D}$ onto regions which is starlike with respect to 1 and symmetric along the real axis. Further, they introduced the subclasses of starlike and convex functions respectively as:

$$\mathcal{S}^*(\varphi) = \left\{ f \in \mathcal{A} : \frac{zf'(z)}{f(z)} \prec \varphi(z), \ (z \in \mathcal{D}) \right\},$$

$$\mathcal{C}(\varphi) = \left\{ f \in \mathcal{A} : 1 + \frac{zf''(z)}{f'(z)} \prec \varphi(z), \ (z \in \mathcal{D}) \right\}.$$

If we choose $\varphi(z) = \frac{1+Az}{1+Bz}$ $(-1 \leq B < A \leq 1)$, then $\mathcal{S}^*[A,B] := \mathcal{S}^*\left(\frac{1+Az}{1+Bz}\right)$ is the family of Janowski starlike functions, see [2]. Further by taking $A = 1 - 2\alpha$ with $0 \leq \alpha < 1$ and $B = -1$, we get the class $\mathcal{S}^*(\alpha) := \mathcal{S}^*[1-2\alpha,-1]$ of starlike functions of order α. Also, the notation $\mathcal{S}^* := \mathcal{S}^*(0)$ represents the familiar class of starlike functions. The subclass $\mathcal{S}_{\mathcal{B}}^* := \mathcal{S}^*(\sqrt{1+z})$ which motivates the researchers was investigated by Sokół et al. [3], containing functions $f \in \mathcal{A}$ such that $zf'(z)/f(z)$ lies in the region bounded by the right-half of the Bernoulli lemniscate given by $|w^2 - 1| < 1$. If we choose $\varphi(z) = 1 + \frac{4}{3}z + \frac{2}{3}z^2$, then the class $\mathcal{S}^*(\varphi)$ coincides with the class $\mathcal{S}_{\mathcal{C}}^*$ studied by Sharma et al. [4], consisting of functions $f \in \mathcal{A}$ such that $zf'(z)/f(z)$ lies in the region bounded by the cardioid given by $(9x^2 + 9y^2 - 18x + 5)^2 - 16(9x^2 + 9y^2 - 6x + 1) = 0$. If we choose $\varphi(z) = 1 + \sin z$, then we get the set $\mathcal{S}_{\sin}^*$, established by Cho et al. [5]. By selecting $\varphi(z) = 1 + \sqrt{2}z + \frac{1}{2}z^2$, we acheive an interesting class $\mathcal{S}_{\mathcal{L}}^*$ containing starlike functions associated with limacon given by $(4x^2 + 4y^2 - 8x - 5)^2 - 8(4x^2 + 4y^2 - 12x - 3) = 0$. The class $\mathcal{S}_{\mathcal{L}}^*$ was introduced in [6]. Further, by choosing some more particular function $\varphi(z)$, we get several interesting subclasses of starlike functions. For some details, see [7–10].

In some recent years, a more intensive approach has been shown by the researchers in quantum calculus (q-calculus) because of its wide spread applications in various branches of sciences particularly in Mathematics and Physics. Among contributors to the study, Jackson was the first who provided basic notions and established the theory of q-calculus [11,12]. The idea of derivative in q-analogue was used for the first time by Ismail et al. [13] to initiate and study the geometry of q-starlike functions. After that, a comprenhensive applications of q-calculus in the field of Geometric Function Theory was contributed by Srivastava in a book chapter (see, for details, [14] (pp. 347 et seq.)) and in the same chapter he also given the usage of q-hypergeometric functions in function theory. The concepts of q-starlikeness was further extended to certain subclasses of starlike functions in q-analogue by Agrawal and Sahoo in [15] (for the recent contributions on this topic, see the work done by Srivastava et al. [16–19]). Also, with help of Hadamard product, the q-analogue of Ruscheweyh operator has been introduced by Kanas and Răducanu [20] and further studied in [21–24]. Many researchers contributed in the development of the theory by introducing certain classes with the help of q-calculus. For some details about these contributions, see [25–28].

Let $q \in (0,1)$ and $z \in \mathcal{D}$ with $z \neq 0$. Then the q-derivative of f is defined by

$$\partial_q f(z) = \frac{f(z) - f(qz)}{z(1-q)}. \tag{3}$$

By the virtue of (1) and (3), we easily calculated that for $n \in \mathbb{N}$ and $z \in \mathcal{D}$

$$\partial_q f(z) = 1 + \sum_{n=2}^{\infty} [n,q] a_n z^{n-1}, \tag{4}$$

where

$$[n,q] := \frac{1-q^n}{1-q} = 1 + \sum_{k=1}^{n-1} q^l, \text{ and } [0,q] = 0.$$

Using the definition of q-derivative, Seoudy and Aouf [29] introduced the class $\mathcal{S}_q^*(\varphi)$. Also, for this class, the familiar Fekete-Szegö problem was obtained by the authors. This class is defined as;

$$\mathcal{S}_q^*(\varphi) = \left\{ f \in \mathcal{A} : \frac{z\partial_q f(z)}{f(z)} \prec \varphi(z), (z \in \mathcal{D}) \right\}.$$

By choosing particular functions instead of the function φ, we obtain several interesting subclasses of starlike functions associated with different image domains. We define few of them as follows.

$$\mathcal{S}^*_{\mathcal{B}_q} = \left\{f \in \mathcal{A} : \frac{z\partial_q f(z)}{f(z)} \prec \phi_{\mathcal{B}}(z),\ (z \in \mathcal{D})\right\},$$

$$\mathcal{S}^*_{\mathcal{C}_q} = \left\{f \in \mathcal{A} : \frac{z\partial_q f(z)}{f(z)} \prec \phi_{\mathcal{C}}(z),\ (z \in \mathcal{D})\right\},$$

$$\mathcal{S}^*_{\mathcal{L}_q} = \left\{f \in \mathcal{A} : \frac{z\partial_q f(z)}{f(z)} \prec \phi_{\mathcal{L}}(z),\ (z \in \mathcal{D})\right\},$$

where the particular functions $\phi_{\mathcal{B}}(z)$, $\phi_{\mathcal{C}}(z)$ and $\phi_{\mathcal{L}}(z)$ are given by

$$\phi_{\mathcal{B}}(z) = \sqrt{1+z},$$
$$\phi_{\mathcal{C}}(z) = 1 + \frac{4}{3}z + \frac{2}{3}z^2,$$
$$\phi_{\mathcal{L}}(z) = 1 + \sqrt{2}z + \frac{1}{2}z^2.$$

Recently, Ali et al. [30] have studied some differential subordinations. More precisely they studied the differential subordination $1 + \alpha z p'(z)/p^n(z) \prec \sqrt{1+z}$ and found that $p(z) \prec \sqrt{1+z}$, where $n = 0,1,2$ for some particular range of α. Similar kind of differential subordinations are also discussed by various authors. They used these results to find sufficient conditions for starlike functions, see [31–36]. Motivated by the above work, we introduce and investigate some q-differential subordinations. In particular, we determine conditions on α so that $1 + \alpha \frac{z\partial_q h(z)}{(h(z))^n}$ are subordinated by Janowski functions and $h(z)$ is subordinated by $1 + \frac{4}{3}z + \frac{2}{3}z^2$, where $n = 0,1,2,3$. Similar results are also obtained for $h(z) \prec 1 + \sqrt{2}z + \frac{1}{2}z^2$, $z \in \mathcal{D}$. We use these results to find sufficient conditions for q-starlike functions associated with cardioid and limacon.

To prove our main results we need the following.

Lemma 1 ([37] (q-Jack's Lemma)). *Let w be analytic in $\mathcal{D}$ with $w(0) = 0$. If w attains its maximum value on the circle $|z| = 1$ at $z_0 = re^{i\theta}$, for $\theta \in [-\pi, \pi]$, then for $0 < q < 1$*

$$z_0 \partial_q w(z_0) = m w(z_0),$$

where m is real and $m \geq 1$.

2. Differential Subordination Related with Cardioid

Theorem 1. *Assume that*

$$|\alpha| \geq \frac{3(A-B)}{2(1-q)(1-|B|)},\quad -1 < B < A \leq 1 \tag{5}$$

and h is an analytic function defined on $\mathcal{D}$ with $h(0) = 1$ satisfying

$$1 + \alpha z \partial_q h(z) \prec \frac{1 + Az}{1 + Bz},\quad z \in \mathcal{D}. \tag{6}$$

In addition, we suppose that

$$1 + \alpha z \partial_q h(z) = \frac{1 + Aw(z)}{1 + Bw(z)},\quad z \in \mathcal{D},$$

where w is an analytic in $\mathcal{D}$ with $w(0) = 0$. Then

$$h(z) \prec 1 + \frac{4}{3}z + \frac{2}{3}z^2,\quad z \in \mathcal{D}.$$

Proof. We define a function

$$p(z) = 1 + \alpha z \partial_q h(z), \tag{7}$$

where p is analytic and $p(0) = 1$. Consider

$$h(z) = 1 + \frac{4}{3}w(z) + \frac{2}{3}w^2(z). \tag{8}$$

To prove our result, it will be enough to show that $|w(z)| < 1$. Using (7) and (8), we obtain

$$p(z) = 1 + \frac{\alpha}{3}z\partial_q w(z) \{4(1 + w(z)) - 2(1-q)z\partial_q w(z)\}.$$

Also

$$\left|\frac{p(z) - 1}{A - Bp(z)}\right| =$$

$$\left|\frac{1 + \frac{\alpha}{3}z\partial_q w(z)\{4 + 4w(z) - 2(1-q)z\partial_q w(z)\} - 1}{A - B\left[1 + \frac{\alpha}{3}z\partial_q w(z)\{4 + 4w(z) - 2(1-q)z\partial_q w(z)\}\right]}\right|$$

$$= \left|\frac{\alpha z \partial_q w(z)\{4 + 4w(z) - 2(1-q)z\partial_q w(z)\}}{3(A-B) + \alpha B z \partial_q w(z)[4 + 4w(z) - 2(1-q)z\partial_q w(z)]}\right|.$$

Suppose that there exists a point $z_0 \in \mathcal{D}$ such that

$$\max_{|z| \leq |z_0|} |w(z)| = |w(z_0)| = 1.$$

Then by using Lemma 1, there exists a number $m \geq 1$ such that $z_0 \partial_q w(z_0) = mw(z_0)$. Suppose $w(z_0) = e^{i\theta}$, $\theta \in [-\pi, \pi]$. Then for $z_0 \in \mathcal{D}$, we have

$$\left|\frac{p(z_0) - 1}{A - Bp(z_0)}\right| = \left|\frac{\alpha m w(z_0)\{4 + 4w(z_0) - 2(1-q)mw(z_0)\}}{3(A-B) + \alpha Bmw(z_0)[4 + 4w(z_0) - 2(1-q)mw(z_0)]}\right|$$

$$\geq \frac{m|\alpha||4 + 4e^{i\theta} - 2(1-q)me^{i\theta}|}{3(A-B) + \alpha m|B||4 + 4e^{i\theta} - 2(1-q)me^{i\theta}|}$$

$$= \frac{m|\alpha|\sqrt{16 + (4 - 2(1-q)m)^2 + 8(4 - 2(1-q)m)\cos\theta}}{3(A-B) + m|\alpha||B|\sqrt{16 + (4 - 2(1-q)m)^2 + 8(4 - 2(1-q)m)\cos\theta}}.$$

Consider the function

$$\Psi(\theta) = \frac{m|\alpha|\sqrt{16 + (4 - 2(1-q)m)^2 + 8(4 - 2(1-q)m)\cos\theta}}{3(A-B) + m|\alpha||B|\sqrt{16 + (4 - 2(1-q)m)^2 + 8(4 - 2(1-q)m)\cos\theta}},$$

for $\theta \in [-\pi, \pi]$. It is clear that Ψ is an even function, therefore we find the minimum value of Ψ when $\theta \in [0, \pi]$. Now

$$\Psi'(\theta) = \frac{-12|\alpha|m(A-B)\{4 - 2(1-q)m\}\sin\theta}{\left(\sqrt{16 + (4 - 2(1-q)m)^2 + 8(4 - 2(1-q)m)\cos\theta}\right)\left(3(A-B) + |\alpha||B|\sqrt{16 + (4 - 2(1-q)m)^2 + 8(4 - 2(1-q)m)\cos\theta}\right)^2}.$$

It is easy to see that $\Psi'(\theta) = 0$ when $\theta = 0, \pi$. Similarly, we can see that $\Psi''(\theta) > 0$, when $\theta = \pi$. Hence $\Psi(\theta) \geq \Psi(\pi)$. Consider the function

$$\Phi(m) = \frac{|\alpha| m \sqrt{16 + (4 - 2(1-q)m)^2 - 8(4 - 2(1-q)m)}}{3(A-B) + |\alpha| |B| \sqrt{16 + (4 - 2(1-q)m)^2 - 8(4 - 2(1-q)m)}}$$

$$= \frac{2|\alpha|(1-q)m^2}{3(A-B) + 2|\alpha||B|(1-q)m^2}$$

Then

$$\Phi'(m) = \frac{12|\alpha|(1-q)(A-B)}{\{3(A-B) + 2m^2|\alpha||B|(1-q)\}^2} > 0.$$

which shows that Φ is an increasing function and it has its minimum value at $m = 1$, so

$$\left| \frac{p(z_0) - 1}{A - Bp(z_0)} \right| \geq \frac{2|\alpha|(1-q)}{3(A-B) + 2|\alpha||B|(1-q)}.$$

Now by (5), we have

$$\left| \frac{p(z_0) - 1}{A - Bp(z_0)} \right| \geq 1,$$

which contradicts (6). Hence $|w(z)| < 1$ and so we get the desired result. □

If we put $p(z) = \frac{z \partial_q f(z)}{f(z)}$ in Theorem 1, we get the following result.

Corollary 1. *Let* $|\alpha| \geq \frac{3(A-B)}{2(1-q)(1-|B|)}$, $-1 < B < A \leq 1$ *and* $f \in \mathcal{A}$ *satisfy the subordination*

$$1 + \alpha z \partial_q \left(\frac{z \partial_q f(z)}{f(z)} \right) \prec \frac{1 + Az}{1 + Bz}, \quad z \in \mathcal{D}. \tag{9}$$

Then $f \in \mathcal{S}^*_{\mathcal{C}_q}$

If we choose $A = 1$ and $B = 0$ in Corollary 1, then we obtain the following result.

Corollary 2. *Let* $|\alpha| \geq \frac{3}{(1-q)}$, $-1 < B < A \leq 1$ *and* $f \in \mathcal{A}$ *satisfy the subordination*

$$1 + \alpha z \partial_q \left(\frac{z \partial_q f(z)}{f(z)} \right) \prec 1 + z, \quad z \in \mathcal{D}. \tag{10}$$

Then $f \in \mathcal{S}^*_{\mathcal{C}_q}$.

Theorem 2. *Assume that*

$$|\alpha| \geq \frac{(A-B)}{2(1-q)(1-|B|)}, \quad -1 < B < A \leq 1 \tag{11}$$

and h is an analytic function defined on $\mathcal{D}$ with $h(0) = 1$ satisfying

$$1 + \alpha \frac{z \partial_q h(z)}{h(z)} \prec \frac{1 + Az}{1 + Bz}, \quad z \in \mathcal{D}. \tag{12}$$

In addition, we suppose that

$$1 + \alpha \frac{z \partial_q h(z)}{h(z)} = \frac{1 + Aw(z)}{1 + Bw(z)}, \quad z \in \mathcal{D},$$

where w is an analytic in $\mathcal{D}$ with $w(0) = 0$. Then

$$h(z) \prec 1 + \frac{4}{3}z + \frac{2}{3}z^2, \quad z \in \mathcal{D}.$$

Proof. We define a function

$$p(z) = 1 + \alpha \frac{z \partial_q h(z)}{h(z)}, \tag{13}$$

where p is analytic and $p(0) = 1$. Consider

$$h(z) = 1 + \frac{4}{3} w(z) + \frac{2}{3} w^2(z). \tag{14}$$

Using (13) and (14), we obtain

$$p(z) = 1 + \frac{\alpha z \partial_q w(z) \{4 + 4w(z) - 2(1-q) z \partial_q w(z)\}}{3\left(1 + \frac{4}{3} w(z) + \frac{2}{3} w^2(z)\right)}.$$

Therefore

$$\left| \frac{p(z) - 1}{A - Bp(z)} \right| =$$

$$\left| \frac{\alpha z \partial_q w(z) \{4 + 4w(z) - 2(1-q) z \partial_q w(z)\}}{\left(3 + 4w(z) + 2(w(z))^2\right)(A - B) + \alpha B \left[(4 + 4w(z) - 2(1-q) z \partial_q w(z)) z \partial_q w(z)\right]} \right|.$$

Hence by applying Lemma 1, we conclude that

$$\left| \frac{p(z_0) - 1}{A - Bp(z_0)} \right| \geq \frac{2|\alpha|(1-q)}{(A-B) + 2|\alpha B|(1-q)}.$$

By using (11), we have

$$\left| \frac{p(z_0) - 1}{A - Bp(z_0)} \right| \geq 1,$$

which contradicts (12). Hence we get the desired result. □

If we put $p(z) = \frac{z \partial_q f(z)}{f(z)}$ in the above theorem, we get the following result.

Corollary 3. *Let* $|\alpha| \geq \frac{(A-B)}{2(1-q)(1-|B|)}$, $-1 < B < A \leq 1$ *and* $f \in \mathcal{A}$ *such that*

$$1 + \alpha z \left(\frac{f(z)}{z \partial_q f(z)} \right) \partial_q \left(\frac{z \partial_q f(z)}{f(z)} \right) \prec \frac{1 + Az}{1 + Bz}, \quad z \in \mathcal{D}. \tag{15}$$

Then $f \in \mathcal{S}^*_{\mathcal{C}_q}$.

Theorem 3. *Assume that*

$$|\alpha| \geq \frac{(A - B)}{6(1-q)(1-|B|)}, \quad -1 < B < A \leq 1 \tag{16}$$

and h is an analytic function defined on $\mathcal{D}$ with $h(0) = 1$ satisfying

$$1 + \alpha \frac{z \partial_q h(z)}{h^2(z)} \prec \frac{1 + Az}{1 + Bz}, \quad z \in \mathcal{D}. \tag{17}$$

In addition, we suppose that

$$1 + \alpha \frac{z \partial_q h(z)}{h^2(z)} = \frac{1 + Aw(z)}{1 + Bw(z)}, \quad z \in \mathcal{D},$$

where w is an analytic in $\mathcal{D}$ with $w(0) = 0$. Then

$$h(z) \prec 1 + \frac{4}{3}z + \frac{2}{3}z^2, \quad z \in \mathcal{D}.$$

Proof. We define a function

$$p(z) = 1 + \alpha \frac{z\partial_q h(z)}{(h(z))^2}, \tag{18}$$

where p is analytic and $p(0) = 1$. Consider

$$h(z) = 1 + \frac{4}{3}w(z) + \frac{2}{3}w^2(z)$$

After some simple calculations, we obtain

$$p(z) = 1 + \frac{3\alpha z \partial_q w(z)\{4 + 4w(z) - 2(1-q)z\partial_q w(z)\}}{(3 + 4w(z) + 2w^2(z))^2}.$$

Therefore

$$\left| \frac{p(z) - 1}{A - Bp(z)} \right| =$$

$$\left| \frac{3\alpha z \partial_q w(z)\{4 + 4w(z) - 2(1-q)z\partial_q w(z)\}}{(3 + 4w(z) + 2w^2(z))^2 (A - B) - 3\alpha Bz\partial_q w(z)\left[(4 + 4w(z) - 2(1-q)z\partial_q w(z))\right]} \right|.$$

Hence by applying Lemma 1, we conclude that

$$\left| \frac{p(z_0) - 1}{A - Bp(z_0)} \right| \geq \frac{6|\alpha|(1-q)}{(A-B) + 6|\alpha B|(1-q)}.$$

Now by using (16), we have

$$\left| \frac{p(z_0) - 1}{A - Bp(z_0)} \right| \geq 1,$$

which contradicts (17). Hence we get the required result. □

If we put $p(z) = \frac{z\partial_q f(z)}{f(z)}$ in the above theorem, we get the following result.

Corollary 4. *Let* $|\alpha| \geq \frac{(A-B)}{6(1-q)(1-|B|)}$, $-1 < B < A \leq 1$ and $f \in \mathcal{A}$ *satisfy*

$$1 + \alpha z \left(\frac{f(z)}{z\partial_q f(z)}\right)^2 \partial_q \left(\frac{z\partial_q f(z)}{f(z)}\right) \prec \frac{1 + Az}{1 + Bz}, \quad z \in \mathcal{D}. \tag{19}$$

Then $f \in \mathcal{S}_{\mathcal{C}_q}^*$.

Theorem 4. *Assume that*

$$|\alpha| \geq \frac{(A-B)}{18(1-q)(1-|B|)}, \quad -1 < B < A \leq 1 \tag{20}$$

and h is an analytic function defined on $\mathcal{D}$ with $h(0) = 1$ satisfying

$$1 + \alpha \frac{z\partial_q h(z)}{h^3(z)} \prec \frac{1 + Az}{1 + Bz}, \quad z \in \mathcal{D}. \tag{21}$$

In addition, we suppose that

$$1 + \alpha \frac{z\partial_q h(z)}{h^3(z)} = \frac{1 + Aw(z)}{1 + Bw(z)}, \quad z \in \mathcal{D},$$

where w is an analytic in $\mathcal{U}$ with $w(0) = 0$. Then

$$h(z) \prec 1 + \frac{4}{3}z + \frac{2}{3}z^2, \quad z \in \mathcal{D}.$$

Proof. Here we define a function

$$p(z) = 1 + \alpha \frac{z\partial_q h(z)}{(h(z))^3}, \quad (22)$$

where p is analytic and $p(0) = 1$. Consider

$$h(z) = 1 + \tfrac{4}{3}w(z) + \tfrac{2}{3}w^2(z).$$

After some simplifications, we obtain

$$p(z) = 1 + \frac{\alpha z \partial_q w(z)\{4 + 4w(z) - 2(1-q)z\partial_q w(z)\}}{3\left(1 + \tfrac{4}{3}w(z) + \tfrac{2}{3}w^2(z)\right)^3}.$$

Therefore

$$\left|\frac{p(z) - 1}{A - Bp(z)}\right| =$$

$$\left|\frac{9\alpha z \partial_q w(z)\{4 + 4w(z) - 2(1-)z\partial_q w(z)\}}{\left(3 + 4w(z) + 2(w(z))^2\right)^3 (A - B) + 9\alpha B z \partial_q w(z)[4 + 4w(z) - 2(1-q)z\partial_q w(z)]}\right|.$$

Hence by Lemma 1, we conclude that

$$\left|\frac{p(z_0) - 1}{A - Bp(z_0)}\right| \geq \frac{18|\alpha|(1-q)}{(A-B) + 18|\alpha B|(1-q)}.$$

Now by using (20), we have

$$\left|\frac{p(z_0) - 1}{A - Bp(z_0)}\right| \geq 1,$$

which is contradiction. We complete the required proof. □

If we put $p(z) = \frac{z\partial_q f(z)}{f(z)}$ in the above theorem, we get the following result.

Corollary 5. *Let* $|\alpha| \geq \frac{(A-B)}{18(1-q)(1-|B|)}$, $-1 < B < A \leq 1$ *and* $f \in \mathcal{A}$ *such that*

$$1 + \alpha z \left(\frac{f(z)}{z\partial_q f(z)}\right)^3 \partial_q \left(\frac{z\partial_q f(z)}{f(z)}\right) \prec \frac{1 + Az}{1 + Bz}, \quad z \in \mathcal{D}. \quad (23)$$

Then $f \in \mathcal{S}^*_{\mathcal{C}_q}$.

Theorem 5. *Assume that*

$$|\alpha| \geq \frac{(A - B)}{2 \cdot 3^{n-1}(1-q)(1-|B|)}, \quad -1 < B < A \leq 1 \quad (24)$$

and h is an analytic function defined on $\mathcal{D}$ with $h(0) = 1$ satisfying

$$1 + \alpha \frac{z\partial_q h(z)}{(h(z))^n} \prec \frac{1 + Az}{1 + Bz}, \quad z \in \mathcal{D}. \tag{25}$$

In addition, we suppose that

$$1 + \alpha \frac{z\partial_q h(z)}{(h(z))^n} = \frac{1 + Aw(z)}{1 + Bw(z)}, \quad z \in \mathcal{D},$$

where w is an analytic in $\mathcal{D}$ with $w(0) = 0$. Then

$$h(z) \prec 1 + \frac{4}{3}z + \frac{2}{3}z^2, \quad z \in \mathcal{D}.$$

Proof. The proof of Theorem 5 is similar to the theorems proved above and so here we choose to omit the details. □

If we put $p(z) = \frac{z\partial_q f(z)}{f(z)}$ in the above theorem, we get the following result.

Corollary 6. Let $|\alpha| \geq \frac{(A-B)}{2 \cdot 3^{n-1}(1-q)(1-|B|)}$, $-1 < B < A \leq 1$ and $f \in \mathcal{A}$ such that

$$1 + \alpha z \left(\frac{f(z)}{z\partial_q f(z)} \right)^n \partial_q \left(\frac{z\partial_q f(z)}{f(z)} \right) \prec \frac{1 + Az}{1 + Bz}, \quad z \in \mathcal{D}. \tag{26}$$

Then $f \in \mathcal{S}^*_{\tilde{C}_q}$.

3. Differential Subordination Related with Limacon

Theorem 6. Assume

$$|\alpha| \geq \frac{2(A-B)}{\sqrt{8 + (1+q)\left(1 + q - 4\sqrt{2}\right)} (1 - |B|)}, \quad -1 < B < A \leq 1 \tag{27}$$

and h is an analytic function defined on $\mathcal{D}$ with $h(0) = 1$ satisfying

$$1 + \alpha \partial_q h(z) \prec \frac{1 + Az}{1 + Bz}, \quad z \in \mathcal{D}. \tag{28}$$

In addition, we suppose that

$$1 + \alpha \partial_q h(z) = \frac{1 + Aw(z)}{1 + Bw(z)}, \quad z \in \mathcal{D}$$

with $w(0) = 0$, then

$$h(z) \prec 1 + \sqrt{2}z + \frac{1}{2}z^2, \quad z \in \mathcal{D}.$$

Proof. We define a function

$$p(z) = 1 + \alpha z \partial_q h(z), \tag{29}$$

where p is analytic and $p(0) = 1$. Consider

$$h(z) = 1 + \sqrt{2}w(z) + \frac{1}{2}w^2(z). \tag{30}$$

To prove our result, it will be enough to show that $|w(z)| < 1$. Using (29) and (30), we obtain

$$p(z) = 1 + \frac{\alpha z \partial_q w(z)}{2} \left\{ 2\sqrt{2} + 2w(z) - (1-q) z \partial_q w(z) \right\}.$$

Also

$$\left| \frac{p(z) - 1}{A - Bp(z)} \right| =$$

$$\left| \frac{1 + \frac{\alpha z \partial_q w(z)}{2} \left\{ 2\sqrt{2} + 2w(z) - (1-q) z \partial_q w(z) \right\} - 1}{A - B \left[1 + \frac{\alpha z \partial_q w(z)}{2} \left\{ 2\sqrt{2} + 2w(z) - (1-q) z \partial_q w(z) \right\} \right]} \right|$$

$$= \left| \frac{\alpha z \partial_q w(z) \left\{ 2\sqrt{2} + 2w(z) - (1-q) z \partial_q w(z) \right\}}{2(A-B) + \alpha B z \partial_q w(z) \left[2\sqrt{2} + 2w(z) - (1-q) z \partial_q w(z) \right]} \right|.$$

Suppose that there exists a point $z_0 \in \mathcal{D}$ such that

$$\max_{|z| \leq |z_0|} |w(z)| = |w(z_0)| = 1.$$

Then by using Lemma 1, there exists a number $m \geq 1$ such that $z_0 \partial_q w(z_0) = m w(z_0)$. Suppose $w(z_0) = e^{i\theta}$, $\theta \in [-\pi, \pi]$. Then for $z_0 \in \mathcal{D}$, we have

$$\left| \frac{p(z_0) - 1}{A - Bp(z_0)} \right| = \left| \frac{\alpha m w(z_0) \left\{ 2\sqrt{2} + 2w(z_0) - (1-q) m w(z_0) \right\}}{2(A-B) + \alpha B m w(z_0) \left[2\sqrt{2} + 2w(z_0) - (1-q) m w(z_0) \right]} \right|$$

$$\geq \frac{m |\alpha| \left| 2\sqrt{2} + 2e^{i\theta} - (1-q) m e^{i\theta} \right|}{2(A-B) + m |\alpha B| \left| 2\sqrt{2} + 2e^{i\theta} - (1-q) m e^{i\theta} \right|}$$

$$= \frac{m |\alpha| \sqrt{8 + (2 - (1-q) m)^2 + 4\sqrt{2}((2 - (1-q) m)) \cos \theta}}{2(A-B) + m |\alpha| B \sqrt{8 + (2 - (1-q) m)^2 + 4\sqrt{2}((2 - (1-q) m)) \cos \theta}}.$$

Consider the function

$$\Theta_1(\theta) = \frac{m |\alpha| \sqrt{8 + (2 - (1-q) m)^2 + 4\sqrt{2}((2 - (1-q) m)) \cos \theta}}{2(A-B) + m |\alpha B| \sqrt{8 + (2 - (1-q) m)^2 + 4\sqrt{2}((2 - (1-q) m)) \cos \theta}}.$$

For $\theta \in [-\pi, \pi]$. It is clear that Θ_1 is an even function, therefore we find the minimum value of Θ_1 when $\theta \in [0, \pi]$. Now

$$\Theta_1'(\theta) = -\frac{4\sqrt{2} m |\alpha| (A-B) a \sin \theta}{\left(\sqrt{8 + a^2 + 4\sqrt{2} a \cos \theta} \right) \left(2(A-B) + m |\alpha B| \sqrt{8 + a^2 + 4\sqrt{2} a \cos \theta} \right)^2},$$

where $a = 2 - (1-q) m$. It is easy to see that $\Theta_1'(\theta) = 0$ when $\theta = 0, \pi$. Similarly, we can see that $\Theta_1''(\theta) > 0$, when $\theta = \pi$. Hence $\Theta_1(\theta) \geq \Theta_1(\pi)$. Now consider the function

$$\Lambda_1(m) = \frac{m |\alpha| \sqrt{8 + (2 - (1-q) m)^2 - 4\sqrt{2}(2 - (1-q) m)}}{2(A-B) + m |\alpha B| \sqrt{8 + (2 - (1-q) m)^2 - 4\sqrt{2}(2 - (1-q) m)}}.$$

Now

$$\Lambda_1'(m) = \frac{4|\alpha|(A-B)\left\{6-4\sqrt{2}+3bm\left(\sqrt{2}-1\right)+b^2m^2\right\}}{\sqrt{8+(2-bm)^2+4\sqrt{2}(bm-2)}\left(\begin{array}{c}2(A-B)+\\ m|\alpha B|\sqrt{8+(2-bm)^2-4\sqrt{2}(2-bm)}\end{array}\right)^2} > 0,$$

where $b = 1 - q$. This shows that Λ is an increasing function and it has its minimum value at $m = 1$, so

$$\left|\frac{p(z_0)-1}{A-Bp(z_0)}\right| \geq \frac{|\alpha|\sqrt{8+(2-(1-q))^2-4\sqrt{2}(2-(1-q))}}{2(A-B)+|\alpha B|\sqrt{8+(2-(1-q))^2-4\sqrt{2}(2-(1-q))}}.$$

Now by (27), we have

$$\left|\frac{p(z_0)-1}{A-Bp(z_0)}\right| \geq 1,$$

which contradicts (28). Hence $|w(z)| < 1$ and so we get the desired result. □

If we put $p(z) = \frac{z\partial_q f(z)}{f(z)}$ in the above theorem, we get the following result.

Corollary 7. *Let* $|\alpha| \geq \frac{2(A-B)}{\sqrt{8+(1+q)\left(1+q-4\sqrt{2}\right)}(1-|B|)}$, $-1 < B < A \leq 1$ *and* $f \in \mathcal{A}$ *satisfy*

$$1 + \alpha \partial_q \left(\frac{z\partial_q f(z)}{f(z)}\right) \prec \frac{1+Az}{1+Bz}, \quad z \in \mathcal{D}. \tag{31}$$

Then $f \in \mathcal{S}^*_{\mathcal{L}_q}$.

Theorem 7. *Assume*

$$|\alpha| \geq \frac{(A-B)\sqrt{28-16\sqrt{2}}}{(1-|B|)\sqrt{8+(1+q)\left(1+q-4\sqrt{2}\right)}}, \quad -1 < B < A \leq 1 \tag{32}$$

and h is an analytic function defined on $\mathcal{D}$ with $h(0) = 1$ satisfying

$$1 + \alpha \frac{z\partial_q h(z)}{h(z)} \prec \frac{1+Az}{1+Bz}. \tag{33}$$

In addition, we suppose that

$$1 + \alpha \frac{z\partial_q h(z)}{h(z)} = \frac{1+Aw(z)}{1+Bw(z)}, \quad z \in \mathcal{D},$$

where w is an analytic in $\mathcal{D}$ with $w(0) = 0$, then

$$h(z) \prec 1 + \sqrt{2}z + \tfrac{1}{2}z^2.$$

Proof. We define a function

$$p(z) = 1 + \alpha \frac{z\partial_q h(z)}{h(z)}, \tag{34}$$

where p is analytic and $p(0) = 1$. Consider

$$h(z) = 1 + \sqrt{2}w(z) + \tfrac{1}{2}w^2(z). \tag{35}$$

Using (34) and (35), we obtain

$$p(z) = 1 + \frac{\alpha z \partial_q w(z)\{2\sqrt{2} + 2w(z) - (1-q)z\partial_q w(z)\}}{2\left(\frac{2+2\sqrt{2}w(z)+w^2(z).}{2}\right)}.$$

Therefore

$$\left|\frac{p(z)-1}{A-Bp(z)}\right| =$$

$$\left|\frac{\alpha z \partial_q w(z)\{2\sqrt{2}+2w(z)-(1-q)z\partial_q w(z)\}}{\left(2+2\sqrt{2}w(z)+w^2(z)\right)(A-B) + \alpha B z \partial_q w(z)\left[2\sqrt{2}+2w(z)-(1-q)z\partial_q w(z)\right]}\right|.$$

Then using the similar method as in Theorem 7, we obtain

$$\left|\frac{p(z_0)-1}{A-Bp(z_0)}\right| \geq \frac{|\alpha|\sqrt{8+(2-(1-q))^2-4\sqrt{2}(2-(1-q))}}{\sqrt{28-16\sqrt{2}}(A-B)+|\alpha B|\sqrt{8+(2-(1-q))^2-4\sqrt{2}(2-(1-q))}}.$$

By using (32), we have

$$\left|\frac{p(z_0)-1}{A-Bp(z_0)}\right| \geq 1,$$

which contradicts (33). Hence we get the desired result. □

If we put $p(z) = \frac{z\partial_q f(z)}{f(z)}$ in the above theorem, we get the following result.

Corollary 8. *Let* $|\alpha| \geq \frac{(A-B)\sqrt{28-16\sqrt{2}}}{(1-|B|)\sqrt{8+(1+q)(1+q-4\sqrt{2})}}$, $-1 < B < A \leq 1$ *and* $f \in \mathcal{A}$ *satisfy*

$$1 + \alpha z \left(\frac{f(z)}{z\partial_q f(z)}\right) \partial_q \left(\frac{z\partial_q f(z)}{f(z)}\right) \prec \frac{1+Az}{1+Bz}. \quad (36)$$

Then $f \in \mathcal{S}^*_{\mathcal{L}_q}$.

Theorem 8. *Assume*

$$|\alpha| \geq \frac{(12-8\sqrt{2})(A-B)}{(1-|B|)\sqrt{8+(1+q)(1+q-4\sqrt{2})}}, \quad -1 < B < A \leq 1 \quad (37)$$

and h is an analytic function defined on $\mathcal{D}$ *with* $h(0) = 1$ *satisfying*

$$1 + \alpha \frac{z\partial_q h(z)}{(h(z))^2} \prec \frac{1+Az}{1+Bz}.$$

In addition, we suppose that

$$1 + \alpha \frac{z\partial_q h(z)}{(h(z))^2} = \frac{1+Aw(z)}{1+Bw(z)}, \quad z \in \mathcal{D},$$

where w is an analytic in $\mathcal{D}$ *with* $w(0) = 0$, *then*

$$h(z) \prec 1 + \sqrt{2}z + \tfrac{1}{2}z^2.$$

Proof. We define a function

$$p(z) = 1 + \alpha \frac{z^2 \partial_q h(z)}{(h(z))^2}, \quad (38)$$

where p is analytic and $p(0) = 1$. Consider

$$h(z) = 1 + \sqrt{2}w(z) + \tfrac{1}{2}w^2(z).$$

After some simple calculations, we obtain

$$p(z) = 1 + \frac{\alpha z \partial_q w(z)\{2\sqrt{2}+2w(z)-(1-q)z\partial_q w(z)\}}{2\left(\frac{2+2\sqrt{2}w(z)+w^2(z)}{2}\right)^2}.$$

Therefore

$$\left|\frac{p(z)-1}{A-Bp(z)}\right| =$$

$$\left|\frac{\alpha z \partial_q w(z)\left\{2\sqrt{2}+2w(z)-(1-q)z\partial_q w(z)\right\}}{\frac{(2+2\sqrt{2}w(z)+w^2(z))^2}{2}(A-B)+\alpha B z \partial_q w(z)\left[\left(2\sqrt{2}+2w(z)-(1-q)z\partial_q w(z)\right)\right]}\right|.$$

Using similar method as in Theorem 7, we obtain

$$\left|\frac{p(z_0)-1}{A-Bp(z_0)}\right| \geq \frac{|\alpha|\sqrt{8+(2-(1-q))^2-4\sqrt{2}(2-(1-q))}}{(12-8\sqrt{2})(A-B)+|\alpha B|\sqrt{8+(2-(1-q))^2-4\sqrt{2}(2-(1-q))}}.$$

Now by using (37), we have

$$\left|\frac{p(z_0)-1}{A-Bp(z_0)}\right| \geq 1,$$

which is contradiction. Hence we get the required result. □

If we put $p(z) = \frac{z\partial_q f(z)}{f(z)}$ in the above theorem, we get the following result.

Corollary 9. *Let* $|\alpha| \geq \frac{(12-8\sqrt{2})(A-B)}{(1-|B|)\sqrt{8+(1+q)(1+q-4\sqrt{2})}}$, $-1 < B < A \leq 1$ *and* $f \in \mathcal{A}$ *such that*

$$1 + \alpha z \left(\frac{f(z)}{z\partial_q f(z)}\right)^2 \partial_q \left(\frac{z\partial_q f(z)}{f(z)}\right) \prec \frac{1+Az}{1+Bz}.$$

Then $f \in \mathcal{S}^*_{\mathcal{L}_q}$.

Theorem 9. *Assume*

$$|\alpha| \geq \frac{(28-16\sqrt{2})^{\frac{3}{2}}(A-B)}{4(1-|B|)\sqrt{8+(1+q)(1+q-4\sqrt{2})}}, \quad -1 < B < A \leq 1 \tag{39}$$

and h is an analytic function defined on $\mathcal{D}$ *with* $h(0) = 1$ *satisfying*

$$1 + \alpha \frac{z^3 \partial_q h(z)}{(h(z))^3} \prec \frac{1+Az}{1+Bz}. \tag{40}$$

In addition, we suppose that

$$1 + \alpha \frac{z^3 \partial_q h(z)}{(h(z))^3} = \frac{1+Aw(z)}{1+Bw(z)}, \quad z \in \mathcal{D},$$

where w is an analytic in $\mathcal{D}$ *with* $w(0) = 0$, *then*

$$h(z) \prec 1 + \sqrt{2}z + \tfrac{1}{2}z^2.$$

Proof. Here we define a function

$$p(z) = 1 + \alpha \frac{z^3 \partial_q h(z)}{(h(z))^3}, \tag{41}$$

where p is analytic and $p(0) = 1$. Consider

$$h(z) = 1 + \sqrt{2} w(z) + \tfrac{1}{2} w^2(z).$$

After some simplifications, we obtain

$$p(z) = 1 + \frac{\alpha z \partial_q w(z) \{2\sqrt{2} + 2w(z) - (q-1) z \partial_q w(z)\}}{2 \left(\frac{2 + 2\sqrt{2} w(z) + w^2(z)}{2} \right)^3}.$$

Therefore

$$\left| \frac{p(z) - 1}{A - B p(z)} \right| = $$

$$\left| \frac{\alpha z \partial_q w(z) \{2\sqrt{2} + 2w(z) - (1-q) z \partial_q w(z)\}}{\frac{(2 + 2\sqrt{2} w(z) + w^2(z))^3}{4}(A-B) + \alpha B z \partial_q w(z) \left[2\sqrt{2} + 2w(z) - (1-q) z \partial_q w(z) \right]} \right|.$$

Therefore by using Lemma 1, we obtain

$$\left| \frac{p(z_0) - 1}{A - B p(z_0)} \right| \geq \frac{|\alpha| \sqrt{(12 - 8\sqrt{2}) + (1-q)^2 + 4(\sqrt{2}-1)(1-q)}}{\frac{(28 - 16\sqrt{2})^{\frac{3}{2}}}{4}(A-B) + |\alpha||B| \left[\sqrt{(12 - 8\sqrt{2}) + (1-q)^2 + 4(\sqrt{2}-1)(1-q)} \right]}.$$

Next by using (39), we have

$$\left| \frac{p(z_0) - 1}{A - B p(z_0)} \right| \geq 1,$$

which is contradiction. Hence, we get the required proof. □

If we put $p(z) = \frac{z \partial_q f(z)}{f(z)}$ in the above theorem, we get the following result.

Corollary 10. *Let* $|\alpha| \geq \dfrac{(28 - 16\sqrt{2})^{\frac{3}{2}}(A-B)}{4(1-|B|)\sqrt{8 + (1+q)(1+q-4\sqrt{2})}}$, $-1 < B < A \leq 1$ *and* $f \in \mathcal{A}$ *satisfy*

$$1 + \alpha z \left(\frac{f(z)}{z \partial_q f(z)} \right)^3 \partial_q \left(\frac{z \partial_q f(z)}{f(z)} \right) \prec \frac{1 + Az}{1 + Bz}.$$

Then $f \in \mathcal{S}^*_{\mathcal{L}_q}$.

4. Conclusions

In this article, we have studied some q-differential subordinations. We have determined conditions on α and

$$1 + \alpha \frac{z \partial_q h(z)}{(h(z))^n} \prec \frac{1 + Az}{1 + Bz} \quad (n = 0, 1, 2, 3), \tag{42}$$

then $h(z) \prec 1 + \tfrac{4}{3} z + \tfrac{2}{3} z^2$. Similar results are also investigated involving the function $1 + \sqrt{2} z + \tfrac{1}{2} z^2$. Further we have deduced sufficiency criterion for q-starlikeness related with cardioid and limacon from our main results. Moreover, by choosing particular functions instead of h, sufficient conditions for other analytic functions can be found.

Author Contributions: The authors have equally contributed to accomplish this research work.

Funding: This paper was supported partially by the Natural Science Foundation of the People's Republic of China under Grants 11561001 and 11271045, the Program for Young Talents of Science and Technology in Universities of Inner Mongolia Autonomous Region under Grant NJYT-18-A14 and the Natural Science Foundation of Inner Mongolia of the People's Republic of China under Grant 2018MS01026.

Conflicts of Interest: The authors agree with the contents of the manuscript, and there are no conflict of interest among the authors.

References

1. Ma, W.C.; Minda, D. A unified treatment of some special classes of univalent functions. In *Proceedings of the Conference on Complex Analysis*; International Press: Cambridge, MA, USA, 1992; pp. 157–169, .
2. Janowski, W. Extremal problems for a family of functions with positive real part and for some related families. *Ann. Pol.* **1970**, *23*, 159–177. [CrossRef]
3. Sokół, J.; Stankiewicz, J. Radius of convexity of some subclasses of strongly starlike functions. *Zesz. Nauk. Politech. Rzesz. Mat.* **1996**, *19*, 101–105.
4. Sharma, K.; Jain, N.K.; Ravichandran, V. Starlike functions associated with a cardioid. *Afr. Mat.* **2016**, *27*, 923–939. [CrossRef]
5. Cho, N.E.; Kumar, V.; Kumar, S.S.; Ravichandran, V. Radius problems for starlike functions associated with the sine function. *Bull. Iran. Math. Soc.* **2019**, *45*, 213–232. [CrossRef]
6. Yunus,Y.; Halim, S.A.; Akbarally, A.B. Subclass of starlike functions associated with a limacon. *AIP Conf. Proc.* **2018**, *1974*, 030023. [CrossRef]
7. Kargar, R.; Ebadian, A.; Sokół, J. On Booth lemniscate and starlike functions. *Anal. Math. Phys.* **2019**, *9*, 143–154. [CrossRef]
8. Kumar, S.; Ravichandran, V. A subclass of starlike functions associated with a rational function. *Southeast Asian Bull. Math.* **2016**, *40*, 199–212.
9. Mendiratta, R.; Nagpal, S.; Ravichandran, V. On a subclass of strongly starlike functions associated with exponential function. *Bull. Malays. Math. Sci. Soc.* **2015**, *38*, 365–386. [CrossRef]
10. Raina, R.K.; Sokół, J. On coefficient estimates for a certain class of starlike functions. *Hacet. J. Math. Stat.* **2015**, *44*, 1427–1433. [CrossRef]
11. Jackson, F.H. On q-functions and a certain difference operator. *Trans. R. Soc. Edinb.* **1909**, *46*, 253–281. [CrossRef]
12. Jackson, F.H. On q-definite integrals. *Q. J. Pure Appl. Math.* **1910**, *41*, 193–203.
13. Ismail, M.E.H.; Merkes, E.; Styer, D. A generalization of starlike functions. *Complex Var. Theory Appl.* **1990**, *14*, 77–84. [CrossRef]
14. Srivastava, H.M. Univalent functions, fractional calculus, and associated generalized hypergeometric functions in Univalent Functions. In *Fractional Calculus, and Their Applications*; Srivastava, H.M., Owa, S., Eds.; Halsted Press (Ellis Horwood Limited): Chichester, UK; Wileya, J. and Sons: New York, NY, USA; Chichester, UK; Brisbane, Australia; Toronto, ON, USA, 1989; pp. 329–354.
15. Agrawal, S.; Sahoo, S.K. A generalization of starlike functions of order α. *Hokkaido Math. J.* **2017**, *46*, 15–27. [CrossRef]
16. Srivastava, H.M.; Khan, B.; Khan, N.; Ahmad, Q.Z. Coeffcient inequalities for q-starlike functions associated with the Janowski functions. *Hokkaido Math. J.* **2019**, *48*, 407–425. [CrossRef]
17. Srivastava, H.M.; Ahmad, Q.Z.; Khan, N.; Khan, B. Hankel and Toeplitz determinants for a subclass of q-starlike functions associated with a general conic domain. *Mathematics* **2019**, *7*, 181. [CrossRef]
18. Srivastava, H.M.; Tahir, M.; Khan, B.; Ahmad, Q.Z.; Khan, N. Some general classes of q-starlike functions associated with the Janowski functions. *Symmetry* **2019**, *11*, 292. [CrossRef]
19. Srivastava, H.M.; Bansal, D. Close-to-convexity of a certain family of q-Mittag-Leffler functions. *J. Non-linear Var. Anal.* **2017**, *1*, 61–69.
20. Kanas, S.; Răducanu, D. Some class of analytic functions related to conic domains. *Math. Slovaca* **2014**, *64*, 1183–1196. [CrossRef]
21. Aldweby, H.; Darus, M. Some subordination results on q-analogue of Ruscheweyh differential operator. *Abstr. Appl. Anal.* **2014**, *2014*, 958563. [CrossRef]

22. Ahmad, K.; Arif, M.; Liu, J.-L. Convolution properties for a family of analytic functions involving q-analogue of Ruscheweyh differential operator. *Turk. J. Math.* **2019**, *43*, 1712–1720. [CrossRef]
23. Mahmmod, S.; Sokół, J. New subclass of analytic functions in conical domain associated with Ruscheweyh q-differential operator. *Results Math.* **2017**, *71*, 1345–1357. [CrossRef]
24. Mohammed, A.; Darus, M. A generalized operator involving the q-hypergeometric function. *Mat. Vesn.* **2013**, *65*, 454–465.
25. Arif, M.; Ahmad, B. New subfamily of meromorphic starlike functions in circular domain involving q-differential operator. *Math. Slovaca* **2018**, *68*, 1049–1056. [CrossRef]
26. Arif, M.; Srivastava, H.M.; Umar, S. Some applications of a q-analogue of the Ruscheweyh type operator for multivalent functions. *Revista de la Real Academia de Ciencias Exactas Físicas y Naturales Serie A Matemáticas* **2019**, *113*, 1211–1221. [CrossRef]
27. Arif, M.; Haq, M.; Liu, J.-L. A subfamily of univalent functions associated with q-analogue of Noor integral operator. *J. Funct. Spaces* **2018**, *2018*, 3818915. [CrossRef]
28. Shi, L.; Khan, Q.; Srivastava,G.; Liu, J.-L.; Arif, M. A study of multivalent q-starlike functions connected with circular domain. *Mathematics* **2019**, *7*, 670. [CrossRef]
29. Seoudy, T.M.; Aouf, M.K. Coefficient estimates of new classes of q-starlike and q-convex functions of complex order. *J. Math. Inequal.* **2016**, *10*, 135–145. [CrossRef]
30. Ali, R.M.; Cho, N.E.; Ravichandran, V.; Kumar, S.S. Differential subordination for functions associated with the lemniscate of Bernoulli. *Taiwan. J. Math.* **2012**, *16*, 1017–1026. [CrossRef]
31. Halim, S.A.; Omar, R. Applications of certain functions associated with lemniscate Bernoulli. *J. Indones. Soc.* **2012**, *18*, 93–99. [CrossRef]
32. Kumar, S.; Ravichandran, V. Subordinations for functions with positive real part. *Complex Anal. Oper. Theory* **2018**, *12*, 1179–1191. [CrossRef]
33. Kumar, S.S.; Kumar, V.; Ravichandran, V.; Cho, N.E. Sufficient conditions for starlike functions associated with the lemniscate of Bernoulli. *J. Inequal. Appl.* **2013**, *2013*, 176. [CrossRef]
34. Paprocki, E.; Sokół, J. The extremal problems in some subclass of strongly starlike functions. *Zesz. Nauk. Politech. Rzesz. Mat.* **1996**, *20*, 89–94.
35. Raza, M.; Sokół, J.; Mushtaq, S. Differential subordinations for analytic functions. *Iran. J. Sci. Technol. Trans. Sci.* **2019**, *43*, 883–890. [CrossRef]
36. Sharma, K.; Ravichandran, V. Applications of subordination theory to starlike functions. *Bull. Iran. Math.* **2016**, *42*, 761–777.
37. Çetinkaya, A.; Polatoglu, Y. q-Harmonic mappings for which analytic part is q-convex functions of complex order. *Hacet. J. Math. Stat.* **2018**, *47*, 813–820. [CrossRef]

© 2019 by the authors. Licensee MDPI, Basel, Switzerland. This article is an open access article distributed under the terms and conditions of the Creative Commons Attribution (CC BY) license (http://creativecommons.org/licenses/by/4.0/).

Article

Properties of Spiral-Like Close-to-Convex Functions Associated with Conic Domains

Hari M. Srivastava [1,2], Nazar Khan [3], Maslina Darus [4], Muhammad Tariq Rahim [3], Qazi Zahoor Ahmad [3,*] and Yousra Zeb [3]

1. Department of Mathematics and Statistics, University of Victoria, Victoria, BC V8W 3R4, Canada
2. Department of Medical Research, China Medical University Hospital, China Medical University, Taichung 40402, Taiwan
3. Department of Mathematics, Abbottabad University of Science and Technology, Abbottabad 22010, Pakistan
4. School of Mathematical Sciences, Faculty of Sciences and Technology, Universiti Kebangsaan Malaysia, Bangi 43600, Selangor, Malaysia
* Correspondence: zahoorqazi5@gmail.com; Tel.: +92-334-96-60-162

Received: 26 June 2019; Accepted: 29 July 2019; Published: 6 August 2019

Abstract: In this paper, our aim is to define certain new classes of multivalently spiral-like, starlike, convex and the varied Mocanu-type functions, which are associated with conic domains. We investigate such interesting properties of each of these function classes, such as (for example) sufficiency criteria, inclusion results and integral-preserving properties.

Keywords: analytic functions; multivalent functions; starlike functions; close-to-convex functions; uniformly starlike functions; uniformly close-to-convex functions; conic domains

MSC: Primary 05A30; 30C45; Secondary 11B65; 47B38

1. Introduction and Motivation

Let $\mathcal{A}(p)$ denote the class of functions of the form:

$$f(z) = z^p + \sum_{n=1}^{\infty} a_{n+p} z^{n+p} \quad (p \in \mathbb{N} = \{1, 2, 3, \cdots\}), \qquad (1)$$

which are analytic and p-valent in the open unit disk:

$$\mathbb{E} = \{z : z \in \mathbb{C} \text{ and } |z| < 1\}.$$

In particular, we write:

$$\mathcal{A}(1) = \mathcal{A}.$$

Furthermore, by $\mathcal{S} \subset \mathcal{A}$, we shall denote the class of all functions that are univalent in $\mathbb{E}$.

The familiar class of p-valently starlike functions in $\mathbb{E}$ will be denoted by $\mathcal{S}^*(p)$, which consists of functions $f \in \mathcal{A}(p)$ that satisfy the following conditions:

$$\Re\left(\frac{zf'(z)}{f(z)}\right) > 0 \qquad (\forall\, z \in \mathbb{E}).$$

One can easily see that:

$$\mathcal{S}^*(1) = \mathcal{S}^*,$$

where $\mathcal{S}^*$ is the well-known class of normalized starlike functions (see [1]).

We denote by $\mathcal{K}$ the class of close-to-convex functions, which consists of functions $f \in \mathcal{A}$ that satisfy the following inequality:

$$\Re\left(\frac{zf'(z)}{g(z)}\right) > 0 \quad (\forall\, z \in \mathbb{E})$$

for some $g \in \mathcal{S}^*$.

For two functions f and g analytic in $\mathbb{E}$, we say that the function f is subordinate to the function g and write as follows:

$$f \prec g \quad \text{or} \quad f(z) \prec g(z),$$

if there exists a Schwarz function w, which is analytic in $\mathbb{E}$ with:

$$w(0) = 0 \quad \text{and} \quad |w(z)| < 1,$$

such that:

$$f(z) = g(w(z)).$$

Furthermore, if the function g is univalent in $\mathbb{E}$, then it follows that:

$$f(z) \prec g(z) \quad (z \in \mathbb{E}) \implies f(0) = g(0) \text{ and } f(\mathbb{E}) \subset g(\mathbb{E}).$$

Next, for a function $f \in \mathcal{A}(p)$ given by (1) and another function $g \in \mathcal{A}(p)$ given by:

$$g(z) = z^p + \sum_{n=2}^{\infty} b_{n+p} z^{n+p} \quad (\forall\, z \in \mathbb{E}),$$

the convolution (or the Hadamard product) of f and g is given by:

$$(f * g)(z) = z^p + \sum_{n=2}^{\infty} a_{n+p} b_{n+p} z^{n+p} = (g * f)(z).$$

The subclass of $\mathcal{A}$ consisting of all analytic functions with a positive real part in $\mathbb{E}$ is denoted by $\mathcal{P}$. An analytic description of $\mathcal{P}$ is given by:

$$h(z) = 1 + \sum_{n=1}^{\infty} c_n z^n \quad (\forall\, z \in \mathbb{E}).$$

Furthermore, if:

$$\Re\{h(z)\} > \rho,$$

then we say that h is in the class $\mathcal{P}(\rho)$. Clearly, one see that:

$$\mathcal{P}(0) = \mathcal{P}.$$

Historically, in the year 1933, Spaček [2] introduced the β-spiral-like functions as follows.

Definition 1. *A function $f \in \mathcal{A}$ is said to be in the class $\mathcal{S}^*(\beta)$ if and only if:*

$$\Re\left(e^{i\beta}\frac{zf'(z)}{f(z)}\right) > 0 \quad (\forall\, z \in \mathbb{E})$$

for:

$$\beta \in \mathbb{R} \quad \text{and} \quad |\beta| < \frac{\pi}{2},$$

where $\mathbb{R}$ is the set of real numbers.

In the year 1967, Libera [3] extended this definition to the class of functions, which are spiral-like of order ρ denoted by $\mathcal{S}_\rho^*(\beta)$ as follows.

Definition 2. *A function $f \in \mathcal{A}$ is said to be in the class $\mathcal{S}_\rho^*(\beta)$ if and only if:*

$$\Re\left(e^{i\beta}\frac{zf'(z)}{f(z)}\right) > \rho \qquad (\forall\, z \in \mathbb{E})$$

$$\left(0 \leqq \rho < 1;\ \beta \in \mathbb{R} \quad \text{and} \quad |\beta| < \frac{\pi}{2}\right),$$

where $\mathbb{R}$ is the set of real numbers.

The above function classes $\mathcal{S}^*(\beta)$ and $\mathcal{S}_\rho^*(\beta)$ have been studied and generalized by different viewpoints and perspectives. For example, in the year 1974, a subclass $S_\beta^\alpha(\rho)$ of spiral-like functions was introduced by Silvia (see [4]), who gave some remarkable properties of this function class. Subsequently, Umarani [5] defined and studied another function class $SC(\alpha, \beta)$ of spiral-like functions. Recently, Noor et al. [6] generalized the works of Silvia [4] and Umarani [5] by defining the class $M(p, \alpha, \beta, \rho)$. Here, in this paper, we define certain new subclasses of spiral-like close-to-convex functions by using the idea of Noor et al. [6] and Umarani [5].

We now recall that Kanas et al. (see [7,8]; see also [9]) defined the conic domains Ω_k ($k \geqq 0$) as follows:

$$\Omega_k = \left\{ u + iv : u > k\sqrt{(u-1)^2 + v^2} \right\}. \tag{2}$$

By using these conic domains Ω_k ($k \geq 0$), they also introduced and studied the corresponding class k-$\mathcal{ST}$ of k-starlike functions (see Definition 3 below).

Moreover, for fixed k, Ω_k represents the conic region bounded successively by the imaginary axis for $(k = 0)$, for $k = 1$ a parabola, for $0 < k < 1$ the right branch of a hyperbola, and for $k > 1$ an ellipse. For these conic regions, the following functions $p_k(z)$, which are given by (3), play the role of extremal functions.

$$p_k(z) = \begin{cases} \dfrac{1+z}{1-z} = 1 + 2z + 2z^2 + \cdots & (k = 0) \\[2mm] 1 + \dfrac{2}{\pi^2}\left(\log\dfrac{1+\sqrt{z}}{1-\sqrt{z}}\right)^2 & (k = 1) \\[2mm] 1 + \dfrac{2}{1-k^2}\sinh^2\left\{\left(\dfrac{2}{\pi}\arccos k\right)\arctan(h\sqrt{z})\right\} & (0 \leq k < 1) \\[2mm] 1 + \dfrac{1}{k^2-1}\sin\left(\dfrac{\pi}{2K(\kappa)}\int_0^{\frac{u(z)}{\sqrt{\kappa}}}\dfrac{dt}{\sqrt{1-t^2}\sqrt{1-\kappa^2 t^2}}\right) + \dfrac{1}{k^2-1} & (k > 1), \end{cases} \tag{3}$$

where:

$$u(z) = \frac{z - \sqrt{\kappa}}{1 - \sqrt{\kappa}z} \qquad (\forall\, z \in \mathbb{E})$$

and $\kappa \in (0,1)$ is chosen such that:

$$k = \cosh\left(\frac{\pi K'(\kappa)}{4K(\kappa)}\right).$$

Here, $K(\kappa)$ is Legendre's complete elliptic integral of the first kind and:

$$K'(\kappa) = K(\sqrt{1-\kappa^2}),$$

that is, $K'(\kappa)$ is the complementary integral of $K(\kappa)$.

These conic regions are being studied and generalized by several authors (see, for example, [10–13]). The class k-$\mathcal{ST}$ is defined as follows.

Definition 3. *A function $f \in \mathcal{A}$ is said to be in the class k-$\mathcal{ST}$ if and only if:*

$$\frac{zf'(z)}{f(z)} \prec p_k(z) \quad (\forall z \in \mathbb{E}; k \geqq 0)$$

or, equivalently,

$$\Re\left(\frac{zf'(z)}{f(z)}\right) > k\left|\frac{zf'(z)}{f(z)} - 1\right|.$$

The class of k-uniformly close-to-convex functions denoted by k-$\mathcal{UK}$ was studied by Acu [14].

Definition 4. *A function $f \in \mathcal{A}$ is said to be in the class k-$\mathcal{UK}$ if and only if:*

$$\Re\left(\frac{zf'(z)}{g(z)}\right) > k\left|\frac{zf'(z)}{g(z)} - 1\right|,$$

where $g \in k$-$\mathcal{ST}$.

In recent years, several interesting subclasses of analytic functions were introduced and investigated from different viewpoints (see, for example, [6,15–20]; see also [21–25]). Motivated and inspired by the recent and current research in the above-mentioned work, we here introduce and investigate certain new subclasses of analytic and p-valent functions by using the concept of conic domains and spiral-like functions as follows.

Definition 5. *Let $f \in \mathcal{A}(p)$. Then, $f \in k$-$\mathcal{K}(p, \lambda)$ for a real number λ with $|\lambda| < \frac{\pi}{2}$ if and only if:*

$$\Re\left(\frac{e^{i\lambda}}{p}\frac{zf'(z)}{\psi(z)}\right) > k\left|\frac{zf'(z)}{\psi(z)} - p\right| + \rho\cos\lambda \quad (k \geqq 0; 0 \leqq \rho < 1)$$

for some $\psi \in \mathcal{S}^*$.

Definition 6. *Let $f \in \mathcal{A}(p)$. Then, $f \in k$-$\mathcal{Q}(p, \lambda)$ for a real λ with $|\lambda| < \frac{\pi}{2}$ if and only if:*

$$\Re\left(\frac{e^{i\lambda}}{p}\frac{zf'(z)}{\psi'(z)}\right) > k\left|\frac{(zf'(z))'}{\psi'(z)} - p\right| + \rho\cos\lambda \quad (k \geqq 0; 0 \leqq \rho < 1)$$

for some $\psi \in \mathcal{C}$.

Definition 7. *Let $f \in \mathcal{A}(p)$ with:*

$$\frac{f'(z)f(z)}{pz} \neq 0$$

and for some real ϕ and λ with $|\lambda| < \frac{\pi}{2}$. Then, $f \in k$-$\mathcal{Q}(\phi, \lambda, \eta, f, \psi)$ if and only if:

$$\Re(\mathcal{M}(\phi, \lambda, \eta, f, \psi)) > k|\mathcal{M}(\phi, \lambda, \eta, f, \psi) - p| + \rho\cos\lambda,$$

where

$$\mathcal{M}(\phi,\lambda,\eta,f,\psi) = (e^{i\lambda} - \phi\cos\lambda)\frac{zf'(z)}{p\psi(z)}$$
$$+ \frac{\phi\cos\lambda}{p-\eta}\left(\frac{(zf'(z))'}{\psi'(z)} - \eta\right) \quad \left(\frac{-1}{2} \leq \eta < 1\right). \tag{4}$$

2. A Set of Lemmas

Each of the following lemmas will be needed in our present investigation.

Lemma 1. *(see [26] p. 70) Let h be a convex function in $\mathbb{E}$ and:*

$$q : \mathbb{E} \implies \mathbb{C} \text{ and } \Re(q(z)) > 0 \quad (z \in \mathbb{E}).$$

If p is analytic in $\mathbb{E}$ with:

$$p(0) = h(0),$$

then:

$$p(z) + q(z)zp'(z) \prec h(z) \text{ implies } p(z) \prec h(z).$$

Lemma 2. *(see [26] p. 195) Let h be a convex function in $\mathbb{E}$ with:*

$$h(0) = 0 \quad \text{and} \quad A > 1.$$

Suppose that $j \geq \frac{4}{h'(0)}$ and that the functions $B(z)$, $C(z)$ and $D(z)$ are analytic in $\mathbb{E}$ and satisfy the following inequalities:

$$\Re\{B(z)\} \geq A + |C(z) - 1| - \Re(C(z) - 1) + jD(z), \quad z \in \mathbb{E}.$$

If p is analytic in $\mathbb{E}$ with:

$$p(z) = 1 + a_1 z + a_2 z^2 + \cdots$$

and the following subordination relation holds true:

$$Az^2 p''(z) + B(z)zp'(z) + C(z)p(z) + D(z) \prec h(z),$$

then:

$$p(z) \prec h(z).$$

3. Main Results and Their Demonstrations

In this section, we will prove our main results.

Theorem 1. *A function $f \in \mathcal{A}$ is in the class k-$\mathcal{Q}(\phi,\lambda,\eta,f,\psi)$ if:*

$$\sum_{n=2}^{\infty} \ddot{U}_n(p,\phi,\lambda,\eta,\xi) < p^2(p-\eta),$$

where:

$$\ddot{U}_n(p,\phi,\lambda,\eta,\xi) = (k+1)\left[(e^{i\lambda} - \phi\cos\lambda)(p-\eta)p + p^4\phi\cos\lambda\right.$$
$$+ (e^{i\lambda} - \phi\cos\lambda)(p-\eta)(n+p)|a_{n+p}| + (n+p)^2|a_{n+p}|$$
$$\left.+ [(np\phi\cos\lambda + p^3(p-\eta)](n+p)|b_{n+p}| + np^2\phi\cos\lambda - p^3(p-\eta)\right]. \tag{5}$$

Proof. Let us assume that the relation (4) holds true. It now suffices to show that:

$$k\left|\mathcal{M}\left(\phi,\lambda,\eta,f,\psi\right)-p\right|-\Re\left|\mathcal{M}\left(\phi,\lambda,\eta,f,\psi\right)-p\right|<1. \qquad (6)$$

We first consider:

$$\left|\mathcal{M}\left(\phi,\lambda,\eta,f,\psi\right)-p\right|$$
$$=\left|\left(e^{i\lambda}-\phi\cos\lambda\right)\frac{zf'(z)}{p\psi(z)}+\frac{\phi\cos\lambda}{(p-\eta)}\left(\frac{(zf'(z))'}{\psi'(z)}-\eta\right)-p\right|$$
$$=\left|\frac{\left(e^{i\lambda}-\phi\cos\lambda\right)(p-\eta)f'(z)}{p(p-\eta)\psi'(z)}+\frac{p\phi\cos\lambda(zf'(z))'}{p(p-\eta)\psi'(z)}\right.$$
$$\left.-\frac{\eta p\phi\cos\lambda\psi'(z)}{p(p-\eta)\psi'(z)}-\frac{p^2(p-\eta)\psi'(z)}{p(p-\eta)\psi'(z)}\right|.$$

Now, by using the series form of the functions f and ψ given by:

$$f(z)=z^p+\sum_{n=2}^{\infty}a_{n+p}z^{n+p}$$

and:

$$\psi(z)=z^p+\sum_{n=2}^{\infty}b_{n+p}z^{n+p}$$

in the above relation, we have:

$$\left|\mathcal{M}\left(\phi,\lambda,\eta,f,\psi\right)-p\right|$$
$$=\left|\frac{\left(e^{i\lambda}-\phi\cos\lambda\right)(p-\eta)(pz^{p-1})+p\phi\cos\lambda(p^2z^{p-1})}{p(p-\eta)\left(pz^{p-1}+\sum_{n=2}^{\infty}(n+p)b_{n+p}z^{n+p-1}\right)}\right.$$
$$\left.+\frac{\sum_{n=2}^{\infty}(n+p)a_{n+p}z^{n+p-1}[\left(e^{i\lambda}-\phi\cos\lambda\right)(p-\eta)+(n+p)]}{p(p-\eta)\left(pz^{p-1}+\sum_{n=2}^{\infty}(n+p)b_{n+p}z^{n+p-1}\right)}-\frac{n\phi\cos\lambda}{(p-\eta)}-p\right|$$
$$\leqq\frac{\left(e^{i\lambda}-\phi\cos\lambda\right)(p-\eta)(p)+p\phi\cos\lambda(p^2)}{p(p-\eta)\left(p+\sum_{n=2}^{\infty}(n+p)\left|b_{n+p}\right|\right)}$$
$$+\frac{\sum_{n=2}^{\infty}(n+p)\left|a_{n+p}\right|\left\{\left(e^{i\lambda}-\phi\cos\lambda\right)(p-\eta)+(n+p)\right\}}{p(p-\eta)\left(p+\sum_{n=2}^{\infty}(n+p)\left|b_{n+p}\right|\right)}-\left\{\frac{n\phi\cos\lambda}{(p-\eta)}+p\right\}.$$

We now see that:

$$k\left|\mathcal{M}\left(\phi,\lambda,\eta,f,\psi\right)-p\right|-\Re\left\{\mathcal{M}\left(\phi,\lambda,\eta,f,\psi\right)-p\right\}$$
$$\leqq (k+1)\left|\mathcal{M}\left(\phi,\lambda,\eta,f,\psi\right)-p\right|$$
$$\leqq (k+1)\left[\frac{\left(e^{i\lambda}-\phi\cos\lambda\right)(p-\eta)(p)+p\phi\cos\lambda(p^2)}{p(p-\eta)\left(p+\sum_{n=2}^{\infty}(n+p)\left|b_{n+p}\right|\right)}\right.$$
$$\left.+\frac{\sum_{n=2}^{\infty}(n+p)\left|a_{n+p}\right|\left[\left(e^{i\lambda}-\phi\cos\lambda\right)(p-\eta)+(n+p)\right]}{p(p-\eta)\left(p+\sum_{n=2}^{\infty}(n+p)\left|b_{n+p}\right|\right)}-\left[\frac{n\phi\cos\lambda}{(p-\eta)}+p\right]\right].$$

The above inequality is bounded above by one, if:

$$(k+1)\left[\left(e^{i\lambda} - \phi\cos\lambda\right)(p-\eta)p\right] + (p\phi\cos\lambda)p^2$$
$$+ \left(\sum_{n=2}^{\infty}(n+p)|a_{n+p}|\right)\left\{(e^{i\lambda} - \phi\cos\lambda)(p-\eta) + (n+p)\right\} - \left[\frac{n\phi\cos\lambda}{(p-\eta)} - p\right]$$
$$\cdot \left\{p(p-\eta)\left(p + \sum_{n=2}^{\infty}(n+p)|b_{n+p}|\right)\right\}$$
$$\leqq p(p-\eta)p + \sum_{n=2}^{\infty}(n+p)|b_{n+p}|.$$

Hence:
$$\sum_{n=2}^{\infty}\ddot{U}_n(p,\phi,\lambda,\eta,\xi) \leqq p^2(p-\eta),$$

where $\ddot{U}_n(p,\phi,\lambda,\eta,\xi)$ is given by (5), which completes the proof of Theorem 1. □

Theorem 2. *A function $f \in \mathcal{A}(p)$ satisfies the condition:*

$$\left|\frac{1}{e^{ij}F(z)} - \frac{1}{2\rho}\right| < \frac{1}{2\rho} \qquad (0 \leqq \rho < 1; j \in \mathbb{R}) \qquad (7)$$

if and only if $f \in 0\text{-}\mathcal{K}(p,\lambda)$, where

$$F(z) = \frac{zf'(z)}{p\psi(z)}.$$

Proof. Suppose that f satisfies (7). We then can write:

$$\left|\frac{2\rho - e^{ij}F(z)}{e^{ij}F(z)2\rho}\right| < \frac{1}{2\rho}$$

$$\iff \left(\left|\frac{2\rho - e^{ij}F(z)}{e^{ij}F(z)2\rho}\right|\right)^2 < \left(\frac{1}{2\rho}\right)^2$$

$$\iff \left(2\rho - e^{ij}F(z)\right)\left(2\rho - \overline{e^{ij}F(z)}\right) < e^{-ij}\overline{F(z)}e^{ij}F(z)$$

$$\iff 4\rho^2 - 2\rho\left[e^{-ij}\overline{F(z)} + e^{ij}F(z)\right] + F(z)\overline{F(z)} < F(z)\overline{F(z)}$$

$$\iff 4\rho^2 - 2\rho\left[e^{-ij}\overline{F(z)} + e^{ij}F(z)\right] < 0$$

$$\iff 2\rho - 2\Re\left[e^{ij}\overline{F(z)}\right] < 0$$

$$\iff \Re\left[e^{ij}F(z)\right] > \rho$$

$$\iff \Re\left(e^{ij}\frac{zf'(z)}{p\psi(z)}\right) > \rho.$$

This completes the proof of Theorem 2. □

Theorem 3. *For $0 \leqq \varphi_1 < \varphi_2$, it is asserted that:*

$$k\text{-}\mathcal{Q}(p,\varphi_2,\lambda,\eta) \subset 0\text{-}\mathcal{Q}(p,\varphi_1,\lambda,\eta).$$

Proof. Let $f(z) \in k\text{-}\mathcal{Q}(p, \varphi_2, \lambda, \eta)$. Then:

$$\frac{1}{p-\eta}\left[\left(e^{i\lambda} - \varphi_1 \cos \lambda\right)(p-\eta)\frac{zf'(z)}{p\psi(z)} + \varphi_1 \cos \lambda \left(\frac{(zf'(z))'}{\psi(z)'} - \eta\right)\right]$$

$$= \frac{\varphi_1}{\varphi_2}\left[\left(e^{i\lambda} - \varphi_2 \cos \lambda\right)\frac{zf'(z)}{p\psi(z)} + \frac{\varphi_2 \cos \lambda}{(p-\eta)}\left(\frac{(zf'(z))'}{p\psi(z)'} - \eta\right)\right]$$

$$- \left(\frac{\varphi_1 - \varphi_2}{\varphi_2}\right)e^{i\lambda}\frac{sf'(z)}{p\psi(z)}$$

$$= \frac{\varphi_1}{\varphi_2}H_1(z) + \left(1 - \frac{\varphi_1}{\varphi_2}\right)H_2(z) = H(z),$$

where:

$$H_1(z) = \left(e^{i\lambda} - \varphi_2 \cos \lambda\right)\frac{zf'(z)}{p\psi(z)} + \frac{\varphi_2 \cos \lambda}{(p-\eta)}\left(\frac{(zf'z)'}{\psi'(z)} - \eta\right) \in \mathcal{P}\left(h_{k,\rho}\right) \subset \mathcal{P}(\rho)$$

and:

$$H_2(z) = e^{i\lambda}\frac{zf'(z)}{p\psi(z)} \in \mathcal{P}(\rho).$$

Since $\mathcal{P}(\rho)$ is a convex set (see [27]), we therefore have $H(z) \in \mathcal{P}(\rho)$. This implies that $f \in 0\text{-}\mathcal{Q}(p, \varphi_1, \lambda, \eta)$. Thus:

$$k\text{-}\mathcal{Q}(p, \varphi_2, \lambda, \eta) \subset 0\text{-}\mathcal{Q}(p, \varphi_1, \lambda, \eta).$$

The proof of Theorem 3 is now completed. □

Theorem 4. *Let $\phi > 0$ and $\lambda < \frac{\pi}{2}$. Then:*

$$k\text{-}\mathcal{Q}(p, \phi, \lambda, \eta, \xi) \subset k\text{-}\mathcal{K}(p, 0, \xi).$$

Proof. Let $f \in k\text{-}\mathcal{Q}(p, \phi, \lambda, \eta, \xi)$, and suppose that:

$$\frac{f'(z)}{\psi'(z)} = p(z), \tag{8}$$

where $p(z)$ is analytic and $p(0) = 1$. Now, by differentiating both sides of (8) with respect to z, we have:

$$\frac{(zf'(z))'}{\psi'(z)} = zp'(z) + p(z)\varepsilon(z), \tag{9}$$

where:

$$\varepsilon(z) = \frac{(z\psi'(z))'}{\psi'(z)}.$$

By using (8) and (9) in (4), we arrive at:

$$\mathcal{M}(\phi, \lambda, \eta, f, \psi) = \left(e^{i\lambda} - \phi \cos \lambda\right)\frac{p(z)}{p} + \frac{\phi \cos \lambda}{p-\eta}\left(zp'(z) + p(z)\varepsilon(z) - \eta\right)$$

$$= \frac{\phi \cos \lambda}{p-\eta}zp'(z) + \left(\frac{e^{i\lambda}}{p} - \frac{\phi \cos \lambda}{p - \varepsilon(z)}\right)\left(\frac{\phi \cos \lambda}{p-\eta}\right)p(z) - \frac{\eta\phi \cos \lambda}{p-\eta}$$

$$= B(z)zp'(z) + C(z)p(z) + D(z), \tag{10}$$

where:

$$B(z) = \frac{\phi \cos \lambda}{p-\eta},$$

$$C(z) = \frac{e^{i\lambda}(p-\eta) - \phi\cos\lambda(p-\eta) + \phi\cos\lambda\varepsilon(z)p}{p(p-\eta)}$$

and:
$$D(z) = \frac{\eta\phi\cos\lambda}{p-\eta}.$$

Now, since $f \in k\text{-}\mathcal{Q}(p,\phi,\lambda,\eta,\xi)$, we have:
$$B(z)zp'(z) + C(z)p(z) + D(z) \prec p_k(z), \tag{11}$$

which, upon replacing $p(z)$ by:
$$p_*(z) = p(z) - 1,$$

and $p_k(z)$ by:
$$p_k^*(z) = p_k(z) - 1,$$

shows that the above subordination in (11) becomes as follows:
$$B(z)zp_x'(z) + C(z)p_x(z) + D_*(z) \prec p_k^*(z), \tag{12}$$

where:
$$D_*(z) = C(z) + D(z) - 1.$$

We now apply Lemma 2 with:
$$A = 0$$

and
$$p_*(z) \prec p_k^*(z).$$

We thus find that:
$$\frac{f'(z)}{\psi'(z)} = p(z) \prec p_k^*(z). \tag{13}$$

This complete the proof of Theorem 4. □

For $f \in \mathcal{A}$, we next consider the integral operator defined by:
$$F(z) = I_m[f] = \frac{m+1}{z^m}\int_0^z t^{m-1}f(t)\,dt. \tag{14}$$

This operator was given by Bernardi [28] in the year 1969. In particular, the operator I_1 was considered by Libera [29]. We prove the following result.

Theorem 5. *Let $f(z) \in k\text{-}\mathcal{Q}(p,\phi,\lambda,\eta,\xi)$. Then, $I_m[f] \in \mathcal{K}(p,0,\xi)$.*

Proof. Let the function $\psi(z)$ be such that:
$$\mathcal{M}(\phi,\lambda,\eta,f,\psi) = \left(e^{i\lambda} - \phi\cos\lambda\right)\frac{zf'(z)}{p\psi(z)} + \frac{\phi\cos\lambda}{(p-\eta)}\left(\frac{(zf'(z))'}{\psi'(z)} - \eta\right).$$

Then, according to [14], the function $G = I_m[f] \in \mathcal{CD}(k,\delta)$. Furthermore, from (14), we deduce that:
$$(1+m)f(z) = (1+m)F(z) + z(F(z))' \tag{15}$$

and:
$$(1+m)g(z) = (1+m)G(z) + z(G(z))'. \tag{16}$$

If we now put:
$$p(z) = \frac{F'(z)}{G'(z)}$$
and:
$$q(z) = \frac{1}{(m+1) + \left(\frac{zG''(z)}{G'(z)}\right)},$$
then, by simple computations, we find that:
$$\frac{f(z)}{\psi(z)} = \frac{(1+m)F'(z) + zF''(z)}{(1+m)G'(z) + zG''(z)}$$
or, equivalently, that:
$$\frac{f'(z)}{\psi'(z)} = p(z) + zp'(z)q(z). \tag{17}$$

We now let:
$$\frac{f'(z)}{\psi'(z)} = p(z) + zp'(z)q(z) = h(z), \tag{18}$$
where the function $h(z)$ is analytic in $\mathbb{E}$ with $h(0) = 1$. Then, by using (18), we have:
$$\frac{(zf(z))'}{\psi'(z)} = zh'(z) + \varepsilon(z)h(z), \tag{19}$$
where:
$$\varepsilon(z) = \frac{(z\psi'(z))'}{\psi'(z)}.$$

Furthermore, by using (18) and (19) in (4), we obtain:
$$\mathcal{M}(\alpha, \beta, \gamma, \lambda, \delta, f) = \left(e^{i\lambda} - \theta \cos\lambda\right) \frac{zf'(z)}{\psi'(z)} + \frac{\phi \cos\lambda}{p - \eta}\left(\frac{(zf'(z))'}{\psi'(z)} - \eta\right)$$
$$= \left(e^{i\lambda} - \theta \cos\lambda\right) + \frac{\phi \cos\lambda}{p - \eta} zh'(z) + [zh'(z) + \varepsilon(z)h(z) - \eta]$$
$$= \frac{\phi \cos\lambda}{p - \eta} zh'(z) + \left(e^{i\lambda} - \phi \cos\lambda + \frac{\phi \cos\lambda}{p - \eta}\right) h(z) - \frac{\eta(\phi \cos\lambda)}{p - \eta}$$
$$= B(z) zh'(z) + C(z) h(z) + D(z),$$
where:
$$B(z) = \frac{\phi \cos\lambda}{p - \eta},$$
$$C(z) = \frac{((p - \eta)e^{i\lambda} - (p - \eta)\phi \cos\lambda + \phi \cos\lambda)}{p - \eta}$$
and:
$$D(z) = \frac{\eta(\phi \cos\lambda)}{p - \eta}.$$

Now, if we apply Lemma 1 with $A = 0$, we get:
$$\frac{f'(z)}{\psi'(z)} = h(z) \prec p_k(z). \tag{20}$$

Furthermore, from (18), we have:
$$p(z) + zp'(z)q(z) \prec p_k(z).$$

By using Lemma 2 on (20), we obtain the desired result. This completes the proof of Theorem 5. □

4. Conclusions

Using the idea of spiral-like and close-to-convex functions, we have introduced Mocanu-type functions associated with conic domains. We have derived some interesting results such as sufficiency criteria, inclusion results, and integral-preserving properties. We have also proven that the our newly-defined function classes are closed under the famous Libera operator.

Author Contributions: conceptualization, H.M.S. and Q.Z.A.; methodology, N.K.; software, M.T.R. and M.D.; validation, H.M.S., M.D. and Y.Z.; formal analysis, H.M.S. and Q.Z.A; investigation, M.D. and M.T.R.; writing–original draft preparation, H.M.S.; and Y.Z writing–review and editing, N.K. and M.D.; visualization, M.T.R.; supervision, H.M.S.; funding acquisition, M.D.

Funding: The third author is partially supported by UKM grant: GUP-2017-064.

Conflicts of Interest: The authors declare that they have no competing interests.

References

1. Duren, P.L. *Univalent Functions*, Grundlehren der Mathematischen Wissenschaften, Band 259; Springer-Verlag: New York, NY, USA; Berlin/Heidelberg, Germany; Tokyo, Japan, 1983.
2. Spaček, L. Prispevek k teorii funkei prostych. *Casŏpis. Pest. Mat.* **1933**, *62*, 12–19.
3. Libera, R. J. Univalent a-spiral functions, *Can. J. Math.* **1967**, *19*, 449-456.
4. Silvia, M.E. On a subclass of spiral-like functions, *Proc. Am. Math. Soc.* **1974**, *44*, 411–420. [CrossRef]
5. Umarani, P. On a subclass of spiral-like functions, *Indian J. Pure Appl. Math.* **1979**, *10*, 1292–1297.
6. Noor, K.I.; Khan, N.; Ahmad, Q.Z. Some properties of multivalent spiral-like functions. *Maejo Int. J. Sci. Technol.* **2018**, *3*, 353–364.
7. Kanas. S.; Wiśniowska, A. Conic regions and k-uniform convexity. *J. Comput. Appl. Math.* **1999**, *105*, 327–336. [CrossRef]
8. Kanas. S.; Wiśniowska, A. Conic domains and starlike functions. *Rev. Roum. Math. Pures Appl.* **2000**, *45*, 647–657.
9. Kanas, S.; Srivastava, H. M.; Linear operators associated with k-uniformly convex functions. *Integral Transforms Spec. Funct.* **2000**, *9*, 121–132. [CrossRef]
10. Kanas, S.; Răducanu, D. Some class of analytic functions related to conic domains. *Math. Slovaca* **2014**, *64*, 1183–1196. [CrossRef]
11. Noor, K.I.; Malik, S.N. On coefficient inequalities of functions associated with conic domains. *Comput. Math. Appl.* **2011**, *62*, 2209–2217. [CrossRef]
12. Shams, S.; Kulkarni, S.R.; Jahangiri, J.M. Classes of uniformly starlike and convex functions. *Int. J. Math. Math. Sci.* **2004**, *55*, 2959–2961. [CrossRef]
13. Srivastava, H.M.; Shanmugam, T.N.; Ramachandran, C.; Sivassurbramanian, S. A new subclass of k-uniformly convex functions with negative coefficients. *J. Inequal. Pure Appl. Math.* **2007**, *8*, 1–14.
14. Acu, M. On a subclass of k-uniformly close to convex functions. *Gen. Math.* **2006**, *14*, 55–64.
15. Arif, M.; Dziok, J.; Raza, M.; Sokól, J. On products of multivalent close-to-star functions, *J. Inequal. Appl.* **2015**, *2015*, 1–14. [CrossRef]
16. Khan, N.; Khan, B.; Ahmad, Q.Z.; Ahmad, S. Some Convolution properties of multivalent analytic functions. *AIMS Math.* **2017**, *2*, 260–268. [CrossRef]
17. Noor, K.I.; Khan, N.; Noor, M.A. On generalized spiral-like analytic functions. *Filomat* **2014**, *28*, 1493–1503. [CrossRef]
18. Raza, M.; Din, M.U.; Malik, S.N. Certain geometric properties of normalized Wright functions. *J. Funct. Spaces* **2016**, *2016*. [CrossRef]
19. Srivastava, H.M.; Eker, S.S. Some applications of a subordination theorem for a class of analytic functions. *Appl. Math. Lett.* **2008**, *21*, 394–399. [CrossRef]
20. Srivastava, H.M.; El-Ashwah, R.M.; Breaz, N. A certain subclass of multivalent functions involving higher-order derivatives. *Filomat* **2016**, *30*, 113–124. [CrossRef]

21. Aldweby, H.; Darus, M. A note on q-integral operators. *Electron. Notes Discret. Math.* **2018**, *67*, 25–30. [CrossRef]
22. Aldweby, H.; Darus, M. On Fekete-Szegö problems for certain subclasses defined by q-derivative. *J. Funct. Spaces* **2017**, 1–5. . [CrossRef]
23. Elhaddad, S.; Aldweby, H.; Darus, M. Some Properties on a class of harmonic univalent functions defined by q-analoque of Ruscheweyh operator. *J. Math. Anal.* **2018**, *9*, 28–35.
24. Hussain, S.; Khan, S.; Zaighum, M.A.; Darus, M. Applications of a q-Sălăgean type operator on multivalent functions. *J. Inequal. Appl.* **2018**, *301*, 1–12. [CrossRef]
25. Rasheed, A.; Hussain, S.; Zaighum, M.A.; Darus, M. Class of analytic function related with uniformly convex and Janowski's functions. *J. Funct. Spaces* **2018**, *2018*. [CrossRef]
26. Miller, S.S.; Mocanu, P.T. *Differential Subordinations: Theory and Applications*; Marcel Dekker: New York, NY, USA, 2000.
27. Kanas. S. Techniques of the differential subordination for domains bounded by conic sections. *Int. J. Math. Math. Sci.* **2003**, *38*, 2389–2400. [CrossRef]
28. Bernardi, S.D. Convex and starlike univalent functions. *Trans. Am. Math. Soc.* **1969**, *135*, 429–446. [CrossRef]
29. Libera, R.J. Some classes of regular univalent functions. *Proc. Am. Math. Soc.* **1965**, *16*, 755–758. [CrossRef]

© 2019 by the authors. Licensee MDPI, Basel, Switzerland. This article is an open access article distributed under the terms and conditions of the Creative Commons Attribution (CC BY) license (http://creativecommons.org/licenses/by/4.0/).

Article

Efficacy of the Post-Exposure Prophylaxis and of the HIV Latent Reservoir in HIV Infection

Carla M. A. Pinto [1,2,*,†], **Ana R. M. Carvalho** [3,†], **Dumitru Baleanu** [4,†] **and Hari M. Srivastava** [5,6,†]

1. School of Engineering, Polytechnic of Porto, Rua Dr António Bernardino de Almeida, 431, 4200-072 Porto, Portugal
2. Centre for Mathematics, University of Porto, Rua do Campo Alegre s/n 4440-452 Porto, Portugal
3. Faculty of Sciences, University of Porto, Rua do Campo Alegre s/n, 4440-452 Porto, Portugal; up200802541@fc.up.pt
4. Department of Mathematics and Computer Sciences, Cankaya University, Balgat, Ankara 0630, Turkey; dumitru@cankaya.edu.tr
5. Department of Mathematics and Statistics, University of Victoria, Victoria, BC V8W 3R4, Canada; harimsri@math.uvic.ca
6. Department of Medical Research, China Medical University Hospital, China Medical University, Taichung 40402, Taiwan
* Correspondence: cap@isep.ipp.pt
† The authors contributed equally to this work.

Received: 17 April 2019; Accepted: 4 June 2019; Published: 5 June 2019

Abstract: We propose a fractional order model to study the efficacy of the Post-Exposure Prophylaxis (PEP) in human immunodeficiency virus (HIV) within-host dynamics, in the presence of the HIV latent reservoir. Latent reservoirs harbor infected cells that contain a transcriptionally silent but reactivatable provirus. The latter constitutes a major difficulty to the eradication of HIV in infected patients. PEP is used as a way to prevent HIV infection after a recent possible exposure to HIV. It consists of the in-take of antiretroviral drugs for, usually, 28 days. In this study, we focus on the dosage and dosage intervals of antiretroviral therapy (ART) during PEP and in the role of the latent reservoir in HIV infected patients. We thus simulate the model for immunologically important parameters concerning the drugs and the fraction of latently infected cells. The results may add important information to clinical practice of HIV infected patients.

Keywords: Post-Exposure Prophylaxis; latent reservoir; HIV infection; fractional order model

1. Introduction

The human immunodeficiency virus (HIV) is a retrovirus, which impairs the host immune system, by destroying preferably the $CD4^+$ T cells. These cells are essential to guarantee immune protection. They do so by helping B cells produce antibodies, inducing macrophages to develop enhanced microbicidal activity, recruiting neutrophils, eosinophils, and basophils to inflammation and infection sites, and, by producing cytokines and chemokines. A number of $CD4^+$ T cells below a given threshold is a synonym of immunodeficiency. The organism is thus vulnerable to a broad set of infections, cancers and other diseases.

HIV occurs in two types: HIV-1 and HIV-2, and is transmitted by the exchange of HIV-infected body fluids, such as blood, semen, and genital secretions. It may also be transmitted from an HIV-infected pregnant woman to her child, during pregnancy, birth, or breastfeeding [1].

HIV is a defying global health threat, responsible for more than 36.7 million infected people worldwide, and more than 35 million deaths, so far. In 2016, the number of deceased from HIV-related causes was estimated at one million. Figures are even more striking since, globally, 1.8 million people become newly infected each year. Access to antiretroviral therapy is crucial to control the virus and to reduce the risk of transmission, providing HIV infected individuals and those at risk, more healthy, long and productive lives. In 2016, nearly half of the adults and children living with HIV had access to treatment. Effective treatment reduces the risk of HIV transmission to non-infected sexual partners by 96% [2]. Other forms of HIV prevention are the Pre-Exposure Prophylaxis (PrEP) and the Post-Exposure Prophylaxis (PEP).

PrEP is the daily in-take of ART to prevent HIV infection in uninfected people. The usual combination of the two HIV drugs, tenofovir and emtricitabine, sold under the name of Truvada, is approved for daily use as PrEP. PrEP is shown to be highly effective for HIV prevention, when taken consistently. WHO recommends PrEP as one of the prevention options, for people at substantial risk of HIV infection (namely injecting drug users, men who have sex with men, and high-risk heterosexual couples), and for HIV-negative women who are pregnant or breastfeeding [2].

PEP consists of the intake of ART, after possible exposure to HIV. It includes counseling, first aid care, HIV testing, and administration of a 28-day course of ART with follow-up care. It is intended to prevent HIV spread in the human body, protecting against being re-exposed to HIV and reducing the chances of HIV transmission [3]. PEP was initially intended for healthcare workers, who had been accidentally exposed to HIV-infected body fluids, through injury with a contaminated syringe, etc. Nowadays, WHO recommends PEP for both health-worker and non-health-workers, for adults and children [2]. PEP should be started immediately after exposure and at most 72 hours after, to enhance the rate of success, since it is not 100% effective [3].

Latent reservoirs consist of a small proportion of resting $CD4^+$ T cells, containing integrated proviral DNA [4–6]. Latent reservoirs are established during the acute phase of HIV infection. These reservoirs may hide out for years in many tissues in the body, namely lymph nodes, seminal fluid, and cerebral spinal fluid. Latent reservoirs can, however, wake up, and release old viral variants in the blood. The mechanism behind this activation is summarized as follows. Proviral genomes are integrated in resting memory $CD4^+$ T cells. Due to the quiescent state of these latent cells, these genomes are not transcribed into mRNA (messenger ribonucleic acid) and translated in protein to become active virus. Nevertheless, when cell activation occurs, then transcription and translation may recommence [7]. This affects the viral dynamics of untreated patients, promoting viral load rebounds. ART can suppress HIV load levels to undetectable values, however, it cannot eliminate the latent reservoir. This is the main challenge to HIV cure.

Considerable research has been found in the literature to describe the effects of HIV prevention strategies. In 2009, Lou et al. [8] study drug dynamics, drug dosages, and therapy strategies in an impulsive model for the dynamics of HIV in the presence of PEP. Authors conclude that the best choice for an infected individual is a safe dose of medication during PEP. Moreover, the side effects of ART should also be taken into consideration in choosing the appropriate therapy. Conway et al. [9] present a stochastic model for the dynamics of HIV, immediately after exposure, and apply drug prophylaxis to understand how it reduces the risk of infection. The authors predict that a two-week PEP regimen may be as effective as the recommended four-week treatment protocol. In 2014, Kim et al. [10] study a model for HIV infection in men who have sex with men (MSM) in South Korea. They simulate the effects of early ART, early diagnosis, PrEP, and combination interventions, on the incidence and prevalence of HIV infection. The authors conclude that PrEP and early diagnosis would be effective ways in reducing HIV incidence in MSM. In 2017, Pinto et al. [11], evaluate the impact of PrEP and screening in the dynamics of HIV in infected patients. The proposed model incorporates condom use, the number of sexual partners, and treatment for HIV. The basic reproduction number is extremely impacted by the efficacy of the screening, pointing to explicit campaigns highlighting screening. The results from the model are fitted to data on the cumulative HIV and AIDS (acquired immunodeficiency syndrome) cases in Portugal.

Fractional Calculus—Short Recap

Fractional Calculus has been a hot research topic in the last few decades. Researchers from distinct scientific areas, theoretical and applied, have studied fractional order models to obtain a deeper understanding of real world phenomena [4,12–17]. Fractional order models are characterized by a 'memory' property, which brings additional information to analyze the systems' dynamical behaviors.

The classical definitions for a derivative of fractional (non-integer) order are the Caputo (C), the Riemann-Liouville (RL) and the Grünwald-Letnikov (GL). Let $(0,t)$ be the interval, instead of (a,t), for simplification. The function $y(\tau)$ is smooth in every interval $(0,t)$, $t \leq T$. The RL definition reads:

$$D_{RL}^\alpha y(t) = \begin{cases} \frac{1}{\Gamma(m-\alpha)} \frac{d^m}{dt^m} \int_0^t \frac{y(\tau)}{(t-\tau)^{\alpha+1-m}}, & m-1 \leq \alpha < m \\ \frac{d^m y(t)}{dt^m} & \alpha = m \end{cases}$$

where Γ is the Euler Gamma function. The Caputo definition is given by:

$$D_C^\alpha y(t) = \begin{cases} \frac{1}{\Gamma(m-\alpha)} \int_0^t \frac{y^m(\tau)}{(t-\tau)^{\alpha+1-m}}, & m-1 \leq \alpha < m \\ \frac{d^m y(t)}{dt^m} & \alpha = m \end{cases}$$

The GL definition is based on finite differences and is equivalent to the RL formula:

$$D_{GL}^\alpha y(t) = \lim_{h \to 0} h^{-\alpha} \sum_{k=0}^n (-1)^k \frac{\Gamma(\alpha+1)}{k! \Gamma(\alpha-k+1)} y(x-kh), \quad nh = x$$

The memory effect in biology/epidemiology/immunology is extremely important, thus the appearance of fractional order models in the study of patterns arising in these models comes as a natural generalization of the integer order models [18–21]. In [20], the authors generalize an integer order model for HIV dynamics to include a fractional order derivative. In Arafa et al. [18] the authors generalize an integer order model for HIV dynamics to include a fractional order derivative. They conclude that the fractional order model provides a better fit to real data from 10 patients than the integer order model. Pinto [4] studies the role of the latent reservoir in the persistence of the latent reservoir and of the plasma viremia in a fractional-order (FO) model for HIV infection. The model assumes that (i) the latently infected cells may undergo bystander proliferation, without active viral production, (ii) the latent cell activation rate decreases with time on ART, and (iii) the productively infected cells' death rate is a function of the infected cell density. The model clarifies the role of the latent reservoir in the persistence of the latent reservoir and of the plasma virus. The non-integer order derivative is associated with distinct velocities in the dynamics of the latent reservoir and of plasma virus. In [12], the authors study the effect of the HIV viral load in a coinfection fractional order model for HIV and HCV (hepatitis C virus) coinfection. HIV has a significant impact on the burden of the coinfection. Moreover, the order of the fractional derivative may pave the way to a better understanding of the individuals' compliance to treatment, the distinct responses of the immune system. The non-integer order derivative adds another degree of freedom to the model. In what concerns drug diffusion in tissues, there are some interesting results in the literature. In [22], the authors propose non-integer order (fractional order) models to represent anomalous diffusion, memory effects and power-law clearance rates, typical of drug uptake and diffusion in a case-study of a drug used for cancer therapy. They conclude that fractional models avoid unbounded accumulation of drugs, seen in the integer order approach, and help to prevent life-threatening side-effects on patients. In 2017 [23], the authors provide a review on pharmacokinetic models and propose their generalizations to fractional orders. The new models account for tissue trapping as well as short- and long-time recirculating effects. The benefits from such approach are twofold: (i) a better understanding of secondary effects on patients under treatment; and (ii) avoidance of unbounded drug accumulation.

With the aforementioned ideas in mind, we outline the paper as follows. In Section 2, we describe the proposed model. We follow with the computation of the reproductive number and the stability of the disease free equilibrium in Section 3. Then, in Section 4, we prove the global stability of the disease free equilibrium. The model is simulated and the corresponding results are discussed in Section 5. Finally, in Section 6, we conclude this work.

2. The Model

The model consists of seven classes: the healthy and susceptible CD4$^+$ T cells, T, the healthy and non-susceptible CD4$^+$ T cells, T_R, the latently infected CD4$^+$ T cells, L, the infected and infectious CD4$^+$ T cells, I, the infected and non-infectious CD4$^+$ T cells, I_R, the HIV virus, V, and the drug concentration in the plasma, R.

CD4$^+$ T cells are produced with rate λ and die with rate μ. These cells are infected by HIV and by infected CD4$^+$ T cells at rates β_1 and β_2, respectively. The healthy T cells are inhibited by drug at rate q. A fraction, η, of infected CD4$^+$ T cells becomes latently infected. The latently infected CD4$^+$ T cells become productively infected at a rate a_L and die with a rate μ_L. The infected CD4$^+$ T cells die with rate a and are inhibited by drug at rate p. The virus are produced by infected CD4$^+$ T cells at rate k and cleared at rate c. The dynamics of the drugs is as follows. For simplicity, we postulate that after taking the drug, the cell, T_R, inhibits infection until it dies. We further assume that drugs are taken at times $t = t_k$, and their effect is instantaneous. The latter results in a system of impulsive differential equations, with condition $\Delta R = \Delta R_k$, where ΔR_k is the dosage. For $t \neq t_k$, the solutions are continuous and obey system (1). The drug, R, is cleared at rate g.

The nonlinear system of fractional differential equations describing the model is given by:

$$
\begin{aligned}
\frac{d^\alpha T}{dt^\alpha} &= \lambda^\alpha - \mu^\alpha T - \beta_1^\alpha TV - \beta_2^\alpha TI - q^\alpha TR \\
\frac{d^\alpha L}{dt^\alpha} &= \eta \beta_1^\alpha TV + \eta \beta_2^\alpha TI - a_L^\alpha L - \mu_L^\alpha L \\
\frac{d^\alpha I}{dt^\alpha} &= (1-\eta)\beta_1^\alpha TV + (1-\eta)\beta_2^\alpha TI + a_L^\alpha L - a^\alpha I - p^\alpha IR \\
\frac{d^\alpha V}{dt^\alpha} &= k^\alpha I - c^\alpha V \\
\frac{d^\alpha T_R}{dt^\alpha} &= q^\alpha TR - d^\alpha T_R \\
\frac{d^\alpha I_R}{dt^\alpha} &= p^\alpha IR - a^\alpha I_R \\
\frac{d^\alpha R}{dt^\alpha} &= R_k^\alpha - g^\alpha R
\end{aligned}
\tag{1}
$$

where the parameter $\alpha \in (0,1]$ is the order of the fractional derivative. The fractional derivative of the proposed model is used in the Caputo sense.

3. Reproduction Number

In this section, we compute the reproduction number of model (1) in the cases of no drug, R_0, and of drug therapy, R_c^d, and the local stability of its disease-free equilibrium. The basic reproduction number is defined as the number of secondary CD4$^+$ T cells infections due to a single infected cell in a completely susceptible population. We start with R_0. We use the next generation method [24].

The disease-free equilibrium of model (1) is given by:

$$
P_0 = (T_0, L_0, I_0, V_0, T_{R_0}, I_{R_0}, R^0) = \left(\frac{\lambda^\alpha}{\mu^\alpha}, 0, 0, 0, 0, 0, 0\right) \tag{2}
$$

Using the notation in [24] on system (1), matrices for the new infection terms, F_1, and the other terms, V_1, are given as follows. The chosen variables of the model are L, I and V and the procedure is identical to [24].

$$F_1 = \begin{pmatrix} 0 & \eta\beta_2^\alpha T_0 & \eta\beta_1^\alpha T_0 \\ 0 & (1-\eta)\beta_2^\alpha T_0 & (1-\eta)\beta_1^\alpha T_0 \\ 0 & 0 & 0 \end{pmatrix}$$

$$V_1 = \begin{pmatrix} a_L^\alpha + \mu_L^\alpha & 0 & 0 \\ -a_L^\alpha & a^\alpha & 0 \\ 0 & -k^\alpha & c^\alpha \end{pmatrix}$$

The associative basic reproduction number R_0 is written as:

$$R_0 = \rho(F_1 V_1^{-1}) = \frac{T_0\left(\beta_1^\alpha k^\alpha + \beta_2^\alpha c^\alpha\right)\left[(1-\eta)\mu_L^\alpha + a_L^\alpha\right]}{a^\alpha c^\alpha\left(a_L^\alpha + \mu_L^\alpha\right)} \qquad (3)$$

where ρ indicates the spectral radius of $F_1 V_1^{-1}$. The local stability of P_0 can be determined using Lemmas 1 and 2.

Lemma 1. *[25] The disease-free equilibrium P_0 of the system (1) is locally asymptotically stable iff all eigenvalues λ_i of the linearization matrix of model (1), satisfy $|arg(\lambda_i)| > \alpha\frac{\pi}{2}$.*

Lemma 2. *The disease-free equilibrium P_0 of the system (1) is unstable if $R_0 > 1$.*

Proof. Let M_1 be given by:

$$M_1 = \begin{pmatrix} -\mu^\alpha & 0 & -\beta_2^\alpha T_0 & -\beta_1^\alpha T_0 \\ 0 & -(a_L^\alpha + \mu_L^\alpha) & \eta\beta_2^\alpha T_0 & \eta\beta_1^\alpha T_0 \\ 0 & a_L^\alpha & (1-\eta)\beta_2^\alpha T_0 - a^\alpha & (1-\eta)\beta_1^\alpha T_0 \\ 0 & 0 & k^\alpha & -c^\alpha \end{pmatrix}$$

Expanding, $\det(\lambda^p I_4 - M_1) = 0$, where I_4 is the 4×4 identity matrix, we have the following equation in terms of λ:

$$(\lambda^p + \mu^\alpha)\left[\lambda^{3p} + \left(a_L^\alpha + \mu_L^\alpha + a^\alpha + c^\alpha - (1-\eta)\beta_2^\alpha T_0\right)\lambda^{2p} + \left(c^\alpha(a_L^\alpha + \mu_L^\alpha + a^\alpha) + (a_L^\alpha + \mu_L^\alpha)a^\alpha\right.\right.$$
$$\left. - (1-\eta)\beta_2^\alpha T_0(c^\alpha + \mu_L^\alpha) - a_L^\alpha \beta_2^\alpha T_0 - k^\alpha(1-\eta)\beta_1^\alpha T_0\right)\lambda^p + (a_L^\alpha + \mu_L^\alpha) + c^\alpha a^\alpha \qquad (4)$$
$$\left. - \beta_2^\alpha T_0 c^\alpha(\mu_L^\alpha(1-\eta) + a_L^\alpha) - \beta_1^\alpha T_0 k^\alpha\left(1-\eta\right)\mu_L^\alpha + a_L^\alpha\right)\right] = 0$$

Now, the arguments of the roots of the equation, $\lambda^p + \mu^\alpha = 0$, are given by:

$$arg(\lambda_j) = \frac{\pi}{p} + j\frac{2\pi}{p} > \frac{\pi}{M} > \frac{\pi}{2M}$$

where $j = 0, 1, ..., (p-1)$.

Using Descartes' rule of signs, we find that there is exactly one sign change of the equation:

$$\lambda^{3p} + \left(a_L^\alpha + \mu_L^\alpha + a^\alpha + c^\alpha - (1-\eta)\beta_2^\alpha T_0\right)\lambda^{2p} + \left(c^\alpha(a_L^\alpha + \mu_L^\alpha + a^\alpha) + (a_L^\alpha + \mu_L^\alpha)a^\alpha\right.$$
$$\left. - (1-\eta)\beta_2^\alpha T_0(c^\alpha + \mu_L^\alpha) - a_L^\alpha \beta_2^\alpha T_0 - k^\alpha(1-\eta)\beta_1^\alpha T_0\right)\lambda^p + (a_L^\alpha + \mu_L^\alpha) + c^\alpha a^\alpha\left[1 - R_0\right] = 0 \qquad (5)$$

for $R_0 > 1$. Thus, there is exactly one positive real root of the aforesaid equation for which the argument is less than $\frac{\pi}{2M}$. As such, we conclude that if $R_0 > 1$ the disease-free equilibrium P_0 of the system (1) is unstable. □

Now, we discuss the dynamics of system (1) with drugs. The disease-free equilibrium of model (1) with drugs is given by:

$$P_1 = (T_1, L_1, I_1, V_1, T_{R_1}, I_{R_1}, R_1) = \left(\frac{\lambda^\alpha}{\mu^\alpha + q^\alpha R^\star}, 0, 0, 0, \frac{q^\alpha T_1 R^\star}{d^\alpha}, 0, R^\star \right) \quad (6)$$

Using the notation in [24] on system (1), matrices for the new infection terms, F_2, and the other terms, V_2, are given by:

$$F_2 = \begin{pmatrix} 0 & \frac{\eta \beta_2^\alpha \lambda^\alpha}{\mu^\alpha + q^\alpha R^\star} & \frac{\eta \beta_1^\alpha \lambda^\alpha}{\mu^\alpha + q^\alpha R^\star} & 0 \\ 0 & \frac{(1-\eta)\beta_2^\alpha \lambda^\alpha}{\mu^\alpha + q^\alpha R^\star} & \frac{(1-\eta)\beta_1^\alpha \lambda^\alpha}{\mu^\alpha + q^\alpha R^\star} & 0 \\ 0 & 0 & 0 & 0 \\ 0 & 0 & 0 & 0 \end{pmatrix}$$

$$V_2 = \begin{pmatrix} a_L^\alpha + \mu_L^\alpha & 0 & 0 & 0 \\ -a_L^\alpha & a^\alpha + p^\alpha R^\star & 0 & 0 \\ 0 & -k^\alpha & c^\alpha & 0 \\ 0 & 0 & 0 & a^\alpha \end{pmatrix}$$

In this case, the basic reproduction number R_c^d is computed to be:

$$R_c^d = \rho(F_2 V_2^{-1}) = \frac{\lambda^\alpha (\beta_1^\alpha k^\alpha + \beta_2^\alpha c^\alpha)[(1-\eta)\mu_L^\alpha + a_L^\alpha]}{(\mu^\alpha + q^\alpha R^\star)(a^\alpha + p^\alpha R^\star)c^\alpha (a_L^\alpha + \mu_L^\alpha)} \quad (7)$$

where ρ indicates the spectral radius of $F_2 V_2^{-1}$. The stability of disease-free equilibrium in the case of the drug therapy, P_1, can be determined using the following lemmas:

Lemma 3. *[25] The disease-free equilibrium P_1 of the system (1) is locally asymptotically stable iff all eigenvalues λ_i of the linearization matrix of model (1), satisfy $|\arg(\lambda_i)| > \alpha \frac{\pi}{2}$.*

Lemma 4. *The disease-free equilibrium P_1 of the system (1) is unstable if $R_c^d > 1$.*

Proof. Let M_2 be given by:

$$M_2 = \begin{pmatrix} -\mu^\alpha - q^\alpha R^\star & 0 & -\frac{\beta_2^\alpha \lambda^\alpha}{\mu^\alpha + q^\alpha R^\star} & -\frac{\beta_1^\alpha \lambda^\alpha}{\mu^\alpha + q^\alpha R^\star} & 0 & 0 & 0 \\ 0 & -(a_L^\alpha + \mu_L^\alpha) & \frac{\eta \beta_2^\alpha \lambda^\alpha}{\mu^\alpha + q^\alpha R^\star} & \frac{\eta \beta_1^\alpha \lambda^\alpha}{\mu^\alpha + q^\alpha R^\star} & 0 & 0 & 0 \\ 0 & a_L^\alpha & \frac{(1-\eta)\beta_2^\alpha \lambda^\alpha}{\mu^\alpha + q^\alpha R^\star} - a^\alpha - p^\alpha R^\star & \frac{(1-\eta)\beta_1^\alpha \lambda^\alpha}{\mu^\alpha + q^\alpha R^\star} & 0 & 0 & 0 \\ 0 & 0 & k^\alpha & -c^\alpha & 0 & 0 & 0 \\ q^\alpha R^\star & 0 & 0 & 0 & -d^\alpha & 0 & \frac{q^\alpha \lambda^\alpha}{\mu^\alpha + q^\alpha R^\star} \\ 0 & 0 & p^\alpha R^\star & 0 & 0 & -a^\alpha & 0 \\ 0 & 0 & 0 & 0 & 0 & 0 & -g^\alpha \end{pmatrix}$$

Expanding, $\det(\lambda^p I_7 - M_2) = 0$, where I_7 is the 7×7 identity matrix, we have the following equation in terms of λ:

$$(\lambda^p + \mu^\alpha + q^\alpha R^\star)(\lambda^p + d^\alpha)(\lambda^p + a^\alpha)(\lambda^p + g^\alpha)$$

$$\left[\lambda^{3p} + \left(a_L^\alpha + \mu_L^\alpha + a^\alpha + p^\alpha R^\star + c^\alpha - (1-\eta)\beta_2^\alpha \frac{\lambda^\alpha}{\mu^\alpha + q^\alpha R^\star}\right)\lambda^{2p} + \right.$$

$$(c^\alpha(a_L^\alpha + \mu_L^\alpha + a^\alpha + p^\alpha R^\star) + (a_L^\alpha + \mu_L^\alpha)(a^\alpha + p^\alpha R^\star)$$

$$\left. -(1-\eta)\beta_2^\alpha \frac{\lambda^\alpha}{\mu^\alpha + q^\alpha R^\star}(c^\alpha + \mu_L^\alpha) - a_L^\alpha \beta_2^\alpha \frac{\lambda^\alpha}{\mu^\alpha + q^\alpha R^\star} - k^\alpha(1-\eta)\beta_1^\alpha \frac{\lambda^\alpha}{\mu^\alpha + q^\alpha R^\star}\right)\lambda^p$$

$$+(a_L^\alpha + \mu_L^\alpha) + c^\alpha(a^\alpha + p^\alpha R^\star)$$

$$\left. -\beta_2^\alpha \frac{\lambda^\alpha}{\mu^\alpha + q^\alpha R^\star}c^\alpha(\mu_L^\alpha(1-\eta) + a_L^\alpha) - \beta_1^\alpha \frac{\lambda^\alpha}{\mu^\alpha + q^\alpha R^\star}k^\alpha((1-\eta)\mu_L^\alpha + a_L^\alpha)\right] = 0 \quad (8)$$

Now, the arguments of the roots of the equation, $\lambda^p + \mu^\alpha + q^\alpha R^\star = 0$, $\lambda^p + d^\alpha = 0$, $\lambda^p + a^\alpha = 0$, and $\lambda^p + g^\alpha = 0$, are given by:

$$\arg(\lambda_j) = \frac{\pi}{p} + j\frac{2\pi}{p} > \frac{\pi}{M} > \frac{\pi}{2M}$$

where $j = 0, 1, .., (p-1)$.

Using Descartes' rule of signs, we find that there is exactly one sign change of the equation:

$$\lambda^{3p} + \left(a_L^\alpha + \mu_L^\alpha + a^\alpha + p^\alpha R^\star + c^\alpha - (1-\eta)\beta_2^\alpha \frac{\lambda^\alpha}{\mu^\alpha + q^\alpha R^\star}\right)\lambda^{2p}$$

$$+ \left(c^\alpha(a_L^\alpha + \mu_L^\alpha + a^\alpha + p^\alpha R^\star) + (a_L^\alpha + \mu_L^\alpha)(a^\alpha + p^\alpha R^\star)\right.$$

$$\left. -(1-\eta)\beta_2^\alpha \frac{\lambda^\alpha}{\mu^\alpha + q^\alpha R^\star}(c^\alpha + \mu_L^\alpha) - a_L^\alpha \beta_2^\alpha \frac{\lambda^\alpha}{\mu^\alpha + q^\alpha R^\star} - k^\alpha(1-\eta)\beta_1^\alpha \frac{\lambda^\alpha}{\mu^\alpha + q^\alpha R^\star}\right)\lambda^p \quad (9)$$

$$+(a_L^\alpha + \mu_L^\alpha) + c^\alpha(a^\alpha + p^\alpha R^\star)\left[1 - R_c^d\right] = 0$$

for $R_c^d > 1$. Thus, there is exactly one positive real root of the aforesaid equation for which the argument is less than $\frac{\pi}{2M}$. As such, we conclude that, if $R_c^d > 1$, the disease-free equilibrium P_1 of the system (1) is unstable. □

4. Global Stability of the Disease-Free Equilibrium

In this section, we compute the global stability of the disease-free equilibrium P_1 of the model (1). Following Castillo & Chavéz [26], we rewrite model (1) as:

$$\frac{d^\alpha X}{dt^\alpha} = F(X, Z)$$
$$\frac{d^\alpha Z}{dt^\alpha} = G(X, Z), \qquad G(X, 0) = 0 \quad (10)$$

where $X = (T, T_R, R)$ and $Z = (L, I, V, I_R)$, with $X \in \mathbf{R}_+^3$ being the number of uninfected and non-susceptible CD4$^+$ T cells and drugs, and $Z \in \mathbf{R}_+^4$ denoting the number of latent and infected CD4$^+$ T cells, virus, and non-infectious CD4$^+$ T cells.

The disease-free equilibrium is written as $U = (X^\star, 0)$, where $X^\star = (T_1, T_{R_1}, R_1) = \left(\frac{\lambda^\alpha}{\mu^\alpha + q^\alpha R^\star}, \frac{q^\alpha T_1 R^\star}{d^\alpha}, R^\star\right)$.

The conditions (H_1) and (H_2) must be met to guarantee the global asymptotic stability of the disease-free equilibrium of the model (1):

$(H_1):$ For $\frac{d^\alpha X}{dt^\alpha} = F(X,0)$, $X^\star$ is globally asymptotically stable

$(H_2):$ $G(X,Z) = AZ - \hat{G}(X,Z)$, $\hat{G} \geq 0$, for $(X,Z) \in Y_1$ (11)

where $A = D_Z G(X^\star, 0)$ can be written in the form $A = M - D$, where $M \geq 0$ ($m_{ij} \geq 0$) and $D > 0$ is a diagonal matrix. Y_1 is the region where the model makes biological sense. If the system (10) satisfies the conditions in (11) the following theorem holds [26].

Theorem 1. *The fixed point $U = (X^\star, 0)$ is a globally asymptotically stable equilibrium of the system (10) provided that $R_c^d < 1$ and that the assumptions in (11) are satisfied.*

Proof. Let

$$F(X,0) = \begin{bmatrix} \lambda^\alpha - \mu^\alpha T - q^\alpha T R \\ q^\alpha TR - d^\alpha T_R \\ R_k^\alpha - g^\alpha R \end{bmatrix} \quad (12)$$

and

$$A = \begin{pmatrix} -(a_L^\alpha + \mu_L^\alpha) & \eta \beta_2^\alpha T_1 & \eta \beta_1^\alpha T_1 & 0 \\ a_L^\alpha & (1-\eta)\beta_2^\alpha T_1 - (a^\alpha + p^\alpha R^\star) & (1-\eta)\beta_1^\alpha T_1 & 0 \\ 0 & k^\alpha & -c^\alpha & 0 \\ 0 & p^\alpha R^\star & 0 & -a^\alpha \end{pmatrix} \quad (13)$$

and

$$\hat{G}(X,Z) = \begin{pmatrix} \hat{G}_1(X,Z) \\ \hat{G}_2(X,Z) \\ \hat{G}_3(X,Z) \\ \hat{G}_4(X,Z) \end{pmatrix} = \begin{pmatrix} \eta \beta_1^\alpha V T_1 \left(1 - \frac{T}{T_1}\right) + \eta \beta_2^\alpha I T_1 \left(1 - \frac{T}{T_1}\right) \\ (1-\eta)\beta_1^\alpha V T_1 \left(1 - \frac{T}{T_1}\right) + (1-\eta)\beta_2^\alpha I T_1 \left(1 - \frac{T}{T_1}\right) \\ 0 \\ 0 \end{pmatrix} \quad (14)$$

All conditions are satisfied, thus U_0 is globally asymptotically stable. □

5. Numerical Results

We simulate the model (1) for different values of the order of the fractional derivative, α and for clinically relevant parameters. We have applied the Predictor–Evaluator–Corrector–Evaluator PECE method of Adams–Bashford–Moulton type [27]. The parameters used in the simulations, based on [8,28], are: $\lambda = 100$ µL^{-1} day$^{-\alpha}$, $\mu = 0.1$ day$^{-\alpha}$, $a = 0.3$ day$^{-\alpha}$, $c = 3$ day$^{-\alpha}$, $k = 210$ day$^{-\alpha}$, $\beta_1 = 1.5 \times 10^{-5}$ day$^{-\alpha}$, $\beta_2 = 1.5 \times 10^{-4}$ day$^{-\alpha}$, $p = 0.1$ µM^{-1} day$^{-\alpha}$, $q = 0.1$ µM^{-1} day$^{-\alpha}$, $g = 2.7726$ day$^{-\alpha}$, $\eta = 0.03$, $a_L = 0.1$ day$^{-\alpha}$, $\mu_L = 4 \times 10^{-3}$ day$^{-\alpha}$, $R_k = 2.5$, $\tau = 0.5$ day$^\alpha$, and the initial conditions are: $T(0) = 1000$, $L(0) = I(0) = T_R(0) = I_R(0) = 0$, $V(0) = 50$ and $R(0) = 2.5$.

In Figures 1 and 2, we consider model (1) without PEP, for different values of the order of the fractional derivative, α. The concentration of CD4$^+$ T cells decreases over time and with α. This suggests that the infection is more severe as α is lowered. This pattern is supported by the graphs in Figure 2, where it is observed a ratio of healthy T cells to total T cells starting with 0.5 for $\alpha = 1.0$, and decreasing for $\alpha = 0.9$ and $\alpha = 0.7$. Moreover, this ratio points to chronic infection, as the disease evolves, for all α.

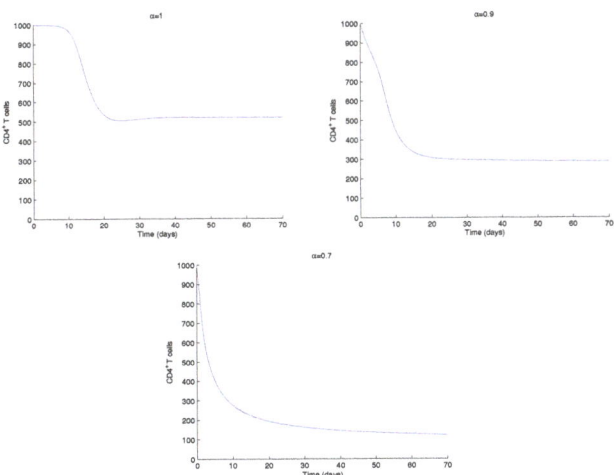

Figure 1. Dynamics of the CD4$^+$ T cells, T, of system (1) without PEP for $\alpha = 1$ (**top left**), $\alpha = 0.9$ (**top right**) and $\alpha = 0.7$ (**bottom**). Parameter values and initial conditions in the text.

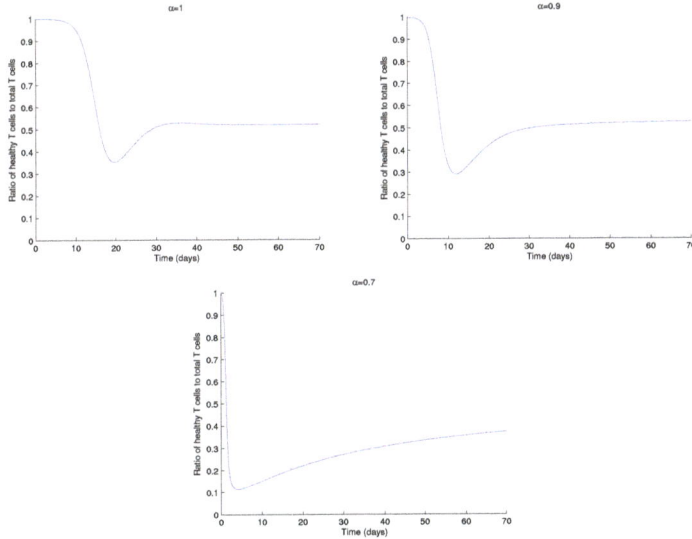

Figure 2. Ratio of healthy CD4$^+$ T cells, T, to total CD4$^+$ T cells, $T + L + I$, of system (1) without PEP for $\alpha = 1$ (**top left**), $\alpha = 0.9$ (**top right**) and $\alpha = 0.7$ (**bottom**). Parameter values and initial conditions in the text.

In Figures 3 and 4, we plot the dynamics of the drug R and of the basic reproduction number R_c^d, for different values of the order of the fractional derivative, α. These figures show that the dosage of the drug is important for controlling HIV infection, since R_c^d varies with R. As R increases, smaller values of R_c^d are observed, which indicate less infection. Moreover, the value of the fractional derivative, α, may also contribute to controlling the severity of the infection, since smaller values of R_c^d are observed with decreasing α.

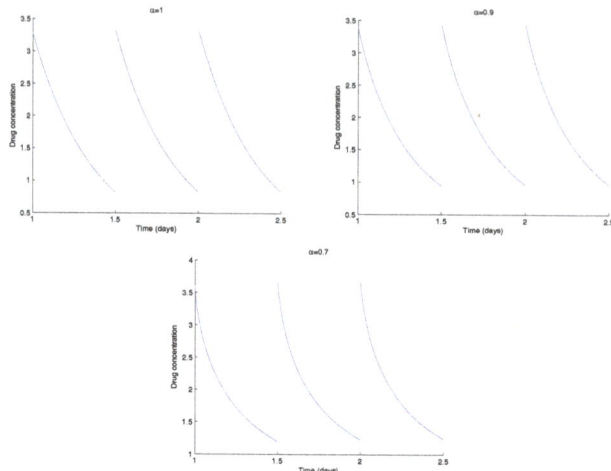

Figure 3. Drug concentration in the plasma, R, given by system (1) for $\alpha = 1$ (**top left**), $\alpha = 0.9$ (**top right**) and $\alpha = 0.7$ (**bottom**). Parameter values and initial conditions in the text.

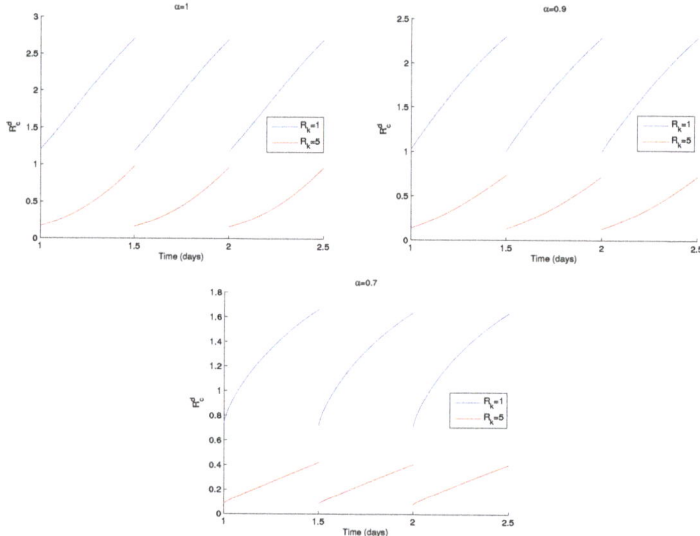

Figure 4. Basic reproduction number R_c^d of system (1) for $\alpha = 1$ (**top left**), $\alpha = 0.9$ (**top right**) and $\alpha = 0.7$ (**bottom**). Parameter values and initial conditions in the text.

Figure 5 depicts the HIV viral load for a dosage $R_k = 5$ and dosing interval $\tau = 0.5$ day, for different values of the order of the fractional derivative, α. As it is shown, the dosage of the drug and the dosing interval are sufficient to control the infection, with the viral load going asymptotically to zero. Similar patterns are seen for all values of α, with higher initial viral load for smaller α, but faster velocity of convergence.

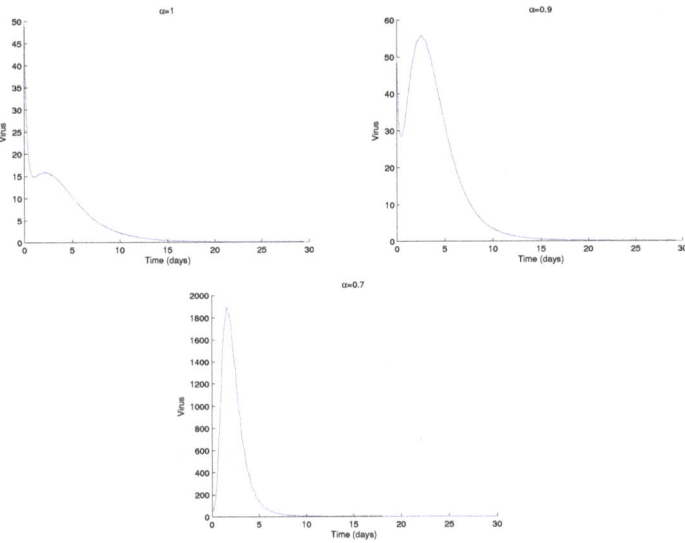

Figure 5. HIV viral-load, V, of the system (1) for $\alpha = 1$ (**top left**), $\alpha = 0.9$ (**top right**) and $\alpha = 0.7$ (**bottom**). Parameter values and initial conditions in the text, except $R_k = 5$.

Figure 6 shows the ratio of the infected CD4$^+$ T cells to total CD4$^+$ T cells in the presence and absence of PEP, with low drug dosage, for different values of the order of the fractional derivative, α. The ratio of infected to total CD4$^+$ T cells is always smaller when patients are under PEP, when compared to the case without treatment.

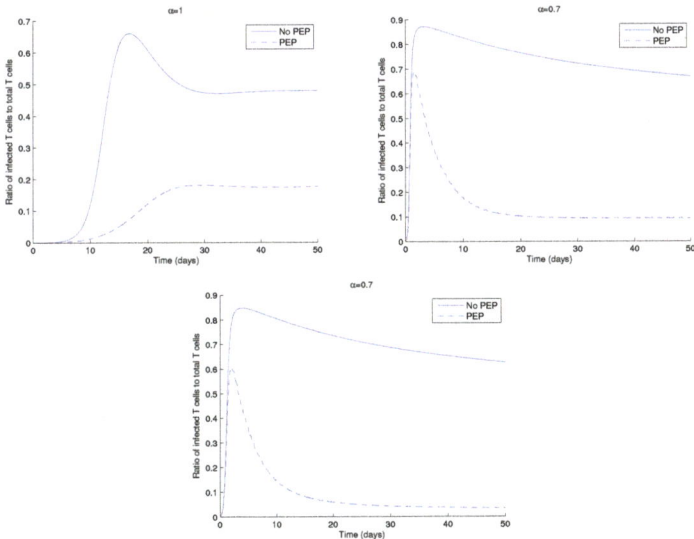

Figure 6. Ratio of infected CD4$^+$ T cells, $I + I_R$, to total CD4$^+$ T cells, $T + L + I + T_R + I_R$ of system (1) for $\alpha = 1$ (**top left**), $\alpha = 0.9$ (**top right**) and $\alpha = 0.7$ (**bottom**). Parameter values and initial conditions in the text, except $R_k = 1$.

In the next figures, we study the effect of different treatment strategies in the dynamics of HIV. We start in Figure 7 with two different treatment strategies: drug perfect adherence and drug therapy breaks. The last strategy consists of intervals (days) in which the therapy is stopped ($\Delta R_k = 0$) followed by intervals where there is perfect drug adherence. Perfect adherence therapy consists of taking a dosage $\Delta R_k = R_k$ for all $t = t_k$. In Figure 7, the drug therapy breaks consist of stopping drug application for two days, followed by drug perfect adherence strategy for five days. It is observed that the elimination of HIV from the body takes longer for drug therapy breaks. This is seen for all α. Moreover, despite a higher initial peak, the asymptotic HIV viral load is reached faster for smaller α.

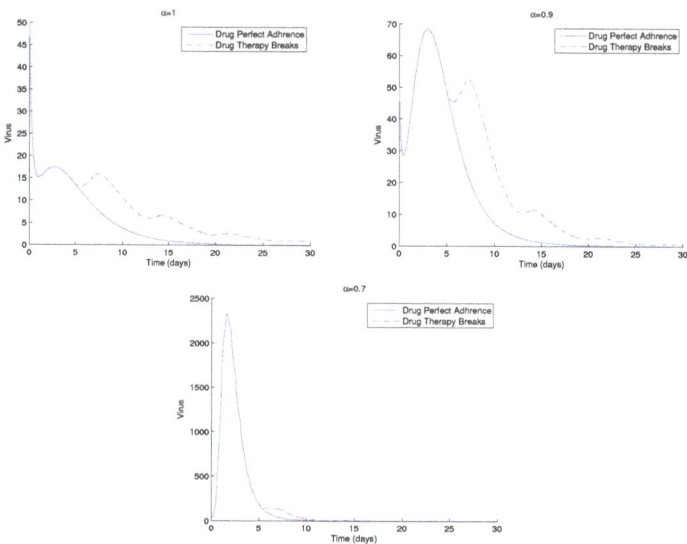

Figure 7. HIV viral load of system (1) for two different therapy strategies and $\alpha = 1$ (**top left**), $\alpha = 0.9$ (**top right**) and $\alpha = 0.7$ (**bottom**). Parameter values and initial conditions are in the text, except $R_k = 4.5$. For more information, see text.

Figure 8 shows another example of the dynamics of the HIV, this time for three distinct treatment strategies: without treatment, drug perfect adherence, and drug therapy breaks. The intervals for the drug therapy breaks are in this case as follows. Five days of no drug administration, which are followed by five more days of perfect drug adherence strategy. The model provides oscillating solutions for the case of drug therapy breaks, as is seen in the figure.

In Figure 9 we show the dynamics of HIV for increasing values of the cell to cell transmission rate, β_2, for three treatment strategies: no treatment, drug perfect adherence and drug therapy breaks, and for varying α. The drug therapy breaks strategy stops drug application for 15 days, followed by another 15 days of perfect drug adherence strategy. We observe higher peaks of the viral load and the corresponding curve, in the case of drug therapy breaks, is between the curves of no treatment and drug perfect adherence. This behaviour is repeated for all α.

The simulations of the model reveal that a combination of sufficient drug dosage and drug frequency may induce better efficacy of PEP. Drug perfect adherence strategy is always better than the other two. Nevertheless, one must think about the side effects of ART, though their toxicity has been reduced as medicine evolves and new treatment options appear.

We now proceed with the simulation of the effect of the latent reservoir in the dynamics of HIV infection, under the conditions of Figure 8. We consider three treatment strategies: without treatment, drug perfect adherence, and drug therapy breaks. The intervals for the drug therapy breaks consist of 10 days. In the first five days, the drug is halted, whereas for the last five days, the perfect drug adherence strategy is applied. The difference from Figure 8 is in the value of η, which represents the proportion of latently infected CD4$^+$ T cells. The value of η is reduced from 0.03 to 0.01. Figure 10 shows slight higher peaks of HIV for $\eta = 0.01$, in particular, for smaller values of α.

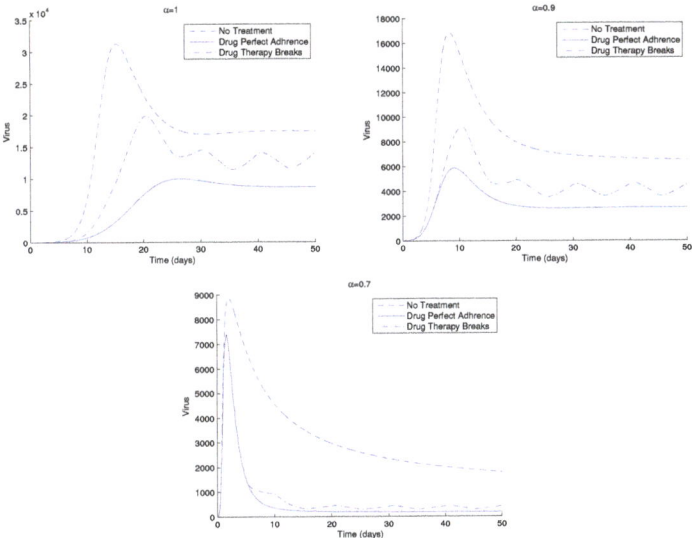

Figure 8. HIV, viral load, V, of system (1) for three different therapy strategies and for $\alpha = 1$ (**top left**), $\alpha = 0.9$ (**top right**) and $\alpha = 0.7$ (**bottom**). Parameter values and initial conditions in the text, except $R_k = 1$. For more information, see text.

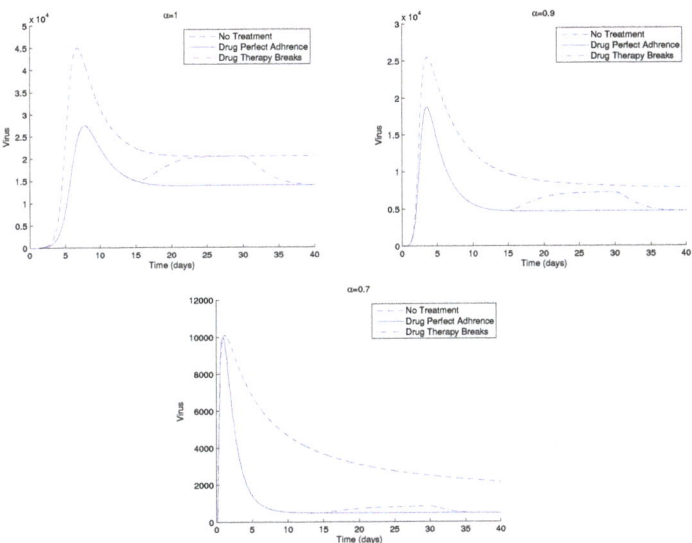

Figure 9. HIV, viral load, V, of system (1) for three different therapy strategies and for $\alpha = 1$ (**top left**), $\alpha = 0.9$ (**top right**) and $\alpha = 0.7$ (**bottom**). Parameter values and initial conditions in the text, except $R_k = 1$ and $\beta_2 = 0.0015$.

In Figure 11, we plot the viral load for two values of η, the fraction of latent infected cells. We note that the asymptotic value of the virus is the same for all α. Nevertheless, there are subtle changes in the dynamics of the virus. In the transient are observed smaller values of HIV viral load for $\eta = 0.03$, whereas in the asymptotic value there is a switch in this behaviour, and higher values of HIV are seen for $\eta = 0.03$. This may be explained as follows. When $\eta = 0.03 > \eta = 0.01$, there are more latently infected cells in the body. If these cells encounter an antigen or are exposed to specific cytokines or chemokines, they become actively infected by proviral transcription. The latter causes viral rebound if a patient stops ART. This happens earlier for smaller values of α.

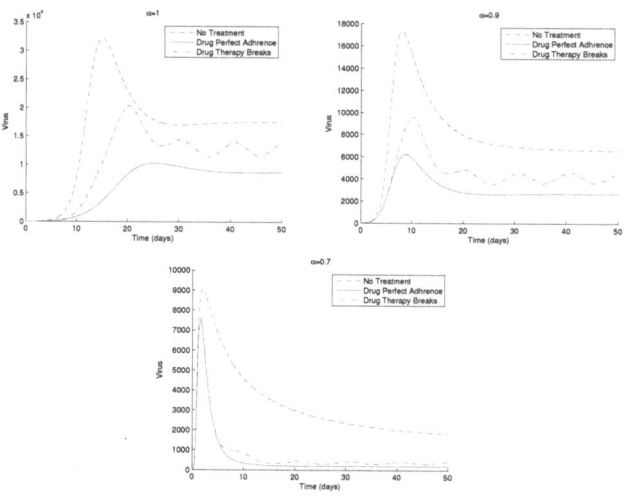

Figure 10. HIV, viral load, V, of system (1) for different therapy strategies and for $\alpha = 1$ (**top left**), $\alpha = 0.9$ (**top right**) and $\alpha = 0.7$ (**bottom**). Parameter values and initial conditions in the text, except $R_k = 1$ and $\eta = 0.01$. For more information, see text.

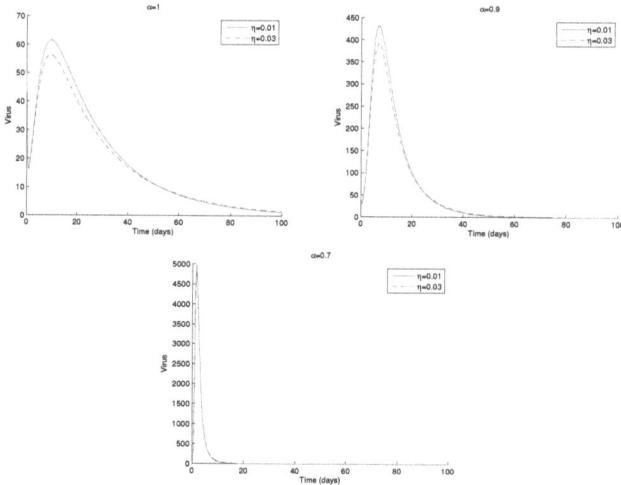

Figure 11. HIV, viral load, V, of system (1) for two values of η, the proportion of latently infected cells, and for $\alpha = 1$ (**top left**), $\alpha = 0.9$ (**top right**) and $\alpha = 0.7$ (**bottom**). Parameter values and initial conditions in the text, except for $R_k = 2.5$. For more information, see text.

6. Conclusions

We propose a model to study the effect of PEP and of the latent reservoir in the dynamics of HIV infection. We find that specific dosages and intervals are extremely important to control the infection. Moreover, we find that the latent reservoir may influence the dynamics of HIV, though slightly. This is understandable from a clinical point of view since the effect of the latent reservoir takes time to be felt and PEP is considered only in the first 28 days after exposure. After that, the person must be evaluated clinically to assess the adequacy of the treatment. The order of the fractional derivative, α, seems to help control the infection in the presence of PEP and increases the severity of infection when there is no PEP. We observe a somewhat 'synergistic' relation between PEP and α. The FO derivative may also help to distinguish other traits (age, immune system response, genetic profile), and this may help to devise better therapeutic regimens, that could improve patients' quality of life, either by diminishing the burden of the therapy or increasing the life span. Moreover, since HIV anti-retroviral therapy (ART) is extremely expensive, an 'optimal' (in the sense of more adjusted to each patient) therapy could also imply a reduction in the economic burden of HIV, especially in poor countries, such as the ones included in sub-Saharan Africa. Future work will focus on deepening these and other issues arising from the model.

Author Contributions: All authors contributed equally to the work reported.

Acknowledgments: The authors were partially funded by the European Regional Development Fund through the program COMPETE and by the Portuguese Government through the FCT - Fundação para a Ciência e a Tecnologia under the project PEst-C/MAT/UI0144/2013. The research of AC was partially supported by an FCT grant with reference SFRH/BD/96816/2013.

Conflicts of Interest: The authors declare no conflict of interest.

References

1. CDC HIV Among Pregnant Women, Infants, and Children. Available online: https://www.cdc.gov/hiv/group/gender/pregnantwomen/index.html (accessed on 2 April 2019).
2. WHO. Available online: http://www.who.int/mediacentre/factsheets/fs360/en/ (accessed on 2 April 2019).
3. CDCP. Available online: https://www.cdc.gov/hiv/basics/pep.html (accessed on 2 April 2019).
4. Pinto, C.M.A. Persistence of low levels of plasma viremia and of the latent reservoir in patients under ART: A fractional-order approach. *Commun. Nonlinear Sci. Numer. Simul.* **2017**, *43*, 251–260. [CrossRef]
5. Rong, L.; Perelson, A.S. Modeling HIV persistence, the latent reservoir, and viral blips. *J. Theor. Biol.* **2009**, *260*, 308–331. [CrossRef] [PubMed]
6. Rong, L.; Perelson, A.S. Modeling latently infected cell activation: viral and latent reservoir persistence, and viral blips in HIV-infected patients on potent therapy. *PLoS Comput. Biol.* **2009**, *5*, e1000533. [CrossRef] [PubMed]
7. Hill, A.L.; Rosenbloom, D.I.S.; Nowak, M.A.; Siliciano, R.F. Insight into treatment of HIV infection from viral dynamics models. *Immunol. Rev.* **2018**, *285*, 9–25. [CrossRef] [PubMed]
8. Lou, J.; Chen, L.; Ruggeri, T. An impulsive differential model on post exposure prophylaxis to HIV-1 exposed individual. *J. Biol. Syst.* **2009**, *17*, 659–683. [CrossRef]
9. Conway, J.M.; Konrad, B.P.; Coombs, D. Stochastic analysis of Pre- and Postexposure prophylaxis against HIV infection. *SIAM J. Appl. Math.* **2013**, *73*, 904–928. [CrossRef]
10. Kim, S.B.; Yoon, M.; Ku, N.S.; Kim, M.H.; Song, J.E.; Ahn, J.Y.; Jeong, S.J.; Kim, C.; Kwon, H.D.; Lee, J.; et al. Mathematical Modeling of HIV Prevention Measures Including Pre-Exposure Prophylaxis on HIV Incidence in South Korea. *PLoS ONE* **2014**, *9*, e90080. [CrossRef] [PubMed]
11. Pinto, C.M.A.; Carvalho, A.R.M. The impact of pre-exposure prophylaxis (PrEP) and screening on the dynamics of HIV. *J. Comput. Appl. Math.* **2018**, *339*, 231–244. [CrossRef]
12. Carvalho, A.R.M.; Pinto, C.M.A.; Baleanu, D. HIV/HCV coinfection model: A fractional-order perspective for the effect of the HIV viral load. *Adv. Differ. Equat.* **2018**, *2018*, 1–22.

13. Kilbas, A.A.; Srivastava, H.M.; Trujillo, J.J. *Theory and Applications of Fractional Differential Equations, North-Holland Mathematical Studies*; Elsevier (North-Holland) Science Publishers: Amsterdam, The Netherlands, 2006; Volume 204.
14. Pinto, C.M.A.; Carvalho, A.R.M. Fractional complex-order model for HIV infection with drug resistance during therapy. *J. Vib. Control* **2016**, *22*, 2222–2239. [CrossRef]
15. Samko, S.; Kilbas, A.; Marichev, O. *Fractional Integrals and Derivatives: Theory and Applications*; Gordon and Breach Science Publishers: London, UK, 1993.
16. Sweilam, N.H.; Abou Hasan, M.M.; Baleanu, D. New studies for general fractional financial models of awareness and trial advertising decisions. *Chaos Solitons Fractals* **2017**, *104*, 772–784. [CrossRef]
17. Téjado, I.; Valério, D.; Pérez, E.; Valério, N. Fractional calculus in economic growth modelling: the Spanish and Portuguese cases. *Int. J. Dyn. Control* **2017**, *5*, 208–222. [CrossRef]
18. Arafa, A.A.M.; Rida, S.Z.; Khalil, M. A fractional-order model of HIV infection: Numerical solution and comparisons with data of patients. *Int. J. Biomath.* **2014**, *7*, 1450036. [CrossRef]
19. Arafa, A.A.M.; Rida, S.Z.; Khalil, M. Fractional modeling dynamics of HIV and CD4+ T-cells during primary infection. *Nonlinear Biomed. Phys.* **2012**, *6*, 1. [CrossRef] [PubMed]
20. Ding, Y.; Ye, H. A fractional-order differential equation model of HIV infection of CD4+ T-cells. *Math. Comput. Model.* **2009**, *50*, 386–392. [CrossRef]
21. Yan, Y.; Kou, C. Stability analysis for a fractional differential model of HIV infection of CD4+ T-cells with time delay. *Math. Comput. Simul.* **2012**, *82*, 1572–1585. [CrossRef]
22. Ionescu, C.; Copot, D.; De Keyser, R. Modelling Doxorubicin effect in various cancer therapies by means of fractional calculus. In Proceedings of the 2016 American Control Conference (ACC), Boston, MA, USA, 6–8 July 2016; pp. 1283–1288.
23. Copot, D.; Magin, R.L.; De Keyser, R.; Ionescu, C. Data-driven modelling of drug tissue trapping using anomalous kinetics. *Chaos Solitons Fractals* **2017**, *102*, 441–446. [CrossRef]
24. Driessche, P.; Watmough, P. Reproduction numbers and sub-threshold endemic equilibria for compartmental models of disease transmission. *Math. Biosci.* **2002**, *180*, 29–48. [CrossRef]
25. Matignon, D. Stability results for fractional differential equations with applications to control processing. *Comput. Eng. Syst. Appl.* **1996**, *2*, 963–968.
26. Castillo-Chavez, C.; Feng, Z.; Huang, W. On the computation of Ro and its role on global stability. In *Mathematical Approaches for Emerging and Reemerging Infectious Diseases: An Introduction*; Kassem, T., Roudenko, S., Castillo-Chavez, C., Eds.; Springer: New York, NY, USA, 2002.
27. Diethelm, K.; Freed, A.D. The Frac PECE subroutine for the numerical solution of differential equations of fractional order. In *Forschung und Wissenschaftliches Rechnen 1998*; Heinzel, S., Plesser, T., Eds.; Gessellschaft fur Wissenschaftliche Datenverarbeitung: Gottingen, Germany, 1999; pp. 57–71.
28. Hadjiandreou, M.M.; Conejeros, R.; Wilson, D.I. Long-term HIV dynamics subject to continuous therapy and structured treatment interruptions. *Chem. Eng. Sci.* **2009**, *64*, 1600–1617. [CrossRef]

© 2019 by the authors. Licensee MDPI, Basel, Switzerland. This article is an open access article distributed under the terms and conditions of the Creative Commons Attribution (CC BY) license (http://creativecommons.org/licenses/by/4.0/).

Article

Impact of Fractional Calculus on Correlation Coefficient between Available Potassium and Spectrum Data in Ground Hyperspectral and Landsat 8 Image

Chengbiao Fu [1,2], Shu Gan [1,*], Xiping Yuan [1], Heigang Xiong [3] and Anhong Tian [1,2]

1. Faculty of Land Resource Engineering, Kunming University of Science and Technology, Kunming 650093, China; fucb@mail.qjnu.edu.cn (C.F.); YXP@kmust.edu.cn (X.Y.); tianah@mail.qjnu.edu.cn (A.T.)
2. College of Information Engineering, Qujing Normal University, Qujing 655011, China
3. College of Applied Arts and Science, Beijing Union University, Beijing 100083, China; heigang@buu.edu.cn
* Correspondence: sgbf@kmust.edu.cn

Received: 1 May 2019; Accepted: 22 May 2019; Published: 28 May 2019

Abstract: As the level of potassium can interfere with the normal circulation process of biosphere materials, the available potassium is an important index to measure the ability of soil to supply potassium to crops. There are rarely studies on the inversion of available potassium content using ground hyperspectral remote sensing and Landsat 8 multispectral satellite data. Pretreatment of saline soil field hyperspectral data based on fractional differential has rarely been reported, and the corresponding relationship between spectrum and available potassium content has not yet been reported. Because traditional integer-order differential preprocessing methods ignore important spectral information at fractional-order, it is easy to reduce the accuracy of inversion model. This paper explores spectral preprocessing effect based on Grünwald–Letnikov fractional differential (order interval is 0.2) between zero-order and second-order. Field spectra of saline soil were collected in Fukang City of Xinjiang. The maximum absolute of correlation coefficient between ground hyperspectral reflectance and available potassium content for five mathematical transformations appears in the fractional-order. We also studied the tendency of correlation coefficient under different fractional-order based on seven bands corresponding to the Landsat 8 image. We found that fractional derivative can significantly improve the correlation, and the maximum absolute of correlation coefficient under five spectral transformations is in Band 2, which is 0.715766 for the band at 467 nm. This study deeply mined the potential information of spectra and made up for the gap of fractional differential for field hyperspectral data, providing a new perspective for field hyperspectral technology to monitor the content of soil available potassium.

Keywords: field spectrum; fractional calculus; desert soil; available potassium; correlation analysis

1. Introduction

Precision agricultural variable fertilizer depends on the understanding of soil nutrient distribution in farmland. Acquiring soil nutrient is the basis for implementing precision agriculture. Available potassium plays an important role in supplying potassium for crops, and it is a necessary nutrient for plant growth and development [1,2]. Excessive potassium content in the soil can result in waste of resources, soil environmental pollution, water pollution, and imbalance of soil nutrient distribution [3]. The rapid and accurate nondestructive determination of soil available potassium content is of great significance for the development of agriculture [4–7]. Traditional laboratory chemical detection methods have the problems of being expensive and time-consuming, while hyperspectral analysis technology has the advantages of convenience, speed, and high precision [8–10]. Visible, near-infrared and mid-infrared spectroscopy technologies have been widely applied in soil science.

In recent years, domestic and foreign scholars have conducted extensive research on soil salinity, moisture, organic matter, and total nitrogen in different types of ecosystems, such as wetlands, forests, grasslands, and farmlands in arid and semi-arid regions [11–15]. There is less research on available potassium content [16]. Liu et al. [17] adopted visible/short-wave near-infrared spectroscopy to measure soil available nitrogen and available potassium. They introduced first-order differential algorithm for spectral pretreatment, and their simulation showed that the model built by least squares support vector machine (LS-SVM) combined with 1-order differential has higher precision. However, the hyperspectral inversion models established for available potassium are mainly constructed based on 1-order or 2-order derivative for spectral reflectance, reciprocal, and logarithm. However, related research points out that traditional integer-order differential transformation ignores the gradual fractional differential information [18,19], especially for high-dimensional data sources such as hyperspectral images with massive information, which may cause some information to be lost or be difficult to extract, and restrict the modeling accuracy.

Fractional calculus theory is a mathematical problem for studying the properties of differential and integral operators of any order and its application. First proposed in 1695, its development is almost in synchronization with the theory of integer-order calculus. However, theoretical research is limited to pure mathematics, and it is not closely related to real life. At the end of last century, with the rapid development of science and technology and the increasing complexity of research issues, fractional calculus has been rapidly developed and applied to many fields [20–23], such as fluid mechanics, viscoelastic mechanics, electrical conduction in biological systems, robot control, chaos phenomena, molecular spectroscopy, etc. At the same time, research in these application fields has also accelerated the development of fractional calculus theory.

In the field of spectral analysis, Schmitt [24] introduced fractional derivative into diffuse reflectance spectroscopy processing and found that it can effectively eliminate baseline drift, shift, etc., and separate overlapping peaks. At the same time, order choice of fractional derivative is more flexible, providing a broader space for band selection. Zheng et al. [25] used Savitzky–Golay (SG) fractional derivative to preprocess near infrared spectra based on corn, wheat, and diesel, and conducted quantitative regression analysis of corresponding properties. They found that fractional prediction effect of non-concentration indicators such as viscosity, density, and hardness was better than that of integer derivatives. Zhang et al. [26] applied fractional differentials to the pretreatment of hyperspectral data and used partial least squares regression (PLSR) to verify the model accuracy of saline soils. The logarithm reciprocal transformation at 1.2-order was an optimal model, showing that fractional differentials could improve model inversion accuracy.

However, these fractional differential studies measure spectral reflectance in an ideal indoor environment, focusing on the research of salinity and organic matter content, and failing to consider field spectrum tests that are in line with actual conditions. At present, the application of fractional differential algorithm in the field of available potassium content is still lacking. Thus, this study collected desert soils located in Fukang City of Xinjiang as research target, and measured field hyperspectral data of soil samples. We explored the effect of Grünwald–Letnikov fractional differential on the pretreatment of field hyperspectral data, and studied the correlation coefficient between available potassium and soil reflectance spectra. The methods used in this study could enrich soil hyperspectral data preprocessing methods, and provide scientific support for local precision agriculture.

2. Experiment Procedure

2.1. Study Area

The research area belongs to middle temperate continental arid climate, which is located in the northern part of the Tianshan Mountains and the southern margin of Junggar Basin (87°44′–88°46′ E,

43°29′–45°45′ N), with an average elevation of 452 nm. The selected research area is not developed and utilized because it is far from the place where people live. It basically maintains the original ecological style. In this study, soil sample data collection was conducted from 9 to 23 May 2017. Five east–west sampling transects with a spacing of 600–800 m were installed from south to north in the study area. Five representative points were selected for each sampling line with a spacing of 300–500 m, collecting a total of 25 soil samples. The location of sample point is shown in Figure 1.

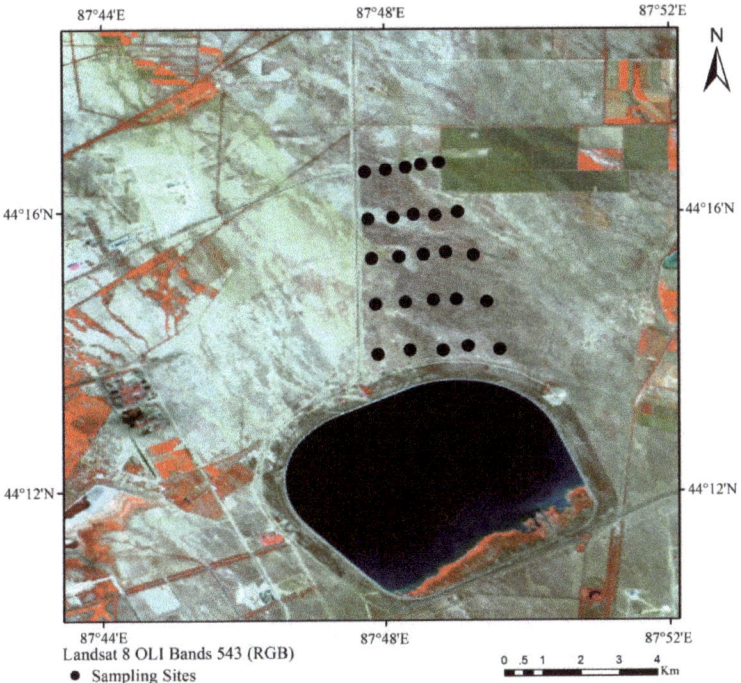

Figure 1. Location of sample point.

2.2. Field Hyperspectral Data Collection

The soil ground spectrum was measured using a portable Field Spec®3Hi-Res spectrometer (Analytica Spectra Devices., Inc., Boulder, CO, USA) with a spectral range of 350–2500 nm. To avoid the adverse effects of weather (e.g., poor sunlight, heavy cloud cover and strong wind, the experiment was conducted at 11:00–15:00 (local time), with little clouds and no wind. Soil sample data collection was conducted from 9 to 23 May 2017. The spectrometer was calibrated on the white board before each acquisition to remove the dark current. The probe with 25° field-of-view was used for spectral measurement, and it was 15 cm vertically above the soil sample. At about 1 cm around each sampling point, five representative sites were selected to collect the surface soil spectrum, and each position was repeatedly measured 10 times. The average of the 50 spectral curves was the measured spectral value of this sampling point. The spectra curves for 25 sampling points were measured in the study area.

2.3. Soil Sample Collection

Hyperspectral data testing and soil sample collection were conducted simultaneously in the same area at locations with flat topography and representative features around sampling point were selected as sampling units. Soil samples were acquired at 0–10 cm depth for the 25 sampling points. The latitude and longitude of the sample points were recorded by a handheld GPS, and they were numbered into bags for the laboratory. The soil available potassium content was tested by chemical professionals at Xinjiang Institute of Ecology and Geography, Chinese Academy of Sciences.

3. Spectral Data Preprocessing Method

Before the qualitative and quantitative analysis, proper pretreatment of spectrum reduces or even eliminates the impact of various non-target factors on the spectrum, and cleans the spectral information. Spectral preprocessing is very important method to establish a good and robust predictive model, and sometimes even plays a decisive role. Common spectral preprocessing methods include removing interference bands, smoothing algorithms, mathematical transformations, and differential algorithms.

3.1. Remove Interference Bands

In this study, the 350–399 nm and 2401–2500 nm bands with low signal-to-noise ratio were removed. At the same time, the bands located in the moisture absorption band have a great influence on the accuracy of spectral inversion, thus the bands of 1355–1410 nm and 1820–1942 nm also needed to be removed.

3.2. Savitzky–Golay Convolution Smoothing

Savitzky–Golay (SG) convolutional smoothing, also known as polynomial smoothing [27], was proposed by Savitzky and Golay. The SG convolution smoothing method is currently a relatively widely used spectral filtering method. The smoothing method combines a least-squares fitting with a moving window. First, a window with an odd number of points is taken. Then, each point of the spectrum in the window is taken as a polynomial. Finally, least square method is used to fit the polynomial coefficient value. The formula is defined as follows:

$$X_{k,\text{smooth}} = \frac{1}{H} \sum_{i=-w}^{+w} x_{k+i} h_i, \qquad (1)$$

where h_i is a smooth coefficient and can be obtained by polynomial fitting. H is a normalization factor and the calculation method is $H = \sum_{i=-w}^{+w} h_i$.

3.3. Fractional Calculus

Gamma function, also called generalized factorial, is often used in the definition and operation of fractional calculus. The integral form defined by the Gamma function is described as

$$\Gamma(z) = \int_0^\infty e^{-t} t^{z-1} dt, \ \text{Re}(z) > 0. \qquad (2)$$

The limit of the definition of gamma function can be expressed as follows

$$\Gamma(z) = \lim_{x \to \infty} \frac{n! n^z}{z(z+1)\cdots(z+n)}. \qquad (3)$$

Grünwald–Letnikov fractional derivative has been generalized from the definition of integer-order derivative. For any real number p, suppose that function f(x) has continuous derivative of m + 1 in the interval [a,t]. Then, p-order derivative for f(x) can be defined as follows:

$$_a D_t^p f(x) = \lim_{h \to 0} \frac{1}{h^p} \sum_{j=0}^{[(t-p)/h]} (-1)^j \binom{p}{j} f(x - jh), \qquad (4)$$

where h is the step size and [(t−p)/h] represents the integer part of (t−p)/h. When p is a positive real number, Equation (4) represents p-order derivative. If p is a negative real number, Equation (4) represents p-order integral.

1-order derivative of function f(x) is defined as

$$f'(x) = \lim_{h \to 0} \frac{f(x+h) - f(x)}{h}. \tag{5}$$

2-order derivative of function f(x) is described as

$$f''(x) = \lim_{h \to 0} \frac{f'(x+h) - f'(x)}{h} = \lim_{h \to 0} \frac{f(x+2h) - 2f(x+h) + f(x)}{h^2}. \tag{6}$$

If the derivative order of function f(x) is raised to higher order of p, then the p-order derivative of function f(x) is expressed as

$$f^{(p)}(x) = \lim_{h \to 0} \frac{1}{h^p} \sum_{m=0}^{p} (-1)^m \binom{p}{m} f(x - mh). \tag{7}$$

If we use Gamma function to replace the binomial coefficient of Equation (7) and extend the derivative order to a non-integer order, we can get the Grünwald–Letnikov fractional derivative in Equation (4). Since the re-sampling interval of ASD (Analytica Spectra Devices) spectrometer was 1 nm, in Equation (4), let h = 1, and then the derivative expression of v-order derivative for function f(x) can be deduced as follows:

$$\frac{d^v f(x)}{dx^v} \approx f(x) + (-v)f(x-1) + \frac{(-v)(-v+1)}{2}f(x-2) + \frac{(-v)(-v+1)(-v+2)}{6}f(x-3) + \ldots + \frac{\Gamma(-v+1)}{n!\Gamma(-v+n+1)}f(x-n). \tag{8}$$

In particular, when v = 1, 2, it is consistent with first-order and second-order derivative formulas of spectrum, respectively. From Equation (8), we can see that fractional derivatives have global and memory characteristics.

3.4. Spectral Mathematical Transformation

Before estimation model of surface parameters based on spectral reflectance is established, it is often necessary to perform nonlinear mathematical transformation for original spectral reflectance (R). The commonly used non-linear mathematical transformations include: root mean square transform ($\sqrt{R}$), reciprocal transform (1/R), logarithmic transformation (lgR), and logarithm reciprocal transformation (1/lgR). The main purpose is that linear relationship between spectral reflectance and surface parameters is transformed into a nonlinear relationship, a relatively simple linear regression analysis is performed to obtain approximately nonlinear results, and various forms of estimation models are established to improve the recognition accuracy. In addition, non-linear transformation can enhance spectral difference to some extent; it is convenient to distinguish the influence on spectrum caused by the difference of surface parameters. Spectral reflectance R and its four kinds of spectral transformation curves are shown in Figure 2.

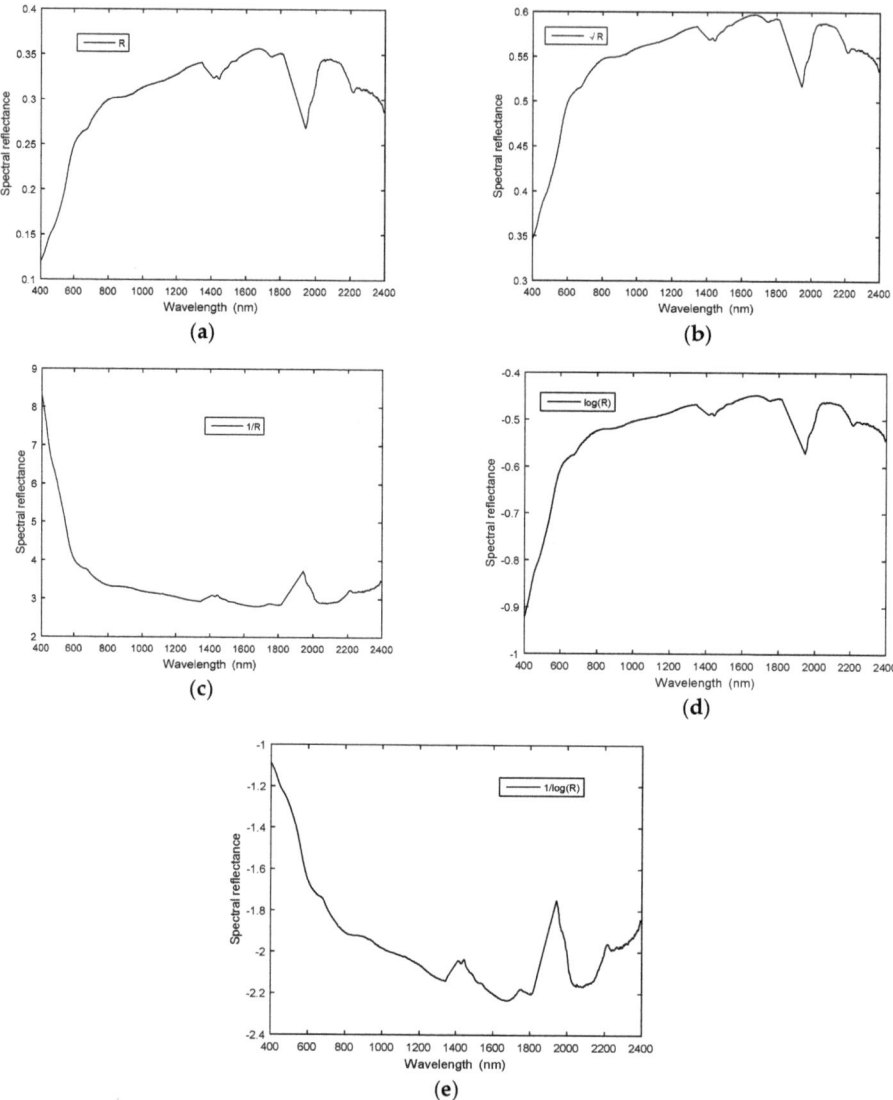

Figure 2. Spectral reflectance of soil and its four mathematical transformation forms: (**a**) R; (**b**) √R; (**c**) 1/R; (**d**) lgR; and (**e**) 1/lgR.

4. Simulation Results

4.1. Differential Calculation of Root Mean Square and Logarithm Reciprocal

To study the effects on spectral data by fractional differentials in detail, starting differential order is 0, termination differential order is 2, and order interval is 0.2. The results of differential calculation in the bands 1450 nm and 1650 nm of soil ground hyperspectral curve for root mean square transformation and logarithm reciprocal transformation are shown in Figure 3. Differential values of two spectral transformations gradually approach 0, as the order slowly ascends from 0-order to 1-order, fractional differential curve gradually approximates the first-order differential curve. When the order is gradually increased from 1-order to 2-order, fractional derivative curve slowly approaches the 2-order differential curve, which verifies the sensitivity of fractional derivative to some extent. In addition, it can be also seen in Figure 3c,d that the derivative value in the band 1450–1550 nm fluctuates greatly, while the derivative value in the band 1550–1650 nm is less fluctuating.

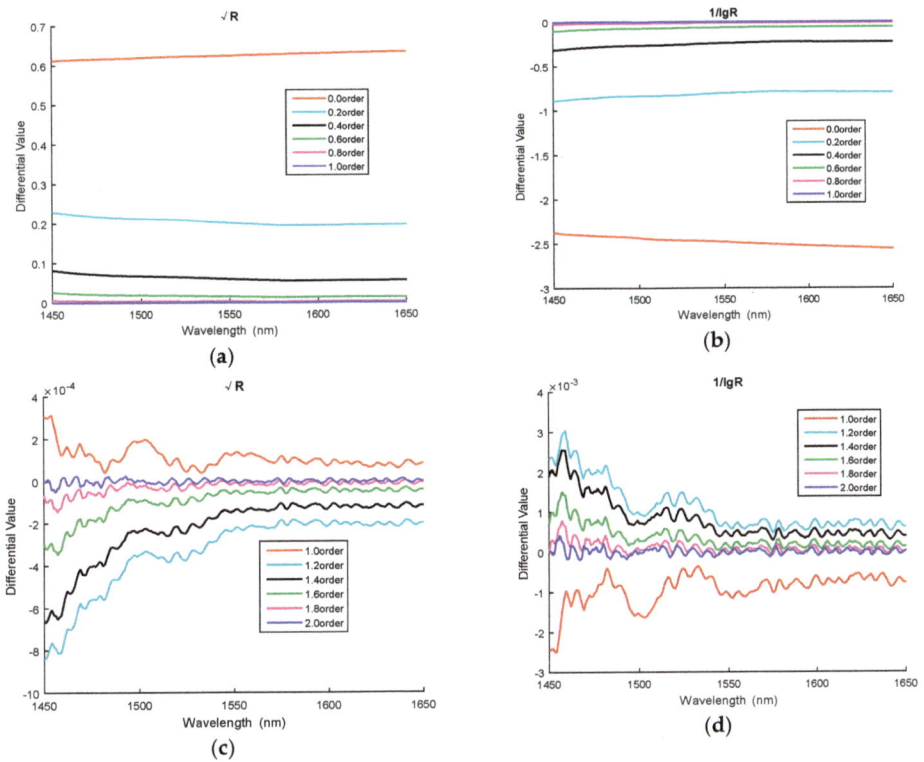

Figure 3. Fractional differential calculation of $\sqrt{R}$ and 1/lgR reflectance at 1450–1650 nm: (**a**) 0-order to 1-order of root mean square; (**b**) 0-order to 1-order of logarithm reciprocal; (**c**) 1-order to 2-order of root mean square; and (**d**) 1-order to 2-order of logarithm reciprocal.

4.2. Trends of Correlation Coefficients for Root Mean Square and Logarithm Reciprocal

Correlation analysis is a key step in spectral data preprocessing. When the correlation coefficient between available potassium and spectral signal passes the significance test, the corresponding band is likely to become the sensitive band, and the band reflectance can be used as the independent variable in the model to establish a reliable predictive model of available potassium content. In this paper, the significance test was carried out at 0.05 level, and the calculus was programmed in Matlab software (MathWorks, Natick, MA, USA) to calculate the correlation between the spectral reflectance and the available potassium content after root mean square and logarithmic inverse transformation, and the differential results between 0-order and 2-order were calculated (at intervals of 0.2). The simulation results are shown in Figures 4 and 5. When differential order gradually increases from zero-order to first-order, the curve of correlation coefficient shows a certain gradual change trend. When the order is increased from 1-order to 2-order, correlation coefficient curve fluctuates greatly, and the gradual change trend is not obvious.

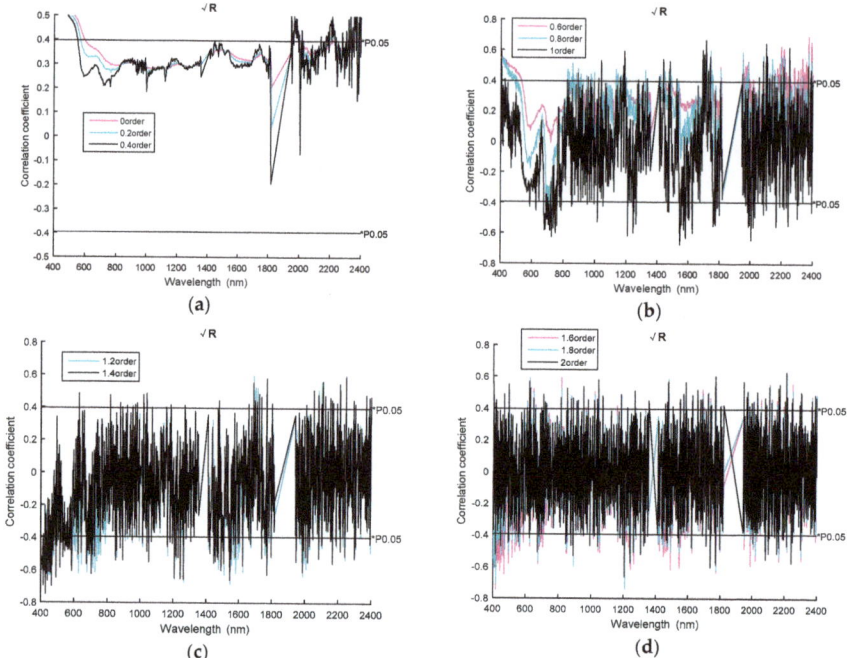

Figure 4. Trends of correlation coefficient for root mean square: (**a**) 0-order to 0.4-order; (**b**) 0.6-order to 1-order; (**c**) 1.2-order to 1.4-order; and (**d**) 1.6-order to 2-order.

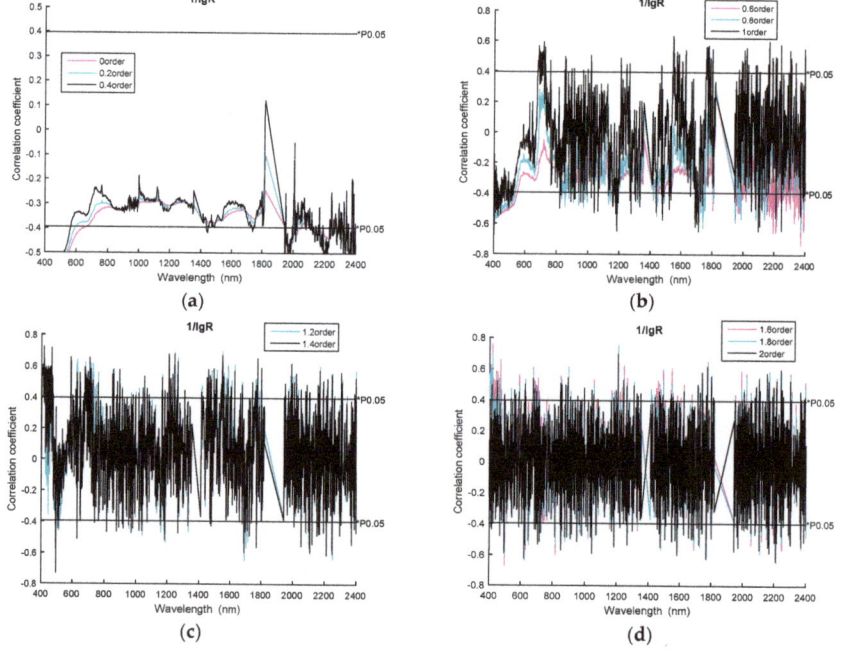

Figure 5. Trends of correlation coefficient for logarithm reciprocal: (**a**) 0-order to 0.4-order; (**b**) 0.6-order to 1-order; (**c**) 1.2-order to 1.4-order; and (**d**) 1.6-order to 2-order.

4.3. Number of Bands that Passed 0.05 Significance Level Test

The number of spectral bands that passed the 0.05 significance level of the correlation coefficient between spectrum and available potassium is shown in Figure 6. There are hundreds of spectral bands passed the 0.05 significance level for these five spectral transformations. Compared with more than 2000 full-band spectra in the range of 350–2500 nm, Figure 6 shows that the preprocessing operation of fractional calculus can reduce the dimensionality of soil hyperspectral data. Overall, the trend of the number for logarithm reciprocal, original spectrum and root mean square is gradually decreasing, and the trend increases first and then decreases for reciprocal and logarithm.

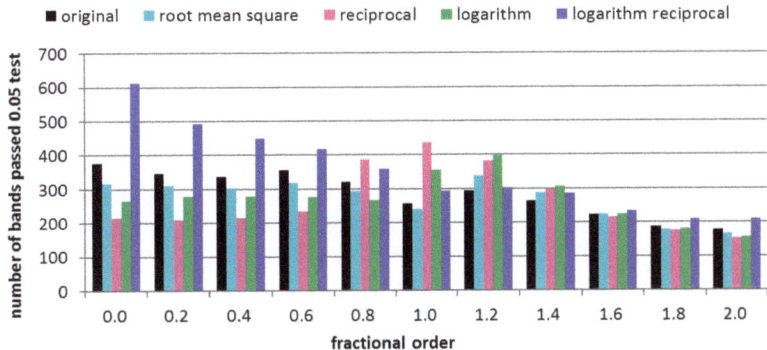

Figure 6. Number of bands that passed the 0.05 test.

4.4. Absolute Maximum Band of Correlation Coefficient under Five Spectral Transformations

Maximum absolute values of correlation coefficients under five different spectral transformations in the 0-order to 2-order range and the corresponding band information are shown in Table 1. The maximum absolute values of correlation coefficients appear in fractional order: when R and $1/\lg R$ are in the 1.6 order, the corresponding band is 416 nm, and the largest absolute of R and $1/\lg R$ are 0.763605 and 0.76218, respectively; when $1/R$ and $\sqrt{R}$ are in the 1.4-order, the corresponding bands are 494 and 430 nm, and the largest absolute of $1/R$ and $\sqrt{R}$ are 0.741574 and 0.750124, respectively; and when $\lg R$ is in the 1.2-order, the corresponding band is 495 nm, and the largest absolute is 0.747359. For first-order differential transformation, $\sqrt{R}$, $1/R$, $\lg R$, and $1/\lg R$ increase the correlation between spectral reflectance of R and available potassium content to some extent. For 0-order and 2-order differential transformation, $1/\lg R$ improves the correlation, while the others reduce the correlation. In addition, the absolute values of correlation coefficients of 1.2-, 1.4-, 1.6-, and 1.8-order for R, $\sqrt{R}$, $\lg R$, and $1/\lg R$ are all greater than 0.7.

Table 1. Bands with the largest absolute values of correlation coefficients under five spectral transformation.

Order	R		√R		1/R		lgR		1/lgR	
	Largest Absolute	Band	Largest Absolute	Band	Largest Absolute	Band	Largest Absolute	Band	Largest Absolute	Band
0	0.560973	405	0.547695	405	0.499854	400	0.532827	405	0.562356	405
0.2	0.570438	2391	0.559709	2391	0.521992	2392	0.547824	2392	0.586224	2391
0.4	0.700301	2390	0.690903	2391	0.655011	2391	0.680045	2391	0.719717	2390
0.6	0.712129	2371	0.697076	2371	0.645772	2390	0.680265	2371	0.734104	2371
0.8	0.632373	2100	0.62877	2100	0.663153	2006	0.623198	2100	0.645372	2371
1	0.668786	1547	0.675601	1547	0.674436	2006	0.674577	1547	0.67265	404
1.2	0.707065	416	0.746563	430	0.674899	495	0.747359	495	0.705747	416
1.4	0.722362	416	0.750124	430	0.741574	494	0.739563	430	0.730641	497
1.6	0.763605	416	0.746466	416	0.689794	1207	0.725154	416	0.762187	416
1.8	0.74986	1207	0.739655	1207	0.701802	416	0.726421	1207	0.75847	1207
2	0.678462	1206	0.664644	1206	0.621093	1740	0.648568	1206	0.69792	1206

4.5. Fractional Derivative Impact on Correlation Coefficient of Landsat 8 Image Bands

To further explain the influence of correlation on partial bands by fractional derivative, seven bands corresponding to Landsat 8 image [28,29] were selected to study the variation trend of the correlation coefficient under different fractional order. The band ranges of Landsat 8 image are shown in Table 2. The seven wavelength bands selected from Landsat 8 are 442 nm in Band 1, 467 nm in Band 2, 587 nm in Band 3, 675 nm in Band 4, 851 nm in Band 5, 1597 nm in Band 6, and 2247 nm in Band 7. The trend of correlation coefficient for the seven selected wavelength bands is shown in Figure 7.

Table 2. Spectral ranges of Landsat 8 image bands.

Band name	Band range (nm)
Band 1 Coastal	433–453
Band 2 Blue	450–515
Band 3 Green	525–600
Band 4 Red	630–680
Band 5 NIR	845–885
Band 6 SWIR1	1560–1660
Band 7 SWIR2	2100–2300

It can be seen in Figure 7 that the correlation coefficient change of R and 1/lgR is opposite to the other three transformations. In Band 7, the correlation coefficients of 1/R and 1/lgR are negative at 0-order to 2-order, and the remaining transformations are positive. In the range of Band 1, Band 2, Band 4, Band 5 and Band 6, the correlation coefficients of 1/R and 1/lgR are negative at 0-order to 0.6-order, and the remaining transformations are positive at 0.0–0.6 order. The correlation coefficients of 1/R and 1/lgR are positive at 1.2-order to 1.8-order, and the remaining transformations are negative at 1.2-order to 1.8-order. In addition, the maximum absolute correlation coefficient of five spectral transformations is in Band 2, which is 0.715766 of 467 nm.

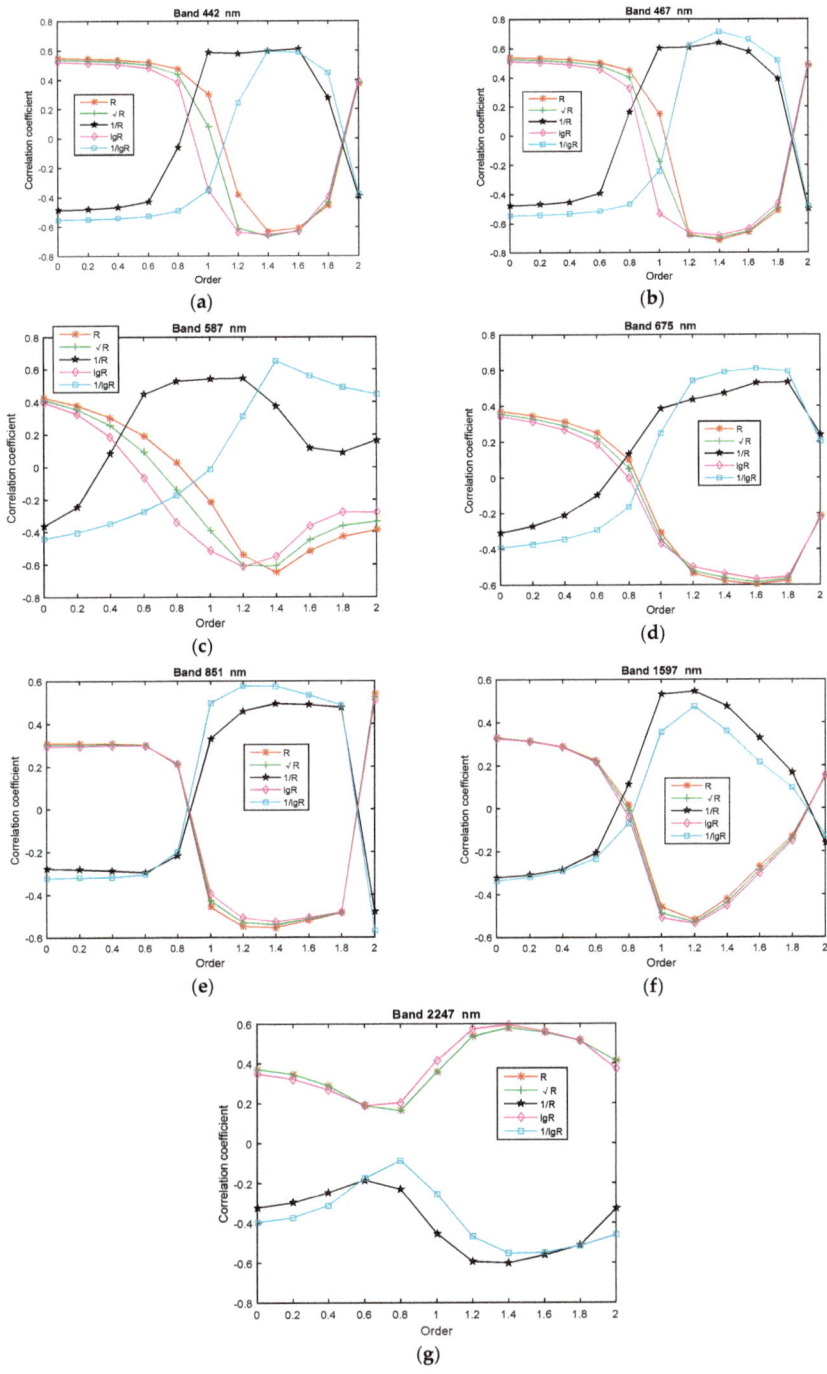

Figure 7. The trend of correlation coefficient for partial bands of Landsat 8 image in each order differential: (**a**) 442 nm; (**b**) 467 nm; (**c**) 587 nm; (**d**) 675 nm ; (**e**) 851 nm ; (**f**) 1597 nm; and (**g**) 2247 nm.

5. Discussion

The integer-order derivative method is widely used in soil spectral signal pretreatment, but its description of the physical model is only an approximation [30,31]. This traditional preprocessing method based on integer-order derivative has obvious shortcomings. One of the main reasons is the defect of integer-order derivative in the numerical calculation process, that is, the integer-order derivative is only related to the information of the points in the differential window. Another main reason is that the fractional derivative has the advantage of "memory" and "non-locality", that is, the fractional order is not only related to the value of the point, but also related to the value of all points before this point. It has been proved that the fractional-order system is more in line with the laws of nature and engineering physics, which can better reflect the performance of the dynamic system, and has a unique historical memory function. Therefore, the fractional derivative model is more accurate than the integer-order derivative model.

In addition to the pretreatment of hyperspectral signals for saline soil between spectral reflectance and salt content, fractional derivatives can also be used to pretreat other types of soil hyperspectral signals between spectral reflectance and nutrient content. For example, Xia et al. [32] used fractional derivative to preprocess the spectrum collected in Ebinur Lake of Xinjiang, China, and the correlation coefficient between electricity conductivity and soil reflectance spectra was analyzed. Results show that fractional derivative details the varying trends of soil reflectance spectra among 0-order to 2-order. Fractional derivative also raises the correlation coefficient between electricity conductivity and soil reflectance spectra for some bands. Hong et al. [33] applied the fractional derivative to analyze the relationship of soil organic matter content and visible and near-infrared spectroscopy. The results show that the highest validation model appears in the 1.5-order derivative combined genetic algorithm. Wang et al. [34] collected 168 sample of soil taken from the coalmine in Eastern Junggar Basin, China. They used fractional derivative to preprocess the hyperspectral data of coalmine soil and PLSR to estimate the soil chromium content. The results show that 1.8-order derivative is the best predictive model, and the ratio of performance to deviation (PRD) is 2.14. Wang et al. [35] used the soil of the Ebinur Lake Wetland National Nature Reserve in Xinjiang as the research object, and used the fractional differential and grey correlation analysis-BP neural network to quantitatively estimate the soil organic matter content. The results show that the 1.2-order model has the highest accuracy and the PRD value is 2.26.

In addition, fractional derivatives are also used to preprocess hyperspectral signals from rubber trees, diesel, tobacco, wheat, corn, and so on. For example, Chen et al. [36] adopted fractional derivative to analyze the near-infrared spectroscopy of nitrogen concentration for natural rubber. The results show that the 0.6-order has the optimal prediction result. Tong et al. [37] adopted SG derivation to analyze the near-infrared spectroscopy of diesel dataset and tobacco dataset. The results show that this method can improve the spectral resolution, and SG derivation combined with competitive adaptive reweighted sampling is the optimal model.

6. Conclusions

Fractional derivative in the field of hyperspectral studies are rarely reported, especially for field-measured ground hyperspectral data. We collected soil samples and hyperspectral data in May 2017. Grünwald–Letnikov fractional derivative was used to analyze the correlation coefficient between the available potassium content and the soil ground hyperspectral data and Landsat 8 multispectral satellite data. Simulation results display that the small difference between spectrum data was clearly described by fractional derivative. The maximum absolute correlation coefficient appeared in the fractional order for the ground hyperspectral data and Landsat 8 multispectral satellite data. Therefore, fractional derivative enriches the pre-processing method of spectral data, provides potential spectrum information, increases the correlation coefficient between spectral reflectance and available potassium content, and provides scientific support for local precision agriculture.

Author Contributions: C.F. designed the research and performed all the modeling. A.T. and H.X. collected the experiment data. S.G., X.Y. and H.X. participated in the data analyses. C.F. and A.T. were involved in drafting and revising the manuscript.

Funding: The authors would like to thank the financial support of National Natural Science Foundation of China (41861054 and 41671198), and Yunnan Province Science and Technology Department and Education Department Project (2017FH001-067 and 2017FH001-117), China.

Conflicts of Interest: The authors declare no conflict of interest.

References

1. Sardans, J.; Penuelas, J. Potassium: A neglected nutrient in global change. *Glob. Ecol. Biogeogr.* **2015**, *24*, 261–275. [CrossRef]
2. Qiu, K.; Xie, Y.; Xu, D.; Pott, R. Ecosystem functions including soil organic carbon, total nitrogen and available potassium are crucial for vegetation recovery. *Sci. Report.* **2018**, *8*, 7607. [CrossRef] [PubMed]
3. Panda, R.; Patra, S.K. Assessment of Suitable Extractants for Predicting Plant-available Potassium in Indian Coastal Soils. *Commun. Soil Sci. Plant Anal.* **2018**, *49*, 1157–1167. [CrossRef]
4. Viscarra Rossel, R.A.; Walvoort, D.J.J.; McBratney, A.B.; Janik, L.J.; Skjemstad, J.O. Visible, near infrared, mid infrared or combined diffuse reflectance spectroscopy for simultaneous assessment of various soil properties. *Geoderma* **2006**, *131*, 59–75. [CrossRef]
5. Jia, S.; Yang, X.; Li, G.; Zhang, J. Quantitatively determination of available phosphorus and available potassium in soil by near infrared spectroscopy combining with recursive partial least squares. *Spectro. Spectr. Anal.* **2015**, *35*, 2516–2520.
6. Sarkar, S.; Patra, S.K. Evaluation of Chemical Extraction Methods for Determining Plant-Available Potassium in Some Soils of West Bengal, India. *Commun. Soil Sci. Plant Anal.* **2017**, *48*, 1008–1019. [CrossRef]
7. Zhang, L.; Zhang, M.; Ren, H.; Pu, P.; Kong, P.; Zhao, H. Comparative investigation on soil nitrate-nitrogen and available potassium measurement capability by using solid-state and PVC ISE. *Comput. Electr. Agric.* **2015**, *112*, 83–91. [CrossRef]
8. O'Rourke, S.M.; Stockmann, U.; Holden, N.M.; McBratney, A.B.; Minasny, B. An assessment of model averaging to improve predictive power of portable vis-NIR and XRF for the determination of agronomic soil properties. *Geoderma* **2016**, *279*, 31–44. [CrossRef]
9. Gholizadeh, A.; Saberioo, M.; Ben-Dor, E.; Boruvka, L. Monitoring of selected soil contaminants using proximal and remote sensing techniques: Background, state-ofthe-art and future perspectives. *Crit. Rev. Environ. Sci. Tech.* **2018**, *48*, 243–278. [CrossRef]
10. Lamine, S.; Petropoulos, G.P.; Brewer, P.A.; Bachari, N.E.; Srivastava, P.K.; Manevski, K.; Kalaitzidis, C.; Macklin, M.G. Heavy Metal Soil Contamination Detection Using Combined Geochemistry and Field Spectroradiometry in the United Kingdom. *Sensors* **2019**, *19*, 762. [CrossRef]
11. Liu, X.; Wang, L.; Chang, Q.; Wang, X.; Shang, Y. Prediction of total nitrogen and alkali hydrolysable nitrogen content in loess using hyperspectral data based on correlation analysis and partial least squares regression. *Chin. J. Appl. Ecol.* **2015**, *26*, 2107–2114.
12. Debaene, G.; Niedźwiecki, J.; Pecio, A.; Żurek, A. Effect of the number of calibration samples on the prediction of several soil propertiesat the farm-scale. *Geoderma* **2014**, *214/215*, 114–125. [CrossRef]
13. Mashimbye, Z.E.; Cho, M.A.; Nell, J.P.; Declercq, W.P.D.; Niekerk, A.V.; Turner, D.P. Model-Based Integrated Methods for Quantitative Estimation of Soil Salinity from Hyperspectral Remote Sensing Data A Case Study of Selected South African Soils. *Pedosphere* **2012**, *22*, 640–649. [CrossRef]
14. Yang, H.; Kuang, B.; Mouazen, A.M. Quantitative analysis of soil nitrogen and carbon at a farm scale using visible and near infrared spectroscopy coupled with wavelength reduction. *Eur. J. Soil Sci.* **2011**, *63*, 410–420. [CrossRef]
15. Fu, Y.; Yang, G.; Wang, J.; Song, X.; Feng, H. Winter wheat biomass estimation based on spectral indices, band depth analysis and partial least squares regression using hyperspectral measurements. *Comput. Electr. Agric.* **2014**, *100*, 51–59. [CrossRef]

16. Wu, Q.; Yang, Y.; Xu, Z.; Jin, Y.; Guo, Y.; Lao, C. Applying Local Neural Network and Visible/Near-Infrared Spectroscopy to Estimating Available Nitrogen, Phosphorus and Potassium in Soil. *Spectro. Spectr. Anal.* **2014**, *34*, 2102–2105.
17. Liu, X.; Liu, J. Based on the LS-SVM Modeling Method Determination of Soil Available N and Available K by Using Near-Infrared Spectroscopy. *Spectro. Spectr. Anal.* **2012**, *32*, 3019–3023.
18. Wang, J.; Tiyip, T.; Ding, J.; Zhang, D.; Liu, W.; Wang, F.; Nigara, T. Desert soil clay content estimation using reflectance spectroscopy preprocessed by fractional derivative. *PLOS ONE* **2017**, *12*, e0184836. [CrossRef]
19. Wang, J.; Ding, J.; Aerzuna, A.; Cai, L. Quantitative estimation of soil salinity by means of different modeling methods and visible-near infrared (VIS-NIR) spectroscopy, Ebinur Lake Wetland, Northwest China. *PeerJ* **2018**. [CrossRef]
20. Kaslik, E.; Sivasundaram, S. Non-Existence of Periodic Solutions in Fractional-Order Dynamical Systems and a Remarkable Difference Between Integer and Fractional-Order Derivatives of Periodic Functions. *Nonlinear Anal. Real World Appl.* **2012**, *13*, 1489–1497. [CrossRef]
21. Mbouna, G.S.; Woafo, P. Dynamics and Synchronization Analysis of Coupled Fractional-Order Nonlinear Electromechanical Systems. *Mech. Res. Commun.* **2012**, *46*, 20–25.
22. Wu, L.; Liu, S.; Yao, L.; Yan, S.; Liu, D. Grey System Model with the Fractional Order Accumulation. *Commun. Nonlinear Sci. Numer. Simul.* **2013**, *18*, 1775–1785. [CrossRef]
23. Galeone, L.; Garrappa, R. Explicit Methods for Fractional Differential Equations and Their Stability Properties. *J. Comput. Appl. Math.* **2009**, *228*, 548–560. [CrossRef]
24. Schmitt, J.M. Fractional Derivative Analysis of Diffuse Reflectance Spectra. *Appl. Spectro.* **1998**, *52*, 840–846. [CrossRef]
25. Zheng, K.; Zhang, X.; Tong, P.; Yao, Y.; Du, Y. Pretreating near infrared spectra with fractional order Savitzky–Golay differentiation (FOSGD). *Chin. Chem. Letters* **2015**, *26*, 293–296. [CrossRef]
26. Zhang, D.; Tiyip, T.; Ding, J.; Zhang, F.; Nurmemet, I.; Kelimu, A.; Wang, J. Quantitative Estimating Salt Content of Saline Soil Using Laboratory Hyperspectral Data Treated by Fractional Derivative. *J. Spectro.* **2016**. [CrossRef]
27. Liu, Y.; Liu, Y.; Chen, Y.; Zhang, Y.; Shi, T.; Wang, J.; Hong, Y.; Fei, T.; Zhang, Y. The Influence of Spectral Pretreatment on the Selection of Representative Calibration Samples for Soil Organic Matter Estimation Using Vis-NIR Reflectance Spectroscopy. *Remote Sens.* **2019**, *11*, 450. [CrossRef]
28. Zhang, K.; Dong, X.; Liu, Z.; Gao, W.; Hu, Z.; Wu, G. Mapping Tidal Flats with Landsat 8 Images and Google Earth Engine: A Case Study of the China's Eastern Coastal Zone circa 2015. *Remote Sens.* **2019**, *11*, 924. [CrossRef]
29. Lima, T.A.; Beuchle, R.; Langner, A.; Grecchi, R.; Griess, V.C.; Frédéric, A. Comparing Sentinel-2 MSI and Landsat 8 OLI Imagery for Monitoring Selective Logging in the Brazilian Amazon. *Remote Sens.* **2019**, *11*, 961. [CrossRef]
30. Nawar, S.; Buddenbaum, H.; Hill, J.; Kozak, J.; Mouazen, A.M. Estimating the soil clay content and organic matter by means of different calibration methods of VIS–NIR diffuse reflectance spectroscopy. *Soil Tillage Res.* **2016**, *155*, 510–522. [CrossRef]
31. Srivastava, R.; Sarkar, D.; Mukhopadhayay, S.S.; Sood, A.; Singh, M.; Nasre, R.A.; Dhale, S.A. Development of hyperspectral model for rapid monitoring of soil organic carbon under precision farming in the Indo-Gangetic Plains of Punjab, India. *J. Indian Soc. Remote Sens.* **2015**, *43*, 751–759. [CrossRef]
32. Xia, N.; Tiyip, T.; Kelimu, A.; Nurmemet, I.; Ding, J.; Zhang, F.; Zhang, D. Influence of Fractional Differential on Correlation Coefficient between EC1:5 and Reflectance Spectra of Saline Soil. *J. Spectro.* **2017**. [CrossRef]
33. Hong, Y.; Chen, Y.; Yu, L.; Liu, Y.; Liu, Y.; Zhang, Y.; Liu, Y.; Cheng, H. Combining Fractional Order Derivative and Spectral Variable Selection for Organic Matter Estimation of Homogeneous Soil Samples by VIS–NIR Spectroscopy. *Remote Sens.* **2018**, *10*, 479. [CrossRef]
34. Wang, J.; Tiyip, T.; Zhang, D. Spectral Detection of Chrominum Content in Desert Soil Based on Fractional Differential. *Trans. Chin. Soc. Agric. Mach.* **2017**, *48*, 152–158.
35. Wang, X.; Zhang, F.; Kung, H.; Johnson, V.C. New methods for improving the remote sensing estimation of soil organic matter content (SOMC) in the Ebinur Lake Wetland National Nature Reserve (ELWNNR) in northwest China. *Remote Sens. Environ.* **2018**, *218*, 104–118. [CrossRef]

36. Chen, K.; Li, C.; Tang, R. Estimation of the nitrogen concentration of rubber tree using fractional calculus augmented NIR spectra. *Ind. Crops Product.* **2017**, *108*, 831–839. [CrossRef]
37. Tong, P.; Du, Y.; Zheng, K.; Wu, T.; Wang, J. Improvement of NIR model by fractional order Savitzky–Golay derivation (FOSGD) coupled with wavelength selection. *Chemom. Intell. Lab. Syst.* **2015**, *143*, 40–48. [CrossRef]

© 2019 by the authors. Licensee MDPI, Basel, Switzerland. This article is an open access article distributed under the terms and conditions of the Creative Commons Attribution (CC BY) license (http://creativecommons.org/licenses/by/4.0/).

Article

Some New Fractional-Calculus Connections between Mittag–Leffler Functions

Hari M. Srivastava [1,2,*], Arran Fernandez [3,4] and Dumitru Baleanu [5,6]

[1] Department of Mathematics and Statistics, University of Victoria, Victoria, BC V8W 3R4, Canada
[2] Department of Medical Research, China Medical University Hospital, China Medical University, Taichung 40402, Taiwan
[3] Department of Applied Mathematics and Theoretical Physics, University of Cambridge, Wilberforce Road, Cambridge CB3 0WA, UK; arran.fernandez@emu.edu.tr
[4] Department of Mathematics, Faculty of Arts and Sciences, Eastern Mediterranean University, Famagusta 99628, TRNC, Mersin-10, Turkey
[5] Department of Mathematics, Cankaya University, Balgat, Ankara 06530, Turkey; dumitru@cankaya.edu.tr
[6] Institute of Space Sciences, 077125 Magurele-Bucharest, Romania
* Correspondence: harimsri@math.uvic.ca

Received: 30 March 2019; Accepted: 20 May 2019; Published: 28 May 2019

Abstract: We consider the well-known Mittag–Leffler functions of one, two and three parameters, and establish some new connections between them using fractional calculus. In particular, we express the three-parameter Mittag–Leffler function as a fractional derivative of the two-parameter Mittag–Leffler function, which is in turn a fractional integral of the one-parameter Mittag–Leffler function. Hence, we derive an integral expression for the three-parameter one in terms of the one-parameter one. We discuss the importance and applications of all three Mittag–Leffler functions, with a view to potential applications of our results in making certain types of experimental data much easier to analyse.

Keywords: fractional integrals; fractional derivatives; Mittag–Leffler functions

MSC: 26A33; 33E12

---

1. Introduction

In fractional calculus, the standard calculus operations of differentiation and integration are generalised to orders beyond the integers: rational, real, and even complex numbers can be used for the order of *differintegration* [1–3]. This area of research is four centuries old, but it has expanded rapidly only in the last fifty years, discovering applications in many fields of science and engineering [4–6]. The most commonly used definition of fractional derivatives and integrals is the Riemann–Liouville one, where fractional integrals are defined by

$$\,^{RL}_a D_x^\alpha f(x) = \frac{1}{\Gamma(-\alpha)} \int_a^x (x-y)^{-\alpha-1} f(y)\,\mathrm{d}y, \qquad \mathrm{Re}(\alpha) < 0, \tag{1}$$

and fractional derivatives are defined by

$$\,^{RL}_a D_x^\alpha f(x) = \frac{\mathrm{d}^m}{\mathrm{d}x^m}\left(\,^{RL}_a D_x^{\alpha-m} f(x)\right), \qquad \mathrm{Re}(\alpha) \geq 0, m := \lfloor \mathrm{Re}(\alpha) \rfloor + 1. \tag{2}$$

Here, $D^\alpha f$ denotes the derivative to order α of a function f, and a is a constant of differintegration. It is important to note that in fractional calculus, derivatives as well as integrals rely on the choice of an arbitrary constant a. This constant is usually set to be either $a = 0$ or $a = -\infty$. To see why both choices are useful, we present the following Lemma which provides two "'natural" differintegration formulae, one requiring $a = 0$ and the other requiring $a = -\infty$. Neither option can be eliminated from the range of possible values for a, if we wish to retain natural expressions for differintegrals of elementary functions.

Lemma 1. *The Riemann–Liouville (RL) differintegrals of power functions and exponential functions, with constant of differintegration $a = 0$ and $a = -\infty$, respectively, are as follows.*

$$^{RL}_0 D_x^\alpha(x^\beta) = \frac{\Gamma(\beta+1)}{\Gamma(\beta-\alpha+1)} x^{\beta-\alpha}, \qquad \alpha, \beta \in \mathbb{C}, \operatorname{Re}(\beta) > -1; \tag{3}$$

$$^{RL}_{-\infty} D_x^\alpha \left(e^{\beta x}\right) = \beta^\alpha e^{\beta x}, \qquad \alpha, \beta \in \mathbb{C}, \beta \notin \mathbb{R}_0^-. \tag{4}$$

In both cases, we define complex power functions using the principal branch with arguments between $-\pi$ and π.

Proof. In both cases, the proof for fractional integrals follows from manipulation and substitution in the integral formula in Equation (1), and then the proof for fractional derivatives is immediate from the definition in Equation (2). For more details, we refer the reader to [1,3]. □

In recent years, many alternative definitions of fractional differintegrals have been proposed. Some of these were motivated by the different real-world systems which can be modelled by different fractional-calculus structures: for example, replacing the power function in Equation (1) by another function to better describe certain types of processes in dynamical systems [7,8]. Others were created by adding extra parameters and levels of generalisation into functions and formulae [9–11].

One particular function which frequently appears in the study of fractional derivatives and integrals [12–14] is the **Mittag–Leffler function**, which in its simplest form is defined by

$$E_\alpha(z) = \sum_{n=0}^{\infty} \frac{z^n}{\Gamma(n\alpha+1)}, \qquad z \in \mathbb{C}. \tag{5}$$

The above function with a single parameter α has also been extended to more general functions defined with two or more parameters, such as the following [15,16].

$$E_{\alpha,\beta}(z) = \sum_{n=0}^{\infty} \frac{z^n}{\Gamma(n\alpha+\beta)}, \qquad z \in \mathbb{C}; \tag{6}$$

$$E_{\alpha,\beta}^\rho(z) = \sum_{n=0}^{\infty} \frac{\Gamma(\rho+n) z^n}{\Gamma(\rho)\Gamma(n\alpha+\beta) n!}, \qquad z \in \mathbb{C}; \tag{7}$$

$$E_{\alpha,\beta}^{\rho,\kappa}(z) = \sum_{n=0}^{\infty} \frac{\Gamma(\rho+\kappa n) z^n}{\Gamma(\rho)\Gamma(n\alpha+\beta) n!}, \qquad z \in \mathbb{C}. \tag{8}$$

It is clear that the following interrelations hold between the above functions:

$$E_{\alpha,\beta}^{\rho,1}(z) = E_{\alpha,\beta}^\rho(z), \qquad E_{\alpha,\beta}^1(z) = E_{\alpha,\beta}(z), \qquad E_{\alpha,1}(z) = E_\alpha(z).$$

Several new models of fractional calculus have used such functions in their definitions, and we mention two of these in particular.

The **AB model**, formulated by Atangana and Baleanu [7] and further studied in [17–20], is defined by replacing the power function in Equation (1) by a one-parameter Mittag–Leffler function of the type in Equation (5):

$$^{ABR}_{a}D^{\alpha}_{x}f(x) = \frac{B(\alpha)}{1-\alpha} \cdot \frac{d}{dx} \int_{a}^{x} E_{\alpha}\left(\frac{-\alpha}{1-\alpha}(x-y)^{\alpha}\right) f(y)\, dy, \qquad 0 < \alpha < 1, \tag{9}$$

$$^{AB}_{a}I^{\alpha}_{x}f(x) = \frac{1-\alpha}{B(\alpha)} f(x) + \frac{\alpha}{B(\alpha)}\, ^{RL}_{a}I^{\alpha}_{x}f(x), \qquad 0 < \alpha < 1. \tag{10}$$

We note that here α is a real variable and not a complex one. All discussion of the AB model in the literature so far has assumed the order of differentiation to be real. The first paper to consider complex-order AB differintegrals is currently in press [21].

The **Prabhakar model**, based on an integral operator defined in 1971 [22] but only later formulated as part of fractional calculus [23,24], is defined by replacing the power function in Equation (1) by a three-parameter Mittag–Leffler function of the type in Equation (7). This model has also been generalised [16] to use a 4-parameter Mittag–Leffler function of the type in Equation (8), and its properties have been explored in many papers (e.g., [25–27]).

One useful application of fractional calculus in pure mathematics has been to find new functional equations and interrelations between various important functions. For example, fractional versions of the product rule and chain rule have given rise to new formulae for assorted special functions [28,29], and fractional differintegration of infinite series has yielded new identities on the Riemann zeta function and its generalisations [30–32].

In the current work, we use these techniques to prove new relationships between the several Mittag–Leffler functions defined above. It is possible to write the three-parameter Mittag–Leffler function as a fractional differintegral of the two-parameter one, as well as writing the two-parameter one as a fractional differintegral of the one-parameter one, and thence to deduce an integral relationship between the one-parameter and three-parameter Mittag–Leffler functions, which suggests a relationship between the AB and Prabhakar models of fractional calculus. Throughout all of this, we use only the classical Riemann–Liouville fractional integrals and derivatives. We also examine the possibility of applications of these results in fields of science such as bioengineering and dielectric relaxation.

This paper is structured as follows. In Section 2, through a number of theorems and propositions, we state the main results concerning relationships between Mittag–Leffler functions. In Section 3, we discuss applications, and, in Section 4, we conclude the article.

2. The Main Results

We first state an important result about fractional differintegration of series, which we need to use in the proofs below.

Lemma 2. *Consider a function S defined by an infinite series*

$$S(x) = \sum_{n=1}^{\infty} S_n(x)$$

which is uniformly convergent on the set $|x - a| \leq K$ for some fixed constants $a \in \mathbb{C}$, $K > 0$. Let $\alpha \in \mathbb{C}$ be a fixed order of differintegration.

1. *If* $\operatorname{Re}(\alpha) < 0$ *(fractional integration), then we have*

$$^{RL}_{a}D^{\alpha}_{x}S(x) = \sum_{n=1}^{\infty}\, ^{RL}_{a}D^{\alpha}_{x}S_n(x), \qquad |x - a| \leq K,$$

and the series on the right-hand side is uniformly convergent on the given region.

2. If $\operatorname{Re}(\alpha) \geq 0$ (fractional differentiation) and the series $\sum_{n=1}^{\infty} {}^{RL}_{a}D^{\alpha}_{x}S_n(x)$ is uniformly convergent on the region $|x - a| \leq K$, then we have

$${}^{RL}_{a}D^{\alpha}_{x}S(x) = \sum_{n=1}^{\infty} {}^{RL}_{a}D^{\alpha}_{x}S_n(x), \qquad |x - a| \leq K.$$

Proof. This is Theorem VIII in [31]. □

Since our work involves analytic functions defined on the complex plane, we find it useful to define the domain

$$\mathbb{D} := \mathbb{C} \setminus \mathbb{R}_0^-, \qquad (11)$$

namely the complex plane slit along a branch cut from the origin. This is used as a domain for various complex power functions and other related functions.

Our first main result is a fractional identity between the Mittag–Leffler functions with two and three parameters as defined by Equations (6) and (7). This is motivated by previous work (e.g., [32]), in which gamma functions that appear in infinite power series can be interpreted as arising from fractional differintegrals. After submitting the paper, we realised that this result was previously proved in [33]. However, our original proof is preserved below.

Proposition 1. *For any $\alpha, \beta, \rho \in \mathbb{C}$ with $\operatorname{Re}(\alpha), \operatorname{Re}(\beta) > 0$, we have:*

$$E^{\rho}_{\alpha,\beta}(z) = \frac{1}{\Gamma(\rho)} {}^{RL}_{0}D^{\rho-1}_{z}\left[z^{\rho-1} E_{\alpha,\beta}(z)\right], \qquad z \in \mathbb{D}. \qquad (12)$$

Proof. First, Lemma 1 tells us that a quotient of gamma functions can very often be interpreted as arising from a fractional differintegral of a power function. In this case, the expression $\frac{\Gamma(\rho+n)}{n!}$ appearing in the coefficients of the series in Equation (7) gives rise to the following:

$$E^{\rho}_{\alpha,\beta}(z) = \sum_{n=0}^{\infty} \frac{\Gamma(\rho+n)z^n}{\Gamma(\rho)\Gamma(n\alpha+\beta)n!} = \sum_{n=0}^{\infty} \frac{1}{\Gamma(n\alpha+\beta)\Gamma(\rho)} \cdot \frac{\Gamma(\rho+n)}{\Gamma(n+1)} z^n$$

$$= \sum_{n=0}^{\infty} \frac{1}{\Gamma(n\alpha+\beta)\Gamma(\rho)} \cdot {}^{RL}_{0}D^{\rho-1}_{z}\left[z^{n+\rho-1}\right] = \frac{1}{\Gamma(\rho)} \sum_{n=0}^{\infty} {}^{RL}_{0}D^{\rho-1}_{z}\left[\frac{z^{n+\rho-1}}{\Gamma(n\alpha+\beta)}\right].$$

Since the series here is uniformly convergent, we can use the result of Lemma 2 to swap the summation with the fractional differintegration. This gives:

$$E^{\rho}_{\alpha,\beta}(z) = \frac{1}{\Gamma(\rho)} \sum_{n=0}^{\infty} {}^{RL}_{0}D^{\rho-1}_{z}\left[\frac{z^{n+\rho-1}}{\Gamma(n\alpha+\beta)}\right] = \frac{1}{\Gamma(\rho)} {}^{RL}_{0}D^{\rho-1}_{z}\left[\sum_{n=0}^{\infty} \frac{z^{n+\rho-1}}{\Gamma(n\alpha+\beta)}\right]$$

$$= \frac{1}{\Gamma(\rho)} {}^{RL}_{0}D^{\rho-1}_{z}\left[z^{\rho-1} \sum_{n=0}^{\infty} \frac{z^n}{\Gamma(n\alpha+\beta)}\right] = \frac{1}{\Gamma(\rho)} {}^{RL}_{0}D^{\rho-1}_{z}\left[z^{\rho-1} E_{\alpha,\beta}(z)\right],$$

and we have the result desired. □

Note that, by setting $\rho = 1$ in Equation (12), we recover the trivial identity $E^1_{\alpha,\beta}(z) = E_{\alpha,\beta}(z)$.

Corollary 1. *For any $\alpha, \rho \in \mathbb{C}$ with $\operatorname{Re}(\alpha) > 0$, we have:*

$$E^{\rho}_{\alpha,1}(z) = \frac{1}{\Gamma(\rho)} {}^{RL}_{0}D^{\rho-1}_{z}\left[z^{\rho-1} E_{\alpha}(z)\right], \qquad z \in \mathbb{D}. \qquad (13)$$

Proof. This follows immediately by setting $\beta = 1$ in Proposition 1. □

Remark 1. *By exactly the same argument as in Proposition 1, we can show that*

$$E^\rho_{\alpha,\beta}(\gamma z) = \frac{1}{\Gamma(\rho)} \, {}^{RL}_0 D^{\rho-1}_z \left[z^{\rho-1} E_{\alpha,\beta}(\gamma z) \right], \qquad z \in \mathbb{D}, \tag{14}$$

for any $\alpha, \beta, \gamma, \rho \in \mathbb{C}$ with $\mathrm{Re}(\alpha), \mathrm{Re}(\beta) > 0$. The proof is as above with an extra factor of γ^n included in each term of the sum.

We have now established a relation between the Mittag–Leffler functions of two and three parameters, and hence a relation between the Mittag–Leffler function of one parameter and the function $E^\rho_{\alpha,1}(z)$. However, in order to find a connection with the AB model, we need to consider not the function $E_\alpha(x)$ but rather the function $E_\alpha\left(\frac{-\alpha}{1-\alpha} x^\alpha\right)$, which appears in the kernel of the definition in Equation (9). To this end, we note the following result, which is seen in (Equation (7.1), [12]) but without reference to fractional calculus.

Proposition 2. *For any $\alpha, \beta, \gamma \in \mathbb{C}$ with $\mathrm{Re}(\alpha), \mathrm{Re}(\beta) > 0$, we have:*

$$E_{\alpha,\beta}(\gamma z^\alpha) = z^{1-\beta} \, {}^{RL}_0 D^{1-\beta}_z \left[E_\alpha(\gamma z^\alpha) \right], \qquad z \in \mathbb{D}. \tag{15}$$

Proof. This time we start from the right-hand side of the desired identity, and use the definition in Equation (5) of the function E_α:

$$z^{1-\beta} \, {}^{RL}_0 D^{1-\beta}_z \left[E_\alpha(\gamma z^\alpha) \right] = z^{1-\beta} \, {}^{RL}_0 D^{1-\beta}_z \left[\sum_{n=0}^\infty \frac{\gamma^n z^{\alpha n}}{\Gamma(n\alpha + 1)} \right].$$

This series is uniformly convergent, thus, by Lemma 2, we can swap the summation and fractional differintegration provided that (at least in the case $0 < \mathrm{Re}(\beta) < 1$) the resulting series also converges uniformly. We swap the operations now and justify this assumption at the end.

$$\begin{aligned} z^{1-\beta} \, {}^{RL}_0 D^{1-\beta}_z \left[E_\alpha(\gamma z^\alpha) \right] &= z^{1-\beta} \sum_{n=0}^\infty {}^{RL}_0 D^{1-\beta}_z \left[\frac{\gamma^n z^{\alpha n}}{\Gamma(n\alpha + 1)} \right] \\ &= z^{1-\beta} \sum_{n=0}^\infty \frac{\Gamma(\alpha n + 1)}{\Gamma(\alpha n + \beta)} \cdot \frac{\gamma^n z^{\alpha n + \beta - 1}}{\Gamma(n\alpha + 1)} \\ &= \sum_{n=0}^\infty \frac{\gamma^n z^{\alpha n}}{\Gamma(\alpha n + \beta)}. \end{aligned}$$

This series converges uniformly, being precisely the series expression in Equation (6) for $E_{\alpha,\beta}(\gamma z^\alpha)$. Thus, our swapping of operations above was justified, and the proof is complete. □

The key point here is that the dependence of the Mittag–Leffler function in Equation (7) on the parameters ρ and β can be encoded by fractional differintegrals. Proposition 1 enables us to interpret the parameter ρ as merely the order of a differintegral, and Proposition 2 enables us to do the same with β.

By combining the results of Proposition 1 and Proposition 2, it is possible to obtain a composite expression for the three-parameter Mittag–Leffler function in Equation (7) in terms of fractional-type integrals, as described by the following theorem.

Theorem 1. *For any $\alpha, \beta, \gamma, \rho \in \mathbb{C}$ with $\mathrm{Re}(\alpha), \mathrm{Re}(\beta) > 0$ and $\mathrm{Re}(\rho) < 1$, we have:*

$$E^\rho_{\alpha,\beta}(\gamma z^\alpha) = \frac{\alpha \sin(\pi \rho)}{\pi} \int_0^z (z^\alpha - u^\alpha)^{-\rho} u^{\alpha \rho - \beta} \, {}^{RL}_0 D^{1-\beta}_u \left[E_\alpha(\gamma u^\alpha) \right] du, \qquad z \in \mathbb{D}. \tag{16}$$

Proof. For $\text{Re}(\rho) < 1$, the fractional differintegrals appearing in Proposition 1 and Corollary 1 are integrals (because $\text{Re}(\rho - 1) < 0$), and so Equation (14) can be rewritten as follows:

$$E_{\alpha,\beta}^{\rho}(\gamma z) = \frac{1}{\Gamma(\rho)} \, {}^{RL}_0 I_z^{1-\rho}\left[z^{\rho-1} E_{\alpha,\beta}(\gamma z)\right] = \frac{1}{\Gamma(\rho)\Gamma(1-\rho)} \int_0^z (z-y)^{-\rho} y^{\rho-1} E_{\alpha,\beta}(\gamma y)\, dy.$$

Making the change of variables $u = y^{1/\alpha}$, and using the reflection formula for the gamma function:

$$E_{\alpha,\beta}^{\rho}(\gamma z) = \frac{\sin(\pi\rho)}{\pi} \int_0^{z^{1/\alpha}} (z-u^\alpha)^{-\rho} u^{\alpha(\rho-1)} E_{\alpha,\beta}(\gamma u^\alpha) \alpha u^{\alpha-1}\, du$$

$$= \frac{\alpha \sin(\pi\rho)}{\pi} \int_0^{z^{1/\alpha}} (z-u^\alpha)^{-\rho} u^{\alpha\rho-1} E_{\alpha,\beta}(\gamma u^\alpha)\, du.$$

Now, we can apply the result of Proposition 2 to the two-parameter Mittag–Leffler function appearing in the integrand of this expression:

$$E_{\alpha,\beta}^{\rho}(\gamma z) = \frac{\alpha \sin(\pi\rho)}{\pi} \int_0^{z^{1/\alpha}} (z-u^\alpha)^{-\rho} u^{\alpha\rho-1}\left[u^{1-\beta} \, {}^{RL}_0 D_u^{1-\beta}\left[E_\alpha(\gamma u^\alpha)\right]\right] du$$

$$= \frac{\alpha \sin(\pi\rho)}{\pi} \int_0^{z^{1/\alpha}} (z-u^\alpha)^{-\rho} u^{\alpha\rho-\beta} \, {}^{RL}_0 D_u^{1-\beta}\left[E_\alpha(\gamma u^\alpha)\right] du.$$

Substituting z^α for z:

$$E_{\alpha,\beta}^{\rho}(\gamma z^\alpha) = \frac{\alpha \sin(\pi\rho)}{\pi} \int_0^{z} (z^\alpha - u^\alpha)^{-\rho} u^{\alpha\rho-\beta} \, {}^{RL}_0 D_u^{1-\beta}\left[E_\alpha(\gamma u^\alpha)\right] du.$$

Note that this can almost, but not quite, be expressed as a composition of two fractional differintegrals. □

Corollary 2. *For any $\alpha, \beta, \gamma, \rho \in \mathbb{C}$ with $\text{Re}(\alpha) > 0$, $\text{Re}(\beta) > 1$, and $\text{Re}(\rho) < 1$, we have:*

$$E_{\alpha,\beta}^{\rho}(\gamma z^\alpha) = \frac{\alpha \sin(\pi\rho)}{\pi \Gamma(\beta-1)} \int_0^z (z^\alpha - u^\alpha)^{-\rho} u^{\alpha\rho-\beta} \int_0^u (u-t)^{\beta-2} E_\alpha(\gamma t^\alpha)\, dt\, du, \qquad z \in \mathbb{D}. \qquad (17)$$

Proof. In this case, we have $\text{Re}(1-\beta) < 0$ and so the fractional differintegral which appears in the integrand of Equation (16) is an *integral*. Thus, we can write

$$^{RL}_0 I_u^{\beta-1}\left[E_\alpha(\gamma u^\alpha)\right] = \frac{1}{\Gamma(\beta-1)} \int_0^u (u-t)^{\beta-2} E_\alpha(\gamma t^\alpha)\, dt,$$

and substitute this into Equation (16) to find:

$$E_{\alpha,\beta}^{\rho}(\gamma z^\alpha) = \frac{\alpha \sin(\pi\rho)}{\pi} \int_0^z (z^\alpha - u^\alpha)^{-\rho} u^{\alpha\rho-\beta} \left[\frac{1}{\Gamma(\beta-1)} \int_0^u (u-t)^{\beta-2} E_\alpha(\gamma t^\alpha)\, dt\right] du,$$

which rearranges to the required result. □

Theorem 2. *The three-parameter Mittag–Leffler function in Equation (7) can be written as an integral transform of the one-parameter Mittag–Leffler function in Equation (5) in the following way:*

$$E_{\alpha,\beta}^{\rho}(\gamma z^\alpha) = \frac{\alpha \sin(\pi\rho)}{\pi \Gamma(\beta-1)} \int_0^z F_{\alpha,\beta,\rho}(t;z) E_\alpha(\gamma t^\alpha)\, dt \qquad z \in \mathbb{D}, \qquad (18)$$

where we assume $\alpha, \beta, \gamma, \rho \in \mathbb{C}$ with $\mathrm{Re}(\alpha) > 0$, $\mathrm{Re}(\beta) > 1$, and $\mathrm{Re}(\rho) < 1$, and where the function F is defined as

$$F_{\alpha,\beta,\rho}(t;z) := \int_t^z (z^\alpha - u^\alpha)^{-\rho} u^{\alpha\rho-\beta}(u-t)^{\beta-2}\, du. \qquad (19)$$

Proof. By Fubini's theorem, it is possible to swap the order of the integrals in Equation (17). We have $0 \leq u \leq z$ and $0 \leq t \leq u$, which after swapping is equivalent to $0 \leq t \leq z$ and $t \leq u \leq z$. We have from Equation (17):

$$\begin{aligned}
E_{\alpha,\beta}^\rho(\gamma z^\alpha) &= \frac{\alpha \sin(\pi\rho)}{\pi\Gamma(\beta-1)} \int_0^z \int_0^u (z^\alpha - u^\alpha)^{-\rho} u^{\alpha\rho-\beta}(u-t)^{\beta-2} E_\alpha(\gamma t^\alpha)\, dt\, du \\
&= \frac{\alpha \sin(\pi\rho)}{\pi\Gamma(\beta-1)} \int_0^z \int_t^z (z^\alpha - u^\alpha)^{-\rho} u^{\alpha\rho-\beta}(u-t)^{\beta-2} E_\alpha(\gamma t^\alpha)\, du\, dt \\
&= \frac{\alpha \sin(\pi\rho)}{\pi\Gamma(\beta-1)} \int_0^z E_\alpha(\gamma t^\alpha) \int_t^z (z^\alpha - u^\alpha)^{-\rho} u^{\alpha\rho-\beta}(u-t)^{\beta-2}\, du\, dt \\
&= \frac{\alpha \sin(\pi\rho)}{\pi\Gamma(\beta-1)} \int_0^z E_\alpha(\gamma t^\alpha) F_{\alpha,\beta,\rho}(t;z)\, dt,
\end{aligned}$$

as required. □

Thus, using Riemann–Liouville fractional calculus, we have forged new connections between the Mittag–Leffler functions of one, two and three parameters. The connection between those of one and three parameters, in particular, may give rise to new formulae linking AB fractional calculus with Prabhakar fractional calculus, in a way more profound than simply writing one as a special case of the other.

3. Applications

As we have already discussed, the various Mittag–Leffler functions are interesting from the point of view of pure mathematical analysis and fractional calculus [34–37] (see also the correction [16,36,38]). However, it is also important to discuss the motivation for studying these functions from the point of view of real-world applications in science and engineering.

The one-parameter Mittag–Leffler function has already discovered many applications via the AB model, and also previously in relaxation models which involve interpolation between exponential and power-law behaviours [39]. In recent years, the two-parameter and three-parameter Mittag–Leffler functions have also been emerging from real experimental data.

A group of biologists and engineers in Cambridge and London have been experimenting with models for cells and tissues, and discovered that their data fit most closely to an operator involving two-parameter Mittag–Leffler functions [40].

The three-parameter Mittag–Leffler function, sometimes called the Prabhakar function, is closely connected with the phenomenon of Havriliak–Negami relaxation [41], and this has been studied also in the context of fractional relaxation [26,42].

In view of these manifold applications of the Mittag–Leffler functions of one, two and three parameters, we believe that our results herein may also discover applications. The one-parameter Mittag–Leffler function is much more elementary and easier to handle than the two- and three-parameter Mittag–Leffler functions. Thus, reducing the latter to the former should mark a major step forward. The physical processes which are modelled using two- and three-parameter Mittag–Leffler functions may now be more easily analysed using only the one-parameter Mittag–Leffler function.

In particular, we note that numerical computation of Mittag–Leffler functions has been a challenging problem for researchers in recent years [43–45]. Naturally, the one-parameter Mittag–Leffler function is the most straightforward to handle. If we can use relations such as those proved in this paper to express the more advanced Mittag–Leffler functions purely in terms of the most basic one, then it may enable much easier computation of the two- and three-parameter Mittag–Leffler functions than before.

4. Conclusions

In this article, we have established new relations between the Mittag–Leffler functions of one, two and three parameters by using Riemann–Liouville fractional calculus. The main results are Proposition 1 (three-parameter Mittag–Leffler in terms of two-parameter Mittag–Leffler), and Theorem 2 (three-parameter Mittag–Leffler in terms of one-parameter Mittag–Leffler), which come from combining Proposition 1 with Proposition 2 (two-parameter Mittag–Leffler in terms of one-parameter Mittag–Leffler). We believe that these results can be applied in the future, to simplify some important physical models that use two- or three-parameter Mittag–Leffler functions, or to provide more efficient computational models for these functions, since the original one-parameter Mittag–Leffler function is much better known and more deeply studied.

Author Contributions: Conceptualisation, H.M.S.; methodology, H.M.S., A.F., and D.B.; formal analysis, A.F.; writing—original draft preparation, A.F.; writing—review and editing, D.B.

Funding: This research received no external funding.

Conflicts of Interest: The authors declare no conflict of interest.

References

1. Miller, K.S.; Ross, B. *An Introduction to the Fractional Calculus and Fractional Differential Equations*; Wiley: New York, NY, USA, 1993.
2. Oldham, K.B.; Spanier, J. *The Fractional Calculus*; Academic Press: San Diego, CA, USA, 1974.
3. Samko, S.G.; Kilbas, A.A.; Marichev, O.I. *Fractional Integrals and Derivatives: Theory and Applications*, Orig. ed.; in Russian, Nauka i Tekhnika, Minsk, 1987; Taylor & Francis: London, UK, 2002.
4. Hilfer, R. (Ed.) *Applications of Fractional Calculus in Physics*; World Scientific: Singapore, 2000.
5. Kilbas, A.A.; Srivastava, H.M.; Trujillo, J.J. *Theory and Applications of Fractional Differential Equations*; Elsevier: Amsterdam, The Netherlands, 2006.
6. Podlubny, I. *Fractional Differential Equations*; Academic Press: San Diego, CA, USA, 1999.
7. Atangana, A.; Baleanu, D. New fractional derivatives with nonlocal and non-singular kernel: Theory and application to heat transfer model. *Therm. Sci.* **2016**, *20*, 763–769. [CrossRef]
8. Caputo, M.; Fabrizio, M. A new Definition of Fractional Derivative without Singular Kernel. *Prog. Fract. Differ. Appl.* **2015**, *1*, 73–85.
9. Çetinkaya, A.; Kıymaz, I.O.; Agarwal, P.; Agarwal, R. A comparative study on generating function relations for generalized hypergeometric functions via generalized fractional operators. *Adv. Differ. Equ.* **2018**, *2018*, 156. [CrossRef]
10. Özarslan, M.A.; Ustaoğlu, C. Some Incomplete Hypergeometric Functions and Incomplete Riemann-Liouville Fractional Integral Operators. *Mathematics* **2019**, *7*, 483. [CrossRef]
11. Özarslan, M.A.; Ustaoğlu, C. Incomplete Caputo fractional derivative operators. *Adv. Differ. Equ.* **2018**, *2018*, 209. [CrossRef]
12. Haubold, H.J.; Mathai, A.M.; Saxena, R.K. Mittag-Leffler functions and their applications. *J. Appl. Math.* **2011**, *2011*, 298628. [CrossRef]
13. Mainardi, F.; Gorenflo, R. On Mittag-Leffler-type functions in fractional evolution processes. *J. Comput. Appl. Math.* **2000**, *118*, 283–299. [CrossRef]
14. Mathai, A.M.; Haubold, H.J. Mittag-Leffler functions and fractional calculus. In *Special Functions for Applied Scientists*; Springer: New York, NY, USA, 2008; pp. 79–134.
15. Gorenflo, R.; Kilbas, A.A.; Mainardi, F.; Rogosin, S.V. *Mittag-Leffler Functions, Related Topics and Applications*; Springer: Berlin, Germany, 2016.
16. Srivastava, H.M.; Tomovski, Ž. Fractional calculus with an integral operator containing a generalized Mittag-Leffler function in the kernel. *Appl. Math. Comput.* **2009**, *211*, 198–210. [CrossRef]

17. Abdeljawad, T.; Baleanu, D. Integration by parts and its applications of a new nonlocal fractional derivative with Mittag-Leffler nonsingular kernel. *J. Nonlinear Sci. Appl.* **2017**, *10*, 1098–1107. [CrossRef]
18. Baleanu, D.; Fernandez, A. On some new properties of fractional derivatives with Mittag-Leffler kernel. *Commun. Nonlinear Sci. Numer. Simul.* **2018**, *59*, 444–462. [CrossRef]
19. Djida, J.-D.; Atangana, A.; Area, I. Numerical Computation of a Fractional Derivative with Non-Local and Non-Singular Kernel. *Math. Model. Nat. Phenom.* **2017**, *12*, 4–13. [CrossRef]
20. Fernandez, A.; Baleanu, D. The mean value theorem and Taylor's theorem for fractional derivatives with Mittag-Leffler kernel. *Adv. Differ. Equ.* **2018**, *2018*, 86. [CrossRef]
21. Fernandez, A. A complex analysis approach to Atangana–Baleanu fractional calculus. *Math. Model. Appl. Sci.* **2019**. [CrossRef]
22. Prabhakar, T.R. A singular integral equation with a generalized Mittag Leffler function in the kernel. *Yokohama Math. J.* **1971**, *19*, 7–15.
23. Garra, R.; Gorenflo, R.; Polito, F.; Tomovski, Ž. Hilfer–Prabhakar derivatives and some applications. *Appl. Math. Comput.* **2014**, *242*, 576–589. [CrossRef]
24. Kilbas, A.A.; Saigo, M.; Saxena, R.K. Generalized Mittag-Leffler function and generalized fractional calculus operators. *Integral Transform. Spec. Funct.* **2004**, *15*, 31–49. [CrossRef]
25. Fernandez, A.; Baleanu, D.; Srivastava, H.M. Series representations for models of fractional calculus involving generalised Mittag-Leffler functions. *Commun. Nonlinear Sci. Numer. Simul.* **2019**, *67*, 517–527. [CrossRef]
26. Garra, R.; Garrappa, R. The Prabhakar or three parameter Mittag-Leffler function: Theory and application. *Commun. Nonlinear Sci. Numer. Simul.* **2018**, *56*, 314–329. [CrossRef]
27. Sandev, T. Generalized Langevin Equation and the Prabhakar Derivative. *Mathematics* **2017**, *5*, 66. [CrossRef]
28. Osler, T.J. Leibniz rule for fractional derivatives generalised and an application to infinite series. *SIAM J. Appl. Math.* **1970**, *18*, 658–674. [CrossRef]
29. Osler, T.J. The fractional derivative of a composite function. *SIAM J. Math. Anal.* **1970**, *1*, 288–293. [CrossRef]
30. Fernandez, A. The Lerch zeta function as a fractional derivative. *arXiv* **2018**, arXiv:1804.07936.
31. Keiper, J.B. Fractional Calculus and Its Relationship to Riemann's Zeta Function. Master's Thesis, Ohio State University, Columbus, OH, USA, 1975; 37p.
32. Lin, S.-D.; Srivastava, H.M. Some families of the Hurwitz–Lerch zeta functions and associated fractional derivative and other integral representations. *Appl. Math. Comput.* **2004**, *154*, 725–733. [CrossRef]
33. Paneva-Konovska, J. Differential and integral relations in the class of multi-index Mittag-Leffler functions. *Fract. Calc. Appl. Anal.* **2018**, *21*, 254–265. [CrossRef]
34. Srivastava, H.M. On an extension of the Mittag-Leffler function. *Yokohama Math. J.* **1968**, *16*, 77–88.
35. Srivastava, H.M. Some families of Mittag-Leffler type functions and associated operators of fractional calculus. *TWMS J. Pure Appl. Math.* **2016**, *7*, 123–145.
36. Tomovski, Ž.; Hilfer, R.; Srivastava, H.M. Fractional and operational calculus with generalized fractional derivative operators and Mittag-Leffler type functions. *Integral Transform. Spec. Funct.* **2010**, *21*, 797–814. [CrossRef]
37. Tomovski, Ž.; Pogány, T.K.; Srivastava, H.M. Laplace type integral expressions for a certain three-parameter family of generalized Mittag-Leffler functions with applications involving complete monotonicity. *J. Frankl. Inst.* **2014**, *351*, 5437–5454. [CrossRef]
38. Srivastava, H.M. Remarks on some families of fractional-order differential equations. *Integral Transform. Spec. Funct.* **2017**, *28*, 560–564. [CrossRef]
39. Metzler, R.; Klafter, J. From stretched exponential to inverse power-law: Fractional dynamics, Cole–Cole relaxation processes, and beyond. *J. Non-Cryst. Solids* **2002**, *305*, 81–87. [CrossRef]
40. Bonfanti, A.; Fouchard, J.; Khalilgharibi, N.; Charras, G.; Kabla, A. A unified rheological model for cells and cellularised materials. **2019**, under review. [CrossRef]
41. Havriliak, S.; Negami, S. A complex plane representation of dielectric and mechanical relaxation processes in some polymers. *Polymer* **1967**, *8*, 161–210. [CrossRef]
42. Garrappa, R. Grünwald–Letnikov operators for fractional relaxation in Havriliak–Negami models. *Commun. Nonlinear Sci. Numer. Simul.* **2016**, *38*, 178–191. [CrossRef]
43. Garrappa, R. Numerical Evaluation of Two and Three Parameter Mittag-Leffler Functions. *SIAM J. Numer. Anal.* **2015**, *53*, 1350–1369.[CrossRef]

44. Seybold, H.; Hilfer, R. Numerical Algorithm for Calculating the Generalized Mittag-Leffler Function. *SIAM J. Numer. Anal.* **2008**, *47*, 69–88. [CrossRef]
45. Valério, D.; Machado, J.T. On the numerical computation of the Mittag-Leffler function. *Commun. Nonlinear Sci. Numer. Simul.* **2014**, *19*, 3419–3424. [CrossRef]

© 2019 by the authors. Licensee MDPI, Basel, Switzerland. This article is an open access article distributed under the terms and conditions of the Creative Commons Attribution (CC BY) license (http://creativecommons.org/licenses/by/4.0/).

Article

Some Incomplete Hypergeometric Functions and Incomplete Riemann-Liouville Fractional Integral Operators

Mehmet Ali Özarslan [1] and Ceren Ustaoğlu [2,*]

[1] Department of Mathematics, Faculty of Arts and Sciences, Eastern Mediterranean University, Famagusta, TRNC, Mersin 10, Turkey; mehmetali.ozarslan@emu.edu.tr
[2] Department of Computer Engineering, Final International University, Toroslar Caddesi, No. 6, Çatalköy, Girne, TRNC, Mersin 10, Turkey
* Correspondence: ceren.ustaoglu@final.edu.tr

Received: 19 March 2019; Accepted: 22 May 2019; Published: 27 May 2019

Abstract: Very recently, the incomplete Pochhammer ratios were defined in terms of the incomplete beta function $B_y(x,z)$. With the help of these incomplete Pochhammer ratios, we introduce new incomplete Gauss, confluent hypergeometric, and Appell's functions and investigate several properties of them such as integral representations, derivative formulas, transformation formulas, and recurrence relations. Furthermore, incomplete Riemann-Liouville fractional integral operators are introduced. This definition helps us to obtain linear and bilinear generating relations for the new incomplete Gauss hypergeometric functions.

Keywords: Gauss hypergeometric function; confluent hypergeometric function; Appell's functions; incomplete fractional calculus; Riemann-Liouville fractional integral; generating functions

1. Introduction and Preliminaries

In recent years, some extensions of the well-known special functions have been considered by several authors (see, for example, [1–9]). The familiar incomplete gamma functions $\gamma(s,x)$ and $\Gamma(s,x)$ are defined by:

$$\gamma(s,x) := \int_0^x t^{s-1} e^{-t} dt \qquad (Re(s) > 0;\ x \geqq 0)$$

and

$$\Gamma(s,x) := \int_x^\infty t^{s-1} e^{-t} dt \qquad (x \geqq 0;\ Re(s) > 0 \text{ when } x = 0),$$

respectively. They satisfy the following decomposition formula:

$$\gamma(s,x) + \Gamma(s,x) = \Gamma(s) \quad (Re(s) > 0). \qquad (1)$$

The function $\Gamma(s)$ and its incomplete versions $\gamma(s,x)$ and $\Gamma(s,x)$ play important roles in the study of analytical solutions of a variety of problems in diverse areas of science and engineering.

The widely-used Pochhammer symbol $(\lambda)_\nu$ ($\lambda, \nu \in \mathbb{C}$) is defined, in general, by:

$$(\lambda)_\nu := \frac{\Gamma(\lambda+\nu)}{\Gamma(\lambda)} = \begin{cases} 1 & (\nu = 0;\ \lambda \in \mathbb{C}\setminus\{0\}) \\ \lambda(\lambda+1)\ldots(\lambda+\nu-1) & (\nu \in \mathbb{N};\ \lambda \in \mathbb{C}) \end{cases} \qquad (2)$$

In terms of the incomplete gamma functions $\gamma(s,x)$ and $\Gamma(s,x)$, the incomplete Pochhammer symbols $(\lambda;x)_\nu$ and $[\lambda;x]_\nu$ $(\lambda,\nu \in \mathbb{C};\ x \geqq 0)$ were defined as follows [10]:

$$(\lambda;x)_\nu := \frac{\gamma(\lambda+\nu,x)}{\Gamma(\lambda)} \quad (\lambda,\nu \in \mathbb{C};\ x \geqq 0) \tag{3}$$

and:

$$[\lambda;x]_\nu := \frac{\Gamma(\lambda+\nu,x)}{\Gamma(\lambda)} \quad (\lambda,\nu \in \mathbb{C};\ x \geqq 0). \tag{4}$$

In view of (1), these incomplete Pochhammer symbols $(\lambda;x)_\nu$ and $[\lambda;x]_\nu$ satisfy the following decomposition relation:

$$(\lambda;x)_\nu + [\lambda;x]_\nu = (\lambda)_\nu \quad (\lambda,\nu \in \mathbb{C};\ x \geqq 0), \tag{5}$$

where $(\lambda)_\nu$ is the Pochhammer symbol given by (2).

The incomplete Gauss hypergeometric functions were defined by means of the incomplete gamma functions as follows [10]:

$$_2\gamma_1\left[\begin{array}{cc}(a,x) & \cdot & b\ ;\\ & c\ ;\end{array} z\right] := \sum_{n=0}^{\infty} \frac{(a;x)_n (b)_n}{(c)_n} \frac{z^n}{n!} \tag{6}$$

and:

$$_2\Gamma_1\left[\begin{array}{cc}(a,x) & \cdot & b\ ;\\ & c\ ;\end{array} z\right] := \sum_{n=0}^{\infty} \frac{[a;x]_n (b)_n}{(c)_n} \frac{z^n}{n!}. \tag{7}$$

After this work, incomplete hypergeometric functions have become a fruitful topic of research in recent years [4,5,9,11–20].

Fractional derivative and integral operators are another important topic of research in recent years. They have found applications in many diverse areas of mathematical, physical, and engineering problems; good summaries of these applications may be found in [21–26] and recently in [27]. The use of fractional derivative operators in obtaining generating relations for some special functions can be found in [6,9,28–30].

In fractional calculus, there are two important differential operators: the Riemann-Liouville and Liouville-Caputo fractional derivatives. In a recent paper [12], which covered work done after the work herein, we introduced incomplete Liouville-Caputo fractional derivative operators and focused on their use in special function theory. For the definitions in [12], we considered the same incomplete Riemann-Liouville integral as in (60) and (61) of this paper, but the operators introduced there were of the Liouville-Caputo type and not of the Riemann-Liouville type like those in the current work. The difference between Liouville-Caputo and Riemann-Liouville is very important for applications to differential equations, because the required initial conditions are of different types between these two cases.

In the present paper, we introduce new incomplete hypergeometric functions with the aid of incomplete Pochhammer ratios and investigate certain properties of them. Moreover, we introduce incomplete Riemann-Liouville fractional integral operators, and we obtain some generating relations for these new incomplete hypergeometric functions with the aid of these new defined operators. The organization of the paper is as follows.

In Section 2, the incomplete Pochhammer ratios are introduced by using the incomplete beta function, and some derivative formulas involving these new incomplete Pochhammer ratios are investigated. In Section 3, new incomplete Gauss hypergeometric functions and confluent hypergeometric functions are introduced with the help of these incomplete Pochhammer ratios, and integral representations, derivative formulas, transformation formulas, and recurrence relations are obtained for them. In Section 4, we define new incomplete Appell's functions $\mathfrak{F}_1[a,b,c;d;x,z;y]$, $\mathfrak{F}_1\{a,b,c;d;x,z;y\}$, $\mathfrak{F}_2[a,b,c;d,e;x,z;y]$, and

$\mathfrak{F}_2\{a,b,c;d,e;x,z;y\}$ and obtain their integral representations. In Section 5, we introduce incomplete Riemann-Liouville fractional integral operators and show that the incomplete Riemann-Liouville fractional integrals of some elementary functions give the new incomplete functions defined in Sections 3 and 4. Finally, in the last section, we obtain linear and bilinear generating relations for the incomplete hypergeometric functions.

2. The Incomplete Pochhammer Ratio

The incomplete beta function is defined by:

$$B_y(x,z) := \int_0^y t^{x-1}(1-t)^{z-1} dt, \ Re(x) > Re(z) > 0, \ 0 \leq y < 1 \tag{8}$$

and can be expressed in terms of the Gauss hypergeometric function:

$$B_y(x,z) := \frac{y^x}{x} {}_2F_1(x, 1-z; 1+x; y). \tag{9}$$

The incomplete Pochhammer ratios $[b,c;y]_n$ and $\{b,c;y\}_n$ are introduced in terms of the incomplete beta function $B_y(x,z)$ as follows [12]:

$$[b,c;y]_n := \frac{B_y(b+n, c-b)}{B(b, c-b)} \tag{10}$$

and:

$$\{b,c;y\}_n := \frac{B_{1-y}(c-b, b+n)}{B(b, c-b)} \tag{11}$$

where $0 \leq y < 1$. They satisfy the following relation:

$$[b,c;y]_n + \{b,c;y\}_n = \frac{(b)_n}{(c)_n}. \tag{12}$$

In view of (9), we have the following relations:

$$[b,c;y]_n := \frac{1}{B(b,c-b)} \frac{y^{b+n}}{b+n} {}_2F_1(b+n, 1-c+b; b+n+1; y) \tag{13}$$

and:

$$\{b,c;y\}_n := \frac{1}{B(b,c-b)} \frac{(1-y)^{c-b}}{c-b} {}_2F_1(c-b, 1-b-n; 1+c-b; 1-y). \tag{14}$$

In the following theorem, we investigate the nth derivatives of the incomplete beta function by means of incomplete Pochhammer ratios.

Theorem 1. *The following derivative formulas hold true:*

$$[b,c;y]_n = \frac{(-1)^n \Gamma(c)}{\Gamma(c-b+n) \Gamma(b)} y^{b+n} \frac{d^n}{dy^n} \left[y^{-b} B_y(b, c-b+n) \right], \tag{15}$$

and:

$$\{b,c;y\}_n = \frac{\Gamma(b+n)}{\Gamma(b+2n)} \frac{1}{B(b,c-b)} (1-y)^{c-b} \frac{d^n}{dy^n} ((1-y)^{-c+b+n} B_{1-y}(c-b-n, b+2n)). \tag{16}$$

Proof. Using (8) and (10), we immediately obtain the following equation:

$$[b,c;y]_n = \frac{y^{b+n}}{B(b,c-b)} \int_0^1 u^{b+n-1}(1-uy)^{c-b-1} du.$$

On the other hand, we have:

$$y^{-b} B_y(b, c-b+n) = \int_0^1 u^{b-1}(1-uy)^{c-b+n-1} du. \tag{17}$$

Taking derivatives n times on both sides of (17) with respect to y, we can obtain a derivative formula for the incomplete beta function $[b,c;y]_n$ asserted by (15). Formula (16) can be proven in a similar way. □

3. The New Incomplete Gauss and Confluent Hypergeometric Functions

In this section, we introduce new incomplete Gauss and confluent hypergeometric functions by:

$$_2\mathfrak{F}_1(a, [b,c;y]; x) := \sum_{n=0}^{\infty} (a)_n [b,c;y]_n \frac{x^n}{n!}, \tag{18}$$

$$_2\mathfrak{F}_1(a, \{b,c;y\}; x) := \sum_{n=0}^{\infty} (a)_n \{b,c;y\}_n \frac{x^n}{n!}, \tag{19}$$

$$_1\mathfrak{F}_1([a,b;y]; x) := \sum_{n=0}^{\infty} [a,b;y]_n \frac{x^n}{n!}, \tag{20}$$

and:

$$_1\mathfrak{F}_1(\{a,b;y\}; x) := \sum_{n=0}^{\infty} \{a,b;y\}_n \frac{x^n}{n!} \tag{21}$$

where $0 \leq y < 1$.

An immediate consequence of (12) and the definitions (18), (19), (20), and (21) is the following decomposition formulas:

$$_2\mathfrak{F}_1(a, [b,c;y]; x) + {}_2\mathfrak{F}_1(a, \{b,c;y\}; x) = {}_2F_1(a,b;c;x) \tag{22}$$

and:

$$_1\mathfrak{F}_1([a,b;y]; x) + {}_1\mathfrak{F}_1(\{a,b;y\}; x) = {}_1F_1(a;b;x). \tag{23}$$

Theorem 2. *The following integral representation holds true:*

$$_2\mathfrak{F}_1(a, [b,c;y], x) = \frac{y^b}{B(b,c-b)} \int_0^1 u^{b-1}(1-uy)^{c-b-1}(1-xuy)^{-a} du, \tag{24}$$

$$Re(c) > Re(b) > 0, \ |\arg(1-x)| < \pi).$$

Proof. Replacing the incomplete Pochhammer ratio $[b,c;y]_n$ in the definition (18) by its integral representation given by (8) and interchanging the order of summation and integral, which is permissible under the conditions given in the hypothesis of the Theorem, we find:

$$_2\mathfrak{F}_1(a, [b,c;y], x) = \frac{1}{B(b,c-b)} \int_0^y t^{b-1}(1-t)^{c-b-1}(1-xt)^{-a} dt, \tag{25}$$

which can be written as follows:

$$_2\mathfrak{F}_1(a,[b,c;y],x) = \frac{y^b}{B(b,c-b)} \int_0^1 u^{b-1}(1-uy)^{c-b-1}(1-xuy)^{-a} du. \qquad (26)$$

□

In a similar way, we have the following theorem:

Theorem 3. *The following integral representation holds true:*

$$_2\mathfrak{F}_1(a,\{b,c;y\},x) = \frac{(1-y)^{c-b}}{B(b,c-b)} \int_0^1 u^{c-b-1}(1-u(1-y))^{b-1}(1-x+xu(1-y))^{-a} du,$$
$$\mathrm{Re}(c) > \mathrm{Re}(b) > 0, |\arg(1-x)| < \pi. \qquad (27)$$

Theorem 4. *The following result holds true:*

$$_2\mathfrak{F}_1(a,[b,c;y],1) = \frac{\Gamma(c)\Gamma(c-a-b)}{\Gamma(c-a)\Gamma(c-b)} - \frac{(1-y)^{c-b-a}y^b}{B(b,c-b)(c-a-b)} \,_2F_1(c-a,1;1+c-b-a;1-y). \qquad (28)$$

Proof. Putting $x = 1$ in (22), we obtain:

$$_2\mathfrak{F}_1(a,[b,c;y],1) = \,_2F_1(a,b;c;1) - \,_2\mathfrak{F}_1(a,\{b,c;1-y\},1) \qquad (29)$$
$$= \frac{\Gamma(c)\Gamma(c-a-b)}{\Gamma(c-a)\Gamma(c-b)} - \frac{(1-y)^{c-b-a}}{B(b,c-b)} \int_0^1 u^{c-b-a-1}(1-u(1-y))^{b-1} du.$$

Using Euler's integral representation for (29), we have:

$$_2\mathfrak{F}_1(a,[b,c;y],1) = \frac{\Gamma(c)\Gamma(c-a-b)}{\Gamma(c-a)\Gamma(c-b)} - \frac{(1-y)^{c-b-a}}{B(b,c-b)(c-b-a)} \,_2F_1(1-b,c-b-a;1+c-b-a;1-y). \qquad (30)$$

Using transformation formula:

$$_2F_1(\alpha,\beta;\gamma;z) = (1-z)^{\gamma-\beta-\alpha} \,_2F_1(\gamma-\alpha,\gamma-\beta;\gamma;z), \qquad (31)$$

in (30), we obtain:

$$_2F_1(1-b,c-b-a;1+c-b-a;1-y) = y^b \,_2F_1(c-a,1;1+c-b-a;1-y). \qquad (32)$$

Considering (32) in (30), we get:

$$_2\mathfrak{F}_1(a,[b,c;y],1) = \frac{\Gamma(c)\Gamma(c-a-b)}{\Gamma(c-a)\Gamma(c-b)} - \frac{(1-y)^{c-b-a}y^b}{B(b,c-b)(c-b-a)} \,_2F_1(c-a,1;1+c-b-a;1-y). \qquad (33)$$

□

Theorem 5. *The following result holds true:*

$$_2\mathfrak{F}_1(a,\{b,c;y\},1) = \frac{\Gamma(c)\Gamma(c-a-b)}{\Gamma(c-a)\Gamma(c-b)} - \frac{(1-y)^{c-b-a}y^b}{B(b,c-b)b} \,_2F_1(c-a,1;b+1;y). \qquad (34)$$

Theorem 6. *The following integral representations hold true:*

$$_1\mathfrak{F}_1([a,b;y],x) = \frac{y^a}{B(a,b-a)} \int_0^1 u^{a-1}(1-uy)^{b-a-1} e^{xuy} du, \quad Re(b) > Re(a) > 0 \qquad (35)$$

and:

$$_1\mathfrak{F}_1(\{a,b;y\},x) = \frac{(1-y)^{b-a}}{B(a,b-a)} \int_0^1 u^{b-a-1}(1-u(1-y))^{a-1} e^{(1-u(1-y))x} du, \quad Re(b) > Re(a) > 0. \qquad (36)$$

Proof. Replacing the incomplete Pochhammer ratio $[a,b;y]_n$ in the definition (20) by its integral representation given by (8), we are led to the desired result (35). Formula (36) can be proven in a similar way. □

Theorem 7. *The following integral representation holds true:*

$$\int_0^1 y^{k-1} {}_2\mathfrak{F}_1(a,[b,c-k;y];x) dy = \frac{1}{k}\left[{}_2F_1(a,b;c-k;x) - \frac{\Gamma(c-k)\Gamma(b+k)}{\Gamma(b)\Gamma(c)} {}_2F_1(a,b+k;c;x)\right], \quad k \in \mathbb{N}. \qquad (37)$$

Proof. It is known that from Euler's formula that:

$$_2F_1(a,b+k;c;x) = \frac{1}{B(b+k,c-b-k)} \int_0^1 y^{b+k-1}(1-y)^{c-b-k-1}(1-xy)^{-a} dy, \quad k \in \mathbb{N}.$$

Taking $u = y^k$ and the remaining part as dv and applying the integration by parts, we get:

$$_2F_1(a,b+k;c;x) = \frac{\Gamma(b)\Gamma(c)}{\Gamma(c-k)\Gamma(b+k)} \left[{}_2F_1(a,b;c-k;x) - k\int_0^1 y^{k-1} {}_2\mathfrak{F}_1(a,[b,c-k;y],x) dy\right].$$

By rearranging the terms, we get the result. □

Corollary 1. *Taking $k=1$ in Theorem 7, we get the following result:*

$$\int_0^1 {}_2\mathfrak{F}_1(a,[b,c-1;y],x) dy = {}_2F_1(a,b;c-1;x) - \frac{b}{c-1} {}_2F_1(a,b+1;c;x). \qquad (38)$$

Theorem 8. *The following integral representation holds true:*

$$\int_0^1 y^{k-1} {}_2\mathfrak{F}_1(a,[b,c;y],x) dy = \frac{1}{k} \frac{\Gamma(c)\Gamma(c-b+k)}{\Gamma(c-b)\Gamma(c+k)} {}_2F_1(a,b;c+k;x). \qquad (39)$$

Proof. It is known that:

$$_2F_1(a,b;c+k;x) = \frac{1}{B(b,c-b+k)} \int_0^1 y^{b-1}(1-y)^{c-b+k-1}(1-xy)^{-a} dy.$$

Taking $u = (1-y)^k$ and the rest as dv and using integration by parts, we get the result. □

Corollary 2. *Taking $k=1$ in Theorem 9, we get the following result:*

$$_2F_1(a,b;c+1;x) = \frac{c}{c-b} \int_0^1 {}_2\mathfrak{F}_1(a,[b,c;y],x) dy. \qquad (40)$$

Theorem 9. *The following derivative formula holds true:*

$$\frac{d^n}{dx^n}(_2\mathfrak{F}_1(a,[b,c;y];x)) = \frac{(a)_n(b)_n}{(c)_n} {_2\mathfrak{F}_1}(a+n,[b+n,c+n;y];x). \tag{41}$$

Proof. Using (25), differentiating on both sides with respect to x, we obtain:

$$\begin{aligned}
\frac{d}{dx}(_2\mathfrak{F}_1(a,[b,c;y];x)) &= \frac{a}{B(b,c-b)}\int_0^y t^b(1-t)^{c-b-1}(1-xt)^{-a-1}dt \\
&= \frac{a}{B(b,c-b)}\int_0^y t^{(b+1)-1}(1-t)^{(c+1)-(b+1)-1}(1-xt)^{-(a+1)}dt \\
&= \frac{ab}{c}\frac{1}{B(b+1,c-b)}\int_0^y t^{(b+1)-1}(1-t)^{(c+1)-(b+1)-1}(1-xt)^{-(a+1)}dt \\
&= \frac{ab}{c} {_2\mathfrak{F}_1}(a+1,[b+1,c+1;y];x)
\end{aligned}$$

which is (41) for $n=1$. The general result follows by the principle of mathematical induction on n. □

Theorem 10. *The following derivative formula holds true:*

$$\frac{d^n}{dx^n}(_1\mathfrak{F}_1([a,b;y];x)) = \frac{(a)_n}{(b)_n} {_1\mathfrak{F}_1}([a+n,b+n;y];x). \tag{42}$$

Theorem 11. *We have the following difference formula for* $_2\mathfrak{F}_1(a,[b,b+h;y];x)$:

$$\frac{b+h-1}{B(b,h)}y^{b-1}(1-y)^{h-1}(1-xy)^{-a} = {_2\mathfrak{F}_1}(a,[b,b+h-1;y];x) + \tag{43}$$
$${_2\mathfrak{F}_1}(a,[b-1,b+h-1;y];x) - ax(b+h-1){_2\mathfrak{F}_1}(a+1,[b,b+h;y];x).$$

Proof. Recalling that the Mellin transform operator is defined by:

$$\mathfrak{M}\{f(t):s\} := \int_0^\infty t^{s-1}f(t)dt,\ \mathrm{Re}(s)>0,$$

we observe that $_2F_1(a,[b,b+h;y];x)$ is the Mellin transform of the function:

$$f(t:x;y,a;h) = H(y-t)(1-t)^{h-1}(1-xt)^{-a},$$

where:

$$H(t) = \begin{cases} 1 & \text{if } t>0 \\ 0 & \text{if } t<0 \end{cases},$$

is the Heaviside unit function. Observing the fact that:

$$_2\mathfrak{F}_1(a,[b,b+h;y];x) := \frac{\mathfrak{M}\{f(t:x;y,a;h):b\}}{B(b,h)}, \tag{44}$$

we can write that:

$$\frac{\partial}{\partial t}(f(t:x;y,a;h)) = -[(y-t)(1-t)^{h-1}(1-xt)^{-a} + (h-1)H(y-t)(1-t)^{h-2}(1-xt)^{-a}] \tag{45}$$
$$+ ax(1-xt)^{-a-1}H(y-t)(1-t)^{h-1},$$

where $\frac{\partial}{\partial t}(H(t)) = \delta(t - t_0)$,

$$\delta(t - t_0) = \begin{cases} \infty & \text{if } t = t_0 \\ 0 & \text{if } t \neq t_0 \end{cases},$$

is the Dirac delta function. Applying the Mellin transform on both sides (45) and using (44) and the fact that:

$$\mathfrak{M}\{f'(t) : x\} = (1-x)\mathfrak{M}\{f(t) : x - 1\},$$

we have:

$$\frac{b+h-1}{B(b,h)} y^{b-1}(1-y)^{h-1}(1-xy)^{-a} = {}_2\mathfrak{F}_1(a, [b, b+h-1; y]; x)$$
$$+ {}_2\mathfrak{F}_1(a, [b-1, b+h-1; y]; x) - ax(b+h-1) {}_2\mathfrak{F}_1(a+1, [b, b+h; y]; x).$$

This completes the proof. □

In the following theorems, we give transformation formulas:

Theorem 12. *The following transformation formula holds true:*

$${}_2\mathfrak{F}_1(a, [\beta, \gamma; y]; z) = (1-z)^{-a} {}_2\mathfrak{F}_1(a, \{\gamma - \beta, \gamma; 1 - y\}; \frac{z}{z-1}), \quad |\arg(1-z)| < \pi. \tag{46}$$

Proof. Using (25), we obtain:

$${}_2\mathfrak{F}_1(a, [\beta, \gamma; y]; z) = \frac{(1-z)^{-a}}{B(\beta, \gamma - \beta)} \int_{1-y}^{1} (1-s)^{\beta-1} s^{\gamma-\beta-1} \left(1 - \frac{z}{z-1} s\right)^{-a} ds. \tag{47}$$

The substitution $s = 1 - t$ in (47) leads to:

$$\begin{aligned}{}_2\mathfrak{F}_1(a, [\beta, \gamma; y]; z) &= \frac{(1-z)^{-a}}{B(\beta, \gamma - \beta)} \int_0^y t^{\beta-1}(1-t)^{\gamma-\beta-1} \left(1 - \frac{z(1-t)}{z-1}\right)^{-a} dt \\ &= (1-z)^{-a} {}_2\mathfrak{F}_1(a, \{\gamma - \beta, \gamma; 1 - y\}; \frac{z}{z-1}).\end{aligned}$$

□

Theorem 13. *The following transformation formula holds true:*

$${}_2\mathfrak{F}_1(a, \{\beta, \gamma; y\}; z) = (1-z)^{-a} {}_2\mathfrak{F}_1(a, [\gamma - \beta, \gamma; 1-y]; \frac{z}{z-1}), \quad |\arg(1-z)| < \pi. \tag{48}$$

Theorem 14. *The following transformation formulas hold true:*

$${}_1\mathfrak{F}_1(\{\alpha, \beta; 1-y\}; z) = e^z \, {}_1\mathfrak{F}_1([\beta - \alpha, \beta; y]; -z) \tag{49}$$

and:

$${}_1\mathfrak{F}_1([\alpha, \beta; y]; z) = e^z \, {}_1\mathfrak{F}_1(\{\beta - \alpha, \beta; 1-y\}; -z). \tag{50}$$

Proof. The proofs of (49) and (50) are direct consequences of Theorem 6. □

4. The Incomplete Appell's Functions

In this section, we introduce the incomplete Appell's functions $\mathfrak{F}_1[a,b,c;d;x,z;y]$, $\mathfrak{F}_1\{a,b,c;d;x,z;y\}$, $\mathfrak{F}_2[a,b,c;d,e;x,z;y]$, and $\mathfrak{F}_2\{a,b,c;d,e;x,z;y\}$ by:

$$\mathfrak{F}_1[a,b,c;d;x,z;y] := \sum_{m,n=0}^{\infty} [a,d;y]_{m+n}(b)_m(c)_n \frac{x^m}{m!}\frac{z^n}{n!}, \quad \max\{|x|,|z|\} < 1 \qquad (51)$$

and:

$$\mathfrak{F}_1\{a,b,c;d;x,z;y\} := \sum_{m,n=0}^{\infty} \{a,d;y\}_{m+n}(b)_m(c)_n \frac{x^m}{m!}\frac{z^n}{n!}, \quad \max\{|x|,|z|\} < 1 \qquad (52)$$

and:

$$\mathfrak{F}_2[a,b,c;d,e;x,z;y] := \sum_{m,n=0}^{\infty} (a)_{m+n}[b,d;y]_m[c,e;y]_n \frac{x^m}{m!}\frac{z^n}{n!}, \quad |x|+|z| < 1 \qquad (53)$$

and:

$$\mathfrak{F}_2\{a,b,c;d,e;x,z;y\} := \sum_{m,n=0}^{\infty} (a)_{m+n}\{b,d;y\}_m\{c,e;y\}_n \frac{x^m}{m!}\frac{z^n}{n!}, \quad |x|+|z| < 1. \qquad (54)$$

Remark 1. *For the reader's convenience, we show how the convergence domains are obtained for the functions defined in* (51)–(54). *We just give the proof of* (51). *The other three definitions can be proven in a similar manner. Considering the absolute value:*

$$\left|\mathfrak{F}_1[a,b,c;d;x,z;y]\right| \leq \sum_{m,n=0}^{\infty} \left|[a,d;y]_{m+n}(b)_m(c)_n \frac{x^m}{m!}\frac{z^n}{n!}\right|$$

$$\leq \sum_{m,n=0}^{\infty} \left|[a,d;y]_{m+n}\right|\left|(b)_m(c)_n \frac{x^m}{m!}\frac{z^n}{n!}\right|$$

$$= \sum_{m,n=0}^{\infty} \left|\frac{1}{B(a,d-a)}\int_0^y t^{a+m+n-1}(1-t)^{d-a-1}dt\right|\left|(b)_m(c)_n \frac{x^m}{m!}\frac{z^n}{n!}\right|$$

$$\leq \sum_{m,n=0}^{\infty} \frac{1}{B(a,d-a)}\int_0^1 \left|t^{a+m+n-1}(1-t)^{d-a-1}dt\right|\left|(b)_m(c)_n \frac{x^m}{m!}\frac{z^n}{n!}\right|$$

$$= \sum_{m,n=0}^{\infty} \frac{B(a+m+n,d-a)}{B(a,d-a)}\left|(b)_m(c)_n \frac{x^m}{m!}\frac{z^n}{n!}\right|,$$

where the final series is the one corresponding to absolute convergence of the series for $F_1(a,b,c;d;x,z)$. Therefore, the series for $\mathfrak{F}_1[a,b,c;d;x,z;y]$ is absolutely convergent under the same conditions as the one for $F_1(a,b,c;d;x,z)$.

We proceed by obtaining the integral representations of the functions $\mathfrak{F}_1[a,b,c;d;x,z;y]$, $\mathfrak{F}_1\{a,b,c;d;x,z;y\}$, $\mathfrak{F}_2[a,b,c;d,e;x,z;y]$, and $\mathfrak{F}_2\{a,b,c;d,e;x,z;y\}$.

Theorem 15. *For the incomplete Appell's functions $\mathfrak{F}_1[a,b,c;d;x,z;y]$ and $\mathfrak{F}_1\{a,b,c;d;x,z;y\}$, we have the following integral representation:*

$$\mathfrak{F}_1[a,b,c;d;x,z;y] = \frac{y^a}{B(a,d-a)}\int_0^1 u^{a-1}(1-uy)^{d-a-1}(1-xuy)^{-b}(1-zuy)^{-c}du, \qquad (55)$$

$$Re(d) > 0,\ Re(a) > 0,\ Re(b) > 0,\ Re(c) > 0,\ |\arg(1-x)| < \pi,\ |\arg(1-z)| < \pi.$$

and:

$$\mathfrak{F}_1\{a,b,c;d;x,z;y\} = \frac{(1-y)^{d-a}}{B(a,d-a)} \times \int_0^1 u^{d-a-1}(1-u(1-y))^{a-1}$$
$$(1-x(1-u(1-y)))^{-b}(1-z(1-u(1-y)))^{-c}du, \; Re(d) > 0,$$
$$Re(a) > 0, \; Re(b) > 0, \; Re(c) > 0, \; |\arg(1-x)| < \pi, \; |\arg(1-z)| < \pi. \quad (56)$$

Proof. Replacing the integral representation for the incomplete beta function, which is given by (8), we find that:

$$\mathfrak{F}_1[a,b,c;d;x,z;y] = \frac{1}{B(a,d-a)} \int_0^y t^{a-1}(1-t)^{d-a-1}(1-xt)^{-b}(1-zt)^{-c}dt,$$

which can be written as:

$$\mathfrak{F}_1[a,b,c;d;x,z;y] = \frac{y^a}{B(a,d-a)} \int_0^1 u^{a-1}(1-uy)^{d-a-1}(1-xuy)^{-b}(1-zuy)^{-c}du;$$

whence the result. Formula (56) can be proven in a similar way. □

Theorem 16. *For the incomplete Appell's functions* $\mathfrak{F}_2[a,b,c;d,e;x,z;y]$ *and* $\mathfrak{F}_2\{a,b,c;d,e;x,z;y\}$*, we have the following integral representation:*

$$\mathfrak{F}_2[a,b,c;d,e;x,z;y] = \frac{y^{b+c}}{B(b,d-b)B(c,e-c)}$$
$$\times \int_0^1 \int_0^1 u^{b-1}(1-uy)^{d-b-1}v^{c-1}(1-vy)^{e-c-1}(1-xuy-zvy)^{-a}dudv,$$
$$Re(d) > Re(a) > Re(b) > Re(c) > Re(m) > 0, \; |\arg(1-x-z)| < \pi. \quad (57)$$

and:

$$\mathfrak{F}_2\{a,b,c;d,e;x,z;y\}$$
$$= \frac{(1-y)^{d-b+e-c}}{B(b,d-b)B(c,e-c)} \int_0^1 \int_0^1 u^{d-b-1}(1-u(1-y))^{b-1}v^{e-c-1}(1-v(1-y))^{c-1}$$
$$(1-x(1-u(1-y))-z(1-v(1-y)))^{-a}dudv,$$
$$Re(d) > 0, \; Re(a) > 0, \; Re(b) > 0, \; Re(c) > 0, \; Re(e) > 0, \; |\arg(1-x-z)| < \pi. \quad (58)$$

Proof. Replacing the integral representation for the incomplete beta function, which is given by (8), we get:

$$\mathfrak{F}_2[a,b,c;d,e;x,z;y] = \frac{1}{B(b,d-b)B(c,e-c)}$$
$$\times \sum_{m,n=0}^{\infty} \int_0^y \int_0^y (a)_{m+n} t^{b+m-1}(1-t)^{d-b-1} s^{c+n-1}(1-s)^{e-c-1} \frac{x^m}{m!} \frac{z^n}{n!} dtds.$$

Considering the fact that the series involved are uniformly convergent and we have a right to interchange the order of summation and integration, we get:

$$\mathfrak{F}_2[a,b,c;d,e;x,z;y] = \frac{1}{B(b,d-b)B(c,e-c)}$$
$$\times \int_0^y \int_0^y t^{b-1}(1-t)^{d-b-1}s^{c-1}(1-s)^{e-c-1}(1-xt-zs)^{-a}dtds,$$
$$= \frac{y^{b+c}}{B(b,d-b)B(c,e-c)}$$
$$\times \int_0^1 \int_0^1 u^{b-1}(1-uy)^{d-b-1}v^{c-1}(1-vy)^{e-c-1}(1-xuy-zvy)^{-a}dudv.$$

Formula (58) can be proven in a similar way. □

5. Incomplete Riemann-Liouville Fractional Integral Operators

In this section, we introduce and investigate the incomplete Riemann-Liouville fractional integral operators. The Riemann-Liouville fractional integral of order μ is defined by:

$$D_z^\mu\{f(z)\} := \frac{1}{\Gamma(-\mu)} \int_0^z f(t)(z-t)^{-\mu-1}dt, \quad Re(\mu) < 0. \tag{59}$$

Now, we define the incomplete Riemann-Liouville fractional integral operators $D_z^\mu[f(z);y]$ and $D_z^\mu\{f(z);y\}$ by:

$$D_z^\mu[f(z);y] := \frac{z^{-\mu}}{\Gamma(-\mu)} \int_0^y f(uz)(1-u)^{-\mu-1}du \tag{60}$$
$$:= \frac{z^{-\mu}y}{\Gamma(-\mu)} \int_0^1 f(ywz)(1-wy)^{-\mu-1}dw, \quad Re(\mu) < 0.$$

and its counterpart is by:

$$D_z^\mu\{f(z);y\} := \frac{z^{-\mu}}{\Gamma(-\mu)} \int_y^1 f(uz)(1-u)^{-\mu-1}du \tag{61}$$
$$:= \frac{z^{-\mu}}{\Gamma(-\mu)} \int_0^{1-y} f((1-t)z)t^{-\mu-1}dt, \quad Re(\mu) < 0.$$

Remark 2. *If $y = 1$, then (60) is equivalent to the standard Riemann-Liouville fractional integral (59). If $y = 0$, then (61) is equivalent to the standard Riemann-Liouville fractional integral (59). Thus, the original definition (59) is a particular case of both types of the incomplete Riemann-Liouville fractional integral.*

We start our investigation by calculating the incomplete fractional integrals of some elementary functions.

Theorem 17. *Let $Re(\lambda) > -1$, $Re(\mu) < 0$. Then:*

$$D_z^\mu[z^\lambda;y] = \frac{B_y(\lambda+1,-\mu)}{\Gamma(-\mu)}z^{\lambda-\mu}. \tag{62}$$

Proof. Using (60) and (8), we get:

$$\begin{aligned} D_z^{\mu}[z^{\lambda};y] &= \frac{z^{-\mu}}{\Gamma(-\mu)} \int_0^y (uz)^{\lambda}(1-u)^{-\mu-1} du \\ &= \frac{B_y(\lambda+1,-\mu)}{\Gamma(-\mu)} z^{\lambda-\mu}; \end{aligned}$$

whence the result. □

Theorem 18. *Let $Re(\lambda) > -1$, $Re(\mu) < 0$. Then:*

$$D_z^{\mu}\{z^{\lambda};y\} = \frac{B_{1-y}(-\mu,\lambda+1)}{\Gamma(-\mu)} z^{-\mu+\lambda}. \tag{63}$$

Theorem 19. *Let $Re(\lambda) > 0$, $Re(\alpha) > 0$, $Re(\mu) < 0$ and $|z| < 1$. Then:*

$$D_z^{\lambda-\mu}[z^{\lambda-1}(1-z)^{-\alpha};y] = \frac{\Gamma(\lambda)}{\Gamma(\mu)} z^{\mu-1} {}_2\mathfrak{F}_1(\alpha,[\lambda,\mu;y];z), \tag{64}$$

and:

$$D_z^{\lambda-\mu}\{z^{\lambda-1}(1-z)^{-\alpha};y\} = \frac{\Gamma(\lambda)}{\Gamma(\mu)} z^{\mu-1} {}_2\mathfrak{F}_1(\alpha,\{\lambda,\mu;y\};z). \tag{65}$$

Proof. Direct calculations yield:

$$\begin{aligned} D_z^{\lambda-\mu}[z^{\lambda-1}(1-z)^{-\alpha};y] &= \frac{z^{\mu-\lambda}}{\Gamma(\mu-\lambda)} \int_0^y (uz)^{\lambda-1}(1-uz)^{-\alpha}(1-u)^{\mu-\lambda-1} du \\ &= \frac{z^{\mu-\lambda}y}{\Gamma(\mu-\lambda)} \int_0^1 (yz)^{\lambda-1} w^{\lambda-1}(1-ywz)^{-\alpha}(1-wy)^{\mu-\lambda-1} dw \\ &= \frac{z^{\mu-1}y^{\lambda}}{\Gamma(\mu-\lambda)} \int_0^1 w^{\lambda-1}(1-ywz)^{-\alpha}(1-wy)^{\mu-\lambda-1} dw. \end{aligned}$$

By (24), we can write:

$$\begin{aligned} D_z^{\lambda-\mu}[z^{\lambda-1}(1-z)^{-\alpha};y] &= \frac{z^{\mu-1}}{\Gamma(\mu-\lambda)} B(\lambda,\mu-\lambda) {}_2\mathfrak{F}_1(\alpha,[\lambda,\mu;y];z) \\ &= \frac{\Gamma(\lambda)}{\Gamma(\mu)} z^{\mu-1} {}_2\mathfrak{F}_1(\alpha,[\lambda,\mu;y];z). \end{aligned}$$

Hence, the proof is completed. Formula (65) can be proven in a similar way. □

Theorem 20. *Let $Re(\lambda) > Re(\mu) > 0$, $Re(\alpha) > 0$, $Re(\beta) > 0$; $|az| < 1$ and $|bz| < 1$. Then:*

$$D_z^{\lambda-\mu}[z^{\lambda-1}(1-az)^{-\alpha}(1-bz)^{-\beta};y] = \frac{\Gamma(\lambda)}{\Gamma(\mu)} z^{\mu-1} \mathfrak{F}_1[\lambda,\alpha,\beta;\mu;az,bz;y], \tag{66}$$

and:

$$D_z^{\lambda-\mu}\{z^{\lambda-1}(1-az)^{-\alpha}(1-bz)^{-\beta};y\} = \frac{\Gamma(\lambda)}{\Gamma(\mu)} z^{\mu-1} \mathfrak{F}_1\{\lambda,\alpha,\beta;\mu;az,bz;y\}. \tag{67}$$

Proof. We have:

$$D_z^{\lambda-\mu}[z^{\lambda-1}(1-az)^{-\alpha}(1-bz)^{-\beta};y]$$

$$= \frac{z^{\mu-\lambda}}{\Gamma(\mu-\lambda)} \int_0^y (uz)^{\lambda-1}(1-auz)^{-\alpha}(1-buz)^{-\beta}(1-u)^{\mu-\lambda-1} du$$

$$= \frac{z^{\mu-\lambda}y}{\Gamma(\mu-\lambda)} \int_0^1 (yw)^{\lambda-1}(z)^{\lambda-1}(1-aywz)^{-\alpha}(1-bywz)^{-\beta}(1-wy)^{\mu-\lambda-1} dw$$

$$= \frac{z^{\mu-1}y^\lambda}{\Gamma(\mu-\lambda)} \int_0^1 w^{\lambda-1}(1-aywz)^{-\alpha}(1-bywz)^{-\beta}(1-wy)^{\mu-\lambda-1} dw.$$

By (55), we can write:

$$D_z^{\lambda-\mu}[z^{\lambda-1}(1-az)^{-\alpha}(1-bz)^{-\beta};y] = \frac{z^{\mu-1}}{\Gamma(\mu-\lambda)} B(\lambda,\mu-\lambda) \mathfrak{F}_1[\lambda,\alpha,\beta;\mu;az,bz;y]$$

$$= \frac{\Gamma(\lambda)}{\Gamma(\mu)} z^{\mu-1} \mathfrak{F}_1[\lambda,\alpha,\beta;\mu;az,bz;y];$$

whence the result. Formula (67) can be proven in a similar way. □

Theorem 21. *Let* $Re(\lambda) > Re(\mu) > 0, Re(\alpha) > 0, Re(\beta) > 0, Re(\gamma) > 0$; $\left|\frac{t}{1-z}\right| < 1$ *and* $|t| + |z| < 1$. *We have:*

$$D_z^{\lambda-\mu}[z^{\lambda-1}(1-z)^{-\alpha} \,_2F_1(\alpha.[\beta,\gamma;y];\frac{t}{1-z});y] = \frac{\Gamma(\lambda)}{\Gamma(\mu)} z^{\mu-1} \mathfrak{F}_2[\alpha,\beta,\lambda;\gamma,\mu;t,z;y], \quad (68)$$

and:

$$D_z^{\lambda-\mu}\{z^{\lambda-1}(1-z)^{-\alpha} \,_2F_1(\alpha.[\beta,\gamma;y];\frac{t}{1-z});y\} = \frac{\Gamma(\lambda)}{\Gamma(\mu)} z^{\mu-1} \mathfrak{F}_2\{\alpha,\beta,\lambda;\gamma,\mu;t,z;y\}. \quad (69)$$

Proof. Using Theorem 19 and (53), we get:

$$D_z^{\lambda-\mu}[z^{\lambda-1}(1-z)^{-\alpha} \,_2F_1(\alpha.[\beta,\gamma;y];\frac{t}{1-z});y]$$

$$= D_z^{\lambda-\mu}[z^{\lambda-1}(1-z)^{-\alpha} \frac{1}{B(\beta,\gamma-\beta)} \sum_{n=0}^{\infty} \frac{(\alpha)_n B_y(\beta+n,\gamma-\beta)}{n!} \left(\frac{t}{1-z}\right)^n;y]$$

$$= \frac{1}{B(\beta,\gamma-\beta)} D_z^{\lambda-\mu}[z^{\lambda-1} \sum_{n=0}^{\infty} (\alpha)_n B_y(\beta+n,\gamma-\beta) \frac{t^n}{n!} (1-z)^{-\alpha-n};y]$$

$$= \frac{1}{B(\beta,\gamma-\beta)} \sum_{m,n=0}^{\infty} B_y(\beta+n,\gamma-\beta) \frac{t^n}{n!} \frac{(\alpha)_n (\alpha+n)_m}{m!} D_z^{\lambda-\mu}[z^{\lambda-1+m};y]$$

$$= \frac{1}{B(\beta,\gamma-\beta)} \sum_{m,n=0}^{\infty} B_y(\beta+n,\gamma-\beta) \frac{t^n}{n!} \frac{(\alpha)_{n+m}}{m!} \frac{B_y(\lambda+m,\mu-\lambda)}{\Gamma(\mu-\lambda)} z^{\mu+m-1}$$

$$= \frac{\Gamma(\lambda)}{\Gamma(\mu)} z^{\mu-1} \mathfrak{F}_2[\alpha,\beta,\lambda;\gamma,\mu;t,z;y].$$

Hence, the proof is complete. Formula (69) can be proven in a similar way. □

6. Generating Functions

Now, we obtain linear and bilinear generating relations for the incomplete hypergeometric functions $_2\mathfrak{F}_1(a, [b, c; y]; x)$ by following the methods described in [2]. We start with the following theorem:

Theorem 22. *For the incomplete hypergeometric functions, we have:*

$$\sum_{n=0}^{\infty} \frac{(\lambda)_n}{n!} {}_2\mathfrak{F}_1(\lambda + n, [\alpha, \beta; y]; z)t^n = (1-t)^{-\lambda} {}_2\mathfrak{F}_1(\lambda, [\alpha, \beta; y]; \frac{z}{1-t}) \quad (70)$$

and:

$$\sum_{n=0}^{\infty} \frac{(\lambda)_n}{n!} {}_2\mathfrak{F}_1(\lambda + n, \{\alpha, \beta; y\}; z)t^n = (1-t)^{-\lambda} {}_2\mathfrak{F}_1(\lambda, \{\alpha, \beta; y\}; \frac{z}{1-t}) \quad (71)$$

where $|z| < \min\{1, |1-t|\}$ *and* $Re(\lambda) > 0$, $Re(\beta) > Re(\alpha) > 0$.

Proof. Considering the elementary identity:

$$[(1-z) - t]^{-\lambda} = (1-t)^{-\lambda} \left[1 - \frac{z}{1-t}\right]^{-\lambda}$$

and expanding the left-hand side, we have for $|t| < |1-z|$ that:

$$(1-z)^{-\lambda} \sum_{n=0}^{\infty} \frac{(\lambda)_n}{n!} \left(\frac{t}{1-z}\right)^n = (1-t)^{-\lambda} \left[1 - \frac{z}{1-t}\right]^{-\lambda}.$$

Now, multiplying both sides of the above equality by $z^{\alpha-1}$ and applying the incomplete fractional integral operator $D_z^{\alpha-\beta}[f(z); y]$ on both sides, we can write:

$$D_z^{\alpha-\beta}\left[\sum_{n=0}^{\infty} \frac{(\lambda)_n}{n!}(1-z)^{-\lambda}\left(\frac{t}{1-z}\right)^n z^{\alpha-1}; y\right] = (1-t)^{-\lambda} D_z^{\alpha-\beta}\left[z^{\alpha-1}\left[1 - \frac{z}{1-t}\right]^{-\lambda}; y\right].$$

Interchanging the order, which is valid for $Re(\alpha) > 0$ and $|t| < |1-z|$, we get:

$$\sum_{n=0}^{\infty} \frac{(\lambda)_n}{n!} D_z^{\alpha-\beta}\left[z^{\alpha-1}(1-z)^{-\lambda-n}; y\right] t^n = (1-t)^{-\lambda} D_z^{\alpha-\beta}\left[z^{\alpha-1}\left[1 - \frac{z}{1-t}\right]^{-\lambda}; y\right].$$

Using Theorem 21, we get the desired result. Formula (71) can be proven in a similar way. □

The following theorem gives another linear generating relation for the incomplete hypergeometric functions.

Theorem 23. *For the incomplete hypergeometric functions, we have:*

$$\sum_{n=0}^{\infty} \frac{(\lambda)_n}{n!} {}_2\mathfrak{F}_1(\rho - n, [\alpha, \beta; y]; z)t^n = (1-t)^{-\lambda} \mathfrak{F}_1[\alpha, \rho, \lambda; \beta; z; \frac{-zt}{1-t}; y] \quad (72)$$

and:

$$\sum_{n=0}^{\infty} \frac{(\lambda)_n}{n!} {}_2\mathfrak{F}_1(\rho - n, \{\alpha, \beta; y\}; z)t^n = (1-t)^{-\lambda} \mathfrak{F}_1\{\alpha, \rho, \lambda; \beta; z; \frac{-zt}{1-t}; y\} \quad (73)$$

where $Re(\lambda) > 0$, $Re(\rho) > 0$, $Re(\beta) > Re(\alpha) > 0$; $|t| < \frac{1}{1+|z|}$.

Proof. Considering:

$$[1-(1-z)t]^{-\lambda} = (1-t)^{-\lambda}\left[1+\frac{zt}{1-t}\right]^{-\lambda}$$

and expanding the left-hand side, we have for $|t| < |1-z|$ that:

$$\sum_{n=0}^{\infty}\frac{(\lambda)_n}{n!}(1-z)^n t^n = (1-t)^{-\lambda}\left[1-\frac{-zt}{1-t}\right]^{-\lambda}.$$

Now, multiplying both sides of the above equality by $z^{\alpha-1}(1-z)^{-\rho}$ and applying the fractional integral operator $D_z^{\alpha-\beta}[f(z);y]$ on both sides, we get:

$$D_z^{\alpha-\beta}\left[\sum_{n=0}^{\infty}\frac{(\lambda)_n}{n!}z^{\alpha-1}(1-z)^{-\rho+n}t^n;y\right] = (1-t)^{-\lambda}D_z^{\alpha-\beta}\left[z^{\alpha-1}(1-z)^{-\rho}\left[1-\frac{-zt}{1-t}\right]^{-\lambda};y\right].$$

Interchanging the order, which is valid for $Re(\alpha) > 0$ and $|zt| < |1-t|$, we get:

$$\sum_{n=0}^{\infty}\frac{(\lambda)_n}{n!}D_z^{\alpha-\beta}\left[z^{\alpha-1}(1-z)^{-(\rho-n)};y\right]t^n = (1-t)^{-\lambda}D_z^{\alpha-\beta}\left[z^{\alpha-1}(1-z)^{-\rho}\left[1-\frac{-zt}{1-t}\right]^{-\lambda};y\right].$$

Using Theorem 21 and 22, we get the desired result. The generating relation (73) can be proven in a similar way. □

Finally, we have the following bilinear generating relation for the incomplete hypergeometric functions.

Theorem 24. *For the incomplete hypergeometric functions, we have:*

$$\sum_{n=0}^{\infty}\frac{(\lambda)_n}{n!} {_2\mathfrak{F}_1}(\gamma,[-n,\delta;y];x) \, {_2\mathfrak{F}_1}(\gamma,[\lambda+n,\beta;y];z)t^n = (1-t)^{-\lambda}\mathfrak{F}_2[\lambda,\alpha,\gamma;\beta,\delta;\frac{z}{1-t},\frac{-xt}{1-t};y] \quad (74)$$

and:

$$\sum_{n=0}^{\infty}\frac{(\lambda)_n}{n!} {_2\mathfrak{F}_1}(\gamma,\{-n,\delta;y\};x) \, {_2\mathfrak{F}_1}(\gamma,\{\lambda+n,\beta;y\};z)t^n = (1-t)^{-\lambda}\mathfrak{F}_2\{\lambda,\alpha,\gamma;\beta,\delta;\frac{z}{1-t},\frac{-xt}{1-t};y\} \quad (75)$$

where $Re(\lambda) > 0$, $Re(\gamma) > 0$, $Re(\beta) > 0$, $Re(\delta) > 0$, $Re(\alpha) > 0$; $|t| < \frac{1-|z|}{1+|x|}$, and $|z| < 1$.

Proof. Replacing t by $(1-x)t$ in (70), multiplying the resulting equality by $x^{\gamma-1}$, and then applying the incomplete fractional integral operator $D_x^{\gamma-\delta}[f(x);y]$, we get:

$$D_x^{\gamma-\delta}\left[\sum_{n=0}^{\infty}\frac{(\lambda)_n}{n!}x^{\gamma-1}\,{_2\mathfrak{F}_1}(\lambda+n,[\alpha,\beta;y];z)(1-x)^n t^n;y\right]$$

$$= D_x^{\gamma-\delta}\left[(1-(1-x)t)^{-\lambda}x^{\gamma-1}\,{_2\mathfrak{F}_1}(\lambda,[\alpha,\beta;y];\frac{z}{1-(1-x)t});y\right].$$

Interchanging the order, which is valid for $|z| < 1$, $\left|\frac{1-x}{1-z}t\right| < 1$ and $\left|\frac{z}{1-t}\right| + \left|\frac{xt}{1-t}\right| < 1$, we can write that:

$$\sum_{n=0}^{\infty} \frac{(\lambda)_n}{n!} D_x^{\gamma-\delta} \left[x^{\gamma-1}(1-x)^n; y \right] {}_2\mathfrak{F}_1(\lambda + n, [\alpha, \beta; y]; z)$$

$$= (1-t)^{-\lambda} D_x^{\gamma-\delta} \left[x^{\gamma-1}\left(1 - \frac{-xt}{1-t}\right) {}_2\mathfrak{F}_1\left(\lambda, [\alpha, \beta; y]; \frac{\frac{z}{1-t}}{1 - \frac{-xt}{1-t}}\right); y \right].$$

Using Theorems 21 and 23, we get (74). The generating relation (75) can be proven in a similar way. □

In the following remark, first of all, we obtained a series formula for the Gauss hypergeometric functions as an application of Theorem 22. Similar results can be obtained for Theorem 23 and 24. Furthermore, we showed that the result obtained in (70) coincides with usual case when $y \to 1^-$.

Remark 3. *Using the relation that is given by (13) in Equation (70), we have:*

$$\frac{1}{B(\alpha, \beta - \alpha)} \sum_{n,k=0}^{\infty} (\lambda)_n (\lambda + n)_k \frac{y^{\alpha+k}}{\alpha + k} {}_2F_1(\alpha + k, 1 - \beta + \alpha; \alpha + k + 1; y) \frac{z^k}{k!} \frac{t^n}{n!} \quad (76)$$

$$= \frac{(1-t)^{-\lambda}}{B(\alpha, \beta - \alpha)} \sum_{m=0}^{\infty} \frac{(\lambda)_m}{m!} \left(\frac{z}{1-t}\right)^m \frac{y^{\alpha+m}}{\alpha + m} {}_2F_1(\alpha + m, 1 - \beta + \alpha; \alpha + m + 1; y) \quad (77)$$

which is a series identity between the Gauss hypergeometric functions. If we take $y = 1$ in the above identity and use the following relation:

$$_2F_1(a, b; c; 1) = \frac{\Gamma(c)\Gamma(c - a - b)}{\Gamma(c - a)\Gamma(c - b)}$$

we obtain:

$$\sum_{n=0}^{\infty} \frac{(\lambda)_n}{n!} {}_2F_1(\lambda + n, \alpha, \beta; z) t^n = (1-t)^{-\lambda} {}_2F_1\left(\lambda, \alpha, \beta; \frac{z}{1-t}\right) \quad (78)$$

7. Conclusions

Recently, in [27], various applications of fractional calculus were exhibited in areas ranging from engineering to life sciences. For the applications of fractional calculus, we should also recommend the references of the paper [27] and, in particular, the book [21].

In the present paper, we introduced the incomplete versions of Riemann-Liouville integral operators. Approaching the problems mentioned in [27] using these incomplete operators may give rise to interesting perspectives on solving these problems. For instance, in a nonlocal fractional process, which occurs on an interval, but whose behavior changes in the middle, it may be useful to consider splitting the domain into subintervals and integrating from both sides separately using incomplete fractional operators.

These operators have already been used to define Liouville-Caputo-type incomplete fractional derivatives in [12]. Furthermore, for the incomplete Riemann-Liouville fractional integrals defined here, their analyticity properties have been investigated in [31]. Some of these, such as a transformation property on the domains of the functions concerned, may also lend themselves well to applications.

Incomplete Pochhammer ratios were defined in (10) and (11) by using the incomplete beta functions. Several properties of these functions were obtained. Incomplete hypergeometric functions were introduced with the help of these incomplete Pochhammer ratios, and certain properties such as integral representations, derivative formulas, transformation formulas, and recurrence relations were investigated.

Furthermore, incomplete Riemann-Liouville fractional integral operators were defined. The incomplete Riemann-Liouville fractional integrals for the some elementary functions were given. Linear and bilinear generating relations for incomplete hypergeometric functions were obtained.

Author Contributions: These authors contributed equally to this work.

Funding: This research received no external funding.

Conflicts of Interest: The authors declare no conflict of interest.

References

1. Chaudhry, M.A.; Qadir, A.; Rafique, M.; Zubair, S.M. Extension of Euler's beta function. *J. Comput. Appl. Math.* **1997**, *78*, 19–32. [CrossRef]
2. Chaudhry, M.A.; Qadir, A.; Srivastava, H.M.; Paris, R.B. Extended hypergeometric and confluent hypergeometric functions. *Appl. Math. Comput.* **2014**, *159*, 589–602. [CrossRef]
3. Chaudhry, M.A.; Zubair, S.M. *On a Class of Incomplete Gamma Functions with Aplications*; Chapman and Hall: Dhahran, Saudi Arabia, 2001.
4. Cho, N.E.; Srivastava, R. Some extended Pochhammer symbols and their applications involving generalized hypergeometric polynomials. *Appl. Math. Comput.* **2014**, *234*, 277–285.
5. Lin, S.; Srivastava, H.M.; Wong, M. Some Applications of Srivastava's Theorem Involving a Certain Family of Generalized and Extended Hypergeometric Polyomials. *Filomat* **2015**, *29*, 1811–1819. [CrossRef]
6. Özarslan, M.A.; Özergin, E. Some generating relations for extended hypergeometric functions via generalized fractional derivative operator. *Math. Comput. Model.* **2010**, *52*, 1825–1833. [CrossRef]
7. Özergin, E.; Özarslan, M.A.; Altin, A. Extension of gamma, beta and hypergeometric functions. *J. Comput. Appl. Math.* **2011**, *235*, 4601–4610.
8. Özergin, E. Some Properties of Hypergeometric Functions. Ph.D. Thesis, Eastern Mediterranean University, North Cyprus, Turkey, 2011.
9. Srivastava, R. Some classes of generating functions associated with a certain family of extended and generalized hypergeometric functions. *Appl. Math. Comput.* **2014**, *243*, 132–137. [CrossRef]
10. Srivastava, H.M.; Chaudry, M.A.; Agarwal, R.P. The incomplete Pochhammer symbols and their applications to hypergeometric and related functions. *Integral Transforms Spec. Funct.* **2012**, *23*, 659–683. [CrossRef]
11. Choi, J.; Parmar, R.K.; Chopra, P. The Incomplete Srivastava's Triple Hypergeometric Functions gamma(H)(B) and Gamma(H)(B). *Filomat* **2016**, *7*, 1779–1787. [CrossRef]
12. Özarslan, M.A.; Ustaoglu, C. Incomplete Caputo fractional derivative operators. *Adv. Differ. Equ.* **2018**, *2018*, 209. [CrossRef]
13. Çetinkaya, A. The incomplete second Appell hypergeometric functions. *Appl. Math. Comput.* **2013**, *219*, 8332–8337. [CrossRef]
14. Sahai, V.; Verma, A. On an extension of the generalized Pochhammer symbol and its applications to hypergeometric functions. *Asian-Eur. J. Math.* **2016**, *9*, 1650064. [CrossRef]
15. Srivastava, H.M.; Saxena, R.K.; Parmar, R.K. Some Families of the Incomplete H-Functions and the Incomplete H-Functions and Associated Integral Transforms and Operators of Fractional Calculus with Applications. *Russ. J. Math. Phys.* **2018**, *25*, 116–138. [CrossRef]
16. Srivastava, H.M.; Agarwal, R.; Jain, S. Integral transform and fractional derivative formulas involving the extended generalized hypergeometric functions and probability distributions. *Math. Methods Appl. Sci.* **2017**, *40*, 255–273. [CrossRef]
17. Srivastava, H.M.; Çetinkaya, A.; Kiymaz, O.I. A certain generalized Pochhammer symbol and its applications to hypergeometric functions. *Appl. Math. Comput.* **2014**, *226*, 484–491. [CrossRef]
18. Srivastava, R.; Cho, N.E. Generating functions for a certain class of incomplete hypergeometric polynomials. *Appl. Math. Comput.* **2012**, *219*, 3219–3225. [CrossRef]

19. Srivastava, R. Some properties of a family of incomplete hypergeometric functions. *Russ. J. Math. Phys.* **2013**, *20*, 121–128. [CrossRef]
20. Srivastava, R. Some generalizations of Pochhammer's symbol and their associated families of hypergeometric functions and hypergeometric polynomials. *Appl. Math. Inf. Sci.* **2013**, *7*, 2195–2206. [CrossRef]
21. Hilfer, R. *Applications of Fractional Calculus in Physics*; World Scientific: Singapore, 2000.
22. Kilbas, A.A.; Srivastava, H.M.; Trujillo, J.J. *Theory and Applications of Fractional Differantial Equations*; Elsevier: Amsterdam, The Netherlands, 2006.
23. Rossikhin, Y.A.; Shitikova, M.V. Application of fractional calculus for dynamic problems of solid mechanics: Novel trends and recent results. *Appl. Mech. Rev.* **2010**, *63*, 010801. [CrossRef]
24. Sabatier, J.; Agrawal, O.P.; Tenreiro Machado, J.A. Advances in fractional calculus. In *Theoretical Developments and Applications in Physics and Engineering*; Springer: Berlin/Heidelberg, Germany, 2007.
25. Spanos, P.D.; Matteo, A.D.; Cheng, Y.; Pirrotta, A.; Li, J. Galerkin Scheme-Based Determination of Survival Probability of Oscillators with Fractional Derivative Element. *J. Appl. Mech.* **2016**, *83*, 121003. [CrossRef]
26. Yang, X.; Baleanu, D.; Srivastava, H.M. *Local Fractional Integral Transforms and Their Applications*; Elsevier/Academic Press: Amsterdam, The Netherlands, 2016.
27. Sun, H.G.; Zhang, Y.; Baleanu, D.; Chen, W.; Chen, Y.Q. A new collection of real world applications of fractional calculus in science and engineering. *Commun. Nonlinear Sci. Numer. Simul.* **2018**, *64*, 213–231. [CrossRef]
28. Lin, S.; Srivastava, H.M.; Yao, J.C. Some classes of of generating relations associated with a family of the generalized Gauss type hypergeometric functions. *Appl. Math. Inform. Sci.* **2015**, *9*, 1731–1738.
29. Srivastava, H.M.; Parmar, R.K.; Chopra, P. A Class of Extended Fractional Derivative Operators and Associated Generating Relations Involving Hypergeometric Functions. *Axioms* **2012**, *1*, 238–258. [CrossRef]
30. Srivastava, H.M.; Manocha, H.L. *A Treatise on Generating Functions*; Ellis Horwood: New York, NY, USA, 1984.
31. Fernandez, A.; Ustaoğlu, C.; Özarslan, M.A. Analytical Development of Incomplete Riemann–Liouville Fractional Calculus. Unpublished work.

© 2019 by the authors. Licensee MDPI, Basel, Switzerland. This article is an open access article distributed under the terms and conditions of the Creative Commons Attribution (CC BY) license (http://creativecommons.org/licenses/by/4.0/).

Article

Logarithmic Coefficients For Univalent Functions Defined by Subordination

Ebrahim Analouei Adegani [1], Nak Eun Cho [2,*] and Mostafa Jafari [3]

[1] Faculty of Mathematical Sciences, Shahrood University of Technology, P.O. Box 316-36155, Shahrood 3619995161, Iran; analoey.ebrahim@gmail.com
[2] Department of Applied Mathematics, College of Natural Sciences, Pukyong National University, Busan 608-737, Korea
[3] Department of Mathematics, Najafabad Branch, Islamic Azad University, Najafabad 8514143131, Iran; mostafajafari83@gmail.com
* Correspondence: necho@pknu.ac.kr

Received: 18 March 2019; Accepted: 30 April 2019 ; Published: 7 May 2019

Abstract: In this work, the bounds for the logarithmic coefficients γ_n of the general classes $\mathcal{S}^*(\varphi)$ and $\mathcal{K}(\varphi)$ were estimated. It is worthwhile mentioning that the given bounds would generalize some of the previous papers. Some consequences of the main results are also presented, noting that our method is more general than those used by others.

Keywords: starlike functions; convex functions; subordination; logarithmic coefficients

MSC: 30C45

---

1. Introduction

Let $\mathcal{H}$ denote the class of analytic functions in the open unit disk $\mathbb{U} := \{z \in \mathbb{C} : |z| < 1\}$ and $\mathcal{A}$ denote the subclass of $\mathcal{H}$ consisting of functions of the form

$$f(z) = z + \sum_{n=2}^{\infty} a_n z^n. \tag{1}$$

Also, let $\mathcal{S}$ be the subclass of $\mathcal{A}$ consisting of all univalent functions in $\mathbb{U}$. Then the logarithmic coefficients γ_n of $f \in \mathcal{S}$ are defined with the following series expansion:

$$\log\left(\frac{f(z)}{z}\right) = 2\sum_{n=1}^{\infty} \gamma_n(f) z^n, \ z \in \mathbb{U}. \tag{2}$$

These coefficients play an important role for various estimates in the theory of univalent functions. Note that we use γ_n instead of $\gamma_n(f)$. The idea of studying the logarithmic coefficients helped Kayumov [1] to solve Brennan's conjecture for conformal mappings.

Recall that we can rewrite (2) in the series form as follows:

$$2\sum_{n=1}^{\infty} \gamma_n z^n = a_2 z + a_3 z^2 + a_4 z^3 + \cdots - \frac{1}{2}[a_2 z + a_3 z^2 + a_4 z^3 + \cdots]^2$$
$$+ \frac{1}{3}[a_2 z + a_3 z^2 + a_4 z^3 + \cdots]^3 + \cdots.$$

Now, considering the coefficients of z^n for $n = 1, 2, 3$, it follows that

$$\begin{cases} 2\gamma_1 = a_2, \\ 2\gamma_2 = a_3 - \dfrac{1}{2}a_2^2, \\ 2\gamma_3 = a_4 - a_2 a_3 + \dfrac{1}{3}a_2^3. \end{cases} \qquad (3)$$

For two functions f and g that are analytic in $\mathbb{U}$, we say that the function f is subordinate to g in $\mathbb{U}$ and write $f(z) \prec g(z)$ if there exists a Schwarz function ω that is analytic in $\mathbb{U}$ with $\omega(0) = 0$ and $|\omega(z)| < 1$ such that

$$f(z) = g(\omega(z)) \quad (z \in \mathbb{U}).$$

In particular, if the function g is univalent in $\mathbb{U}$, then $f \prec g$ if and only if $f(0) = g(0)$ and $f(\mathbb{U}) \subseteq g(\mathbb{U})$.

Using subordination, different subclasses of starlike and convex functions were introduced by Ma and Minda [2], in which either of the quantity $\frac{zf'(z)}{f(z)}$ or $1 + \frac{zf''(z)}{f'(z)}$ is subordinate to a more general superordinate function. To this aim, they considered an analytic univalent function φ with positive real part in $\mathbb{U}$. $\varphi(\mathbb{U})$ is symmetric respecting the real axis and starlike considering $\varphi(0) = 1$ and $\varphi'(0) > 0$. They defined the classes consisting of several well-known classes as follows:

$$\mathcal{S}^*(\varphi) := \left\{ f \in \mathcal{S} : \frac{zf'(z)}{f(z)} \prec \varphi(z), \, z \in \mathbb{U} \right\},$$

and

$$\mathcal{K}(\varphi) := \left\{ f \in \mathcal{S} : 1 + \frac{zf''(z)}{f'(z)} \prec \varphi(z), \, z \in \mathbb{U} \right\}.$$

For example, the classes $\mathcal{S}^*(\varphi)$ and $\mathcal{K}(\varphi)$ reduce to the classes $\mathcal{S}^*[A, B] := \dfrac{1 + Az}{1 + Bz}$ and $\mathcal{K}[A, B] := \dfrac{1 + Az}{1 + Bz}$ of the well-known Janowski starlike and Janowski convex functions for $-1 \leq B < A \leq 1$, respectively. By replacing $A = 1 - 2\alpha$ and $B = -1$ where $0 \leq \alpha < 1$, we conclude the classes $\mathcal{S}^*(\alpha)$ and $\mathcal{K}(\alpha)$ of the starlike functions of order α and convex functions of order α, respectively. In particular, $\mathcal{S}^* := \mathcal{S}^*(0)$ and $\mathcal{K} := \mathcal{K}(0)$ are the class of starlike functions and of convex functions in the unit disk $\mathbb{U}$, respectively. The Koebe function $k(z) = z/(1-z)^2$ is starlike but not convex in $\mathbb{U}$. Thus, every convex function is starlike but not conversely; however, each starlike function is convex in the disk of radius $2 - \sqrt{3}$.

Lately, several researchers have subsequently investigated similar problems in the direction of the logarithmic coefficients, the coefficient problems, and differential subordination [3–11], to mention a few. For example, the rotation of Koebe function $k(z) = z(1 - e^{i\theta})^{-2}$ for each θ has logarithmic coefficients $\gamma_n = e^{i\theta n}/n$, $n \geq 1$. If $f \in \mathcal{S}$, then by using the Bieberbach inequality for the first equation of (3) it concludes $|\gamma_1| \leq 1$ and by utilizing the Fekete–Szegö inequality for the second equation of (3), (see [12] (Theorem 3.8)),

$$|\gamma_2| = \frac{1}{2}\left|a_3 - \frac{1}{2}a_2^2\right| \leq \frac{1}{2}(1 + 2e^{-2}) = 0.635\cdots.$$

It was shown in [12] (Theorem 4) that the logarithmic coefficients γ_n of every function $f \in \mathcal{S}$ satisfy

$$\sum_{n=1}^{\infty} |\gamma_n|^2 \leq \frac{\pi^2}{6},$$

and the equality is attained for the Koebe function. For $f \in \mathcal{S}^*$, the inequality $|\gamma_n| \leq 1/n$ holds but is not true for the full class $\mathcal{S}$, even in order of magnitude (see [12] (Theorem 8.4)). In 2018, Ali and Vasudevarao [3] and Pranav Kumar and Vasudevarao [6] obtained the logarithmic coefficients γ_n for certain subclasses of close-to-convex functions. Nevertheless, the problem of the best upper bounds for the logarithmic coefficients of univalent functions for $n \geq 3$ is presumably still a concern.

Based on the results presented in previous research, in the current study, the bounds for the logarithmic coefficients γ_n of the general classes $\mathcal{S}^*(\varphi)$ and $\mathcal{K}(\varphi)$ were estimated. It is worthwhile mentioning that the given bounds in this paper would generalize some of the previous papers and that many new results are obtained, noting that our method is more general than those used by others. The following lemmas will be used in the proofs of our main results.

For this work, let Ω represent the class of all analytic functions ω in $\mathbb{U}$ that equips with conditions $\omega(0) = 0$ and $|\omega(z)| < 1$ for $z \in \mathbb{U}$. Such functions are called Schwarz functions.

Lemma 1. *[13] (p. 172) Assume that ω is a Schwarz function so that $\omega(z) = \sum_{n=1}^{\infty} p_n z^n$. Then*

$$|p_1| \leq 1, \quad |p_n| \leq 1 - |p_1|^2 \quad n = 2, 3, \ldots.$$

Lemma 2. *[14] Let $\psi, \varpi \in \mathcal{H}$ be any convex univalent functions in $\mathbb{U}$. If $f(z) \prec \psi(z)$ and $g(z) \prec \varpi(z)$, then $f(z) * g(z) \prec \psi(z) * \varpi(z)$ where $f, g \in \mathcal{H}$.*

We observe that in the above lemma, nothing is assumed about the normalization of ψ and ϖ, and "$*$" represents the Hadamard (or convolution) product.

Lemma 3. *[12,15] (Theorem 6.3, p. 192; Rogosinski's Theorem II (i)) Let $f(z) = \sum_{n=1}^{\infty} a_n z^n$ and $g(z) = \sum_{n=1}^{\infty} b_n z^n$ be analytic in $\mathbb{U}$, and suppose that $f \prec g$ where g is univalent in $\mathbb{U}$. Then*

$$\sum_{k=1}^{n} |a_k|^2 \leq \sum_{k=1}^{n} |b_k|^2, \quad n = 1, 2, \ldots.$$

Lemma 4. *[12,15] (Theorem 6.4 (i), p. 195; Rogosinski's Theorem X) Let $f(z) = \sum_{n=1}^{\infty} a_n z^n$ and $g(z) = \sum_{n=1}^{\infty} b_n z^n$ be analytic in $\mathbb{U}$, and suppose that $f \prec g$ where g is univalent in $\mathbb{U}$. Then*

(i) *If g is convex, then $|a_n| \leq |g'(0)| = |b_1|$, $n = 1, 2, \ldots$.*
(ii) *If g is starlike (starlike with respect to 0), then $|a_n| \leq n|g'(0)| = n|b_1|$, $n = 2, 3, \ldots$.*

Lemma 5. *[16] If $\omega(z) = \sum_{n=1}^{\infty} p_n z^n \in \Omega$, then for any real numbers q_1 and q_2, the following sharp estimate holds:*

$$|p_3 + q_1 p_1 p_2 + q_2 p_1^3| \leq H(q_1; q_2),$$

where

$$H(q_1; q_2) = \begin{cases} 1 & \text{if } (q_1, q_2) \in D_1 \cup D_2 \cup \{(2,1)\}, \\ |q_2| & \text{if } (q_1, q_2) \in \cup_{k=3}^{7} D_k, \\ \frac{2}{3}(|q_1|+1)\left(\frac{|q_1|+1}{3(|q_1|+1+q_2)}\right)^{\frac{1}{2}} & \text{if } (q_1, q_2) \in D_8 \cup D_9, \\ \frac{q_2}{3}\left(\frac{q_1^2-4}{q_1^2-4q_2}\right)\left(\frac{q_1^2-4}{3(q_2-1)}\right)^{\frac{1}{2}} & \text{if } (q_1, q_2) \in D_{10} \cup D_{11} \setminus \{(2,1)\}, \\ \frac{2}{3}(|q_1|-1)\left(\frac{|q_1|-1}{3(|q_1|-1-q_2)}\right)^{\frac{1}{2}} & \text{if } (q_1, q_2) \in D_{12}. \end{cases}$$

While the sets D_k, $k = 1, 2, \ldots, 12$ are defined as follows:

$$D_1 = \left\{ (q_1, q_2) : |q_1| \leq \frac{1}{2}, |q_2| \leq 1 \right\},$$

$$D_2 = \left\{ (q_1, q_2) : \frac{1}{2} \leq |q_1| \leq 2, \frac{4}{27}\left((|q_1|+1)^3\right) - (|q_1|+1) \leq |q_2| \leq 1 \right\},$$

$$D_3 = \left\{ (q_1, q_2) : |q_1| \leq \frac{1}{2}, |q_2| \leq -1 \right\},$$

$$D_4 = \left\{ (q_1, q_2) : |q_1| \geq \frac{1}{2}, |q_2| \leq -\frac{2}{3}(|q_1|+1) \right\},$$

$$D_5 = \{(q_1, q_2) : |q_1| \leq 2, |q_2| \geq 1\},$$

$$D_6 = \left\{ (q_1, q_2) : 2 \leq |q_1| \leq 4, |q_2| \geq \frac{1}{12}(q_1^2 + 8) \right\},$$

$$D_7 = \left\{ (q_1, q_2) : |q_1| \geq 4, |q_2| \geq \frac{2}{3}(|q_1| - 1) \right\},$$

$$D_8 = \left\{ (q_1, q_2) : \frac{1}{2} \leq |q_1| \leq 2, -\frac{2}{3}(|q_1|+1) \leq q_2 \leq \frac{4}{27}\left((|q_1|+1)^3\right) - (|q_1|+1) \right\},$$

$$D_9 = \left\{ (q_1, q_2) : |q_1| \geq 2, -\frac{2}{3}(|q_1|+1) \leq q_2 \leq \frac{2|q_1|(|q_1|+1)}{q_1^2 + 2|q_1| + 4} \right\},$$

$$D_{10} = \left\{ (q_1, q_2) : 2 \leq |q_1| \leq 4, \frac{2|q_1|(|q_1|+1)}{q_1^2 + 2|q_1| + 4} \leq q_2 \leq \frac{1}{12}(q_1^2 + 8) \right\},$$

$$D_{11} = \left\{ (q_1, q_2) : |q_1| \geq 4, \frac{2|q_1|(|q_1|+1)}{q_1^2 + 2|q_1| + 4} \leq q_2 \leq \frac{2|q_1|(|q_1|-1)}{q_1^2 - 2|q_1| + 4} \right\},$$

$$D_{12} = \left\{ (q_1, q_2) : |q_1| \geq 4, \frac{2|q_1|(|q_1|-1)}{q_1^2 - 2|q_1| + 4} \leq q_2 \leq \frac{2}{3}(|q_1| - 1) \right\}.$$

2. Main Results

Throughout this paper, we assume that φ is an analytic univalent function in the unit disk $\mathbb{U}$ satisfying $\varphi(0) = 1$ such that it has series expansion of the form

$$\varphi(z) = 1 + B_1 z + B_2 z^2 + B_3 z^3 + \cdots, \quad B_1 \neq 0. \tag{4}$$

Theorem 1. *Let the function $f \in \mathcal{S}^*(\varphi)$. Then the logarithmic coefficients of f satisfy the inequalities:*

(i) If φ is convex, then

$$|\gamma_n| \leq \frac{|B_1|}{2n}, \quad n \in \mathbb{N}, \tag{5}$$

$$\sum_{n=1}^{k} |\gamma_n|^2 \leq \frac{1}{4} \sum_{n=1}^{k} \frac{|B_n|^2}{n^2}, \quad k \in \mathbb{N}, \tag{6}$$

and

$$\sum_{n=1}^{\infty} |\gamma_n|^2 \leq \frac{1}{4} \sum_{n=1}^{\infty} \frac{|B_n|^2}{n^2}. \tag{7}$$

(ii) If φ is starlike with respect to 1, then

$$|\gamma_n| \leq \frac{|B_1|}{2}, \quad n \in \mathbb{N}. \tag{8}$$

All inequalities in (5), (7), and (8) are sharp such that for any $n \in \mathbb{N}$, there is the function f_n given by $\frac{z f_n'(z)}{f_n(z)} = \varphi(z^n)$ and the function f given by $\frac{z f'(z)}{f(z)} = \varphi(z)$, respectively.

Proof. Suppose that $f \in \mathcal{S}^*(\varphi)$. Then considering the definition of $\mathcal{S}^*(\varphi)$, it follows that

$$z\frac{d}{dz}\left(\log\left(\frac{f(z)}{z}\right)\right) = \frac{zf'(z)}{f(z)} - 1 \prec \varphi(z) - 1 =: \phi(z), \; z \in \mathbb{U},$$

which according to the logarithmic coefficients γ_n of f given by (1), concludes

$$\sum_{n=1}^{\infty} 2n\gamma_n z^n \prec \phi(z), \; z \in \mathbb{U}.$$

Now, for the proof of inequality (5), we assume that φ is convex in $\mathbb{U}$. This implies that $\phi(z)$ is convex with $\phi'(0) = B_1$, and so by Lemma 4(i) we get

$$2n|\gamma_n| \leq |\phi'(0)| = |B_1|, \; n \in \mathbb{N},$$

and concluding the result.

Next, for the proof of inequality (6), we define $h(z) := \frac{f(z)}{z}$, which is an analytic function, and it satisfies the relation

$$\frac{zh'(z)}{h(z)} = \frac{zf'(z)}{f(z)} - 1 \prec \phi(z), \; z \in \mathbb{U}, \tag{9}$$

as ϕ is convex in $\mathbb{U}$ with $\phi(0) = 0$.

On the other hand, it is well known that the function (see [17])

$$b_0(z) = \log\left(\frac{1}{1-z}\right) = \sum_{n=1}^{\infty} \frac{z^n}{n}$$

belongs to the class $\mathcal{K}$, and for $f \in \mathcal{H}$,

$$f(z) * b_0(z) = \int_0^z \frac{f(t)}{t} dt. \tag{10}$$

Now, by Lemma 2 and from (9), we obtain

$$\frac{zh'(z)}{h(z)} * b_0(z) \prec \phi(z) * b_0(z).$$

Considering (10), the above relation becomes

$$\log\left(\frac{f(z)}{z}\right) \prec \int_0^z \frac{\phi(t)}{t} dt.$$

In addition, it has been proved in [18] that the class of convex univalent functions is closed under convolution. Therefore, the function $\int_0^z \frac{\phi(t)}{t} dt$ is convex univalent. In addition, the above relation considering the logarithmic coefficients γ_n of f given by (1) is equivalent to

$$\sum_{n=1}^{\infty} 2\gamma_n z^n \prec \sum_{n=1}^{\infty} \frac{B_n z^n}{n}.$$

Applying Lemma 3, from the above subordination this gives

$$4\sum_{n=1}^k |\gamma_n|^2 \leq \sum_{n=1}^k \frac{|B_n|^2}{n^2},$$

which yields the inequality in (6). Supposing that $k \to \infty$, we deduce that

$$4 \sum_{n=1}^{\infty} |\gamma_n|^2 \le \sum_{n=1}^{\infty} \frac{|B_n|^2}{n^2},$$

and it concludes the inequality (7).

Finally, we suppose that φ is starlike with respect to 1 in $\mathbb{U}$, which implies $\phi(z)$ is starlike, and thus by Lemma 4(ii), we obtain

$$2n|\gamma_n| \le n|\phi'(0)| = n|B_1|, \quad n \in \mathbb{N},$$

This implies the inequality in (8).

For the sharp bounds, it suffices to use the equality

$$z \frac{d}{dz}\left(\log\left(\frac{f(z)}{z} \right) \right) = \frac{zf'(z)}{f(z)} - 1,$$

and so these results are sharp in inequalities (5), (6), and (8) such that for any $n \in \mathbb{N}$, there is the function f_n given by $\frac{zf'_n(z)}{f_n(z)} = \varphi(z^n)$ and the function f given by $\frac{zf'(z)}{f(z)} = \varphi(z)$, respectively. This completes the proof. □

In the following corollaries, we obtain the logarithmic coefficients γ_n for two subclasses $\mathcal{S}^*(\alpha + (1-\alpha)e^z)$ and $\mathcal{S}^*(\alpha + (1-\alpha)\sqrt{1+z})$, which were defined by Khatter et al. in [19], and $\alpha + (1-\alpha)e^z$ and $\alpha + (1-\alpha)\sqrt{1+z}$ are the convex univalent functions in $\mathbb{U}$. For $\alpha = 0$, these results reduce to the logarithmic coefficients γ_n for the subclasses $\mathcal{S}^*(e^z)$ and $\mathcal{S}^*(\sqrt{1+z})$ (see [20,21]).

Corollary 1. *For $0 \le \alpha < 1$, let the function $f \in \mathcal{S}^*(\alpha + (1-\alpha)e^z)$. Then the logarithmic coefficients of f satisfy the inequalities*

$$|\gamma_n| \le \frac{1-\alpha}{2n}, \quad n \in \mathbb{N}$$

and

$$\sum_{n=1}^{\infty} |\gamma_n|^2 \le \frac{1}{4} \sum_{n=1}^{\infty} \frac{(1-\alpha)^2/(n!)^2}{n^2}.$$

These results are sharp such that for any $n \in \mathbb{N}$, there is the function f_n given by $\frac{zf'_n(z)}{f_n(z)} = \alpha + (1-\alpha)e^{z^n}$ and the function f given by $\frac{zf'(z)}{f(z)} = \alpha + (1-\alpha)e^z$.

Corollary 2. *For $0 \le \alpha < 1$, let the function $f \in \mathcal{S}^*(\alpha + (1-\alpha)\sqrt{1+z})$. Then the logarithmic coefficients of f satisfy the inequalities*

$$|\gamma_n| \le \frac{1-\alpha}{4n}, \quad n \in \mathbb{N}$$

and

$$\sum_{n=1}^{\infty} |\gamma_n|^2 \le \frac{1}{4} \sum_{n=1}^{\infty} \frac{((1-\alpha)(\frac{1}{2})_n)^2}{n^2}.$$

These results are sharp such that for any $n \in \mathbb{N}$, there is the function f_n given by $\frac{zf'_n(z)}{f_n(z)} = \alpha + (1-\alpha)\sqrt{1+z^n}$ and the function f given by $\frac{zf'(z)}{f(z)} = \alpha + (1-\alpha)\sqrt{1+z}$.

The following corollary concludes the logarithmic coefficients γ_n for a subclass $\mathcal{S}^*(1+\sin z)$ defined by Cho et al. in [22], in which considering the proof of Theorem 1 and Corollary 1, the convexity radius for $q_0(z) = 1 + \sin z$ is given by $r_0 \approx 0.345$.

Corollary 3. Let the function $f \in \mathcal{S}^*(1+\sin z)$ where $q_0(z)$ is a convex univalent function for $r_0 \approx 0.345$ in $\mathbb{U}$. Then the logarithmic coefficients of f satisfy the inequalities

$$|\gamma_n| \leq \frac{1}{2n}, \quad n \in \mathbb{N}$$

and

$$\sum_{n=1}^{\infty} |\gamma_n|^2 \leq \frac{1}{4} \sum_{n=1}^{\infty} \frac{1}{((2n+1)!n)^2}.$$

These results are sharp such that for any $n \in \mathbb{N}$, there is the function f_n given by $\frac{z f_n'(z)}{f_n(z)} = q_0(z^n)$ and the function f given by $\frac{z f'(z)}{f(z)} = q_0(z)$.

In the following result, we get the logarithmic coefficients γ_n for a subclass $\mathcal{S}^*(p_k(z))$ defined by Kanas and Wisniowska in [23] (see also [24,25]), in which

$$p_k(z) = 1 + P_1(k)z + P_2(k)z^2 + \cdots,$$

where $p_k(z)$ is a convex univalent function in $\mathbb{U}$ and

$$P_1(k) = \begin{cases} \frac{2A^2}{1-k^2} & \text{if } 0 \leq k < 1, \\ \frac{8}{\pi^2} & \text{if } k = 1, \\ \frac{\pi^2}{4\kappa^2(t)(k^2-1)(1+t)\sqrt{t}} & \text{if } k > 1. \end{cases}$$

$A = \frac{2}{\pi} \arccos k$ and $\kappa(t)$ is the complete elliptic integral of the first kind.

Corollary 4. For $0 \leq k < \infty$, let the function $f \in \mathcal{S}^*(p_k(z))$. Then the logarithmic coefficients of f satisfy the inequalities

$$|\gamma_n| \leq \frac{P_1(k)}{2n}, \quad n \in \mathbb{N}.$$

This result is sharp such that for any $n \in \mathbb{N}$, there is the function f_n given by $\frac{z f_n'(z)}{f_n(z)} = p_k(z^n)$.

The following result concludes the logarithmic coefficients γ_n for a subclass $\mathcal{S}^*\left(\sqrt{2} - (\sqrt{2} - 1)\sqrt{\frac{1-z}{1+2(\sqrt{2}-1)z}}\right)$ defined by Mendiratta et al. in [26], in which

$$\varphi_0(z) = \sqrt{2} - (\sqrt{2}-1)\sqrt{\frac{1-z}{1+2(\sqrt{2}-1)z}} = 1 + \frac{5-3\sqrt{2}}{2}z + \frac{71-51\sqrt{2}}{8}z^2 + \cdots,$$

where φ_0 is a convex univalent function in $\mathbb{U}$.

Corollary 5. Let the function $f \in \mathcal{S}^*\left(\sqrt{2} - (\sqrt{2} - 1)\sqrt{\frac{1-z}{1+2(\sqrt{2}-1)z}}\right)$. Then the logarithmic coefficients of f satisfy the inequalities

$$|\gamma_n| \leq \frac{5-3\sqrt{2}}{4n}, \quad n \in \mathbb{N}.$$

This result is sharp such that for any $n \in \mathbb{N}$, there is the function f_n given by $\frac{z f_n'(z)}{f_n(z)} = \varphi_0(z^n)$.

The following results conclude the logarithmic coefficients γ_n for two subclasses $\mathcal{S}^*(z+\sqrt{1+z^2})$ and $\mathcal{S}^*\left(1+\frac{z}{(1-\alpha z^2)}\right)$ defined by Krishna Raina and Sokół in [27] and Kargar et al. in [28], where

$$z+\sqrt{1+z^2} = 1+z+\sum_{n=1}^{\infty}\binom{\frac{1}{2}}{n}^2 z^{2n},$$

and

$$1+\frac{z}{(1-\alpha z^2)} = 1+z+\sum_{n=1}^{\infty}\alpha^n z^{2n+1}, \quad (0\leq \alpha <1),$$

respectively. These functions are univalent and starlike with respect to 1 in $\mathbb{U}$.

Corollary 6. *Let the function $f \in \mathcal{S}^*(z+\sqrt{1+z^2})$. Then the logarithmic coefficients of f satisfy the inequalities*

$$|\gamma_n| \leq \frac{1}{2}, \quad n \in \mathbb{N}.$$

This result is sharp such that for any $n \in \mathbb{N}$, there is the function f_n given by $\frac{zf_n'(z)}{f_n(z)} = z^n + \sqrt{1+z^{2n}}$.

Corollary 7. *Let the function $f \in \mathcal{S}^*\left(1+\frac{z}{(1-\alpha z^2)}\right)$, where $0\leq \alpha <1$. Then the logarithmic coefficients of f satisfy the inequalities*

$$|\gamma_n| \leq \frac{1}{2}, \quad n \in \mathbb{N}.$$

This result is sharp such that for any $n \in \mathbb{N}$, there is the function f_n given by $\frac{zf_n'(z)}{f_n(z)} = 1+\frac{z}{(1-\alpha z^{2n})}$.

Remark 1. 1. Letting

$$\varphi(z) = \frac{1+Az}{1+Bz}$$
$$= 1+(A-B)z - B(A-B)z^2 + B^2(A-B)z^3 + \cdots$$
$$= 1+ \begin{cases} \frac{A-B}{B}\sum_{n=1}^{\infty}(-1)^{n-1}B^n z^n, & \text{if } B \neq 0 \\ Az, & \text{if } B = 0, \end{cases} \quad (-1\leq B < A \leq 1),$$

which is convex univalent in $\mathbb{U}$ in Theorem 1, then we get the results obtained by Ponnusamy et al. [7] (Theorem 2.1 and Corollary 2.3).

2. For $A = e^{i\alpha}(e^{i\alpha} - 2\beta\cos\alpha)$, where $\beta \in [0,1)$ and $\alpha \in (-\pi/2, \pi/2)$ in the above expression, then we get the results obtained by Ponnusamy et al. [7] (Theorem 2.5).

3. Taking

$$\varphi(z) = \left(\frac{1+z}{1-z}\right)^\alpha = 1+2\alpha z + 2\alpha^2 z^2 + \frac{8\alpha^3+4\alpha}{6}z^3 + \cdots$$
$$= 1+\sum_{n=1}^{\infty}A_n(\alpha)z^n, \quad (0<\alpha \leq 1),$$

which is convex univalent in $\mathbb{U}$, and $A_n(\alpha) = \sum_{k=1}^{n}\binom{n-1}{k-1}\binom{\alpha}{k}2^k$ in Theorem 1, then we get the results obtained by Ponnusamy et al. [7] (Theorem 2.6).

4. Setting

$$\varphi(z) = 1 + \frac{\beta - \alpha}{\pi} i \log\left(\frac{1 - e^{2\pi i \frac{1-\alpha}{\beta-\alpha} z}}{1 - z}\right) = 1 + \sum_{n=1}^{\infty} C_n z^n, \quad (\alpha > 1, \beta < 1),$$

which is convex univalent in $\mathbb{U}$, and $C_n = \frac{\beta - \alpha}{n\pi} i\left(1 - e^{2n\pi i \frac{1-\alpha}{\beta-\alpha}}\right)$ in Theorem 1, then we get the results obtained by Kargar [5] (Theorems 2.2 and 2.3).

5. Letting

$$\varphi(z) = 1 + \frac{1}{2i \sin \delta} \log\left(\frac{1 + z e^{i\delta}}{1 + z e^{-i\delta}}\right) = 1 + \sum_{n=1}^{\infty} D_n z^n, \quad (\pi/2 \leq \delta < \pi),$$

which is convex univalent in $\mathbb{U}$, and $D_n = \frac{(-1)^{n-1} \sin n\delta}{n \sin \delta}$ in Theorem 1, then we get the results obtained by Kargar [5] (Theorems 2.5 and 2.6).

6. Letting

$$\varphi(z) = \left(\frac{1 + cz}{1 - z}\right)^{(\alpha_1 + \alpha_2)/2} = 1 + \sum_{n=1}^{\infty} \lambda_n z^n, \quad \left(0 < \alpha_1, \alpha_2 \leq 1, c = e^{\pi i \theta}, \theta = \frac{\alpha_2 - \alpha_1}{\alpha_2 + \alpha_1}\right),$$

which is convex univalent in $\mathbb{U}$, and

$$\lambda_n = \sum_{k=1}^{n} \binom{n-1}{k-1}\binom{(\alpha_1 + \alpha_2)/2}{k}(1 + c)^k$$

in Theorem 1, then we get the results obtained for $|\gamma_n|$ by Kargar et al. [29] (Theorem 3.1). Moreover, for $\alpha_1 = \alpha_2 = \beta$, we get the result presented by Thomas in [30] (Theorem 1).

7. Let the function $f \in \mathcal{K}\left(1 - \frac{cz}{1-z}\right) = \mathcal{K}(1 - cz - cz^2 - cz^3 + \ldots)$, where $c \in (0, 1]$. It is equivalent to

$$\operatorname{Re}\left(1 + \frac{zf''(z)}{f'(z)}\right) < 1 + \frac{c}{2}.$$

Then we have (see e.g., [31] (Theorem 1))

$$\frac{zf'(z)}{f(z)} \prec \frac{(1 + c)(1 - z)}{1 + c - z},$$

where $\frac{(1 + c)(1 - z)}{1 + c - z}$ is a convex univalent function in $\mathbb{U}$, and

$$\frac{(1 + c)(1 - z)}{1 + c - z} = 1 - \frac{c}{c + 1} z - \frac{c}{(c + 1)^2} z^2 + \cdots = 1 - c \sum_{n=1}^{\infty} \frac{z^n}{(1 + c)^n}.$$

Thus, applying Theorem 1, we get the results obtained by Obradović et al. [4] (Theorem 2 and Corollary 2).

Theorem 2. *Let the function $f \in \mathcal{K}(\varphi)$. Then the logarithmic coefficients of f satisfy the inequalities*

$$|\gamma_1| \leq \frac{|B_1|}{4}, \tag{11}$$

$$|\gamma_2| \leq \begin{cases} \dfrac{|B_1|}{12} & \text{if} \quad |4B_2 + B_1^2| \leq 4|B_1| \\ \dfrac{|4B_2 + B_1^2|}{48} & \text{if} \quad |4B_2 + B_1^2| > 4|B_1| \end{cases}, \qquad (12)$$

and if B_1, B_2, and B_3 are real values,

$$|\gamma_3| \leq \dfrac{|B_1|}{24} H(q_1; q_2), \qquad (13)$$

where $H(q_1; q_2)$ is given by Lemma 5, $q_1 = \dfrac{B_1 + \frac{4B_2}{B_1}}{2}$, and $q_2 = \dfrac{B_2 + \frac{2B_3}{B_1}}{2}$. The bounds (11) and (12) are sharp.

Proof. Let $f \in \mathcal{K}(\varphi)$. Then by the definition of the subordination, there is a $\omega \in \Omega$ with $\omega(z) = \sum_{n=1}^{\infty} c_n z^n$ so that

$$\begin{aligned} 1 + \dfrac{zf''(z)}{f'(z)} &= \varphi(\omega(z)) \\ &= 1 + B_1 c_1 z + (B_1 c_2 + B_2 c_1^2) z^2 + (B_1 c_3 + 2c_1 c_2 B_2 + B_3 c_1^3) z^3 + \cdots. \end{aligned}$$

From the above equation, we get that

$$\begin{cases} 2a_2 = B_1 c_1 \\ 6a_3 - 4a_2^2 = B_1 c_2 + B_2 c_1^2 \\ 12a_4 - 18a_2 a_3 + 8a_2^3 = B_1 c_3 + 2c_1 c_2 B_2 + B_3 c_1^3. \end{cases} \qquad (14)$$

By substituting values a_n (n = 1, 2, 3) from (14) in (3), we have

$$\begin{cases} 2\gamma_1 = \dfrac{B_1 c_1}{2} \\ 2\gamma_2 = \dfrac{8B_1 c_2 + (8B_2 + 2B_1^2) c_1^2}{48} \\ 2\gamma_3 = \dfrac{B_1}{12} \left[c_3 + \dfrac{B_1 + \frac{4B_2}{B_1}}{2} c_1 c_2 + \dfrac{B_2 + \frac{2B_3}{B_1}}{2} c_1^3 \right]. \end{cases}$$

Firstly, for γ_1, by applying Lemma 1 we get $|\gamma_1| \leq \dfrac{|B_1|}{4}$, and this bound is sharp for $|c_1| = 1$. Next, applying Lemma 1 for γ_2, we have

$$\begin{aligned} |\gamma_2| &\leq \dfrac{4|B_1|(1 - |c_1|^2) + |4B_2 + B_1^2||c_1|^2}{48} \\ &= \dfrac{4|B_1| + [|4B_2 + B_1^2| - 4|B_1|]|c_1|^2}{48} \\ &\leq \begin{cases} \dfrac{4|B_1|}{48} & \text{if} \quad |4B_2 + B_1^2| \leq 4|B_1| \\ \dfrac{|4B_2 + B_1^2|}{48} & \text{if} \quad |4B_2 + B_1^2| > 4|B_1|. \end{cases} \end{aligned}$$

These bounds are sharp for $c_1 = 0$ and $|c_1| = 1$, respectively.

Finally, using Lemma 5 for γ_3, we obtain

$$2|\gamma_3| \leq \frac{|B_1|}{12}\left|c_3 + \frac{B_1 + \frac{4B_2}{B_1}}{2}c_1c_2 + \frac{B_2 + \frac{2B_3}{B_1}}{2}c_1^3\right| \leq H(q_1; q_2),$$

where $q_1 = \frac{B_1 + \frac{4B_2}{B_1}}{2}$ and $q_2 = \frac{B_2 + \frac{2B_3}{B_1}}{2}$. Therefore, this completes the proof. □

Remark 2. *1. Letting*

$$\varphi(z) = 1 + \frac{cz}{1-z}$$
$$= 1 + cz + cz^2 + cz^3 + \ldots \quad (c \in (0, 3])$$

in Theorem 2, (for $|\gamma_3|$ with respect to D_6) then we get the results obtained by Ponnusamy et al. [7] (Theorem 2.7 and Corollary 2.8).

2. Taking

$$\varphi(z) = 1 - \frac{cz}{1-z}$$
$$= 1 - cz - cz^2 - cz^3 + \ldots \quad (c \in (0, 1])$$

in Theorem 2, (for $|\gamma_3|$ respect to D_2) then we get the results obtained by Ponnusamy et al. [7] (Theorem 2.10).

Author Contributions: All authors contributed equally.

Funding: The authors would like to express their gratitude to the referees for many valuable suggestions regarding the previous version of this paper. This research was supported by the Basic Science Research Program through the National Research Foundation of Korea (NRF) funded by the Ministry of Education, Science, and Technology (No. 2016R1D1A1A09916450).

Conflicts of Interest: The authors declare no conflict of interest.

References

1. Kayumov, I.R. On Brennan's conjecture for a special class of functions. *Math. Notes* **2005**, *78*, 498–502. [CrossRef]
2. Ma, W.C.; Minda, D. A unified treatment of some special classes of univalent functions. In *Proceedings of the Conference on Complex Analysis (Tianjin, 1992)*; Internat Press: Cambridge, MA, USA, 1992; pp. 157–169.
3. Ali, M.F.; Vasudevarao, A. On logarithmic coefficients of some close-to-convex functions. *Proc. Am. Math. Soc.* **2018**, *146*, 1131–1142. [CrossRef]
4. Obradović, M.; Ponnusamy, S.; Wirths, K.-J. Logarithmic coeffcients and a coefficient conjecture for univalent functions. *Monatsh. Math.* **2018**, *185*, 489–501. [CrossRef]
5. Kargar, R. On logarithmic coefficients of certain starlike functions related to the vertical strip. *J. Anal.* **2018**, 1–11. [CrossRef]
6. Kumar, U.P.; Vasudevarao, A. Logarithmic coefficients for certain subclasses of close-to-convex functions. *Monatsh. Math.* **2018**, *187*, 543–563. [CrossRef]
7. Ponnusamy, S.; Sharma, N.L.; Wirths, K.J. Logarithmic Coefficients of the Inverse of Univalent Functions. *Results Math.* **2018**, *73*, 160. [CrossRef]
8. Ponnusamy, S.; Sharma, N.L.; Wirths, K.J. Logarithmic coefficients problems in families related to starlike and convex functions. *J. Aust. Math. Soc.* **2018**, 1–20. [CrossRef]
9. Srivastava, H.M.; Hussain, S.; Raziq, A.; Raza, M. The Fekete-Szegö functional for a subclass of analytic functions associated with quasi-subordination. *Carpathian J. Math.* **2018**, *34*, 103–113.

10. Srivastava, H.M.; Prajapati, A.; Gochhayat, P. Third-order differential subordination and differential superordination results for analytic functions involving the Srivastava-Attiya operator. *Appl. Math. Inf. Sci.* **2018**, *12*, 469–481. [CrossRef]
11. Srivastava, H.M.; Rǎducanu, D.; Zaprawa, P. A certain subclass of analytic functions defined by means of differential subordination. *Filomat* **2016**, *30*, 3743–3757. [CrossRef]
12. Duren, P.L. *Univalent Functions*; Springer: New York, NY, USA; Berlin/Heidelberg, Germany; Tokyo, Japan, 1983.
13. Nehari, Z. *Conformal Mapping*; McGraw-Hill: New York, NY, USA, 1952.
14. Ruscheweyh, S.; Stankiewicz, J. Subordination under convex univalent function. *Bull. Pol. Acad. Sci. Math.* **1985**, *33*, 499–502.
15. Rogosinski, W. On the coefficients of subordinate functions. *Proc. Lond. Math. Soc.* **1943**, *48*, 48–82. [CrossRef]
16. Prokhorov, D.V.; Szynal, J. Inverse coefficients for $(\alpha;\beta)$-convex functions. *Ann. Univ. Mariae Curie-Sklodowska Sect. A* **1984**, *35*, 125–143.
17. Ruscheweyh, S. New criteria for univalent functions. *Proc. Am. Math. Soc.* **1975**, *49*, 109–115. [CrossRef]
18. Ruscheweyh, S.; Sheil-Small, T. Hadamard product of schlicht functions and the Pòyla Schoenberg conjecture. *Comment. Math. Helv.* **1973**, *48*, 119–135. [CrossRef]
19. Khatter, K.; Ravichandran, V.; Kumar, S.S. Starlike functions associated with exponential function and the lemniscate of Bernoulli. *Rev. Real Acad. Cienc. Exactas Físicas Nat. Ser. A Mat.* **2019**, *113*, 233–253. [CrossRef]
20. Mendiratta, R.; Nagpal, S.; Ravichandran, V. On a subclass of strongly starlike functions associated with exponential function. *Bull. Malays. Math. Sci. Soc.* **2015**, *38*, 365–386. [CrossRef]
21. Sokół, J.; Stankiewicz, J. Radius of convexity of some subclasses of strongly starlike functions. *Zeszyty Nauk. Politech. Rzeszowskiej Mat.* **1996**, *19*, 101–105.
22. Cho, N.E.; Kumar, V.; Kumar, S.S.; Ravichandran, V. Radius problems for starlike functions associated with the sine function. *Bull. Iran. Math. Soc.* **2019**, *45*, 213–232. [CrossRef]
23. Kanas, S.; Wiśniowska, A. Conic regions and k-uniform convexity. *J. Comput. Appl. Math.* **1999**, *105*, 327–336. [CrossRef]
24. Kanas, S. Coefficient estimates in subclasses of the Caratheodory class related to conical domains. *Acta Math. Univ. Comen.* **2005**, *74*, 149–161.
25. Kanas, S.; Srivastava, H.M. Linear operators associated with k-uniformly convex functions. *Integral Transform. Spec. Funct.* **2000**, *9*, 121–132. [CrossRef]
26. Mendiratta, R.; Nagpal, S.; Ravichandran, V. A subclass of starlike functions associated with left-half of the lemniscate of Bernoulli. *Int. J. Math.* **2014**, *25*, 1450090. [CrossRef]
27. Raina, R.K.; Sokół, J. Some properties related to a certain class of starlike functions. *C. R. Acad. Sci. Paris Ser. I* **2015**, *353*, 973–978. [CrossRef]
28. Kargar, R.; Ebadian, A.; Sokół, J. On Booth lemniscate and starlike functions. *Anal. Math. Phys.* **2019**, *9*, 143–154. [CrossRef]
29. Kargar, R.; Sokół, J.; Mahzoon, H. Some properties of a certain subclass of strongly starlike functions. *arXiv* **2018**, arXiv:1811.01271.
30. Thomas, D.K. On the coefficients of strongly starlike functions. *Indian J. Math.* **2016**, *58*, 135–146.
31. Jovanović, I.; Obradovixcx, M. A note on certain classes of univalent functions. *Filomat* **1995**, *9*, 69–72.

© 2019 by the authors. Licensee MDPI, Basel, Switzerland. This article is an open access article distributed under the terms and conditions of the Creative Commons Attribution (CC BY) license (http://creativecommons.org/licenses/by/4.0/).

Article

Third-Order Hankel and Toeplitz Determinants for Starlike Functions Connected with the Sine Function

Hai-Yan Zhang [1], Rekha Srivastava [2,*] and Huo Tang [1,*]

[1] School of Mathematics and Statistics, Chifeng University, Chifeng 024000, China; cfxyzhhy@163.com
[2] Department of Mathematics and Statistics, University of Victoria, Victoria, BC V8W 3R4, Canada
* Correspondence: rekhas@math.uvic.ca (R.S.); thth2009@163.com (H.T.)

Received: 26 March 2019; Accepted: 25 April 2019; Published: 6 May 2019

Abstract: Let $\mathcal{S}_s^*$ be the class of normalized functions f defined in the open unit disk $\mathbb{D} = \{z : |z| < 1\}$ such that the quantity $\frac{zf'(z)}{f(z)}$ lies in an eight-shaped region in the right-half plane and satisfying the condition $\frac{zf'(z)}{f(z)} \prec 1 + \sin z$ $(z \in \mathbb{D})$. In this paper, we aim to investigate the third-order Hankel determinant $H_3(1)$ and Toeplitz determinant $T_3(2)$ for this function class $\mathcal{S}_s^*$ associated with sine function and obtain the upper bounds of the determinants $H_3(1)$ and $T_3(2)$.

Keywords: starlike function; Toeplitz determinant; Hankel determinant; sine function; upper bound

MSC: 30C45; 30C50; 30C80

---

1. Introduction

Let $\mathcal{A}$ denote the class of functions f which are analytic in the open unit disk $\mathbb{D} = \{z : |z| < 1\}$ of the form

$$f(z) = z + a_2 z^2 + a_3 z^3 + \cdots \quad (z \in \mathbb{D}) \tag{1}$$

and let $\mathcal{S}$ denote the subclass of $\mathcal{A}$ consisting of univalent functions.

Suppose that $\mathcal{P}$ denotes the class of analytic functions p normalized by

$$p(z) = 1 + c_1 z + c_2 z^2 + c_3 z^3 + \cdots$$

and satisfying the condition

$$\Re(p(z)) > 0 \quad (z \in \mathbb{D}).$$

We easily see that, if $p(z) \in \mathcal{P}$, then a Schwarz function $\omega(z)$ exists with $\omega(0) = 0$ and $|\omega(z)| < 1$, such that (see [1])

$$p(z) = \frac{1 + w(z)}{1 - w(z)} \quad (z \in \mathbb{D}).$$

Very recently, Cho et al. [2] introduced the following function class $\mathcal{S}_s^*$, which are associated with sine function:

$$\mathcal{S}_s^* := \left\{ f \in \mathcal{A} : \frac{zf'(z)}{f(z)} \prec 1 + \sin z \quad (z \in \mathbb{D}) \right\}, \tag{2}$$

where "$\prec$" stands for the subordination symbol (for details, see [3]) and also implies that the quantity $\frac{zf'(z)}{f(z)}$ lies in an eight-shaped region in the right-half plane.

The qth Hankel determinant for $q \geq 1$ and $n \geq 1$ of functions f was stated by Noonan and Thomas [4] as

$$H_q(n) = \begin{vmatrix} a_n & a_{n+1} & \cdots & a_{n+q-1} \\ a_{n+1} & a_{n+2} & \cdots & a_{n+q} \\ \vdots & \vdots & & \vdots \\ a_{n+q-1} & a_{n+q} & \cdots & a_{n+2q-2} \end{vmatrix} \quad (a_1 = 1).$$

This determinant has been considered by several authors, for example, Noor [5] determined the rate of growth of $H_q(n)$ as $n \to \infty$ for functions $f(z)$ given by Equation (1) with bounded boundary and Ehrenborg [6] studied the Hankel determinant of exponential polynomials.

In particular, we have

$$H_3(1) = \begin{vmatrix} a_1 & a_2 & a_3 \\ a_2 & a_3 & a_4 \\ a_3 & a_4 & a_5 \end{vmatrix} \quad (n = 1, q = 3).$$

Since $f \in \mathcal{S}$, $a_1 = 1$,

$$H_3(1) = a_3(a_2 a_4 - a_3^2) - a_4(a_4 - a_2 a_3) + a_5(a_3 - a_2^2).$$

We note that $|H_2(1)| = |a_3 - a_2^2|$ is the well-known Fekete-Szego functional (see, for example, [7–9]).

On the other hand, Thomas and Halim [10] defined the symmetric Toeplitz determinant $T_q(n)$ as follows:

$$T_q(n) = \begin{vmatrix} a_n & a_{n+1} & \cdots & a_{n+q-1} \\ a_{n+1} & a_n & \cdots & a_{n+q} \\ \vdots & \vdots & & \vdots \\ a_{n+q-1} & a_{n+q} & \cdots & a_n \end{vmatrix} \quad (n \geq 1, q \geq 1).$$

The Toeplitz determinants are closely related to Hankel determinants. Hankel matrices have constant entries along the reverse diagonal, whereas Toeplitz matrices have constant entries along the diagonal. For a good summary of the applications of Toeplitz matrices to the wide range of areas of pure and applied mathematics, we can refer to [11].

As a special case, when $n = 2$ and $q = 3$, we have

$$T_3(2) = \begin{vmatrix} a_2 & a_3 & a_4 \\ a_3 & a_2 & a_3 \\ a_4 & a_3 & a_2 \end{vmatrix}.$$

In recent years, many authors studied the second-order Hankel determinant $H_2(2)$ and the third-order Hankel determinant $H_3(1)$ for various classes of functions (the interested readers can see, for instance, [12–25]). However, apart from the work in [10,21,26,27], there appears to be little literature dealing with Toeplitz determinants. Inspired by the aforementioned works, in this paper, we aim to investigate the third-order Hankel determinant $H_3(1)$ and Toeplitz determinant $T_3(2)$ for the above function class $\mathcal{S}_s^*$ associated with sine function, and obtain the upper bounds of the above determinants.

2. Main Results

To prove our desired results, we need the following lemmas.

Lemma 1. *If $p(z) \in \mathcal{P}$, then exists some x, z with $|x| \leq 1$(see [28]), $|z| \leq 1$, such that*

$$2c_2 = c_1^2 + x(4 - c_1^2),$$

$$4c_3 = c_1^3 + 2c_1 x(4 - c_1^2) - (4 - c_1^2)c_1 x^2 + 2(4 - c_1^2)(1 - |x|^2)z.$$

Lemma 2. *Let $p(z) \in \mathcal{P}$ (see [29]), then*

$$|c_n| \leq 2, \ n = 1, 2, \cdots.$$

We now state and prove the main results of our present investigation.

Theorem 1. *If the function $f(z) \in \mathcal{S}_s^*$ and of the form Equation (1), then*

$$|a_2| \leq 1, \ |a_3| \leq \frac{1}{2}, \ |a_4| \leq \frac{5}{9}, \ |a_5| \leq \frac{47}{72}. \tag{3}$$

Proof. Since $f(z) \in \mathcal{S}_s^*$, according to subordination relationship, so there exists a Schwarz function $\omega(z)$ with $\omega(0) = 0$ and $|\omega(z)| < 1$, such that

$$\frac{zf'(z)}{f(z)} = 1 + \sin(\omega(z)).$$

Now,

$$\frac{zf'(z)}{f(z)} = \frac{z + \sum_{n=2}^{\infty} na_n z^n}{z + \sum_{n=2}^{\infty} a_n z^n}$$

$$= (1 + \sum_{n=2}^{\infty} na_n z^{n-1})[1 - a_2 z + (a_2^2 - a_3)z^2 - (a_2^3 - 2a_2 a_3 + a_4)z^3$$

$$+ (a_2^4 - 3a_2^2 a_3 + 2a_2 a_4 + a_3^2 - a_5)z^4 + \cdots]$$

$$= 1 + a_2 z + (2a_3 - a_2^2)z^2 + (a_2^3 - 3a_2 a_3 + 3a_4)z^3$$

$$+ (4a_5 - a_2^4 + 4a_2^2 a_3 - 4a_2 a_4 - 2a_3^2)z^4 + \cdots. \tag{4}$$

Define a function

$$p(z) = \frac{1 + \omega(z)}{1 - \omega(z)} = 1 + c_1 z + c_2 z^2 + \cdots.$$

Clearly, we have $p(z) \in \mathcal{P}$ and

$$\omega(z) = \frac{p(z) - 1}{1 + p(z)} = \frac{c_1 z + c_2 z^2 + c_3 z^3 + \cdots}{2 + c_1 z + c_2 z^2 + c_3 z^3 + \cdots}. \tag{5}$$

On the other hand,

$$1 + \sin(\omega(z)) = 1 + \frac{1}{2}c_1 z + (\frac{c_2}{2} - \frac{c_1^2}{4})z^2 + (\frac{5c_1^3}{48} + \frac{c_3 - c_1 c_2}{2})z^3$$

$$+ (\frac{c_4}{2} + \frac{5c_1^2 c_2}{16} - \frac{c_2^2}{4} - \frac{c_1 c_3}{2} - \frac{c_1^4}{32})z^4 + \cdots. \tag{6}$$

Comparing the coefficients of z, z^2, z^3, z^4 between Equations (4) and (6), we obtain

$$a_2 = \frac{c_1}{2}, \; a_3 = \frac{c_2}{4}, \; a_4 = \frac{c_3}{6} - \frac{c_1 c_2}{24} - \frac{c_1^3}{144}, \; a_5 = \frac{c_4}{8} - \frac{c_1 c_3}{24} + \frac{5 c_1^4}{1152} - \frac{c_1^2 c_2}{192} - \frac{c_2^2}{32}. \tag{7}$$

By using Lemma 2, we thus know that

$$|a_2| \leq 1, \; |a_3| \leq \frac{1}{2}, \; |a_4| \leq \frac{5}{9}, \; |a_5| \leq \frac{47}{72}.$$

The proof of Theorem 1 is completed. □

Theorem 2. *If the function $f(z) \in \mathcal{S}_s^*$ and of the form in Equation (1), then we have*

$$|a_3 - a_2^2| \leq \frac{1}{2}. \tag{8}$$

Proof. According to Equation (7), we have

$$|a_3 - a_2^2| = \left| \frac{c_2}{4} - \frac{c_1^2}{4} \right|.$$

By applying Lemma 1, we get

$$|a_3 - a_2^2| = \left| \frac{x(4 - c_1^2)}{8} - \frac{c_1^2}{8} \right|.$$

Let $|x| = t$, $t \in [0, 1]$, $c_1 = c$, $c \in [0, 2]$. Then, using the triangle inequality, we obtain

$$|a_3 - a_2^2| \leq \frac{t(4 - c^2)}{8} + \frac{c^2}{8}.$$

Suppose that

$$F(c, t) = \frac{t(4 - c^2)}{8} + \frac{c^2}{8},$$

then $\forall t \in (0, 1)$, $\forall c \in (0, 2)$,

$$\frac{\partial F}{\partial t} = \frac{4 - c^2}{8} > 0,$$

which shows that $F(c, t)$ is an increasing function on the closed interval $[0,1]$ about t. Therefore, the function $F(c, t)$ can get the maximum value at $t = 1$, that is, that

$$\max F(c, t) = F(c, 1) = \frac{(4 - c^2)}{8} + \frac{c^2}{8} = \frac{1}{2}.$$

Thus, obviously,

$$|a_3 - a_2^2| \leq \frac{1}{2}.$$

The proof of Theorem 2 is thus completed. □

Theorem 3. *If the function $f(z) \in \mathcal{S}_s^*$ and of the form in Equation (1), then we have*

$$|a_2 a_3 - a_4| \leq \frac{1}{3}. \tag{9}$$

Proof. From Equation (7), we have

$$|a_2 a_3 - a_4| = \left| \frac{c_1 c_2}{8} + \frac{c_1^3}{144} - \frac{c_3}{6} + \frac{c_1 c_2}{24} \right|$$
$$= \left| \frac{c_1 c_2}{6} - \frac{c_3}{6} + \frac{c_1^3}{144} \right|.$$

Now, in view of Lemma 1, we get

$$|a_2 a_3 - a_4| = \left| \frac{7c_1^3}{144} + \frac{(4 - c_1^2) c_1 x^2}{24} - \frac{(4 - c_1^2)(1 - |x|^2) z}{12} \right|.$$

Let $|x| = t$, $t \in [0, 1]$, $c_1 = c$, $c \in [0, 2]$. Then, using the triangle inequality, we deduce that

$$|a_2 a_3 - a_4| \leq \frac{7c^3}{144} + \frac{(4 - c^2) c t^2}{24} + \frac{(4 - c^2)(1 - t^2)}{12}.$$

Assume that

$$F(c, t) = \frac{7c^3}{144} + \frac{(4 - c^2) c t^2}{24} + \frac{(4 - c^2)(1 - t^2)}{12}.$$

Therefore, we have, $\forall t \in (0, 1)$, $\forall c \in (0, 2)$

$$\frac{\partial F}{\partial t} = \frac{(4 - c^2) t (c - 2)}{12} < 0,$$

namely, $F(c, t)$ is an decreasing function on the closed interval $[0,1]$ about t. This implies that the maximum value of $F(c, t)$ occurs at $t = 0$, which is

$$\max F(c, t) = F(c, 0) = \frac{(4 - c^2)}{12} + \frac{7c^3}{144}.$$

Define

$$G(c) = \frac{(4 - c^2)}{12} + \frac{7c^3}{144},$$

clearly, the function $G(c)$ has a maximum value attained at $c = 0$, also which is

$$|a_2 a_3 - a_4| \leq G(0) = \frac{1}{3}.$$

The proof of Theorem 3 is completed. □

Theorem 4. *If the function $f(z) \in S_s^*$ and of the form in Equation (1), then we have*

$$|a_2 a_4 - a_3^2| \leq \frac{1}{4}. \tag{10}$$

Proof. Suppose that $f(z) \in S_s^*$, then from Equation (7), we have

$$|a_2 a_4 - a_3^2| = \left| \frac{c_1 c_3}{12} - \frac{c_1^2 c_2}{48} + \frac{c_1^4}{48} - \frac{c_2^2}{16} \right|.$$

Now, in terms of Lemma 1, we obtain

$$|a_2 a_4 - a_3^2| = \left| \frac{c_1 c_3}{12} - \frac{c_1^2 c_2}{48} - \frac{c_1^4}{288} - \frac{c_2^2}{16} \right|$$

$$= \left| -\frac{5 c_1^4}{576} - \frac{x^2 c_1^2 (4 - c_1^2)}{48} - \frac{x^2 (4 - c_1^2)^2}{64} + \frac{c_1 (4 - c_1^2)(1 - |x|^2) z}{24} \right|.$$

Let $|x| = t$, $t \in [0,1]$, $c_1 = c$, $c \in [0,2]$. Then, using the triangle inequality, we get

$$|a_2 a_4 - a_3^2| \le \frac{t^2 c^2(4-c^2)}{48} + \frac{(1-t^2)c(4-c^2)}{24} + \frac{t^2(4-c^2)^2}{64} + \frac{5c^4}{576}.$$

Putting

$$F(c,t) = \frac{t^2 c^2(4-c^2)}{48} + \frac{(1-t^2)c(4-c^2)}{24} + \frac{t^2(4-c^2)^2}{64} + \frac{5c^4}{576},$$

then, $\forall t \in (0,1)$, $\forall c \in (0,2)$, we have

$$\frac{\partial F}{\partial t} = \frac{t(c^2 - 8c + 12)(4-c^2)}{96} > 0,$$

which implies that $F(c,t)$ increases on the closed interval $[0,1]$ about t. That is, that $F(c,t)$ have a maximum value at $t = 1$, which is

$$\max F(c,t) = F(c,1) = \frac{c^2(4-c^2)}{48} + \frac{(4-c^2)^2}{64} + \frac{5c^4}{576}.$$

Setting

$$G(c) = \frac{c^2(4-c^2)}{48} + \frac{(4-c^2)^2}{64} + \frac{5c^4}{576},$$

then we have

$$G'(c) = \frac{c(4-c^2)}{24} - \frac{c^3}{24} - \frac{c(4-c^2)}{16} + \frac{5c^3}{144}.$$

If $G'(c) = 0$, then the root is $c = 0$. In addition, since $G''(0) = -\frac{1}{12} < 0$, so the function $G(c)$ can take the maximum value at $c = 0$, which is

$$|a_2 a_4 - a_3^2| \le G(0) = \frac{1}{4}.$$

The proof of Theorem 4 is completed. □

Theorem 5. *If the function* $f(z) \in \mathcal{S}_s^*$ *and of the form in Equation (1), then we have*

$$|a_2^2 - a_3^2| \le \frac{5}{4}. \tag{11}$$

Proof. Suppose that $f(z) \in \mathcal{S}_s^*$, then, by using Equation (7), we have

$$|a_2^2 - a_3^2| = \left|\frac{c_1^2}{4} - \frac{c_2^2}{16}\right|.$$

Next, according to Lemma 1, we obtain

$$|a_2^2 - a_3^2| = \left|\frac{c_1^2}{4} - \frac{c_2^2}{16}\right|$$

$$= \left|\frac{c_1^2}{4} - \frac{c_1^4}{64} - \frac{xc_1^2(4-c_1^2)}{32} - \frac{x^2(4-c_1^2)^2}{64}\right|.$$

Let $|x| = t$, $t \in [0,1]$, $c_1 = c$, $c \in [0,2]$. Then, by applying the triangle inequality, we get

$$|a_2^2 - a_3^2| \le \frac{c^2}{4} + \frac{c^4}{64} + \frac{tc^2(4-c^2)}{32} + \frac{t^2(4-c^2)^2}{64}.$$

Taking
$$F(c,t) = \frac{c^2}{4} + \frac{c^4}{64} + \frac{tc^2(4-c^2)}{32} + \frac{t^2(4-c^2)^2}{64}.$$

Then, $\forall t \in (0,1)$, $\forall c \in (0,2)$, we have
$$\frac{\partial F}{\partial t} = \frac{c^2(4-c^2)}{32} + \frac{t(4-c^2)^2}{32} > 0,$$

which implies that $F(c,t)$ increases on the closed interval $[0,1]$ about t. Namely, the maximum value of $F(c,t)$ attains at $t = 1$, which is
$$\max F(c,t) = F(c,1) = \frac{c^2}{4} + \frac{c^4}{64} + \frac{c^2(4-c^2)}{32} + \frac{(4-c^2)^2}{64}.$$

Let
$$G(c) = \frac{c^2}{4} + \frac{c^4}{64} + \frac{c^2(4-c^2)}{32} + \frac{(4-c^2)^2}{64},$$

then
$$G'(c) = \frac{c}{2} > 0, \; \forall c \in (0,2).$$

Therefore, the function $G(c)$ is an increasing function on the closed interval $[0,2]$ about c, and thus $G(c)$ has a maximum value attained at $c = 2$, which is
$$|a_2^2 - a_3^2| \leq G(2) = \frac{5}{4}.$$

The proof of Theorem 5 is completed. □

Theorem 6. *If the function $f(z) \in \mathcal{S}_s^*$ and of the form in Equation (1), then we have*
$$|a_2 a_3 - a_3 a_4| \leq \frac{13}{12}. \tag{12}$$

Proof. Assume that $f(z) \in \mathcal{S}_s^*$, then from Equation (7), we obtain
$$|a_2 a_3 - a_3 a_4| = \left|\frac{c_1 c_2}{8} + \frac{c_1^3 c_2}{576} - \frac{c_2 c_3}{24} + \frac{c_1 c_2^2}{96}\right|.$$

Now, by using Lemma 1, we see that
$$|a_2 a_3 - a_3 a_4| = \left|\frac{c_1 c_2}{8} + \frac{c_1^3 c_2}{576} - \frac{c_2 c_3}{24} + \frac{c_1 c_2^2}{96}\right|$$
$$= \left|\frac{c_1^3}{16} - \frac{c_1^5}{576} - \frac{11 x c_1^3(4-c_1^2)}{1152} + \frac{x c_1 (4-c_1^2)}{16} + \frac{x^2 c_1 (4-c_1^2)[c_1^2 + x(4-c_1^2)]}{192} + \frac{c_1 x^2 (4-c_1^2)^2}{128} + \frac{(1-|x|^2) z (4-c_1^2)[x(4-c_1^2) + c_1^2]}{96}\right|.$$

If we let $|x| = t$, $t \in [0,1]$, $c_1 = c$, $c \in [0,2]$, then, using the triangle inequality, we have
$$|a_2 a_3 - a_3 a_4| \leq \frac{c^3}{16} + \frac{c^5}{576} + \frac{11 t c^3 (4-c^2)}{1152} + \frac{t(4-c^2)}{8} + \frac{t^2[c^2 + t(4-c^2)](4-c^2)}{96} + \frac{t^2(4-c^2)^2}{64} + \frac{(4-c^2)[t(4-c^2) + c^2]}{96}.$$

Setting
$$F(c,t) = \frac{c^3}{16} + \frac{c^5}{576} + \frac{11 t c^3 (4-c^2)}{1152} + \frac{t(4-c^2)}{8} + \frac{t^2[c^2 + t(4-c^2)](4-c^2)}{96} + \frac{t^2(4-c^2)^2}{64} + \frac{(4-c^2)[t(4-c^2) + c^2]}{96}.$$

Then, we easily see that, $\forall t \in (0,1)$, $\forall c \in (0,2)$,
$$\frac{\partial F}{\partial t} = \frac{11 c^3 (4-c^2)}{1152} + \frac{(4-c^2)}{8} + \frac{t[c^2 + t(4-c^2)](4-c^2)}{48} + \frac{t^2(4-c^2)^2}{96} + \frac{t(4-c^2)^2}{32} + \frac{(4-c^2)^2}{96} > 0,$$

which implies that $F(c,t)$ is an increasing function on the closed interval [0,1] about t. That is, that the maximum value of $F(c,t)$ occurs at $t = 1$, which is

$$\max F(c,t) = F(c,1) = \frac{c^3}{16} + \frac{c^5}{576} + \frac{11c^3(4-c^2)}{1152} + \frac{(4-c^2)}{8} + \frac{(4-c^2)}{24} + \frac{(4-c^2)^2}{64} + \frac{(4-c^2)}{24}.$$

Taking

$$G(c) = \frac{c^3}{16} + \frac{c^5}{576} + \frac{11c^3(4-c^2)}{1152} + \frac{(4-c^2)}{8} + \frac{(4-c^2)}{24} + \frac{(4-c^2)^2}{64} + \frac{(4-c^2)}{24},$$

then

$$G'(c) = \frac{3c^2}{16} + \frac{5c^4}{576} + \frac{11c^2(4-c^2)}{384} - \frac{11c^4}{576} - \frac{c(4-c^2)}{16} - \frac{c}{12},$$

$$G''(c) = \frac{3c}{8} + \frac{5c^3}{144} + \frac{11c(4-2c^2)}{192} - \frac{11c^3}{144} - \frac{(4-c^2)}{16} + \frac{c^2}{8} - \frac{1}{12}.$$

We easily find that $c = 0$ is the root of the function $G'(c) = 0$, since $G''(0) < 0$, which implies that the function $G(c)$ can reach the maximum value at $c = 0$, also which is

$$|a_2 a_3 - a_3 a_4| \leq G(0) = \frac{13}{12}.$$

The proof of Theorem 6 is completed. □

Theorem 7. *If the function $f(z) \in \mathcal{S}_s^*$ and of the form in Equation (1), then we have*

$$|H_3(1)| \leq \frac{275}{432} \approx 0.637. \tag{13}$$

Proof. Since

$$H_3(1) = a_3(a_2 a_4 - a_3^2) - a_4(a_4 - a_2 a_3) + a_5(a_3 - a_2^2),$$

by applying the triangle inequality, we get

$$|H_3(1)| \leq |a_3||a_2 a_4 - a_3^2| + |a_4||a_4 - a_2 a_3| + |a_5||a_3 - a_2^2|. \tag{14}$$

Now, substituting Equations (3), (8), (9) and (10) into Equation (14), we easily obtain the desired assertion (Equation (13)). □

Theorem 8. *If the function $f(z) \in \mathcal{S}_s^*$ and of the form in Equation (1), then we have*

$$|T_3(2)| \leq \frac{139}{72} \approx 1.931. \tag{15}$$

Proof. Because

$$T_3(2) = a_2(a_2^2 - a_3^2) - a_3(a_2 a_3 - a_3 a_4) + a_4(a_3^2 - a_2 a_4),$$

by using the triangle inequality, we obtain

$$|T_3(2)| \leq |a_2||a_2^2 - a_3^2| + |a_3||a_2 a_3 - a_3 a_4| + |a_4||a_3^2 - a_2 a_4|. \tag{16}$$

Next, from Equations (3), (10), (11) and (12), we immediately get the desired assertion (Equation (15)). □

Finally, we give two examples to illustrate our results obtained.

Example 1. If we take the function $f(z) = e^z - 1 = z + \sum_{n=2}^{\infty} \frac{z^n}{n!} \in \mathcal{S}_s^*$, then we obtain

$$|H_3(1)| \leq |a_3||a_2a_4 - a_3^2| + |a_4||a_4 - a_2a_3| + |a_5||a_3 - a_2^2|$$

$$= \frac{1}{3!} \times |\frac{1}{2!} \times \frac{1}{4!} - \frac{1}{3!} \times \frac{1}{3!}| + \frac{1}{4!} \times |\frac{1}{4!} - \frac{1}{2!} \times \frac{1}{3!}| + \frac{1}{5!} \times |\frac{1}{3!} - \frac{1}{2!} \times \frac{1}{2!}|$$

$$\approx 0.004 < 0.637.$$

Example 2. If we set the function $f(z) = -\log(1-z) = z + \sum_{n=2}^{\infty} \frac{z^n}{n} \in \mathcal{S}_s^*$, then we get

$$|T_3(2)| \leq |a_2||a_2^2 - a_3^2| + |a_3||a_2a_3 - a_3a_4| + |a_4||a_3^2 - a_2a_4|$$

$$= \frac{1}{2} \times |\frac{1}{2} \times \frac{1}{2} - \frac{1}{3} \times \frac{1}{3}| + \frac{1}{3} \times |\frac{1}{2} \times \frac{1}{3} - \frac{1}{3} \times \frac{1}{4}| + \frac{1}{4} \times |\frac{1}{3} \times \frac{1}{3} - \frac{1}{2} \times \frac{1}{4}|$$

$$\approx 0.107 < 1.931.$$

Author Contributions: conceptualization, H.T. and H.-Y.Z.; methodology, H.T. and H.-Y.Z.; software, H.-Y.Z.; validation, H.-Y.Z, R.S. and H.T.; formal analysis, R.S.; investigation, H.T.; resources, H.T.; data curation, H.T.; writing-original draft preparation, H.-Y.Z.; writing—review and editing, H.T. and R.S.; visualization, R.S.; supervision, H.T. and R.S.; project administration, H.T.; funding acquisition, H.T.

Funding: This research was funded by the Natural Science Foundation of the People's Republic of China under Grants 11561001 and 11271045, the Program for Young Talents of Science and Technology in Universities of Inner Mongolia Autonomous Region under Grant NJYT-18-A14, the Natural Science Foundation of Inner Mongolia of the People's Republic of China under Grant 2018MS01026, the Higher School Foundation of Inner Mongolia of the People's Republic of China under Grants NJZY17300 and NJZY18217 and the Natural Science Foundation of Chifeng of Inner Mongolia.

Conflicts of Interest: The authors declare no conflict of interest.

References

1. Srivastava, H.M.; Owa, S. *Current Topics in Analytic Function Theory*; World Scientific Publishing Company: Singapore, 1992.
2. Cho, N.E.; Kumar, V.; Kumar, S.S.; Ravichandran, V. Radius problems for starlike functions associated with the Sine function. *Bull. Iran. Math. Soc.* **2019**, *45*, 213–232. [CrossRef]
3. Miller, S.S.; Mocanu, P.T. *Differential Subordinations: Theory and Applications, Series on Monographs and Textbooks in Pure and Applied Mathematics, No. 225*; Marcel Dekker Incorporated: New York, NY, USA, 2000.
4. Noonan, J.W.; Thomas, D.K. On the second Hankel determinant of areally mean *p*-valent functions. *Trans. Am. Math. Soc.* **1976**, *223*, 337–346.
5. Noor, K.I. Hankel determinant problem for the class of functions with bounded boundary rotation. *Rev. Roumaine Math. Pure Appl.* **1983**, *28*, 731–739.
6. Ehrenborg, R. The Hankel determinant of exponential polynomials. *Am. Math. Mon.* **2000**, *107*, 557–560. [CrossRef]
7. Fekete, M.; Szegö, G. Eine benberkung uber ungerada schlichte funktionen. *J. Lond. Math. Soc.* **1933**, *8*, 85–89. [CrossRef]
8. Koepf, W. On the Fekete-Szego problem for close-to-convex functions. *Proc. Am. Math. Soc.* **1987**, *101*, 89–95.
9. Koepf, W. On the Fekete-Szego problem for close-to-convex functions II. *Arch. Math.* **1987**, *49*, 420–433. [CrossRef]
10. Thomas, D.K.; Halim, S.A. Toeplitz matrices whose elements are the coefficients of starlike and close-to-convex functions. *Bull. Malays. Math. Sci. Soc.* **2017**. [CrossRef]
11. Ye, K.; Lim, L.-H. Every matrix is a product of Toeplitz matrices. *Found. Comput. Math.* **2016**, *16*, 577–598. [CrossRef]
12. Babalola, K.O. On $H_3(1)$ Hankel determinant for some classes of univalent functions. *Inequal. Theory Appl.* **2010**, *6*, 1–7.

13. Bansal, D. Upper bound of second Hankel determinant for a new class of analytic functions. *Appl. Math. Lett.* **2013**, *26*, 103–107. [CrossRef]
14. Bansal, D.; Maharana, S.; Prajapat, J.K. Third order Hankel determinant for certain univalent functions. *J. Korean Math. Soc.* **2015**, *52*, 1139–1148. [CrossRef]
15. Caglar, M.; Deniz, E.; Srivastava, H.M. Second Hankel determinant for certain subclasses of bi-univalent functions. *Turk. J. Math.* **2017**, *41*, 694–706. [CrossRef]
16. Janteng, A.; Halim, S.; Darus, M. Coefficient inequality for a function whose derivative has a positive real part. *J. Inequal. Pure Appl. Math.* **2006**, *50*, 5.
17. Janteng, A.; Halim, S.A.; Darus, M. Hankel determinant for starlike and convex functions. *Int. J. Math. Anal.* **2007**, *13*, 619–625.
18. Lee, S.K.; Ravichandran, V.; Subramaniam, S. Bounds for the second Hankel determinant of certain univalent functions. *J. Inequal. Appl.* **2013**, *281*, 17. [CrossRef]
19. Mahmood, S.; Srivastava, H.M.; Khan, N.; Ahmad, Q.Z.; Khan, B.; Ali, I. Upper bound of the third Hankel determinant for a subclass of q-starlike functions. *Symmetry* **2019**, *11*, 1–13. [CrossRef]
20. Raza, M.; Malik, S.N. Upper bound of the third Hankel determinant for a class of analytic functions related with lemniscate of Bernoulli. *J. Inequal. Appl.* **2013**, *2013*, 1–8. [CrossRef]
21. Srivastava, H.M.; Ahmad, Q.Z.; Khan, N.; Khan, B. Hankel and Toeplitz determinants for a subclass of q-starlike functions associated with a general conic domain. *Mathematics* **2019**, *7*, 1–15. [CrossRef]
22. Srivastava, H.M.; Altinkaya, S.; Yalcin, S. Hankel determinant for a subclass of bi-univalent functions defined by using a symmetric q-derivative operator. *Filomat* **2018**, *32*, 503–516. [CrossRef]
23. Zaprawa, P. Third Hankel determinants for subclasses of univalent functions. *Med. J. Math.* **2017**. [CrossRef]
24. Zhang, H.-Y.; Tang, H.; Ma, L.-N. Upper bound of third Hankel determinant for a class of analytic functions. Pure Appl. Math. **2017**, *33*, 211–220. (In Chinese)
25. Zhang, H.-Y.; Tang, H.; Niu, X.-M. Third-order Hankel determinant for certain class of analytic functions related with exponential function. *Symmetry* **2018**, *10*, 1–8. [CrossRef]
26. Ali, M.F.; Thomas, D.K.; Vasudevarao, A. Toeplitz determinants whose elements are the coefficients of analytic and univalent functions. *Bull. Aust. Math. Soc.* **2018**, *97*, 253–264. [CrossRef]
27. Radhika, V.; Sivasubramanian, S.; Murugusundaramoorthy, G.; Jahangiri, J.M. Toeplitz matrices whose elements are the coefficients of functions with bounded boundary rotation. *J. Complex Anal.* **2016**, *4960704*, 4. [CrossRef]
28. Libera, R.J.; Zlotkiewicz, E.J. Coefficient bounds for the inverse of a function with derivative in P. *Proc. Am. Math. Soc.* **1983**, *87*, 251–257. [CrossRef]
29. Pommerenke, C. *Univalent Functions*; Vandenhoeck and Ruprecht: Gottingen, Germany, 1975.

© 2019 by the authors. Licensee MDPI, Basel, Switzerland. This article is an open access article distributed under the terms and conditions of the Creative Commons Attribution (CC BY) license (http://creativecommons.org/licenses/by/4.0/).

Article

Unique Existence Result of Approximate Solution to Initial Value Problem for Fractional Differential Equation of Variable Order Involving the Derivative Arguments on the Half-Axis

Shuqin Zhang [1,*] and Lei Hu [2]

[1] Department of Mathematics, China University of Mining and Technology Beijing, Ding No. 11 Xueyuan Road, Haidian District, Beijing 100083, China
[2] School of Science, Shandong Jiaotong University, Jinan 250023, China; huleimath@163.com
* Correspondence: zsqjk@163.com

Received: 29 December 2018; Accepted: 15 March 2019; Published: 20 March 2019

Abstract: The semigroup properties of the Riemann–Liouville fractional integral have played a key role in dealing with the existence of solutions to differential equations of fractional order. Based on some results of some experts', we know that the Riemann–Liouville variable order fractional integral does not have semigroup property, thus the transform between the variable order fractional integral and derivative is not clear. These judgments bring us extreme difficulties in considering the existence of solutions of variable order fractional differential equations. In this work, we will introduce the concept of approximate solution to an initial value problem for differential equations of variable order involving the derivative argument on half-axis. Then, by our discussion and analysis, we investigate the unique existence of approximate solution to this initial value problem for differential equation of variable order involving the derivative argument on half-axis. Finally, we give examples to illustrate our results.

Keywords: variable order fractional derivative; initial value problem; fractional differential equations; piecewise constant functions; approximate solution

1. Introduction

In this paper, we will observe and study the unique existence of approximate solution to the following initial value problem of variable order

$$\begin{cases} D_{0+}^{p(t)}x(t) = f(t, x, D_{0+}^{q(t)}x), 0 < t < +\infty, \\ x(0) = 0, \end{cases} \quad (1)$$

where $0 < q(t) < p(t) < 1$, $f(t, x, D_{0+}^{q(t)}x)$ are given real functions, and $D_{0+}^{p(t)}$, $D_{0+}^{q(t)}$ denote derivatives of variable order $p(t)$ and $q(t)$ defined by

$$D_{0+}^{p(t)}x(t) = \frac{d}{dt}\int_0^t \frac{(t-s)^{-p(t)}}{\Gamma(1-p(t))}x(s)ds, t > 0. \quad (2)$$

$$D_{0+}^{q(t)}x(t) = \frac{d}{dt}\int_0^t \frac{(t-s)^{-q(t)}}{\Gamma(1-q(t))}x(s)ds, \ t > 0,$$

and $\frac{1}{\Gamma(1-p(t))}\int_0^t(t-s)^{-p(t)}x(s)ds$ is integral of variable order $1-p(t)$ for function $x(t)$, for details, please refer to [1].

The operators of variable order, which fall into a more complex category, are the derivatives and integrals whose orders are the functions of certain variables. There are several definitions of variable order fractional integrals and derivatives. The following are several definitions of variable order fractional integrals and derivatives, which can be found in [2]. Let $-\infty < a < b < \infty$.

Definition 1. *Let $p : [a,b] \to (0, +\infty)$, the left Riemann–Liouville fractional integral of order $\alpha(t)$ for function $x(t)$ are defined as the following two types*

$$I_{a+}^{\alpha(t)} x(t) = \int_a^t \frac{(t-s)^{\alpha(t)-1}}{\Gamma(\alpha(t))} x(s) ds, \quad t > a, \tag{3}$$

$$I_{a+}^{\alpha(t)} x(t) = \int_a^t \frac{(t-s)^{\alpha(s)-1}}{\Gamma(\alpha(s))} x(s) ds, \quad t > a. \tag{4}$$

Definition 2. *Let $\alpha : [a,b] \to (n-1,n]$ (n is a natural number), the left Riemann–Liouville fractional derivative of order $\alpha(t)$ for function $x(t)$ are defined as the following two types*

$$D_{a+}^{\alpha(t)} x(t) = \left(\frac{d}{dt}\right)^n \int_a^t \frac{(t-s)^{n-\alpha(t)-1}}{\Gamma(n-\alpha(t))} x(s) ds, \quad t > a, \tag{5}$$

$$D_{a+}^{\alpha(t)} x(t) = \left(\frac{d}{dt}\right)^n \int_a^t \frac{(t-s)^{n-\alpha(s)-1}}{\Gamma(n-\alpha(s))} x(s) ds, \quad t > a. \tag{6}$$

Definition 3. *Let $\alpha : [a,b] \to (n-1,n]$ (n is a natural number), the left Caputo fractional derivative of order $\alpha(t)$ for function $x(t)$ are defined as the following two types*

$$^C D_{a+}^{\alpha(t)} x(t) = \int_a^t \frac{(t-s)^{n-\alpha(t)-1}}{\Gamma(n-\alpha(t))} x^{(n)}(s) ds, \quad t > a, \tag{7}$$

$$^C D_{a+}^{\alpha(t)} x(t) = \int_a^t \frac{(t-s)^{n-\alpha(s)-1}}{\Gamma(n-\alpha(s))} x^{(n)}(s) ds, \quad t > a. \tag{8}$$

The problems denoted by the operator of variable order are apparently more complicated than the ones denoted by the operator of constant order. Recently, some authors have considered the applications of derivatives of variable order in various sciences such as anomalous diffusion modeling, mechanical applications, multi-fractional Gaussian noises. Among these, there have been many works dealing with numerical methods for some class of variable order fractional differential equations, for instance, [1–20].

We notice that, if the order $p(t)$ is a constant function q, then the Riemann–Liouville variable order fractional derivatives and integrals are the Riemann–Liouville fractional derivative and integral, respectively [21]. We know there are some important properties as following. Let $-\infty < b < \infty$.

Lemma 1. *[21] The Riemann–Liouville fractional integral defined for function $x(t) \in L(0,b)$ exists almost everywhere.*

Lemma 2. *[21] The equality $I_{0+}^{\gamma} I_{0+}^{\delta} x(t) = I_{0+}^{\delta} I_{0+}^{\gamma} x(t) = I_{0+}^{\gamma+\delta} x(t)$, $0 < \gamma < 1, 0 < \delta < 1$ holds for $x \in L(0,b)$.*

Lemma 2 is semigroup property for the Riemann–Liouville fractional integral, which is very crucial in obtaining the following Lemmas 3–5. In other words, without Lemma 2, one could not have Lemmas 3–5, for details, please refer to [21].

Lemma 3. *[21] The equality $D_{0+}^{\gamma} I_{0+}^{\gamma} x(t) = x(t), 0 < \gamma < 1$ holds for $x \in L(0,b)$.*

Lemma 4. *[21] Let $0 < \alpha < 1$, then the differential equation*

$$D_{0+}^{\alpha} x = 0, t > 0$$

has solution

$$x(t) = ct^{\alpha-1}, c \in R.$$

Lemma 5. *[21] Let $0 < \alpha < 1$, $x \in L(0,b)$, $D_{0+}^{\alpha} x \in L(0,b)$. Then the following equality holds*

$$I_{0+}^{\alpha} D_{0+}^{\alpha} x(t) = x(t) + ct^{\alpha-1}, c \in R.$$

These properties play a very important role in considering the existence of the solutions of differential equations for the Riemann–Liouville fractional derivative, for details, please refer to [22–26]. However, from [15–18], for general functions $h(t), g(t)$, we notice that the semigroup property does not hold, i.e., $I_{a+}^{h(t)} I_{a+}^{g(t)} \neq I_{a+}^{h(t)+g(t)}$. Thus, it brings us extreme difficulties, that we cannot get these properties like Lemmas 3–5 for the variable order fractional operators (integral and derivative). Without these properties for variable order fractional derivative and integral, we can hardly consider the existence of solutions of differential equations for variable order derivative by means of nonlinear functional analysis (for instance, some fixed point theorems).

In [18], by means of Banach contraction principle, we considered the uniqueness result of solutions to initial value problems of differential equations of variable order

$$\begin{cases} D_{0+}^{q(t)} x(t) = f(t,x), 0 < t \leq T, \\ x(0) = 0, \end{cases} \tag{9}$$

where $0 < T < +\infty$, $D_{0+}^{q(t)}$ denotes derivative of variable order defined by (2), and $q : [0,T] \to (0,1]$ is a piecewise constant function with partition $P = \{[0, T_1], (T_1, T_2], (T_2, T_3], \cdots, (T_{N^*-1}, T]\}$ (N^* is a given natural number) of the finite interval $[0, T]$, i.e.,

$$q(t) = \sum_{k=1}^{N^*} q_k I_k(t), t \in [0, T],$$

where $0 < q_k \leq 1, k = 1, 2, \cdots, N^*$ are constants, and I_k is the indicator of the interval $[T_{k-1}, T_k]$, $k = 1, 2, \cdots, N^*$(here $T_0 = 0, T_{N^*} = T$), that is $I_k = 1$ for $t \in [T_{k-1}, T_k]$, $I_k = 0$ for elsewhere.

In this paper, we will consider the existence of solutions to the problem (1) for variable orders $p(t), q(t)$ are not piecewise constants. Based on some analysis, we will introduce the concept of approximate solution to the problem (1). Then, according to our discussion and analysis, we explore the unique existence of the approximate solution of the problem (1).

This paper is organized as follows. In Section 2, we provide some facts to the variable order integral and derivative through several examples. Also, we state some results which will play a very important role in obtaining our main results. In Section 3, we set forth our main result. Finally, two examples are given.

2. Some Preliminaries on Approximate Solution

In this section, we give some preliminaries on approximate solutions to the initial value problem (1). First of all, we use an example to illustrate the claim: for general function $p(t), q(t)$, the Riemann–Liouville variable order fractional integral does not have the semigroup property.

Example 1. *Let $p(t) = \frac{t}{6} + \frac{1}{3}$, $q(t) = \frac{t}{4} + \frac{1}{4}$, $f(t) = 1, 0 \leq t \leq 3$. Now, we calculate $I_{0+}^{p(t)} I_{0+}^{q(t)} f(t)|_{t=1}$ and $I_{0+}^{p(t)+q(t)} f(t)|_{t=1}$ which are defined in (3).*

For $1 \leq t \leq 3$, we have

$$I_{0+}^{p(t)} I_{0+}^{q(t)} f(t) = \int_0^t \frac{(t-s)^{\frac{t}{6}+\frac{1}{3}-1}}{\Gamma(\frac{t}{6}+\frac{1}{3})} \int_0^s \frac{(s-\tau)^{\frac{s}{4}+\frac{1}{4}-1}}{\Gamma(\frac{s}{4}+\frac{1}{4})} d\tau ds$$

$$= \int_0^t \frac{(t-s)^{\frac{t}{6}-\frac{2}{3}} s^{\frac{1}{4}+\frac{s}{4}}}{\Gamma(\frac{t}{6}+\frac{1}{3})\Gamma(\frac{5}{4}+\frac{s}{4})} ds$$

$$= \int_0^1 \frac{(t-s)^{\frac{t}{6}-\frac{2}{3}} s^{\frac{1}{4}+\frac{s}{4}}}{\Gamma(\frac{t}{6}+\frac{1}{3})\Gamma(\frac{5}{4}+\frac{s}{4})} ds + \int_1^t \frac{(t-s)^{\frac{t}{6}-\frac{2}{3}} s^{\frac{1}{4}+\frac{s}{4}}}{\Gamma(\frac{t}{6}+\frac{1}{3})\Gamma(\frac{5}{4}+\frac{s}{4})} ds.$$

We set $M_1 = \max_{1 \leq t \leq 3} |\frac{1}{\Gamma(p(t))}|$ and $M_2 = \max_{1 \leq s \leq 3} |\frac{1}{\Gamma(\frac{5}{4}+\frac{s}{4})}|$. For $1 \leq t \leq 3$, it holds

$$\left| \int_1^t \frac{(t-s)^{\frac{t}{6}-\frac{2}{3}} s^{\frac{1}{4}+\frac{s}{4}}}{\Gamma(\frac{t}{6}+\frac{1}{3})\Gamma(\frac{5}{4}+\frac{s}{4})} ds \right| = \left| \int_1^t 3^{\frac{t}{6}-\frac{2}{3}} (\frac{t-s}{3})^{\frac{t}{6}-\frac{2}{3}} \frac{s^{\frac{1}{4}+\frac{s}{4}}}{\Gamma(\frac{t}{6}+\frac{1}{3})\Gamma(\frac{5}{4}+\frac{s}{4})} ds \right|$$

$$\leq M_1 M_2 \int_1^t 3^{\frac{1}{2}-\frac{2}{3}} (\frac{t-s}{3})^{\frac{1}{6}-\frac{2}{3}} sds$$

$$\leq M_1 M_2 \int_1^t 3^{\frac{1}{3}} (t-s)^{-\frac{1}{2}} 3 ds$$

$$= 2 \times 3^{\frac{4}{3}} M_1 M_2 (t-1)^{\frac{1}{2}},$$

hence, we have

$$\left[\int_1^t \frac{(t-s)^{\frac{t}{6}-\frac{2}{3}} s^{\frac{1}{4}+\frac{s}{4}}}{\Gamma(\frac{t}{6}+\frac{1}{3})\Gamma(\frac{5}{4}+\frac{s}{4})} ds \right]_{t=1} = 0.$$

So, we get

$$I_{0+}^{p(t)} I_{0+}^{q(t)} f(t)|_{t=1} = \int_0^1 \frac{(1-s)^{-\frac{1}{2}} s^{\frac{1}{4}+\frac{s}{4}}}{\Gamma(\frac{1}{2})\Gamma(\frac{5}{4}+\frac{s}{4})} ds \approx 1.063$$

and

$$I_{0+}^{p(t)+q(t)} f(t)|_{t=1} = \int_0^1 \frac{(1-s)^{p(1)+q(1)-1}}{\Gamma(p(1)+q(1))} ds = \int_0^1 ds = 1.$$

Therefore,

$$I_{0+}^{p(t)} I_{0+}^{q(t)} f(t)|_{t=1} \neq I_{0+}^{p(t)+q(t)} f(t)|_{t=1}.$$

Without the semigroup property of the Riemann–Liouville variable order fractional integral, we can assure that the variable order fractional integration operator of non-constant continuous functions $p(t)$ for $x(t)$ does not have the properties like Lemmas 3–5. Consequently, we cannot transform differential equations of variable order into an integral equation.

Let $L[x(t);s]$, $L[I_{0+}^{p(t)} x(t);s]$, $L[D_{0+}^{p(t)} x(t);s]$ denote the Laplace transforms of functions $x(t)$, $I_{0+}^{p(t)} x(t)$ and $D_{0+}^{p(t)} x(t)$. We have not found out the explicit connection between $L[x(t);s]$ and $L[I_{0+}^{p(t)} x(t);s]$, as a result, we have not found out the explicit connection between $L[x(t);s]$ and $L[D_{0+}^{p(t)} x(t);s]$.

Example 2. Let $p(t) = \frac{1}{\sqrt{t+1}}, t \geq 0$. We consider the Laplace transforms of functions $t (t \geq 0)$ and $I_{0+}^{p(t)} t (t \geq 0)$ defined in (3). We can know that

$$L[t;s] = \int_0^\infty e^{-st} t dt = \frac{1}{s^2}, \tag{10}$$

$$L[I_{0+}^{(t+1)^{-\frac{1}{2}}}t;s] = \int_0^\infty e^{-st} \int_0^t \frac{(t-\tau)^{(t+1)^{-\frac{1}{2}}-1}}{\Gamma((t+1)^{-\frac{1}{2}})} \tau d\tau dt$$

$$= \int_0^\infty e^{-st} \int_\tau^\infty \frac{(t-\tau)^{(t+1)^{-\frac{1}{2}}-1}}{\Gamma((t+1)^{-\frac{1}{2}})} \tau dt d\tau$$

$$= \int_0^\infty e^{-s(\tau+r)} \int_0^\infty \frac{r^{(\tau+r+1)^{-\frac{1}{2}}-1}}{\Gamma((\tau+r+1)^{-\frac{1}{2}})} \tau dr d\tau$$

$$= \int_0^\infty e^{-s\tau} \tau \int_0^\infty e^{-sr} \frac{r^{(\tau+r+1)^{-\frac{1}{2}}-1}}{\Gamma((\tau+r+1)^{-\frac{1}{2}})} dr d\tau. \qquad (11)$$

By (10) or (11), we do not get the explicit connection between $L[t;s]$ and $L[I_{0+}^{(t+1)^{-\frac{1}{2}}}t;s]$.

In view of this example, the definition of variable order fractional derivative and the connection between the Laplace transforms of function $x(t)$ and its derivative $x'(t)$, we cannot obtain the Laplace transform formula for variable order fractional derivatives (2). Based on these facts, we cannot get the explicit expression of the solutions for the problem (1).

Throughout this paper, we assume that

(A_1) Let $p : [0, +\infty) \to (0, 1)$ and $q : [0, +\infty) \to (0, 1)$ be continuous functions, $q(t) < p(t)$ for all $t \in [0, +\infty)$, and that $p(t), q(t)$ satisfy

$$\lim_{t \to +\infty} p(t) = \rho_1, \lim_{t \to +\infty} q(t) = \rho_2, 0 < \rho_1, \rho_2 < 1. \qquad (12)$$

The following result is necessary in our next analysis of main result.

Lemma 6. *Let condition (A_1) hold. Then there exist positive constant T, natural number n^* and intervals $[0, T_1], (T_1, T_2], \cdots, (T_{n^*-1}, T] \ (T, +\infty)(n^* \in N)$ and functions $\alpha : [0, +\infty) \to (0, 1)$ and $\beta : [0, +\infty) \to (0, 1)$ defined by*

$$\alpha(t) = \sum_{k=1}^{n^*} p_k I_k(t) + \rho_1 I_T(t), \ t \in [0, +\infty), \qquad (13)$$

$$\beta(t) = \sum_{k=1}^{n^*} q_k I_k(t) + \rho_2 I_T(t), \ t \in [0, +\infty), \qquad (14)$$

where $p_k, q_k \in (0, 1)$, $I_k(t)$ is the indicator of the interval $[T_{k-1}, T_k]$ $(k = 1, 2, \cdots, n^*$, here $T_0 = 0, T_{n^*} = T)$, i.e., $I_k(t) = 1$ for $t \in [T_{k-1}, T_k]$, $I_k(t) = 0$ for t lying in elsewhere; $I_T(t)$ is the indicator of interval $(T, +\infty)$, i.e., $I_T(t) = 1$ for $t \in (T, +\infty)$, $I_T(t) = 0$ for t lying in elsewhere, such that for arbitrary small $\varepsilon > 0$,

$$|p(t) - \alpha(t)| < \varepsilon, |q(t) - \beta(t)| < \varepsilon, \ 0 \le t < +\infty. \qquad (15)$$

Proof. By (12), for $\forall \varepsilon > 0$, there exist $\overline{T}_1, \overline{T}_2 > 0$, such that

$$|p(t) - \rho_1| < \varepsilon, t > \overline{T}_1; |p(t) - \rho_2| < \varepsilon, t > \overline{T}_2.$$

Let $T = \max\{\overline{T}_1, \overline{T}_2\}$, then, for $\forall \varepsilon > 0$, we have that

$$|p(t) - \rho_1| < \varepsilon, |p(t) - \rho_2| < \varepsilon, t > T. \qquad (16)$$

We know that $p : [0, T] \to (0, 1), q : [0, T] \to (0, 1)$ are continuous functions. Since $p(t)$ is right continuous at point 0, then, for arbitrary small $\varepsilon > 0$, there is $\delta_{01} > 0$ such that

$$|p(t) - p(0)| < \varepsilon, \text{ for } 0 \le t \le \delta_{01}.$$

Since $q(t)$ is right continuous at point 0, then, for arbitrary small $\varepsilon > 0$, there is $\delta_{02} > 0$ such that

$$|q(t) - q(0)| < \varepsilon, \text{ for } 0 \leq t \leq \delta_{02}.$$

Then for arbitrary small $\varepsilon > 0$, takeing $\delta_0 = \min\{\delta_{01}, \delta_{02}\}$, it holds

$$|p(t) - p(0)| < \varepsilon, \ |q(t) - q(0)| < \varepsilon, \text{ for } 0 \leq t \leq \delta_0. \tag{17}$$

We take point $\delta_0 \doteq T_1$ (if $T_1 < T$, we consider continuities of $p(t), q(t)$ at point T_1, otherwise, we end this procedure). Since $p(t)$ is right continuous at point T_1, so, for arbitrary small $\varepsilon > 0$, there is $\delta_{11} > 0$ such that

$$|p(t) - p(T_1)| < \varepsilon, \text{ for } T_1 \leq t \leq T_1 + \delta_{11},$$

Since $q(t)$ is right continuous at point T_1, then, for arbitrary small $\varepsilon > 0$, there is $\delta_{12} > 0$ such that

$$|q(t) - q(T_1)| < \varepsilon, \text{ for } T_1 \leq t \leq T_1 + \delta_{12}.$$

Hence, for arbitrary small $\varepsilon > 0$, taking $\delta_1 = \min\{\delta_{11}, \delta_{12}\}$, it holds

$$|p(t) - p(T_1)| < \varepsilon, \ |q(t) - q(T_1)| < \varepsilon, \text{ for } T_1 \leq t \leq T_1 + \delta_1. \tag{18}$$

We take point $T_1 + \delta_1 \doteq T_2$ (if $T_2 < T$, we consider continuities of $p(t), q(t)$ at point T_2, otherwise, we end this procedure). Since $p(t)$ is right continuous at point T_2, so, for arbitrary small $\varepsilon > 0$, there is $\delta_{21} > 0$ such that

$$|p(t) - p(T_2)| < \varepsilon, \text{ for } T_2 \leq t \leq T_2 + \delta_{21}.$$

Since $q(t)$ is right continuous at point T_2, so, for arbitrary small $\varepsilon > 0$, there is $\delta_{22} > 0$ such that

$$|q(t) - q(T_2)| < \varepsilon, \text{ for } T_2 \leq t \leq T_2 + \delta_{22}.$$

Thus, for arbitrary small $\varepsilon > 0$, taking $\delta_2 = \min\{\delta_{21}, \delta_{22}\}$, it holds

$$|p(t) - p(T_2)| < \varepsilon, \ |q(t) - q(T_2)| < \varepsilon, \text{ for } T_2 \leq t \leq T_2 + \delta_2. \tag{19}$$

We take point $T_2 + \delta_2 \doteq T_3$ (if $T_3 < T$, we consider continuities of $p(t), q(t)$ at point T_3, otherwise, we end this procedure). Since $p(t)$ is right continuous at point T_3, so, for arbitrary small $\varepsilon > 0$, there is $\delta_{31} > 0$ such that

$$|p(t) - p(T_3)| < \varepsilon, \text{ for } T_3 \leq t \leq T_3 + \delta_{31},$$

Since $q(t)$ is right continuous at point T_3, so, for arbitrary small $\varepsilon > 0$, there is $\delta_{32} > 0$ such that

$$|q(t) - q(T_3)| < \varepsilon, \text{ for } T_3 \leq t \leq T_3 + \delta_{32}.$$

Therefore, for arbitrary small $\varepsilon > 0$, taking $\delta_3 = \min\{\delta_{31}, \delta_{32}\}$, it holds

$$|p(t) - p(T_3)| < \varepsilon, \ |q(t) - q(T_3)| < \varepsilon, \text{ for } T_3 \leq t \leq T_3 + \delta_3. \tag{20}$$

Since $[0, T]$ is a finite interval, then, continuing this analysis procedure, we could obtain that there exist $\delta_{n^*-2} > 0, \delta_{n^*-1} > 0$ ($n^* \in N$) such that $T_{n^*-2} + \delta_{n^*-2} \doteq T_{n^*-1} < T$, $T_{n^*-1} + \delta_{n^*-1} \geq T$, such that for arbitrary small $\varepsilon > 0$, it holds

$$|p(t) - p(T_{n^*-1})| < \varepsilon, |q(t) - q(T_{n^*-1})| < \varepsilon \text{ for } T_{n^*-1} \leq t \leq T, \tag{21}$$

From (16)–(21), we could let

$$p(0) \doteq p_1, p(T_1) \doteq p_2, p(T_2) \doteq p_3, p(T_3) \doteq p_4, \cdots, p(T_{n^*-1}) \doteq p_{n^*},$$

$$q(0) \doteq q_1, q(T_1) \doteq q_2, q(T_2) \doteq q_3, q(T_3) \doteq q_4, \cdots, q(T_{n^*-1}) \doteq q_{n^*}.$$

Thus, we define functions $\alpha, \beta : [0, +\infty) \to (0, 1)$ as following

$$\alpha(t) = \begin{cases} p_1, & t \in [0, T_1], \\ p_2, & t \in (T_1, T_2], \\ \vdots & \\ p_{n^*}, & t \in (T_{n^*-1}, T], \\ \rho_1, & t \in (T, +\infty), \end{cases} \qquad \beta(t) = \begin{cases} q_1, & t \in [0, T_1], \\ q_2, & t \in (T_1, T_2], \\ \vdots & \\ q_{n^*}, & t \in (T_{n^*-1}, T], \\ \rho_2, & t \in (T, +\infty). \end{cases}$$

Hence, from the previous arguments, for arbitrary small $\varepsilon > 0$, we have

$$\begin{cases} |p(t) - p_1| < \varepsilon, |q(t) - q_1| < \varepsilon, \text{ for } t \in [0, T_1], \\ |p(t) - p_2| < \varepsilon, |q(t) - q_2| < \varepsilon, \text{ for } t \in (T_1, T_2], \\ \vdots \\ |p(t) - p_{n^*}| < \varepsilon, |q(t) - q_{n^*}| < \varepsilon, \text{ for } t \in (T_{n^*-1}, T], \\ |p(t) - \rho_1| < \varepsilon, |q(t) - \rho_2| < \varepsilon, \text{ for } t \in (T, +\infty). \end{cases} \quad (22)$$

Thus, we complete this proof. □

The following example illustrates that the semigroup property of the variable order fractional integral does not holds for the piecewise constant functions $p(t)$ and $q(t)$ defined in the same partition of finite interval $[a, b]$.

Example 3. Let $p(t) = \begin{cases} 4, & 0 \leq t \leq 1, \\ 3, & 1 < t \leq 4, \end{cases}$ $q(t) = \begin{cases} 3, & 0 \leq t \leq 1, \\ 2, & 1 < t \leq 4, \end{cases}$ and $f(t) = 1, 0 \leq t \leq 4$. We'll verify $I_{0+}^{p(t)} I_{0+}^{q(t)} f(t)|_{t=3} \neq I_{0+}^{p(t)+q(t)} f(t)|_{t=3}$, here, the variable order fractional integral is defined in (3). For $1 \leq t \leq 4$, we have

$$I_{0+}^{p(t)} I_{0+}^{q(t)} f(t)$$

$$= \int_0^1 \frac{(t-s)^{p(t)-1}}{\Gamma(p(t))} \int_0^s \frac{(s-\tau)^{3-1}}{\Gamma(3)} d\tau ds + \int_1^t \frac{(t-s)^{p(t)-1}}{\Gamma(p(t))} \int_0^s \frac{(s-\tau)^{2-1}}{\Gamma(2)} d\tau ds$$

$$= \int_0^1 \frac{(t-s)^{p(t)-1} s^3}{6\Gamma(p(t))} ds + \int_1^t \frac{(t-s)^{p(t)-1} s^2}{2\Gamma(2)\Gamma(p(t))} ds,$$

thus, we have

$$I_{0+}^{p(t)} I_{0+}^{q(t)} f(t)|_{t=3} = \int_0^1 \frac{(3-s)^2 s^3}{6\Gamma(3)} ds + \int_1^3 \frac{(3-s)^2 s^2}{2\Gamma(2)\Gamma(3)} ds = \frac{245}{144}.$$

$$I_{0+}^{p(t)+q(t)}f(t)|_{t=3} = \int_0^3 \frac{(3-s)^{p(3)+q(3)-1}}{\Gamma(p(3)+q(3))}ds = \frac{3^{3+2}}{\Gamma(1+3+2)} = \frac{81}{40}.$$

Therefore, we obtain

$$I_{0+}^{p(t)}I_{0+}^{q(t)}f(t)|_{t=3} \neq I_{0+}^{p(t)+q(t)}f(t)|_{t=3},$$

which implies that the semigroup property of the variable order fractional integral does not hold for the piecewise constant functions $p(t)$ and $q(t)$ defined in the same partition $[0,1],(1,4]$ of finite interval $[0,4]$.

Lemma 7. *[10] Suppose $\beta > 0$, $a(t)$ is a nonnegative nondecreasing function locally integrable on $0 \leq t < L$ (some $L \leq +\infty$) and $g(t)$ is a nonnegative nondecreasing continuous function defined on $0 \leq t < L$, $g(t) \leq M$ (constant), and suppose $u(t)$ is nonnegative and locally integrable on $0 \leq t < L$ with*

$$u(t) \leq a(t) + g(t)\int_0^t (t-s)^{\beta-1}u(s)ds$$

on this interval. Then

$$u(t) \leq a(t)E_\beta(g(t)\Gamma(\beta)t^\beta), 0 \leq t < L,$$

where E_β is the Mittag–Leffler function defined by $E_\beta(z) = \sum_{k=0}^{\infty}\frac{z^k}{\Gamma(k\beta+1)}$.

3. Existence of Approximate Solution

According to the previous arguments, we do not transform the problem (1) into an integral equation. Here, we consider the unique existence of approximate solution of the problem (1). In this section, we present our main results.

Now we make the following assumptions:

(A_2) $f : [0,+\infty) \times R^2 \to R$ be a continuous function, and there exist positive constants $\lambda > \{\rho_1,\rho_2\}$, $c_1, c_2 > 0$ satisfying

$$\frac{c_1}{\Gamma(1+\rho_1)} + \frac{c_2}{\Gamma(1+\rho_1-\rho_2)} < 1,$$

such that

$$|f(t,(1+t^\lambda)x_1,(1+t^\lambda)y_1) - f(t,(1+t^\lambda)x_2,(1+t^\lambda)y_2)| \leq c_1|x_1-x_2| + c_2|y_1-y_2|, \quad (23)$$

where ρ_1, ρ_2 are the constants in $(A1)$.

(A_3) $f(t,0,0)(t \in (0,+\infty))$ satisfies

$$\lim_{t \to +\infty} \frac{1}{1+t^\lambda}\int_0^t (t-s)^{\rho_1-\rho_2-1}|f(s,0,0)|ds = 0.$$

Let B_i denote the Banach spaces defined as

$$B_i = \{x|x \in C[0,T_i]\}$$

with the norm

$$\|x\|_{B_i} = \max_{t \in [0,T_i]}|x(t)|, \quad (24)$$

where T_i is the constant obtained in Lemma 6, $i = 1,\cdots,n^*(T_{n^*} = T)$. Let

$$E = \left\{x\Big|x \in C[0,+\infty), \sup_{t \geq 0}\frac{|x(t)|}{1+t^\lambda} < \infty\right\}$$

with the norm
$$\|x\|_E = \sup_{t \geq 0} \frac{|x(t)|}{1+t^\lambda}, \tag{25}$$

where $\lambda > \{\rho_1, \rho_2\}$. Then, by the same arguments as in Lemma 2.2 of [22], we know that $(E, \|\cdot\|_E)$ is a Banach space, here we omit this proof.

Now, we consider the following initial value problem
$$\begin{cases} D_{0+}^{\alpha(t)} x(t) = f(t, x, D_{0+}^{\beta(t)} x), 0 < t < +\infty, \\ x(0) = 0, \end{cases} \tag{26}$$

where $\alpha(t), \beta(t)$ are defined in (13) and (14).

In order to obtain our main results, we start off by carrying on essential analysis to the equation of (26).

By (13) and (14), we get

$$\int_0^t \frac{(t-s)^{-\alpha(t)}}{\Gamma(1-\alpha(t))} x(s) ds = \sum_{k=1}^{n^*} I_k(t) \int_0^t \frac{(t-s)^{-p_k}}{\Gamma(1-p_k)} x(s) ds + I_T(t) \int_0^t \frac{(t-s)^{-\rho_1}}{\Gamma(1-\rho_1)} x(s) ds,$$

$$\int_0^t \frac{(t-s)^{-\beta(t)}}{\Gamma(1-\beta(t))} x(s) ds = \sum_{k=1}^{n^*} I_k(t) \int_0^t \frac{(t-s)^{-q_k}}{\Gamma(1-q_k)} x(s) ds + I_T(t) \int_0^t \frac{(t-s)^{-\rho_2}}{\Gamma(1-\rho_2)} x(s) ds \doteq h_{\beta,x}(t),$$

So, the equation of (26) can be written by

$$\frac{d}{dt} \Big(\sum_{k=1}^{n^*} I_k(t) \int_0^t \frac{(t-s)^{-p_k}}{\Gamma(1-p_k)} x(s) ds + I_T(t) \int_0^t \frac{(t-s)^{-\rho_1}}{\Gamma(1-\rho_1)} x(s) ds \Big) = f(t, x, \frac{d}{dt} h_{\beta,x}(t)), \ 0 < t < +\infty. \tag{27}$$

Then, Equation (27) in the interval $(0, T_1]$ can be written by

$$\frac{d}{dt} \int_0^t \frac{(t-s)^{-p_1}}{\Gamma(1-p_1)} x(s) ds = D_{0+}^{p_1} x(t) = f(t, x, D_{0+}^{q_1} x), 0 < t \leq T_1. \tag{28}$$

The Equation (27) in the interval $(T_1, T_2]$ can be written by

$$\frac{d}{dt} \int_0^t \frac{(t-s)^{-p_2}}{\Gamma(1-p_2)} x(s) ds = f(t, x, \frac{d}{dt} \int_0^t \frac{(t-s)^{-q_2}}{\Gamma(1-q_2)} x(s) ds), \ T_1 < t \leq T_2. \tag{29}$$

The Equation (27) in the interval $(T_2, T_3]$ can be written by

$$\frac{d}{dt} \int_0^t \frac{(t-s)^{-p_3}}{\Gamma(1-p_3)} x(s) ds = f(t, x, \frac{d}{dt} \int_0^t \frac{(t-s)^{-q_3}}{\Gamma(1-q_3)} x(s) ds), \ T_2 < t \leq T_3. \tag{30}$$

The Equation (27) in the interval $(T_{i-1}, T_i], i = 4, 5, \cdots, n^*$ ($T_{n^*} = T$) can be written by

$$\frac{d}{dt} \int_0^t \frac{(t-s)^{-p_i}}{\Gamma(1-p_i)} x(s) ds = f(t, x, \frac{d}{dt} \int_0^t \frac{(t-s)^{-q_i}}{\Gamma(1-q_i)} x(s) ds), \ T_{i-1} < t \leq T_i. \tag{31}$$

The Equation (27) in the interval $(T, +\infty)$ can be written by

$$\frac{d}{dt} \int_0^t \frac{(t-s)^{-\rho_1}}{\Gamma(1-\rho_1)} x(s) ds = f(t, x, \frac{d}{dt} \int_0^t \frac{(t-s)^{-\rho_2}}{\Gamma(1-\rho_2)} x(s) ds), \ T < t < +\infty. \tag{32}$$

Now, we present the definition of a solution to the problem (26), which is crucial in our work.

Definition 4. *We say the problem (26) exists one unique solution, if there are unique functions $u_i(t)$, $i = 1, 2, \cdots, n^*$, such that $u_1 \in C[0, T_1]$ satisfying Equation (28) and $u_1(0) = 0$; $u_2 \in C[0, T_2]$ satisfying Equation (29) and $u_2(0) = 0$; $u_3 \in C[0, T_3]$ satisfying Equation (30) and $u_3(0) = 0$; $u_i \in C[0, T_i]$ satisfying Equation (31) and $u_i(0) = 0$ ($i = 4, 5, \cdots, n^*$)($T_{n^*} = T$); $u_T \in C[0, +\infty)$ satisfying Equation (32) and $u_T(0) = 0$.*

The following is the definition of approximate solution of the problem (1).

Definition 5. *If there exist $T > 0$, natural number $n^* \in N$ and intervals $[0, T_1], (T_1, T_2], \cdots, (T_{n^*-1}, T]$, $(T, +\infty)$ and functions defined in Equations (13) and (14), such that the problem (26) exists one unique solution, then, we say this solution of the problem (26) is one unique approximate solution of the problem (1).*

Our main result is as follows.

Theorem 1. *Let conditions $(A_1), (A_2), (A_3)$ hold, then the problem (1) exists one unique approximate solution.*

Proof of Theorem 1. From Definitions 4 and 5 and Lemma 6, we only need to consider the unique existence of solution of the problem (26). According to the above analysis, equation of problem (26) can be written as the Equation (27). So Equation (26) in the interval $(0, T_1]$ can be written as (28). Applying operator $I_{0+}^{p_1}$ to both sides of (28), by Lemma 5, we have

$$x(t) = ct^{p_1-1} + \frac{1}{\Gamma(p_1)} \int_0^t (t-s)^{p_1-1} f(s, x(s), D_{0+}^{q_1} x(s)) ds, \quad 0 < t \leq T_1.$$

By $x(0) = 0$ and the assumption of function f, we get $c = 0$, that is

$$x(t) = \frac{1}{\Gamma(p_1)} \int_0^t (t-s)^{p_1-1} f(s, x(s), D_{0+}^{q_1} x(s)) ds, \quad 0 \leq t \leq T_1. \tag{33}$$

Let $D_{0+}^{q_1} x(t) = y(t)$, then, according to $x(0) = 0$ and Lemma 5, we get that

$$x(t) = I_{0+}^{q_1} y(t),$$

hence we will consider existence of solution to integral equation as following

$$y(t) = \frac{1}{\Gamma(p_1 - q_1)} \int_0^t (t-s)^{p_1-q_1-1} f(s, I_{0+}^{q_1} y(s), y(s)) ds, \quad 0 \leq t \leq T_1. \tag{34}$$

Obviously, if $y^* \in B_1 = C[0, T_1]$ is a solution of (34), then, applying operator $I_{0+}^{q_1}$ on both sides of (34), from Lemma 2, it holds

$$I_{0+}^{q_1} y^*(t) = I_{0+}^{q_1} I_{0+}^{p_1-q_1} f(t, I_{0+}^{q_1} y^*(t), y^*(t)) = I_{0+}^{p_1} f(t, I_{0+}^{q_1} y^*(t), y^*(t)), \quad 0 \leq t \leq T_1,$$

let

$$I_{0+}^{q_1} y^*(t) = x^*(t), 0 \leq t \leq T,$$

as a result, we have that

$$x^*(t) = I_{0+}^{p_1} f(t, x^*(t), D_{0+}^{q_1} x^*(t)), \quad 0 \leq t \leq T_1,$$

that is, $x^* \in B_1 = C[0, T_1]$ is a solution of (33), thus, we know that $x^* \in B_1 = C[0, T_1]$ is a solution of Equation (28) with zero initial value condition.

Define operator $F : B_1 \to B_1$ by

$$Fy(t) = \frac{1}{\Gamma(p_1 - q_1)} \int_0^t (t-s)^{p_1-q_1-1} f(s, I_{0+}^{q_1} y(s), y(s)) ds, \quad 0 \le t \le T_1. \tag{35}$$

From the continuity of function f and the standard arguments, we know that the operator $F : B_1 \to B_1$ is well defined. Let $M = \max_{0 \le t \le T} |f(t,0,0)|$. Let Ω_1 be a bounded, convex and closed subset of B_1 defined by

$$\Omega_1 = \{y | y \in B_1; |y(t)| \le K_1 e^{R_1^2 t^{p_1-q_1}}, 0 \le t \le T_1\},$$

where

$$K_1 = \frac{2MT_1^{p_1-q_1}}{\Gamma(1+p_1-q_1)},$$

$R_1 \in N$ satisfying

$$R_1 > \{1, \left(\frac{2d_1(1+T_1^{p_1-q_1})}{p_1-q_1}\right)^{\frac{1}{p_1-q_1}}\},$$

here $d_1 = \frac{1}{\Gamma(p_1-q_1)} \left[\frac{c_1 T_1^{q_1}}{\Gamma(1+q_1)} + c_2\right]$ (c_1, c_2 are the constants appearing in condition (A_2)).

By the analogy way as in [23], we could verify that $F : \Omega_1 \to \Omega_1$ is well defined. In fact, for $y \in \Omega_1$, since

$$|I_{0+}^{q_1} y(s)| \le \frac{1}{\Gamma(q_1)} \int_0^s (s-\tau)^{q_1-1} |y(\tau)| d\tau$$

$$\le \frac{K_1}{\Gamma(q_1)} \int_0^s (s-\tau)^{q_1-1} e^{R_1^2 \tau^{p_1-q_1}} d\tau$$

$$\le \frac{K_1}{\Gamma(q_1)} \int_0^s (s-\tau)^{q_1-1} e^{R_1^2 s^{p_1-q_1}} d\tau$$

$$= \frac{K_1}{\Gamma(1+q_1)} s^{q_1} e^{R_1^2 s^{p_1-q_1}}$$

$$\le \frac{K_1 T_1^{q_1}}{\Gamma(1+q_1)} e^{R_1^2 s^{p_1-q_1}}.$$

Now, $y \in \Omega_1$, by estimations above and (A_2), we get

$|Fy(t)|$

$$\le \frac{1}{\Gamma(p_1-q_1)} \int_0^t (t-s)^{p_1-q_1-1} |f(s, I_{0+}^{q_1} y(s), y(s))| ds$$

$$= \frac{1}{\Gamma(p_1-q_1)} \int_0^t (t-s)^{p_1-q_1-1} |f(s, I_{0+}^{q_1} y(s), y(s)) - f(s,0,0) + f(s,0,0)| ds$$

$$\le \frac{MT_1^{p_1-q_1}}{\Gamma(1+p_1-q_1)} + \frac{1}{\Gamma(p_1-q_1)} \int_0^t (t-s)^{p_1-q_1-1} (c_1 \frac{|I_{0+}^{q_1} y(s)|}{1+s^\lambda} + c_2 \frac{|y(s)|}{1+s^\lambda}) ds$$

$$\le \frac{K_1}{2} + \frac{1}{\Gamma(p_1-q_1)} \int_0^t (t-s)^{p_1-q_1-1} (c_1 |I_{0+}^{q_1} y(s)| + c_2 |y(s)|) ds$$

$$\leq \quad \frac{K_1}{2} + \frac{1}{\Gamma(p_1-q_1)} \int_0^t (t-s)^{p_1-q_1-1}\left(\frac{K_1 c_1 T_1^{q_1}}{\Gamma(1+q_1)} e^{R_1^2 s^{p_1-q_1}} + c_2 K_1 e^{R_1^2 s^{p_1-q_1}}\right) ds$$

$$\leq \quad \frac{K_1}{2} + K_1 d_1 \left[\sum_{i=1}^{R_1-1} \int_{\frac{(i-1)t}{R_1}}^{\frac{it}{R_1}} (t-s)^{p_1-q_1-1} e^{R_1^2 s^{p_1-q_1}} ds\right.$$

$$\left. + \int_{\frac{(R_1-1)t}{R_1}}^{t} (t-s)^{p_1-q_1-1} e^{R_1^2 s^{p_1-q_1}} ds\right]$$

$$\leq \quad \frac{K_1}{2} + K_1 d_1 \left[\sum_{i=1}^{R_1-1} \int_{\frac{(i-1)t}{R_1}}^{\frac{it}{R_1}} R_1^{1-p_1+q_1}(R_1-i)^{p_1-q_1-1} t^{p_1-q_1-1} e^{R_1^2 s^{p_1-q_1}} ds\right.$$

$$\left. + \int_{\frac{(R_1-1)t}{R_1}}^{t} (t-s)^{p_1-q_1-1} e^{R_1^2 t^{p_1-q_1}} ds\right]$$

$$\leq \quad \frac{K_1}{2} + K_1 d_1 \left[\sum_{i=1}^{R_1-1} \int_{\frac{(i-1)t}{R_1}}^{\frac{it}{R_1}} R_1^{1-p_1+q_1} t^{p_1-q_1-1} e^{R_1^2 s^{p_1-q_1}} ds\right.$$

$$\left. + \int_{\frac{(R_1-1)t}{R_1}}^{t} (t-s)^{p_1-q_1-1} e^{R_1^2 t^{p_1-q_1}} ds\right]$$

$$= \quad \frac{K_1}{2} + K_1 d_1 R_1^{1-p_1+q_1} \int_0^{\frac{(R_1-1)t}{R_1}} t^{p_1-q_1-1} e^{R_1^2 s^{p_1-q_1}} ds + \frac{K_1 d_1 R_1^{q_1-p_1} T_1^{p_1-q_1}}{p_1-q_1} e^{R_1^2 t^{p_1-q_1}}$$

$$\leq \quad \frac{K_1}{2} + K_1 d_1 R_1^{1-p_1+q_1} \int_0^{\frac{(R_1-1)t}{R_1}} s^{p_1-q_1-1} e^{R_1^2 s^{p_1-q_1}} ds + \frac{K_1 d_1 R_1^{q_1-p_1} T_1^{p_1-q_1}}{p_1-q_1} e^{R_1^2 t^{p_1-q_1}}$$

$$\leq \quad \frac{K_1}{2} + \frac{K_1 d_1 R_1^{1-p_1+q_1}}{R_1^2 (p_1-q_1)} e^{R_1^2 (\frac{(R_1-1)t}{R_1})^{p_1-q_1}} + \frac{K_1 d_1 R_1^{q_1-p_1} T_1^{p_1-q_1}}{p_1-q_1} e^{R_1^2 t^{p_1-q_1}}$$

$$\leq \quad \frac{K_1}{2} + \frac{K_1 d_1 R_1^{-1-p_1+q_1}}{p_1-q_1} e^{R_1^2 t^{p_1-q_1}} + \frac{K_1 d_1 R_1^{q_1-p_1} T_1^{p_1-q_1}}{p_1-q_1} e^{R_1^2 t^{p_1-q_1}}$$

$$\leq \quad \frac{K_1}{2} e^{R_1^2 t^{p_1-q_1}} + \frac{K_1 d_1 (1+T_1^{p_1-q_1})}{p_1-q_1} R_1^{q_1-p_1} e^{R_1^2 t^{p_1-q_1}}$$

$$\leq \quad \frac{K_1}{2} e^{R_1^2 t^{p_1-q_1}} + \frac{K_1}{2} e^{R_1^2 t^{p_1-q_1}} = K_1 e^{R_1^2 t^{p_1-q_1}},$$

which implies that $F: \Omega_1 \to \Omega_1$ is well defined. By the standard arguments, we could know that $F: \Omega_1 \to \Omega_1$ is a completely operator. Hence, the Schauder fixed point theorem assures that operator F has at least one fixed point $y_1(t) \in \Omega_1$. Obviously, $y_1(0) = 0$. Now, we will verify the uniqueness of solution to the integral Equation (34). We notice that: for $0 \leq s \leq t \leq T_1$, if $0 \leq t - s \leq 1$, then $(t-s)^{p_1-1} \leq (t-s)^{p_1-q_1-1}$; if $t - s \geq 1$, then $(t-s)^{p_1-q_1-1} \leq (t-s)^{p_1-1}$. As a result, we take

$$\max\{(t-s)^{p_1-1}, (t-s)^{p_1-q_1-1}\} \doteq (t-s)^{\alpha-1},$$

where α denotes p_1 or $p_1 - q_1$. Now, let $u_1(t), u_2(t)$ zre two solutions of the integral Equation (34), by expression above and (A_2), we get

$$|u_1(t) - u_2(t)|$$

$$\leq \quad \frac{1}{\Gamma(p_1-q_1)} \int_0^t (t-s)^{p_1-q_1-1}\left(c_1 \frac{|I_{0+}^{q_1}(u_1(s) - u_2(s))|}{1+s^\lambda} + c_2 \frac{|u_1(s) - u_2(s)|}{1+s^\lambda}\right) ds$$

$$\leq \frac{1}{\Gamma(p_1-q_1)}\int_0^t (t-s)^{p_1-q_1-1}(c_1|I_{0+}^{q_1}(u_1(s)-u_2(s))|+c_2|u_1(s)-u_2(s)|)ds$$

$$\leq \frac{c_1}{\Gamma(p_1-q_1)\Gamma(q_1)}\int_0^t (t-s)^{p_1-q_1-1}\int_0^s (s-\tau)^{q_1-1}|u_1(\tau)-u_2(\tau)|d\tau ds$$

$$+\frac{c_2}{\Gamma(p_1-q_1)}\int_0^t (t-s)^{p_1-q_1-1}|u_1(s)-u_2(s)|ds$$

$$= \frac{c_1}{\Gamma(p_1-q_1)\Gamma(q_1)}\int_0^t \int_\tau^t (t-s)^{p_1-q_1-1}(s-\tau)^{q_1-1}|u_1(\tau)-u_2(\tau)|dsd\tau$$

$$+\frac{c_2}{\Gamma(p_1-q_1)}\int_0^t (t-s)^{p_1-q_1-1}|u_1(s)-u_2(s)|ds$$

$$= \frac{c_1}{\Gamma(p_1)}\int_0^t (t-\tau)^{p_1-1}|u_1(\tau)-u_2(\tau)|d\tau + \frac{c_2}{\Gamma(p_1-q_1)}\int_0^t (t-s)^{p_1-q_1-1}|u_1(s)-u_2(s)|ds$$

$$\leq \frac{c_1}{\Gamma(p_1)}\int_0^t (t-\tau)^{\alpha-1}|u_1(\tau)-u_2(\tau)|d\tau + \frac{c_2}{\Gamma(p_1-q_1)}\int_0^t (t-s)^{\alpha-1}|u_1(s)-u_2(s)|ds$$

$$= [\frac{c_1}{\Gamma(p_1)}+\frac{c_2}{\Gamma(p_1-q_1)}]\int_0^t (t-\tau)^{\alpha-1}|u_1(\tau)-u_2(\tau)|d\tau,$$

by Lemma 7, we obtain that $u_1(t) = u_2(t)$, $0 \leq t \leq T_1$, this assures the uniqueness of solution of (34). As a result, by some arguments above, $x_1(t) = I_{0+}^{q_1}y_1(t)$ is one unique solution of the Equation (28) with zero initial value condition.

Also, we have obtained that the Equation (27) in the interval $(T_1, T_2]$ can be written by (29). In order to consider the existence result of solutions to (29), we may discuss the following equation defined on interval $(0, T_2]$

$$\frac{d}{dt}\int_0^t \frac{(t-s)^{-p_2}x(s)}{\Gamma(1-p_2)}ds = D_{0+}^{p_2}x(t) = f(t,x,\frac{d}{dt}\int_0^t \frac{(t-s)^{-q_2}x(s)}{\Gamma(1-q_2)}ds) = f(t,x,D_{0+}^{q_2}x). \quad (36)$$

It is clear that if function $x \in C[0, T_2]$ satisfies the Equation (36), then $x(t)$ must satisfy the Equation (29). In fact, if $x^* \in C[0, T_2]$ with $x^*(0) = 0$ is a solution of the Equation (36) with initial value condition $x(0) = 0$, that is

$$D_{0+}^{p_2}x^*(t)$$

$$= \frac{d}{dt}\int_0^t \frac{(t-s)^{-p_2}x^*(s)}{\Gamma(1-p_2)}ds$$

$$= f(t,x^*(t),D_{0+}^{q_2}x^*(t)) = f(t,x^*(t),\frac{d}{dt}\int_0^t \frac{(t-s)^{-q_2}x^*(s)}{\Gamma(1-q_2)}ds), \quad 0 < t \leq T_2; \ x^*(0) = 0.$$

Hence, from the equality above, we have that $x^* \in C[0, T_2]$ with $x^*(0) = 0$ satisfies the equation

$$\frac{d}{dt}\int_0^t \frac{(t-s)^{-p_2}x^*(s)}{\Gamma(1-p_2)}ds = f(t,x^*(t),\frac{d}{dt}\int_0^t \frac{(t-s)^{-q_2}x^*(s)}{\Gamma(1-q_2)}ds), \quad T_1 \leq t \leq T_2,$$

which means the function $x^* \in C[0, T_2]$ with $x^*(0) = 0$ is a solution of the Equation (29).

Based on this fact, we consider the existence of solutions to the Equation (36) with initial value condition $x(0) = 0$.

Now, applying operator $I_{0+}^{p_2}$ on both sides of (36), by Lemma 5, we have

$$x(t) = ct^{p_2-1} + \frac{1}{\Gamma(p_2)} \int_0^t (t-s)^{p_2-1} f(s, x(s), D_{0+}^{q_2} x(s)) ds, \quad 0 < t \leq T_2.$$

By initial value condition $x(0) = 0$, we have $c = 0$, that is

$$x(t) = \frac{1}{\Gamma(p_2)} \int_0^t (t-s)^{p_2-1} f(s, x(s), D_{0+}^{q_2} x(s)) ds, \quad 0 \leq t \leq T_2. \tag{37}$$

Let $D_{0+}^{q_2} x(t) = y(t)$, then, according to $x(0) = 0$ and Lemma 5, we get that

$$x(t) = I_{0+}^{q_2} y(t),$$

hence we will consider existence of solution to integral equation as following

$$y(t) = \frac{1}{\Gamma(p_2 - q_2)} \int_0^t (t-s)^{p_2-q_2-1} f(s, I_{0+}^{q_2} y(s), y(s)) ds, \quad 0 \leq t \leq T_2. \tag{38}$$

Obviously, if $y^* \in B_2 = C[0, T_2]$ is a solution of (38), then, by (38) and Lemma 2, it holds

$$I_{0+}^{q_2} y^*(t) = I_{0+}^{q_2} I_{0+}^{p_2-q_2} f(t, I_{0+}^{q_2} y^*(t), y^*(t)) = I_{0+}^{p_2} f(t, I_{0+}^{q_2} y^*(t), y^*(t)), \quad 0 \leq t \leq T_2,$$

let

$$I_{0+}^{q_2} y^*(t) = x^*(t), 0 \leq t \leq T_2,$$

as a result, we have that

$$x^*(t) = I_{0+}^{p_2} f(t, x^*(t), D_{0+}^{q_2} x^*(t)), \quad 0 \leq t \leq T_2,$$

that is, $x^* \in B_2 = C[0, T_2]$ is a solution of (37), hence, $x^* \in B_2 = C[0, T_2]$ is a solution of Equation (29) with zero initial value condition.

Define operator $F : B_2 \to B_2$ by

$$Fy(t) = \frac{1}{\Gamma(p_2 - q_2)} \int_0^t (t-s)^{p_2-q_2-1} f(s, I_{0+}^{q_2} y(s), y(s)) ds, \quad 0 \leq t \leq T_2.$$

From the continuity of function f and the standard arguments, we know that the operator $F : B_1 \to B_2$ is well defined. Let Ω_2 be a bounded, convex and closed subset of B_2 defined by

$$\Omega_2 = \{y | y \in B_2; |y(t)| \leq K_2 e^{R_2^2 t^{p_2-q_2}}, 0 \leq t \leq T_2\},$$

where

$$K_2 = \frac{2MT_2^{p_2-q_2}}{\Gamma(1 + p_2 - q_2)},$$

$R_2 \in N$ satisfying

$$R_2 > \{1, (\frac{2d_2(1 + T_2^{p_2-q_2})}{p_2 - q_2})^{\frac{1}{p_2-q_2}}\},$$

here $d_2 = \frac{1}{\Gamma(p_2-q_2)} \left[\frac{c_1 T_2^{q_2}}{\Gamma(1+q_2)} + c_2 \right]$ (c_1, c_2 are the constants appearing in condition (A_2)). By the same arguments above, there exists $y_2 \in \Omega_2$ such that $x_2(t) = I_{0+}^{q_2} y_2(t)$ is one unique solution of the Equation (29) with zero initial value condition.

In a similar way, for $i = 3, \cdots, n^*$, we get that the Equation (31) defined on $(T_{i-1}, T_i]$ ($T_{n^*} = T$) has one solution $x_i(t) \in \Omega_i \subset B_i$ with $x_i(0) = 0$, where

$$\Omega_i = \{y | y \in B_i; |y(t)| \leq K_i e^{R_i^2 t^{p_i - q_i}}, 0 \leq t \leq T_i\},$$

$$K_i = \frac{2MT_i^{p_i - q_i}}{\Gamma(1 + p_i - q_i)},$$

$R_i \in N$ satisfying

$$R_i > \left\{1, \left(\frac{2d_i(1 + T_i^{p_i - q_i})}{p_i - q_i}\right)^{\frac{1}{p_i - q_i}}\right\},$$

here $d_i = \frac{1}{\Gamma(p_i - q_i)} \left[\frac{c_1 T_i^{q_i}}{\Gamma(1 + q_i)} + c_2\right]$ (c_1, c_2 are the constants appearing in condition (A_2)), $i = 3, 4, \cdots, n^*$, $T_{n^*} = T$.

Finally, we get that the Equation (27) in the interval $(T, +\infty)$ can be written by (32). In order to consider the existence result of solutions to (32), we may discuss the following equation defined on interval $(0, +\infty)$

$$\frac{d}{dt} \int_0^t \frac{(t-s)^{-\rho_1}}{\Gamma(1-\rho_1)} x(s) ds = D_{0+}^{\rho_1} x(t) = f(t, x, D_{0+}^{\rho_2} x), \quad 0 < t < +\infty. \tag{39}$$

We see that, if function $x \in C[0, +\infty)$ satisfies the Equation (39), then $x(t)$ must satisfy the Equation (32). In fact, if $x^* \in C[0, +\infty)$ with $x^*(0) = 0$ is a solution of the Equation (39) with initial value condition $x(0) = 0$, that is

$$D_{0+}^{\rho_1} x^*(t) = \frac{d}{dt} \int_0^t \frac{(t-s)^{-\rho_1} x^*(s)}{\Gamma(1-\rho_1)} ds = f(t, x^*(t), D_{0+}^{\rho_2} x^*)$$

$$= f(t, x^*(t), \frac{d}{dt} \int_0^t \frac{(t-s)^{-\rho_2} x^*(s)}{\Gamma(1-\rho_2)} ds), \quad 0 < t < +\infty; \quad x^*(0) = 0.$$

Hence, from the equality above, we have $x^* \in C[0, +\infty)$ with $x^*(0) = 0$ satisfying the equation

$$\frac{d}{dt} \int_0^t \frac{(t-s)^{-\rho_1} x(s)}{\Gamma(1-\rho_1)} ds = f(t, x(t), \frac{d}{dt} \int_0^t \frac{(t-s)^{-\rho_2} x(s)}{\Gamma(1-\rho_2)} ds), \quad T < t < +\infty,$$

which means the function $x^* \in C[0, +\infty)$ with $x^*(0) = 0$ is a solution of the Equation (32).

Based on this fact, we will consider the existence of solutions to the Equation (39) with initial value condition $x(0) = 0$.

Now, applying operator $I_{0+}^{\rho_1}$ on both sides of (39), by Lemma 5, we have that

$$x(t) = ct^{\rho_1 - 1} + \frac{1}{\Gamma(\rho_1)} \int_0^t (t-s)^{\rho_1 - 1} f(s, x(s), D_{0+}^{\rho_2} x(s)) ds, \quad 0 < t < +\infty.$$

By initial value condition $x(0) = 0$, we have $c = 0$, that is

$$x(t) = \frac{1}{\Gamma(\rho_1)} \int_0^t (t-s)^{\rho_1 - 1} f(s, x(s), D_{0+}^{\rho_2} x(s)) ds, \quad 0 \leq t < +\infty. \tag{40}$$

Similar to arguments above, we let $D_{0+}^{\rho_2} x(t) = y(t)$, then, according to $x(0) = 0$ and Lemma 5, we get that

$$x(t) = I_{0+}^{\rho_2} y(t),$$

hence we will consider existence of solution to integral equation as following

$$y(t) = \frac{1}{\Gamma(\rho_1 - \rho_2)} \int_0^t (t-s)^{\rho_1-\rho_2-1} f(s, I_{0+}^{\rho_2} y(s), y(s)) ds, \quad 0 \le t < +\infty. \tag{41}$$

Obviously, if $y^* \in E$ is a solution of (41), then, by (41) and Lemma 2, it holds

$$I_{0+}^{\rho_2} y^*(t) = I_{0+}^{\rho_2} I_{0+}^{\rho_1-\rho_2} f(t, I_{0+}^{\rho_2} y^*(t), y^*(t)) = I_{0+}^{\rho_1} f(t, I_{0+}^{\rho_2} y^*(t), y^*(t)), \quad 0 \le t < +\infty.$$

Let

$$I_{0+}^{\rho_2} y^*(t) = x^*(t), 0 \le t < +\infty.$$

As a result, we have that

$$x^*(t) = I_{0+}^{\rho_1} f(t, x^*(t), D_{0+}^{\rho_2} x^*(t)), \quad 0 \le t < +\infty,$$

that is, $x^* \in E$ is a solution of (40), hence, $x^* \in E$ is a solution of Equation (32) with zero initial value condition.

Defining operator $F : E \to E$ as follows

$$Fy(t) = \frac{1}{\Gamma(\rho_1 - \rho_2)} \int_0^t (t-s)^{\rho_1-\rho_2-1} f(s, I_{0+}^{\rho_2} y(s), y(s)) ds, 0 \le t < +\infty.$$

To get the operator $F : E \to E$ is well defined. First, we verify that $Fy \in C[0, +\infty)$ for $x \in E$. In fact, for the case of $t_0 \in (0, +\infty)$, take $t > t_0, t - t_0 < 1$, then

$$(t_0 - s)^{\rho_1 - 1} > (t - s)^{\rho_1 - 1}, 0 \le s < t_0.$$

Now, for $y \in E$, it holds

$$\frac{|I_{0+}^{\rho_2} y(s)|}{1 + s^\lambda} \le \frac{\int_0^s (s-\tau)^{\rho_2-1} |y(\tau)| d\tau}{\Gamma(\rho_2)(1+s^\lambda)}$$

$$\le \frac{\int_0^s (s-\tau)^{\rho_2-1} (1+\tau^\lambda) \|y\|_E d\tau}{\Gamma(\rho_2)(1+s^\lambda)}$$

$$\le \frac{\int_0^s (s-\tau)^{\rho_2-1} (1+s^\lambda) \|y\|_E d\tau}{\Gamma(\rho_2)(1+s^\lambda)}$$

$$= \frac{\|y\|_E s^{\rho_2}}{\Gamma(1+\rho_2)},$$

thus, for $y \in E$, we have

$$|Fy(t)| \le \frac{1}{\Gamma(\rho_1 - \rho_2)} \int_0^{t_0} ((t_0-s)^{\rho_1-\rho_2-1} - (t-s)^{\rho_1-\rho_2-1})(c_1 \frac{|I_{0+}^{\rho_2} y(s)|}{1+s^\lambda} + c_2 \frac{|y(s)|}{1+s^\lambda}) ds$$

$$+ \frac{1}{\Gamma(\rho_1 - \rho_2)} \int_{t_0}^t (t-s)^{\rho_1-\rho_2-1} (c_1 \frac{|I_{0+}^{\rho_2} y(s)|}{1+s^\lambda} + c_2 \frac{|y(s)|}{1+s^\lambda}) ds$$

$$+ \frac{1}{\Gamma(\rho_1 - \rho_2)} \int_0^{t_0} ((t_0-s)^{\rho_1-\rho_2-1} - (t-s)^{\rho_1-\rho_2-1}) |f(s,0,0)| ds$$

$$+ \frac{1}{\Gamma(\rho_1 - \rho_2)} \int_{t_0}^t (t-s)^{\rho_1-\rho_2-1} |f(s,0,0)| ds$$

$$\leq \frac{\|y\|_E}{\Gamma(\rho_1-\rho_2)}\int_0^{t_0}((t_0-s)^{\rho_1-\rho_2-1}-(t-s)^{\rho_1-\rho_2-1})\left(c_1\frac{s^{\rho_2}}{\Gamma(1+\rho_2)}+c_2\right)ds$$

$$+\frac{\|y\|_E}{\Gamma(\rho_1-\rho_2)}\int_{t_0}^{t}(t-s)^{\rho_1-\rho_2-1}\left(c_1\frac{s^{\rho_2}}{\Gamma(1+\rho_2)}+c_2\right)ds$$

$$+\frac{\max_{0\leq t\leq t_0+1}|f(t,0,0)|}{\Gamma(\rho_1-\rho_2)}\int_0^{t_0}((t_0-s)^{\rho_1-\rho_2-1}-(t-s)^{\rho_1-\rho_2-1})ds$$

$$+\frac{\max_{0\leq t\leq t_0+1}|f(t,0,0)|}{\Gamma(\rho_1-\rho_2)}\int_{t_0}^{t}(t-s)^{\rho_1-\rho_2-1}ds.$$

We will consider the four terms above, respectively. For $0<\eta<\rho_1-\rho_2$, it is easy to show that

$$\int_0^t (t-s)^{\rho_1-\rho_2-1}s^\eta ds = \frac{\Gamma(1+\eta)\Gamma(\rho_1-\rho_2)t^{\rho_1-\rho_2+\eta}}{\Gamma(1+\rho_1-\rho_2+\eta)}.$$

Hence, for any given $\varepsilon>0$, there exists a $\delta_1>0$, such that, when $0\leq t_0\leq \delta_1$, it holds that

$$\frac{c_1\|y\|_E}{\Gamma(\rho_1-\rho_2)\Gamma(1+\rho_2)}\int_0^{t_0}(t_0-s)^{\rho_1-\rho_2-1}s^{\rho_2}ds < \frac{\varepsilon}{4}, \frac{c_2\|y\|_E}{\Gamma(\rho_2-\rho_2)}\int_0^{t_0}(t_0-s)^{\rho_1-\rho_2-1}ds < \frac{\varepsilon}{4}. \quad (42)$$

Moreover, we get

$$\int_{\delta_1}^{t_0}((t_0-s)^{\rho_1-\rho_2-1}-(t-s)^{\rho_1-\rho_2-1})s^{\rho_2}ds$$

$$\leq t_0^{\rho_2}\int_{\delta_1}^{t_0}((t_0-s)^{\rho_1-\rho_2-1}-(t-s)^{\rho_1-\rho_2-1})ds$$

$$= \frac{t_0^{\rho_2}}{\rho_1-\rho_2}((t_0-\delta_1)^{\rho_1-\rho_2}-(t-\delta_1)^{\rho_1-\rho_2}+(t-t_0)^{\rho_1-\rho_2})$$

$$\leq \frac{t_0^{\rho_2}}{\rho_1-\rho_2}(t-t_0)^{\rho_1-\rho_2},$$

$$\int_{\delta_1}^{t_0}((t_0-s)^{\rho_1-\rho_2-1}-(t-s)^{\rho_1-\rho_2-1})ds \leq \frac{1}{\rho_1-\rho_2}(t-t_0)^{\rho_1-\rho_2},$$

hence, we know that there exists $\delta_2>0$ such that for $0<t-t_0<\delta_2$, we have

$$\frac{c_1\|y\|_E}{\Gamma(\rho_1-\rho_2)\Gamma(1+\rho_2)}\int_{\delta_1}^{t_0}((t_0-s)^{\rho_1-\rho_2-1}-(t-s)^{\rho_1-\rho_2-1})s^{\rho_2}ds < \frac{\varepsilon}{4},$$

$$\frac{c_2\|y\|_E}{\Gamma(\rho_2-\rho_2)}\int_{\delta_1}^{t_0}((t_0-s)^{\rho_1-\rho_2-1}-(t-s)^{\rho_1-\rho_2-1})ds < \frac{\varepsilon}{4},$$

together with (42), it leads to

$$\int_0^{t_0}((t_0-s)^{\rho_1-\rho_2-1}-(t-s)^{\rho_1-\rho_2-1})\left(\frac{c_1\|y\|_E s^{\rho_2}}{\Gamma(\rho_1-\rho_2)\Gamma(1+\rho_2)}+\frac{c_2\|y\|_E}{\Gamma(\rho_2-\rho_2)}\right)ds < \varepsilon.$$

By the direct calculation, we have

$$\int_{t_0}^{t}(t-s)^{\rho_1-\rho_2-1}s^{\rho_2}ds \leq (t_0+1)^{\rho_2}\frac{(t-t_0)^{\rho_1-\rho_2}}{\rho_1-\rho_2},$$

$$\int_{t_0}^{t}(t-s)^{\rho_1-\rho_2-1}ds \le \frac{(t-t_0)^{\rho_1-\rho_2}}{\rho_1-\rho_2},$$

which implies that there exists $\delta_3 > 0$ such that for $0 < t - t_0 < \delta_3$, we get

$$\int_{t_0}^{t}(t-s)^{\rho_1-\rho_2-1}\left(\frac{c_1\|y\|_E s^{\rho_2}}{\Gamma(\rho_1-\rho_2)\Gamma(1+\rho_2)} + \frac{c_2\|y\|_E}{\Gamma(\rho_2-\rho_2)}\right)ds < \varepsilon.$$

By the same arguments, we get that these estimations still hold for the last two terms above. Hence, we obtain $Fx(t)$ is continuous on point t_0. In view of the arbitrariness of t_0, we have $Fx \in C(0,+\infty)$.

For the case of $t_0 = 0$, by (A_2), for $y \in E$, take $t < 1$, then

$$|Fy(t)| = \left|\frac{1}{\Gamma(\rho_1-\rho_2)}\int_0^t (t-s)^{\rho_1-\rho_2-1}f(s, I_{0+}^{\rho_2}y(s), y(s))ds\right|$$

$$\le \frac{\|y\|_E}{\Gamma(\rho_1-\rho_2)}\int_0^t (t-s)^{\rho_1-\rho_2-1}(c_1\frac{s^{\rho_2}}{\Gamma(1+\rho_2)} + c_2)ds$$

$$+ \frac{\max_{0\le t\le 1}|f(t,0,0)|}{\Gamma(\rho_1-\rho_2)}\int_0^t (t-s)^{\rho_1-\rho_2-1}ds,$$

From the previous arguments, we could know that $Fy(t)$ is continuous on point 0. As a result, we have $Fy \in C[0,+\infty)$ for $x \in E$.

By the similar arguments, for $y \in E$, by (A_2), we have

$$\left|\frac{Fy(t)}{1+t^\lambda}\right| \le \frac{1}{\Gamma(\rho_1-\rho_2)(1+t^\lambda)}\int_0^t (t-s)^{\rho_1-\rho_2-1}(c_1\frac{|I_{0+}^{\rho_2}y(s)|}{1+s^\lambda} + c_2\frac{|y(s)|}{1+s^\lambda})ds$$

$$+ \frac{1}{\Gamma(\rho_1-\rho_2)(1+t^\lambda)}\int_0^t (t-s)^{\rho_1-\rho_2-1}|f(s,0,0)|ds$$

$$\le \frac{\|y\|_E}{\Gamma(\rho_1-\rho_2)(1+t^\lambda)}\int_0^t (t-s)^{\rho_1-\rho_2-1}(c_1\frac{s^{\rho_2}}{\Gamma(1+\rho_2)} + c_2)ds$$

$$+ \frac{1}{\Gamma(\rho_1-\rho_2)(1+t^\lambda)}\int_0^t (t-s)^{\rho_1-\rho_2-1}|f(s,0,0)|ds$$

$$= \frac{\|y\|_E}{1+t^\lambda}[\frac{c_1 t^{\rho_1}}{\Gamma(1+\rho_1)} + \frac{c_2 t^{\rho_1-\rho_2}}{\Gamma(1+\rho_1-\rho_2)}]$$

$$+ \frac{1}{\Gamma(\rho_1-\rho_2)(1+t^\lambda)}\int_0^t (t-s)^{\rho_1-\rho_2-1}|f(s,0,0)|ds,$$

according to these estimations and (A_2), we ge that $\lim_{t\to+\infty}\frac{Fy(t)}{1+t^\lambda} = 0$. Hence, $F: E \to E$ is well defined.

Now, for $x, y \in E$, by a similar way, we get

$$\frac{|Fx(t)-Fy(t)|}{1+t^\lambda}$$

$$\le \frac{1}{\Gamma(\rho_1-\rho_2)}\int_0^t (t-s)^{\rho_1-\rho_2-1}(c_1\frac{I_{0+}^{\rho_2}|x(s)-y(s)|}{1+s^\lambda} + c_2\frac{|x(s)-y(s)|}{1+s^\lambda})ds$$

$$\le \frac{\|x-y\|_E}{1+t^\lambda}[\frac{c_1 t^{\rho_1}}{\Gamma(1+\rho_1)} + \frac{c_2 t^{\rho_1-\rho_2}}{\Gamma(1+\rho_1-\rho_2)}]$$

$$\le [\frac{c_1}{\Gamma(1+\rho_1)} + \frac{c_2}{\Gamma(1+\rho_1-\rho_2)}]\|x-y\|_E,$$

which implies that the operator $F : E \to E$ is a contraction operator, so the Banach contraction principle assures that the operator F has a unique fixed point $y_T(t) \in E$. According to some arguments above, we obtain that $x_T(t) = I_{0+}^{\rho_2} y_T(t)$ is one unique solution of the Equation (32) with zero initial value condition. Thus, according to Definition 5, we obtain that the problem (1) has one unique approximate solution. □

Example 4. *Now, we consider the initial value problem as following*

$$\begin{cases} D_{0+}^{\frac{1}{2}+\frac{t}{200(1+t^2)}} x(t) = \frac{\Gamma(\frac{3}{2})x^4}{12(1+t^2)^4(1+x^4)} + \frac{\Gamma(\frac{7}{6})(D_{0+}^{\frac{1}{3}+\frac{t}{600(1+t^2+t^3)}} x)^2}{12(1+t^2)^2(1+(D_{0+}^{\frac{1}{3}+\frac{t}{600(1+t^2+t^3)}} x)^2)}, 0 < t < +\infty, \\ x(0) = 0. \end{cases} \quad (43)$$

We let

$$p(t) = \frac{1}{2} + \frac{t}{200(1+t^2)}, q(t) = \frac{1}{3} + \frac{t}{600(1+t^2+t^3)}, 0 \leq t < +\infty,$$

$$f(t, x(t), y(t)) = \frac{\Gamma(\frac{3}{2})x^4(t)}{12(1+t^2)^4(1+x^4(t))} + \frac{\Gamma(\frac{7}{6})y^2(t)}{12(1+t^2)^2(1+y^2(t))}, 0 < t < +\infty, x(t), y(t) \in R.$$

Obviously, we get $\lim_{t \to +\infty} p(t) = \frac{1}{2}$ and $\lim_{t \to +\infty} q(t) = \frac{1}{3}$, thus, p satisfies (A_1) with $\rho_1 = \frac{1}{2}$, $\rho_2 = \frac{1}{3}$. That $f(t, 0, 0) = 0$. In addition, for all $0 \leq t < +\infty, x(t), y(t) \in R$, from the differentiation mean theorem, we get

$$|f(t, (1+t^2)x_1, (1+t^2)y_1) - f(t, (1+t^2)x_2, (1+t^2)y_2)|$$

$$\leq \frac{\Gamma(\frac{3}{2})}{12} \left| \frac{x_1^4}{1+(1+t^2)^4 x_1^4} - \frac{x_2^4}{1+(1+t^2)^4 x_2^4} \right|$$

$$+ \frac{\Gamma(\frac{7}{6})}{12} \left| \frac{y_1^2(t)}{1+(1+t^2)^2 y_1^2} - \frac{y_2^2}{1+(1+t^2)^2 y_2^2} \right|$$

$$\leq \frac{\Gamma(\frac{3}{2})}{3} |x_1 - x_2| + \frac{\Gamma(\frac{7}{6})}{3} |y_1 - y_1|,$$

which implies that f satisfies (A_2) with $c_1 = \frac{\Gamma(\frac{3}{2})}{3}, c_2 = \frac{\Gamma(\frac{7}{6})}{3}$, which satisfies

$$\frac{c_1}{\Gamma(1+\rho_1)} + \frac{c_2}{\Gamma(1+\rho_1-\rho_2)}$$

$$= \frac{\Gamma(\frac{3}{2})}{3} \frac{1}{\Gamma(1+\frac{1}{2})} + \frac{\Gamma(\frac{7}{6})}{3} \frac{1}{\Gamma(1+\frac{1}{2}-\frac{1}{3})}$$

$$= \frac{2}{3} < 1.$$

For given arbitrary small $\varepsilon = \frac{1.1}{100}$, there exists $T = \frac{22}{\varepsilon} = 2000$, such that

$$|p(t) - \frac{1}{2}| = \frac{t}{200(1+t^2)} < \frac{1}{t} \leq \frac{1}{T} = \frac{\varepsilon}{22} < \varepsilon, t \geq T,$$

$$|q(t) - \frac{1}{3}| = \frac{t}{600(1+t^2+t^3)} < \frac{1}{t} \leq \frac{1}{T} = \frac{\varepsilon}{22} < \varepsilon, t \geq T.$$

Now, we consider function $p(t)$ restricted on interval $[0, T] = [0, 2000]$. By the right continuity of function $p(t)$ at point 0, for $\varepsilon = \frac{1.1}{100}$, taking $\delta_0 = 2$, when $0 \leq t \leq \delta_0 = 2$, we have

$$|p(t) - p(0)| = \left|\frac{t}{200(1+t^2)}\right| \leq \frac{t}{200} < \frac{\delta_0}{200} = \frac{1}{100} < \frac{1.1}{100} = \varepsilon.$$

$$|q(t) - q(0)| = \left|\frac{t}{600(1+t^2+t^3)}\right| \leq \frac{t}{600} < \frac{\delta_0}{200} = \frac{1}{100} < \frac{1.1}{100} = \varepsilon.$$

We get $t_1 = \delta_0 = 2$. By the right continuity of functions $p(t)$, $q(t)$ at the point t_1, for $\varepsilon = \frac{1.1}{100}$, taking $\delta_1 = 2$, when $0 \leq t - t_1 \leq \delta_1$, by differential mean value theorem, we have

$$|p(t) - p(t_1)| = \left|\frac{t}{200(1+t^2)} - \frac{t_1}{200(1+t_1^2)}\right|$$

$$\leq \left|\frac{1-\xi^2}{200(1+\xi^2)^2}\right| |t - t_1|$$

$$\leq \frac{1+\xi^2}{200(1+\xi^2)^2} |t - t_1|$$

$$\leq \frac{1}{200}|t - t_1|$$

$$< \frac{\delta_1}{200} = \frac{1}{100} < \frac{1.1}{100} = \varepsilon,$$

$$|q(t) - q(t_1)| = \left|\frac{t}{600(1+t^2+t^3)} - \frac{t_1}{600(1+t_1^2+t_1^3)}\right|$$

$$\leq \left|\frac{1-\eta^2-2\eta^3}{600(1+\eta^2+\eta^3)^2}\right| |t - t_1|$$

$$\leq \left|\frac{1+\eta^2+2\eta^3}{600(1+\eta^2+\eta^3)^2}\right| |t - t_1|$$

$$\leq \frac{3}{600}|t - t_1|$$

$$< \frac{\delta_1}{200} = \frac{1}{100} < \frac{1.1}{100} = \varepsilon,$$

where $t_1 < \xi < t$, $t_1 < \eta < t$. We let $t_2 = t_1 + \delta_1 = 4$. By the right continuity of function $p(t)$ at point t_2, for $\varepsilon = \frac{1.1}{100}$, taking $\delta_2 = 2$, when $0 \leq t - t_1 \leq \delta_2$, by the same reasons above, we have

$$|p(t) - p(t_1)| = \left|\frac{t}{200(1+t^2)} - \frac{t_2}{20(1+t_2^2)}\right| < \frac{\delta_2}{200} = \frac{1}{100} < \frac{1.1}{100} = \varepsilon,$$

$$|q(t) - q(t_1)| = \left|\frac{t}{600(1+t^2+t^3)} - \frac{t_2}{600(1+t_2^2+t^3)}\right| < \frac{\delta_2}{200} = \frac{1}{100} < \frac{1.1}{100} = \varepsilon,$$

Continuing this procession, from $t_{n-1} = 2(n-1) < 2000$, $t_n = t_{n-1} + \delta_{n-1} = 2(n-1) + 2 = 2000$, we get $n = 1000$. Thus, let

$$p_1 \doteq p(0) = \frac{1}{2}, p_2 \doteq p(t_1) = p(2) = \frac{1}{2} + \frac{2}{200(1+4)},$$

$$p_3 \doteq p(t_2) = p(4) = \frac{1}{2} + \frac{4}{200 \times (1+16)}, \cdots, p_{1000} = p(t_{999}) = p(1998) = \frac{1}{2} + \frac{1998}{200 \times (1+1998^2)}.$$

$$q_1 \doteq q(0) = \frac{1}{3}, q_2 \doteq q(t_1) = q(2) = \frac{1}{3} + \frac{2}{600(1+4+8)},$$

$$q_3 \doteq q(t_2) = q(4) = \frac{1}{3} + \frac{4}{600 \times (1+16+64)}, \cdots,$$

$$q_{1000} = q(t_{999}) = q(1998) = \frac{1}{3} + \frac{1998}{600 \times (1+1998^2+1998^3)}.$$

As a result, we get intervals $[0,2], (2,4], \cdots, (1998, 2000], (2000, +\infty)$ and function $\alpha(t)$ defined by

$$\alpha(t) = \begin{cases} p_1 = \frac{1}{2}, & \text{for } t \in [0,2], \\ p_2 = \frac{1}{2} + \frac{2}{200 \times (1+4)}, & \text{for } t \in (2,4], \\ p_3 = \frac{1}{2} + \frac{4}{200 \times (1+16)}, & \text{for } t \in (4,6], \\ \cdots, \\ p_{1000} = \frac{1}{2} + \frac{1998}{200 \times (1+1998^2)}, & \text{for } t \in (1998, 2000] \\ p_1 = \frac{1}{2}, & \text{for } t \in (2000, +\infty). \end{cases} \quad (44)$$

$$\beta(t) = \begin{cases} q_1 = \frac{1}{3}, & \text{for } t \in [0,2], \\ q_2 = \frac{1}{3} + \frac{2}{600 \times (1+4+8)}, & \text{for } t \in (2,4], \\ q_3 = \frac{1}{3} + \frac{4}{600 \times (1+16+64)}, & \text{for } t \in (4,6], \\ \cdots, \\ q_{1000} = \frac{1}{3} + \frac{1998}{6000 \times (1+1998^2+1998^3)}, & \text{for } t \in (1998, 2000] \\ p_2 = \frac{1}{3}, & \text{for } t \in (2000, +\infty). \end{cases}$$

By Definitions 4 and 5 and the arguments of Theorem 1, the problem (43) has one unique approximate solution.

Remark 1. *From Lemma 6 and Definition 5, we may take arbitrary small ε, such that the problem (43) has one unique approximate solution. This means that the proximity is very high.*

Example 5. *Finally, we calculate the approximate solution of the following initial value problem for linear equation*

$$D_{0+}^{\frac{1}{2}+\frac{t}{200(1+t^2)}} x(t) = t^{\frac{1}{4}}, x(0) = 0, 0 < t < +\infty, \quad (45)$$

According to analysis in Example 4, we get intervals $[0,2], (2,4], \cdots, (1998, 2000], (2000, +\infty)$ and function $\alpha(t)$ defined in (44). By Definitions 4 and 5, we calculate out the approximate solution of the problem (45) as following

$$\begin{cases} x_1(t) = \frac{\Gamma(\frac{5}{4})}{\Gamma(\frac{7}{4})}t^{\frac{3}{4}} \in C[0,2], \\ x_2(t) = \frac{\Gamma(\frac{5}{4})}{\Gamma(\frac{7}{4}+\frac{2}{200\times(1+4)})}t^{\frac{3}{4}+\frac{2}{200\times(1+4)}} \in C[0,4], \\ x_3(t) = \frac{\Gamma(\frac{5}{4})}{\Gamma(\frac{7}{4}+\frac{4}{200\times(1+16)})}t^{\frac{3}{4}+\frac{4}{200\times(1+16)}} \in C[0,6], \\ \ldots, \\ x_{1000} = \frac{\Gamma(\frac{5}{4})}{\Gamma(\frac{7}{4}+\frac{1998}{200\times(1+1998^2)})}t^{\frac{3}{4}+\frac{1998}{200\times(1+1998^2)}} \in C[0,2000], \\ x_{2000}(t) = \frac{\Gamma(\frac{5}{4})}{\Gamma(\frac{7}{4})}t^{\frac{3}{4}} \in C[0,+\infty). \end{cases}$$

Remark 2. *By the characters of variable order derivative, we cannot get accurate solution of the problem (45). Hence, the approximate solution given by us is significative.*

4. Conclusions

In this paper, we have obtained the unique existence result of approximate solution of initial value problem for fractional differential equation of variable order involving with the variable order derivative defined on the half-axis. Through discussing the characters of variable order calculus(integral and derivative), we introduce the concept of approximate solution to the problem. Based on our discussion and analysis, using the fixed point theorem, we have found the unique existence results. As applications, two examples are presented to illustrate the main results. The issue of the existence and qualitative analysis of approximate solution of initial value problems for fractional differential equation of variable order is interesting. In the future, we will consider the existence and qualitative analysis of approximate solution of initial value problem for singular fractional differential equation of variable order.

Author Contributions: All authors contributed equally in writing this article. All authors read and approved the final manuscript.

Funding: This research is funded by the National Natural Science Foundation of China (11671181). This research is also supported by Doctor Science Foundation of Shandong Jiaotong University and Shandong Jiaotong University "Climbing" Research Innovation Team Program.

Conflicts of Interest: The authors declare no conflict of interest.

References

1. Valério, D.; da Costa, J.S. Variable-order fractional derivative and their numerical approximations. *Signal Process.* **2011**, *91*, 470–483. [CrossRef]
2. Tavares, D.; Almeida, R.; Torres, D.F.M. Caputo derivatives of fractional variable order: Numerical approximations. *Commun. Nonlinear Sci. Numer. Simul.* **2016**, *35*, 69–87. [CrossRef]
3. Razminia, A.; Dizaji, A.F.; Majd, V.J. Solution existence for non-autonomous variable-order fractional differential equations. *Math. Comput. Model.* **2012**, *55*, 1106–1117. [CrossRef]
4. Alikhanov, A.A. Boundary value problems for the equation of the variable order in differential and difference settings. *Appl. Math. Comput.* **2012**, *219*, 3938–3946. [CrossRef]
5. Moghaddam, B.P.; Machado, J.A.T.; Behforooz, H. An integro quadratic spline approach for a class of variable-order fractional initial value problems. *Chaos Solitons Fractals* **2017**, *102*, 354–360. [CrossRef]
6. Zúniga-Aguilar, C.J.; Romero-Ugalde, H.M.; Gómez-Aguilar, J.F.; Escobar-Jiménez, R.F.; Valtierra-Rodríguez, M. Solving fractional differential equations of variable-order involving operator with Mittag-Leffler kernel using artifical neural networks. *Chaos Solitons Fractals* **2017**, *103*, 382–403. [CrossRef]

7. Chen, C.M.; Liu, F.; Anh, V.; Turner, I. Numberical schemes with high spatial accuracy for a variable-order anomalous subdiffusion equation. *SIAM J. Sci. Comput.* **2010**, *32*, 1740–1760. [CrossRef]
8. Sierociuk, D.; Malesza, W.; Macias, M. Derivation, interpretation, and analog modelling of fractional variable order derivative definition. *Appl. Math. Model.* **2015**, *39*, 3876–3888. [CrossRef]
9. Sun, H.; Chen, W.; Wei, H.; Chen, Y. A comparative study of constant-order and variable-order fractional models in characterizing memory property of systems. *Eur. Phys. J. Spec. Top.* **2011**, *193*, 185–192. [CrossRef]
10. Ye, H.; Gao, J.; Ding, Y. A generalized Gronwall inequality and its application to a fractional differential equation. *J. Math. Anal. Appl.* **2007**, *328*, 1075–1081. [CrossRef]
11. Sousa, J.V.D.C.; de Oliveira, E.C. Two new fractional derivatives of variable order with non-singular kernel and fractional differential equation. *Comput. Appl. Math.* **2018**, *37*, 5375–5394. [CrossRef]
12. Gómez-Aguilar, J.F. Analytical and numerical solutions of nonlinear alcoholism model via variable-order fractional differential equations. *Phys. A* **2018**, *494*, 52–57. [CrossRef]
13. Yang, J.; Yao, H.; Wu, B. An efficient numberical method for variable order fractional functional differential equation. *Appl. Math. Lett.* **2018**, *76*, 221–226. [CrossRef]
14. Hajipour, M.; Jajarmi, A.; Baleanu, D.; Sun, H. On an accurate discretization of a variable-order fractional reaction-diffusion equation. *Commun. Nonlinear Sci. Numer. Simul.* **2019**, *69*, 119–133. [CrossRef]
15. Samko, S.G. Fractional integration and differentiation of variable order. *Anal. Math.* **1995**, *21*, 213–236. [CrossRef]
16. Samko, S.G.; Boss, B. Integration and differentiation to a variable fractional order. *Integral Transform. Spec. Funct.* **1993**, *1*, 277–300. [CrossRef]
17. Zhang, S.; Li, S.S.; Hu, L. The existeness and uniqueness result of solutions to initial value problems of nonlinear diffusion equations involving with the conformable variable derivative. *Rev. R. Acad. Cienc. Exactas Fís. Nat. Ser. A Math. RACSAM* **2018**. [CrossRef]
18. Zhang, S. The uniqueness result of solutions to initial value problem of differential equations of variable-order. *Rev. R. Acad. Cienc. Exactas Fís. Nat. Ser. A Math. RACSAM* **2018**, *112*, 407–423. [CrossRef]
19. Malesza, W.; Macias, M.; Sierociuk, D. Analysitical solution of fractional variable order differential equations. *J. Comput. Appl. Math.* **2019**, *348*, 214–236. [CrossRef]
20. Kian, Y.; Sorsi, E.; Yamamoto, M. On time-fractional diffusion equations with space-dependent variable order. *Ann. Henri Poincaré* **2018**, *19*, 3855–3881. [CrossRef]
21. Kilbas, A.A.; Srivastava, H.M.; Trujillo, J.J. *Theory and Applications of Fractional Differential Equations*; Elsevier: Amsterdam, The Netherlands, 2006.
22. Kou, C.; Zhou, H.; Yan, Y. Existence of solutions of initial value problems for nonlinear fractional differential equations on the half-axis. *Nonlinear Anal.* **2011**, *74*, 5975–5986. [CrossRef]
23. Deng, J.; Deng, Z. Existence of solutions of initial value problems for nonlinear fractional differential equations. *Appl. Math. Lett.* **2014**, *32*, 6–12. [CrossRef]
24. Agarwal, R.P.; Benchohra, M.; Hamani, S. Boundary value problems for fractional differential equations. *Georgian Math. J.* **2009**, *16*, 401–411.
25. Dong, X.; Bai, Z.; Zhang, S. Positive solutions to boundary value problems of p-Laplacian with fractional derivative. *Bound. Value Probl.* **2017**, *5*, 1–15. [CrossRef]
26. Bai, Z.; Zhang, S.; Sun, S.; Chun, Y. Monotone iterative method for a class of fractional differential equations. *Electron. J. Differ. Equ.* **2016**, *6*, 1–8.

© 2019 by the authors. Licensee MDPI, Basel, Switzerland. This article is an open access article distributed under the terms and conditions of the Creative Commons Attribution (CC BY) license (http://creativecommons.org/licenses/by/4.0/).

Article

Random Coupled Hilfer and Hadamard Fractional Differential Systems in Generalized Banach Spaces

Saïd Abbas [1], Nassir Al Arifi [2], Mouffak Benchohra [3,4] and Yong Zhou [5,6,*]

[1] Laboratory of Mathematics, Geometry, Analysis, Control and Applications, Tahar Moulay University of Saïda, P.O. Box 138, EN-Nasr, Saïda 20000, Algeria; said.abbas@univ-saida.dz or abbasmsaid@yahoo.fr
[2] College of Science, Geology and Geophysics Department, King Saud University, Riyadh 11451, Saudi Arabia; nalarifi@ksu.edu.sa
[3] Laboratory of Mathematics, Djillali Liabes University of Sidi Bel-Abbès, P.O. Box 89, Sidi Bel-Abbès 22000, Algeria; benchohra@yahoo.com
[4] Department of Mathematics, College of Science, King Saud University, P.O. Box 2455, Riyadh 11451, Saudi Arabia
[5] Faculty of Mathematics and Computational Science, Xiangtan University, Xiangtan 411105, China
[6] Nonlinear Analysis and Applied Mathematics (NAAM) Research Group, Faculty of Science, King Abdulaziz University, Jeddah 21589, Saudi Arabia
* Correspondence: yzhou@xtu.edu.cn

Received: 30 January 2019; Accepted: 15 March 2019; Published: 20 March 2019

Abstract: This article deals with some existence and uniqueness result of random solutions for some coupled systems of Hilfer and Hilfer–Hadamard fractional differential equations with random effects. Some applications are made of generalizations of classical random fixed point theorems on generalized Banach spaces.

Keywords: fractional differential systems; mixed Riemann–Liouville integral; mixed Hadamard integral; Hilfer derivative; Hadamard derivative; coupled system; random solution

MSC: 26A33; 34G25

---

1. Introduction

Fractional calculus is an extension of the ordinary differentiation and integration to arbitrary non-integer order. In recent years, this theory has become an important object of investigations due to its demonstrated applications in different areas of physics and engineering (see, for example, [1,2] and the references therein). In particular, time fractional differential equations are used when attempting to describe transport processes with long memory. Recently, the study of time fractional ordinary and partial differential equations has received great attention from many researchers, both in theory and in applications; we refer the reader to the monographs of Abbas et al. [3–5], Samko et al. [6], and Kilbas et al. [7], and the papers [8–14] and the references therein. On the other hand, the existence of solutions of initial and boundary value problems for fractional differential equations with the Hilfer fractional derivative have started to draw attention. For the related works, see for example [1,15–20] and the references therein.

Functional differential equations with random effects are differential equations with a stochastic process in their vector field [21–25]. They play a fundamental role in the theory of random dynamical systems.

Consider the following coupled system of Hilfer fractional differential equations:

$$\begin{cases} (D_0^{\alpha_1,\beta_1} u)(t,w) = f_1(t, u(t,w), v(t,w), w) \\ (D_0^{\alpha_2,\beta_2} v)(t,w) = f_2(t, u(t,w), v(t,w), w) \end{cases} \quad ; t \in I := [0,T], \ w \in \Omega, \tag{1}$$

with the following initial conditions:

$$\begin{cases} (I_0^{1-\gamma_1} u)(0,w) = \phi_1(w) \\ (I_0^{1-\gamma_2} v)(0,w) = \phi_2(w) \end{cases} \quad ; w \in \Omega, \tag{2}$$

where $T > 0$, $\alpha_i \in (0,1)$, $\beta_i \in [0,1]$, $(\Omega, \mathcal{A})$ is a measurable space, $\gamma_i = \alpha_i + \beta_i - \alpha_i\beta_i$, $\phi_i : \Omega \to \mathbb{R}^m$, $f_i : I \times \mathbb{R}^m \times \mathbb{R}^m \times \Omega \to \mathbb{R}^m$; $i = 1,2$, are given functions, $I_0^{1-\gamma_i}$ is the left-sided mixed Riemann–Liouville integral of order $1 - \gamma_i$, and $D_0^{\alpha_i,\beta_i}$ is the generalized Riemann–Liouville derivative (Hilfer) operator of order α_i and type β_i: $i = 1,2$. Next, we discuss the following coupled system of Hilfer–Hadamard fractional differential equations:

$$\begin{cases} (^H D_1^{\alpha_1,\beta_1} u)(t,w) = g_1(t, u(t,w), v(t,w), w) \\ (^H D_1^{\alpha_2,\beta_2} v)(t,w) = g_2(t, u(t,w), v(t,w), w) \end{cases} \quad ; t \in [1,T], \ w \in \Omega, \tag{3}$$

with the following initial conditions:

$$\begin{cases} (^H I_1^{1-\gamma_1} u)(1,w) = \psi_1(w) \\ (^H I_1^{1-\gamma_2} v)(1,w) = \psi_2(w) \end{cases} \quad ; w \in \Omega, \tag{4}$$

where $T > 1$, $\alpha_i \in (0,1)$, $\beta_i \in [0,1]$, $\gamma_i = \alpha_i + \beta_i - \alpha_i\beta_i$, $\psi_i : \Omega \to \mathbb{R}^m$, $g_i : [1,T] \times \mathbb{R}^m \times \mathbb{R}^m \times \Omega \to \mathbb{R}^m$; $i = 1,2$ are given functions, $\mathbb{R}^m$; $m \in \mathbb{N}^*$, $^H I_1^{1-\gamma_i}$ is the left-sided mixed Hadamard integral of order $1 - \gamma_i$, and $^H D_1^{\alpha_i,\beta_i}$ is the Hilfer–Hadamard fractional derivative of order α_i and type β_i; $i = 1,2$.

2. Preliminaries

We denote by C; the Banach space of all continuous functions from I into $\mathbb{R}^m$ with the supremum (uniform) norm $\|\cdot\|_\infty$. As usual, $AC(I)$ denotes the space of absolutely continuous functions from I into $\mathbb{R}^m$. By $L^1(I)$, we denote the space of Lebesgue-integrable functions $v : I \to \mathbb{R}^m$ with the norm:

$$\|v\|_1 = \int_0^T \|v(t)\| dt.$$

By $C_\gamma(I)$ and $C_\gamma^1(I)$, we denote the weighted spaces of continuous functions defined by:

$$C_\gamma(I) = \{w : (0,T] \to \mathbb{R}^m : t^{1-\gamma} w(t) \in C\},$$

with the norm:

$$\|w\|_{C_\gamma} := \sup_{t \in I} \|t^{1-\gamma} w(t)\|,$$

and:

$$C_\gamma^1(I) = \{w \in C : \frac{dw}{dt} \in C_\gamma\},$$

with the norm:

$$\|w\|_{C_\gamma^1} := \|w\|_\infty + \|w'\|_{C_\gamma}.$$

Furthermore, by $\mathcal{C} := C_{\gamma_1} \times C_{\gamma_2}$, we denote the product weighted space with the norm:

$$\|(u,v)\|_\mathcal{C} = \|u\|_{C_{\gamma_1}} + \|v\|_{C_{\gamma_2}}.$$

Now, we give some definitions and properties of fractional calculus.

Definition 1. *[4,6,7] The left-sided mixed Riemann–Liouville integral of order $r > 0$ of a function $w \in L^1(I)$ is defined by:*
$$(I_0^r w)(t) = \frac{1}{\Gamma(r)} \int_0^t (t-s)^{r-1} w(s) ds; \text{ for a.e. } t \in I,$$
where $\Gamma(\cdot)$ is the (Euler's) Gamma function.

Notice that for all $r, r_1, r_2 > 0$ and each $w \in C$, we have $I_0^r w \in C$, and:
$$(I_0^{r_1} I_0^{r_2} w)(t) = (I_0^{r_1 + r_2} w)(t); \text{ for a.e. } t \in I.$$

Definition 2. *[4,6,7] The Riemann–Liouville fractional derivative of order $r \in (0,1]$ of a function $w \in L^1(I)$ is defined by:*
$$\begin{aligned}(D_0^r w)(t) &= \left(\frac{d}{dt} I_0^{1-r} w\right)(t) \\ &= \frac{1}{\Gamma(1-r)} \frac{d}{dt} \int_0^t (t-s)^{-r} w(s) ds; \text{ for a.e. } t \in I.\end{aligned}$$

Let $r \in (0,1]$, $\gamma \in [0,1)$ and $w \in C_{1-\gamma}(I)$. Then, the following expression leads to the left inverse operator as follows.
$$(D_0^r I_0^r w)(t) = w(t); \text{ for all } t \in (0, T].$$

Moreover, if $I_0^{1-r} w \in C_{1-\gamma}^1(I)$, then the following composition is proven in [6]:
$$(I_0^r D_0^r w)(t) = w(t) - \frac{(I_0^{1-r} w)(0^+)}{\Gamma(r)} t^{r-1}; \text{ for all } t \in (0, T].$$

Definition 3. *[4,6,7] The Caputo fractional derivative of order $r \in (0,1]$ of a function $w \in L^1(I)$ is defined by:*
$$\begin{aligned}(^c D_0^r w)(t) &= \left(I_0^{1-r} \frac{d}{dt} w\right)(t) \\ &= \frac{1}{\Gamma(1-r)} \int_0^t (t-s)^{-r} \frac{d}{ds} w(s) ds; \text{ for a.e. } t \in I.\end{aligned}$$

In [1], R.Hilfer studied applications of a generalized fractional operator having the Riemann–Liouville and the Caputo derivatives as specific cases (see also [17,19]).

Definition 4. *(Hilfer derivative). Let $\alpha \in (0,1)$, $\beta \in [0,1]$, $w \in L^1(I)$, and $I_0^{(1-\alpha)(1-\beta)} w \in AC(I)$. The Hilfer fractional derivative of order α and type β of w is defined as:*
$$(D_0^{\alpha,\beta} w)(t) = \left(I_0^{\beta(1-\alpha)} \frac{d}{dt} I_0^{(1-\alpha)(1-\beta)} w\right)(t); \text{ for a.e. } t \in I. \tag{5}$$

Property 1. *Let $\alpha \in (0,1)$, $\beta \in [0,1]$, $\gamma = \alpha + \beta - \alpha\beta$, and $w \in L^1(I)$.*

1. The operator $(D_0^{\alpha,\beta} w)(t)$ can be written as:
$$(D_0^{\alpha,\beta} w)(t) = \left(I_0^{\beta(1-\alpha)} \frac{d}{dt} I_0^{1-\gamma} w\right)(t) = \left(I_0^{\beta(1-\alpha)} D_0^\gamma w\right)(t); \text{ for a.e. } t \in I.$$

Moreover, the parameter γ satisfies:

$$\gamma \in (0,1], \; \gamma \geq \alpha, \; \gamma > \beta, \; 1-\gamma < 1-\beta(1-\alpha).$$

2. The generalization (5) for $\beta = 0$, coincides with the Riemann–Liouville derivative and for $\beta = 1$ with the Caputo derivative.

$$D_0^{\alpha,0} = D_0^{\alpha}, \text{ and } D_0^{\alpha,1} = {}^c D_0^{\alpha}.$$

3. If $D_0^{\beta(1-\alpha)} w$ exists and is in $L^1(I)$, then:

$$(D_0^{\alpha,\beta} I_0^{\alpha} w)(t) = (I_0^{\beta(1-\alpha)} D_0^{\beta(1-\alpha)} w)(t); \text{ for a.e. } t \in I.$$

Furthermore, if $w \in C_\gamma(I)$ and $I_0^{1-\beta(1-\alpha)} w \in C_\gamma^1(I)$, then:

$$(D_0^{\alpha,\beta} I_0^{\alpha} w)(t) = w(t); \text{ for a.e. } t \in I.$$

4. If $D_0^{\gamma} w$ exists and is in $L^1(I)$, then:

$$(I_0^{\alpha} D_0^{\alpha,\beta} w)(t) = (I_0^{\gamma} D_0^{\gamma} w)(t) = w(t) - \frac{I_0^{1-\gamma}(0^+)}{\Gamma(\gamma)} t^{\gamma-1}; \text{ for a.e. } t \in I.$$

Corollary 1. *Let $h \in C_\gamma(I)$. Then, the Cauchy problem:*

$$\begin{cases} (D_0^{\alpha,\beta} u)(t) = h(t); \; t \in I, \\ (I_0^{1-\gamma} u)(t)|_{t=0} = \phi, \end{cases}$$

has the following unique solution:

$$u(t) = \frac{\phi}{\Gamma(\gamma)} t^{\gamma-1} + (I_0^{\alpha} h)(t).$$

Let $\beta_{\mathbb{R}^m}$ be the σ-algebra of Borel subsets of $\mathbb{R}^m$. A mapping $v: \Omega \to \mathbb{R}^m$ is said to be measurable if for any $B \in \beta_{\mathbb{R}^m}$; one has:

$$v^{-1}(B) = \{w \in \Omega : v(w) \in B\} \subset \mathcal{A}.$$

Definition 5. *Let $\mathcal{A} \times \beta_{\mathbb{R}^m}$ be the direct product of the σ-algebras $\mathcal{A}$ and $\beta_{\mathbb{R}^m}$ those defined in Ω and $\mathbb{R}^m$, respectively. A mapping $T: \Omega \times \mathbb{R}^m \to \mathbb{R}^m$ is called jointly measurable if for any $B \in \beta_{\mathbb{R}^m}$, one has:*

$$T^{-1}(B) = \{(w,v) \in \Omega \times E : T(w,v) \in B\} \subset \mathcal{A} \times \beta_{\mathbb{R}^m}.$$

Definition 6. *A function $T: \Omega \times \mathbb{R}^m \to \mathbb{R}^m$ is called jointly measurable if $T(\cdot, u)$ is measurable for all $u \in \mathbb{R}^m$ and $T(w, \cdot)$ is continuous for all: $w \in \Omega$.*

A random operator is a mapping $T: \Omega \times \mathbb{R}^m \to \mathbb{R}^m$ such that $T(w, u)$ is measurable in w for all $u \in \mathbb{R}^m$, and it expressed as $T(w)u = T(w,u)$; we also say that $T(w)$ is a random operator on $\mathbb{R}^m$. The random operator $T(w)$ on E is called continuous (resp. compact, totally bounded, and completely continuous) if $T(w, u)$ is continuous (resp. compact, totally bounded, and completely continuous) in u for all $w \in \Omega$. The details of completely continuous random operators in Banach spaces and their properties appear in Itoh [26].

Definition 7. *[27] Let $\mathcal{P}(Y)$ be the family of all nonempty subsets of Y and C be a mapping from Ω into $\mathcal{P}(Y)$. A mapping $T: \{(w,y) : w \in \Omega, \; y \in C(w)\} \to Y$ is called a random operator with stochastic domain C if C*

is measurable (i.e., for all closed $A \subset Y$, $\{w \in \Omega, C(w) \cap A \neq \emptyset\}$ is measurable), and for all open $D \subset Y$ and all $y \in Y$, $\{w \in \Omega : y \in C(w), T(w,y) \in D\}$ is measurable. T will be called continuous if every $T(w)$ is continuous. For a random operator T, a mapping $y : \Omega \to Y$ is called a random (stochastic) fixed point of T if for P-almost all $w \in \Omega$, $y(w) \in C(w)$ and $T(w)y(w) = y(w)$, and for all open $D \subset Y$, $\{w \in \Omega : y(w) \in D\}$ is measurable.

Definition 8. *A function $f : I \times \mathbb{R}^m \times \Omega \to \mathbb{R}^m$ is called random Carathéodory if the following conditions are satisfied:*

(i) *The map $(t,w) \to f(x,y,u,w)$ is jointly measurable for all $u \in \mathbb{R}^m$ and*
(ii) *The map $u \to f(t,u,w)$ is continuous for all $t \in I$ and $w \in \Omega$.*

Let $x,, y \in \mathbb{R}^m$ with $x = (x_1, x_2, \ldots, x_m)$, $y = (y_1, y_2, \ldots, y_m)$.

By $x \leq y$, we mean $x_i \leq y_i$; $i = 1, \ldots, m$. Also $|x| = (|x_1|, |x_2|, \ldots, |x_m|)$, $\max(x,y) = (\max(x_1, y_1), \max(x_2, y_2), \ldots, \max(x_m, y_m))$, and $\mathbb{R}_+^m = \{x \in \mathbb{R}^m : x_i \in \mathbb{R}_+, i = 1, \ldots, m\}$. If $c \in \mathbb{R}$, then $x \leq c$ means $x_i \leq c$; $i = 1, \ldots, m$.

Definition 9. *Let X be a nonempty set. By a vector-valued metric on X, we mean a map $d : X \times X \to \mathbb{R}^m$ with the following properties:*

(i) $d(x,y) \geq 0$ *for all $x, y \in X$, and if $d(x,y) = 0$, then $x = y$;*
(ii) $d(x,y) = d(y,x)$ *for all $x, y \in X$;*
(iii) $d(x,z) \leq d(x,y) + d(y,z)$ *for all $x, y, z \in X$.*

We call the pair (X, d) a generalized metric space with $d(x,y) := \begin{pmatrix} d_1(x,y) \\ d_2(x,y) \\ \vdots \\ d_m(x,y) \end{pmatrix}$.

Notice that d is a generalized metric space on X if and only if d_i; $i = 1, \ldots, m$ are metrics on X. For $r = (r_1, \ldots, r_m) \in \mathbb{R}^m$ and $x_0 \in X$, we will denote by:

$$B_r(x_0) := \{x \in X : d(x_0, x) < r\} = \{x \in X : d_i(x_0, x) < r_i; i = 1, \ldots, m\}$$

the open ball centered in x_0 with radius r and:

$$\overline{B}_r(x_0) := \{x \in X : d(x_0, x) \leq r\} = \{x \in X : d_i(x_0, x) \leq r_i; i = 1, \ldots, m\}$$

the closed ball centered in x_0 with radius r. We mention that for generalized metric spaces, the notations of open, closed, compact, convex sets, convergence, and Cauchy sequence are similar to those in usual metric spaces.

Definition 10. *[28,29] A square matrix of real numbers is said to be convergent to zero if and only if its spectral radius $\rho(M)$ is strictly less than one. In other words, this means that all the eigenvalues of M are in the open unit disc, i.e., $|\lambda| < 1$; for every $\lambda \in \mathbb{C}$ with $\det(M - \lambda I) = 0$; where I denotes the unit matrix of $M_{m \times m}(\mathbb{R})$.*

Example 1. *The matrix $A \in M_{2 \times 2}(\mathbb{R})$ defined by:*

$$A = \begin{pmatrix} a & b \\ c & d \end{pmatrix},$$

converges to zero in the following cases:

(1) $b = c = 0$, $a, d > 0$, and $\max\{a, d\} < 1$.
(2) $c = 0$, $a, d > 0$, $a + d < 1$, and $-1 < b < 0$.
(3) $a + b = c + d = 0$, $a > 1$, $c > 0$, and $|a - c| < 1$.

In the sequel, we will make use of the following random fixed point theorems:

Theorem 1. *[23–25] Let $(\Omega, \mathcal{F})$ be a measurable space, X a real separable generalized Banach space, and $F : \Omega \times X \to X$ a continuous random operator, and let $M(w) \in \mathcal{M}_{n \times n}(\mathbb{R}_+)$ be a random variable matrix such that for every $w \in \Omega$, the matrix $M(w)$ converges to zero and:*

$$d(F(w, x_1), F(w, x_2)) \leq M(w) d(x_1, x_2); \text{ for each } x_1, x_2 \in X \text{ and } w \in \Omega,$$

then there exists a random variable $x : \Omega \to X$ that is the unique random fixed point of F.

Theorem 2. *[23–25] Let $(\Omega, \mathcal{F})$ be a measurable space, X be a real separable generalized Banach space, and $F : \Omega \times X \to X$ be a completely continuous random operator. Then, either:*

(i) *the random equation $F(w, x) = x$ has a random solution, i.e., there is a measurable function $x : \Omega \to X$ such that $F(w, x(w)) = x(w)$ for all $w \in \Omega$ or*
(ii) *the set $M = \{x : \Omega \to X \text{ is measurable} : \lambda(w) F(w, x) = x\}$ is unbounded for some measurable function $\lambda : \Omega \to X$ with $0 < \lambda(w) < 1$ on Ω.*

Furthermore, we will use the following Gronwall lemma:

Lemma 1. *[23] Let $u : I \to [0, \infty)$ be a real function and $u(\cdot)$ a nonnegative, locally-integrable function on I. Assume that there exist constants $c > 0$ and $r < 1$ such that:*

$$u(t) \leq v(t) + c \int_0^t \frac{u(s)}{(t-s)^r} ds,$$

then, there exists a constant $K := K(r)$ such that:

$$u(t) \leq v(t) + cK \int_0^t \frac{v(s)}{(t-s)^r} ds,$$

for every $t \in I$.

3. Coupled Hilfer Fractional Differential Systems

In this section, we are concerned with the existence and uniqueness results of the system (1) and (2).

Definition 11. *By a solution of the problem (1) and (2), we mean coupled measurable functions $(u, v) \in C_{\gamma_1} \times C_{\gamma_2}$, which satisfy the Equation (1) on I, and the conditions $(I_0^{1-\gamma_1} u)(0^+) = \phi_1$, and $(I_0^{1-\gamma_2} v)(0^+) = \phi_2$.*

The following hypotheses will be used in the sequel.

(H_1) The functions f_i; $i = 1, 2$ are Carathéodory.
(H_2) There exist measurable functions $p_i, q_i : \Omega \to (0, \infty)$; $i = 1, 2$ such that:

$$\|f_i(t, u_1, v_1) - f_i(t, u_2, v_2)\| \leq p_i(w) \|u_1 - u_2\| + q_i(w) \|v_1 - v_2\|;$$

for a.e. $t \in I$, and each $u_i, v_i \in \mathbb{R}^m$, $i = 1, 2$.
(H_3) There exist measurable functions $a_i, b_i : \Omega \to (0, \infty)$; $i = 1, 2$ such that:

$$\|f_i(t, u, v)\| \leq a_i(w) \|u\| + b_i(w) \|v\|; \text{ for a.e. } t \in I, \text{ and each } u, v \in \mathbb{R}^m.$$

First, we prove an existence and uniqueness result for the coupled system (1)–(2) by using Banach's random fixed point theorem in generalized Banach spaces.

Theorem 3. *Assume that the hypotheses (H_1) and (H_2) hold. If for every $w \in \Omega$, the matrix:*

$$M(w) := \begin{pmatrix} \frac{T^{\alpha_1}}{\Gamma(1+\alpha_1)} p_1(w) & \frac{T^{\alpha_1}}{\Gamma(1+\alpha_1)} q_1(w) \\ \frac{T^{\alpha_2}}{\Gamma(1+\alpha_2)} p_2(w) & \frac{T^{\alpha_2}}{\Gamma(1+\alpha_2)} q_2(w) \end{pmatrix}$$

converges to zero, then the coupled system (1) and (2) has a unique random solution.

Proof. Define the operators $N_1 : \mathcal{C} \times \Omega \to \mathcal{C}_{\gamma_1}$ and $N_2 : \mathcal{C} \times \Omega \to \mathcal{C}_{\gamma_2}$ by:

$$(N_1(u,v))(t,w) = \frac{\phi_1(w)}{\Gamma(\gamma_1)} t^{\gamma_1-1} + \int_0^t (t-s)^{\alpha_1-1} \frac{f(s,u(s,w),v(s,w),w)}{\Gamma(\alpha_1)} ds, \tag{6}$$

and:

$$(N_2(u,v))(t,w) = \frac{\phi_2(w)}{\Gamma(\gamma_2)} t^{\gamma_2-1} + \int_0^t (t-s)^{\alpha_2-1} \frac{f(s,u(s,w),v(s,w),w)}{\Gamma(\alpha_2)} ds. \tag{7}$$

Consider the operator $N : \mathcal{C} \times \Omega \to \mathcal{C}$ defined by:

$$(N(u,v))(t,w) = ((N_1(u,v))(t,w), (N_2(u,v))(t,w)). \tag{8}$$

Clearly, the fixed points of the operator N are random solutions of the system (1) and (2).

Let us show that N is a random operator on $\mathcal{C}$. Since f_i; $i = 1, 2$ are Carathéodory functions, then $w \to f_i(t,u,v,w)$ are measurable maps. We concluded that the maps:

$$w \to (N_1(u,v))(t,w) \text{ and } w \to (N_2(u,v))(t,w),$$

are measurable. As a result, N is a random operator on $\mathcal{C} \times \Omega$ into $\mathcal{C}$. We show that N satisfies all conditions of Theorem 1.

For any $w \in \Omega$ and each $(u_1, v_1), (u_2, v_2) \in \mathcal{C}$, and $t \in I$, we have:

$$\|t^{1-\gamma_1}(N_1(u_1,v_1))(t,w) - t^{1-\gamma_1}(N_1(u_2,v_2))(t,w)\|$$
$$\leq \frac{t^{1-\gamma_1}}{\Gamma(\alpha_1)} \int_0^t (t-s)^{\alpha_1-1} \|f_1(s,u_1(s,w),v_1(s,w),w) - f_1(s,u_2(s,w),v_2(s,w),w)\| ds$$
$$\leq \frac{t^{1-\gamma_1}}{\Gamma(\alpha_1)} \int_0^t (t-s)^{\alpha_1-1} (p_1(w)\|u_1(s,w) - v_1(s,w)\|$$
$$+ q_1(w)\|u_2(s,w) - v_2(s,w)\|) ds$$
$$\leq \frac{1}{\Gamma(\alpha_1)} \int_0^t (t-s)^{\alpha_1-1} (p_1(w) s^{1-\gamma_1} \|u_1(s,w) - v_1(s,w)\|$$
$$+ q_1(w) s^{1-\gamma_1} \|u_2(s,w) - v_2(s,w)\|) ds$$
$$\leq \frac{p_1(w)\|u_1(\cdot,w) - v_1(\cdot,w)\|_{\mathcal{C}_{\gamma_1}} + q_1(w)\|u_2(\cdot,w) - v_2(\cdot,w)\|_{\mathcal{C}_{\gamma_2}}}{\Gamma(\alpha_1)}$$
$$\times \int_0^t (t-s)^{\alpha_1-1} ds$$
$$\leq \frac{T^{\alpha_1}}{\Gamma(1+\alpha_1)} (p_1(w)\|u_1(\cdot,w) - v_1(\cdot,w)\|_{\mathcal{C}_{\gamma_1}} + q_1(w)\|u_2(\cdot,w) - v_2(\cdot,w)\|_{\mathcal{C}_{\gamma_2}}).$$

Then,

$$\|(N_1(u_1,v_1))(\cdot,w) - (N_1(u_2,v_2))(\cdot,w)\|_{\mathcal{C}_{\gamma_1}}$$

$$\leq \frac{T^{\alpha_1}}{\Gamma(1+\alpha_1)}(p_1(w)\|u_1(\cdot,w)-v_1(\cdot,w)\|_{C_{\gamma_1}}+q_1(w)\|u_2(\cdot,w)-v_2(\cdot,w)\|_{C_{\gamma_2}}).$$

Furthermore, for any $w \in \Omega$ and each $(u_1,v_1), (u_2,v_2) \in \mathcal{C}$, and $t \in I$, we get:

$$\|(N_2(u_1,v_1))(\cdot,w)-(N_2(u_2,v_2))(\cdot,w)\|_{C_{\gamma_2}}$$
$$\leq \frac{T^{\alpha_2}}{\Gamma(1+\alpha_2)}(p_2(w)\|u_1(\cdot,w)-v_1(\cdot,w)\|_{C_{\gamma_1}}+q_2(w)\|u_2(\cdot,w)-v_2(\cdot,w)\|_{C_{\gamma_2}}).$$

Thus,

$$d((N(u_1,v_1))(\cdot,w),(N(u_2,v_2))(\cdot,w)) \leq M(w)d((u_1(\cdot,w),v_1(\cdot,w)),(u_2(\cdot,w),v_2(\cdot,w))),$$

where:

$$d((u_1(\cdot,w),v_1(\cdot,w)),(u_2(\cdot,w),v_2(\cdot,w))) = \begin{pmatrix} \|u_1(\cdot,w)-v_1(\cdot,w)\|_{C_{\gamma_1}} \\ \|u_2(\cdot,w)-v_2(\cdot,w)\|_{C_{\gamma_2}} \end{pmatrix}.$$

Since for every $w \in \Omega$, the matrix $M(w)$ converges to zero, then Theorem 1 implies that the operator N has a unique fixed point, which is a random solution of system (1) and (2). □

Now, we prove an existence result for the coupled system (1) and (2) by using the random nonlinear alternative of the Leray–Schauder type in generalized Banach space.

Theorem 4. *Assume that the hypotheses* (H_1) *and* (H_3) *hold. Then, the coupled system (1) and (2) has at least one random solution.*

Proof. We show that the operator $N : \mathcal{C} \times \Omega \to \mathcal{C}$ defined in (8) satisfies all conditions of Theorem 2. The proof will be given in four steps.

Step 1. $N(\cdot,\cdot,w)$ *is continuous.*

Let $(u_n,v_n)_n$ be a sequence such that $(u_n,v_n) \to (u,v) \in \mathcal{C}$ as $n \to \infty$. For any $w \in \Omega$ and each $t \in I$, we have:

$$\|t^{1-\gamma_1}(N_1(u_n,v_n))(t,w)-t^{1-\gamma_1}(N_1(u,v))(t,w)\|$$
$$\leq \frac{t^{1-\gamma_1}}{\Gamma(\alpha_1)}\int_0^t (t-s)^{\alpha_1-1}\|f_1(s,u_n(s,w),v_n(s,w),w)-f_1(s,u(s,w),v(s,w),w)\|ds$$
$$\leq \frac{1}{\Gamma(\alpha_1)}\int_0^t (t-s)^{\alpha_1-1}s^{1-\gamma_1}\|f_1(s,u_n(s,w),v_n(s,w),w)-f_1(s,u(s,w),v(s,w),w)\|ds$$
$$\leq \frac{T^{\alpha_1}}{\Gamma(1+\alpha_1)}\|f_1(\cdot,u_n(\cdot,w),v_n(\cdot,w),w)-f_1(\cdot,u(\cdot,w),v(\cdot,w),w)\|_{C_{\gamma_1}}.$$

Since f_1 is Carathéodory, we have:

$$\|(N_1(u_n,v_n))(\cdot,w)-(N_1(u,v))(\cdot,w)\|_{C_{\gamma_1}} \to 0 \text{ as } n \to \infty.$$

On the other hand, for any $w \in \Omega$ and each $t \in I$, we obtain:

$$\|t^{1-\gamma_2}(N_2(u_n,v_n))(t,w)-t^{1-\gamma_2}(N_2(u,v))(t,w)\|$$
$$\leq \frac{T^{\alpha_2}}{\Gamma(1+\alpha_2)}\|f_2(\cdot,u_n(\cdot,w),v_n(\cdot,w),w)-f_2(\cdot,u(\cdot,w),v(\cdot,w),w)\|_{C_{\gamma_2}}.$$

Furthermore, from the fact that f_2 is Carathéodory, we get:

$$\|(N_2(u_n,v_n))(\cdot,w)-(N_2(u,v))(\cdot,w)\|_{C_{\gamma_2}} \to 0 \text{ as } n \to \infty.$$

Hence, $N(\cdot, \cdot, w)$ is continuous.

Step 2. $N(\cdot, \cdot, w)$ maps bounded sets into bounded sets in $\mathcal{C}$.

Let $R > 0$, and set:
$$B_R := \{(\mu, \nu) \in \mathcal{C} : \|\mu\|_{\mathcal{C}_{\gamma_1}} \leq R, \|\nu\|_{\mathcal{C}_{\gamma_2}} \leq R\}.$$

For any $w \in \Omega$ and each $(u,v) \in B_R$ and $t \in I$, we have:

$$\begin{aligned}
\|t^{1-\gamma_1}(N_1(u,v))(t,w)\| &\leq \frac{\|\phi_1(w)\|}{\Gamma(\gamma_1)} + \frac{t^{1-\gamma_1}}{\Gamma(\alpha_1)} \int_0^t (t-s)^{\alpha_1-1} \|f_1(s, u(s,w), v(s,w), w)\| ds \\
&\leq \frac{\|\phi_1(w)\|}{\Gamma(\gamma_1)} \\
&\quad + \frac{1}{\Gamma(\alpha_1)} \int_0^t (t-s)^{\alpha_1-1} s^{1-\gamma_1} (a_1(w)\|u(s,w)\| + b_1(w)\|v(s,w)\|) ds \\
&\leq \frac{\|\phi_1(w)\|}{\Gamma(\gamma_1)} + \frac{R}{\Gamma(\alpha_1)} \int_0^t (t-s)^{\alpha_1-1} s^{1-\gamma_1}(a_1(w) + b_1(w)) ds \\
&\leq \frac{\|\phi_1(w)\|}{\Gamma(\gamma_1)} + \frac{(a_1(w) + b_1(w))T^{\alpha_1}}{\Gamma(1+\alpha_1)} \\
&:= \ell_1.
\end{aligned}$$

Thus,
$$\|(N_1(u,v))(\cdot,w)\|_{\mathcal{C}_{\gamma_1}} \leq \ell_1.$$

Furthermore, for any $w \in \Omega$ and each $(u,v) \in B_R$ and $t \in I$, we get:

$$\begin{aligned}
\|(N_2(u,v))(\cdot,w)\|_{\mathcal{C}_{\gamma_2}} &\leq \frac{\|\phi_2(w)\|}{\Gamma(\gamma_2)} + \frac{(a_2(w) + b_2(w))T^{\alpha_2}}{\Gamma(1+\alpha_)} \\
&:= \ell_2.
\end{aligned}$$

Hence,
$$\|(N(u,v))(\cdot,w)\|_{\mathcal{C}} \leq (\ell_1, \ell_2) := \ell.$$

Step 3. $N(\cdot, \cdot, w)$ maps bounded sets into equicontinuous sets in $\mathcal{C}$.

Let B_R be the ball defined in Step 2. For each $t_1, t_2 \in I$ with $t_1 \leq t_2$ and any $(u,v) \in B_R$ and $w \in \Omega$, we have:

$$\begin{aligned}
&\|t_1^{1-\gamma_1}(N_1(u,v))(t_1,w) - t_2^{1-\gamma_1}(N_1(u,v))(t_2,w)\| \\
&\leq \frac{t_2^{1-\gamma_1}}{\Gamma(\alpha_1)} \int_{t-1}^{t_2} (t_2-s)^{\alpha_1-1} \|f_1(s, u(s,w), v(s,w), w)\| ds \\
&\leq \frac{T^{\alpha_1}}{\Gamma(1+\alpha_1)} (t_2-t_1)^{\alpha_1} (a_1(w)\|u(\cdot,w)\|_{\mathcal{C}_{\gamma_1}} + b_1(w)\|v(\cdot,w)\|_{\mathcal{C}_{\gamma_2}}) \\
&\leq \frac{RT^{\alpha_1}(a_1(w) + b_1(w))}{\Gamma(1+\alpha_1)} (t_2-t_1)^{\alpha_1} \\
&\to 0 \text{ as } t_1 \to t_2.
\end{aligned}$$

Furthermore, we get:

$$\begin{aligned}
&\|t_1^{1-\gamma_2}(N_2(u,v))(t_1,w) - t_2^{1-\gamma_2}(N_2(u,v))(t_2,w)\| \\
&\leq \frac{RT^{\alpha_{12}}(a_2(w) + b_2(w))}{\Gamma(1+\alpha_2)} (t_2-t_1)^{\alpha_2}
\end{aligned}$$

$\to 0$ as $t_1 \to t_2$.

As a consequence of Steps 1–3, with the Arzela–Ascoli theorem, we conclude that $N(\cdot,\cdot,w)$ maps B_R into a precompact set in $\mathcal{C}$.

Step 4. The set $E(w)$ consisting of $(u(\cdot,w),v(\cdot,w)) \in \mathcal{C}$ such that $(u(\cdot,w),v(\cdot,w)) = \lambda(w)(N((u,v))(\cdot,w)$ for some measurable function $\lambda : \Omega \to (0,1)$ is bounded in $\mathcal{C}$.

Let $(u(\cdot,w),v(\cdot,w)) \in \mathcal{C}$ such that $(u(\cdot,w),v(\cdot,w)) = \lambda(w)(N((u,v))(\cdot,w)$. Then, $u(\cdot,w) = \lambda(w)(N_1((u,v))(\cdot,w)$ and $v(\cdot,w) = \lambda(w)(N_2((u,v))(\cdot,w)$. Thus, for any $w \in \Omega$ and each $t \in I$, we have:

$$\|t^{1-\gamma_1}u(t,w)\| \leq \frac{\|\phi_1(w)\|}{\Gamma(\gamma_1)} + \frac{t^{1-\gamma_1}}{\Gamma(\alpha_1)}\int_0^t(t-s)^{\alpha_1-1}\|f_1(s,u(s,w),v(s,w),w)\|ds$$

$$\leq \frac{\|\phi_1(w)\|}{\Gamma(\gamma_1)}$$
$$+ \frac{1}{\Gamma(\alpha_1)}\int_0^t(t-s)^{\alpha_1-1}s^{1-\gamma_1}(a_1(w)\|u(s,w\| + b_1(w)\|v(s,w\|)ds.$$

Furthermore, we get:

$$\|t^{1-\gamma_2}v(t,w)\| \leq \frac{\|\phi_2(w)\|}{\Gamma(\gamma_2)}$$
$$+ \frac{1}{\Gamma(\alpha_2)}\int_0^t(t-s)^{\alpha_2-1}s^{1-\gamma_2}(a_2(w)\|u(s,w\| + b_2(w)\|v(s,w\|)ds.$$

Hence, we obtain:

$$\|t^{1-\gamma_1}u(t,w)\| + \|t^{1-\gamma_2}v(t,w)\| \leq a + bc\int_0^t(t-s)^{\alpha-1}(\|s^{1-\gamma_1}u(s,w\| + \|s^{1-\gamma_2}v(s,w\|)ds,$$

where:

$$a := \frac{\|\phi_1(w)\|}{\Gamma(\gamma_1)} + \frac{\|\phi_2(w)\|}{\Gamma(\gamma_2)},\ b := \frac{1}{\Gamma(\alpha_1)} + \frac{1}{\Gamma(\alpha_2)},$$
$$c := \max\{a_1(w)+a_2(w), b_1(w)+b_2(w)\},\ \alpha := \max\{\alpha_1,\alpha_2\}.$$

Lemma 1 implies that there exists $\rho := \rho(\alpha) > 0$ such that:

$$\|t^{1-\gamma_1}u(t,w)\| + \|t^{1-\gamma_2}v(t,w)\| \leq a + abc\rho\int_0^t(t-s)^{\alpha-1}ds$$
$$\leq \frac{a + abc\rho T^\alpha}{\alpha}$$
$$= L.$$

This gives:
$$\|u(\cdot,w)\|_{C_{\gamma_1}} + \|v(\cdot,w)\|_{C_{\gamma_2}} \leq L.$$

Hence:
$$\|(u(\cdot,w),v(\cdot,w))\|_{\mathcal{C}} \leq L.$$

This shows that the set $E(w)$ is bounded. As a consequence of Steps 1–4 together with Theorem 2, we can conclude that N has at least one fixed point in B_R, which is a solution for the system (1) and (2). □

4. Coupled Hilfer–Hadamard Fractional Differential Systems

Now, we are concerned with the coupled system (3) and (4). Set $C := C([1,T])$, and denote the weighted space of continuous functions defined by:

$$C_{\gamma,\ln}([1,T]) = \{w(t) : (\ln t)^{1-\gamma}w(t) \in C\},$$

with the norm:

$$\|w\|_{C_{\gamma,\ln}} := \sup_{t \in [1,T]} |(\ln t)^{1-r}w(t)|.$$

Furthermore, by $C_{\gamma_1,\gamma_2,\ln}([1,T]) := C_{\gamma_1,\ln}([1,T]) \times C_{\gamma_2,\ln}([1,T])$, we denote the product weighted space with the norm:

$$\|(u,v)\|_{C_{\gamma_1,\gamma_2,\ln}([1,T])} = \|u\|_{C_{\gamma_1,\ln}} + \|v\|_{C_{\gamma_2,\ln}}.$$

Let us recall some definitions and properties of Hadamard fractional integration and differentiation. We refer to [7] for a more detailed analysis.

Definition 12. *[7] (Hadamard fractional integral) The Hadamard fractional integral of order $q > 0$ for a function $g \in L^1([1,T])$ is defined as:*

$$(^H I_1^q g)(x) = \frac{1}{\Gamma(q)} \int_1^x \left(\ln \frac{x}{s}\right)^{q-1} \frac{g(s)}{s} ds,$$

provided the integral exists.

Example 2. *Let $0 < q < 1$. Let $g(x) = \ln x$, $x \in [0,e]$. Then:*

$$(^H I_1^q g)(x) = \frac{1}{\Gamma(2+q)} (\ln x)^{1+q}; \text{ for a.e. } x \in [0,e].$$

Set:

$$\delta = x\frac{d}{dx}, \ q > 0, \ n = [q] + 1,$$

and:

$$AC_\delta^n := \{u : [1,T] \to E : \delta^{n-1}[u(x)] \in AC(I)\}.$$

Definition 13. *[7] The Hadamard fractional derivative of order $q > 0$ applied to the function $w \in AC_\delta^n$ is defined as:*

$$(^H D_1^q w)(x) = \delta^n (^H I_1^{n-q} w)(x).$$

In particular, if $q \in (0,1]$, then:

$$(^H D_1^q w)(x) = \delta(^H I_1^{1-q} w)(x).$$

Example 3. *Let $0 < q < 1$. Let $w(x) = \ln x$, $x \in [0,e]$. Then:*

$$(^H D_1^q w)(x) = \frac{1}{\Gamma(2-q)} (\ln x)^{1-q}, \text{ for a.e. } x \in [0,e].$$

It has been proven (see, e.g., Kilbas [30], Theorem 4.8) that in the space $L^1(I)$, the Hadamard fractional derivative is the left-inverse operator to the Hadamard fractional integral, i.e.:

$$(^H D_1^q)(^H I_1^q w)(x) = w(x).$$

From [7], we have:
$$({}^H I_1^q)({}^H D_1^q w)(x) = w(x) - \frac{({}^H I_1^{1-q} w)(1)}{\Gamma(q)} (\ln x)^{q-1}.$$

The Caputo–Hadamard fractional derivative is defined in the following way:

Definition 14. *The Caputo–Hadamard fractional derivative of order $q > 0$ applied to the function $w \in AC_\delta^n$ is defined as:*
$$({}^{Hc} D_1^q w)(x) = ({}^H I_1^{n-q} \delta^n w)(x).$$

In particular, if $q \in (0,1]$, then:
$$({}^{Hc} D_1^q w)(x) = ({}^H I_1^{1-q} \delta w)(x).$$

Definition 15. *Let $\alpha \in (0,1)$, $\beta \in [0,1]$, $\gamma = \alpha + \beta - \alpha\beta$, $w \in L^1(I)$, and ${}^H I_1^{(1-\alpha)(1-\beta)} w \in AC(I)$. The Hilfer–Hadamard fractional derivative of order α and type β applied to the function w is defined as:*
$$\begin{aligned}({}^H D_1^{\alpha,\beta} w)(t) &= \left({}^H I_1^{\beta(1-\alpha)} ({}^H D_1^\gamma w)\right)(t) \\ &= \left({}^H I_1^{\beta(1-\alpha)} \delta({}^H I_1^{1-\gamma} w)\right)(t); \text{ for a.e. } t \in [1,T].\end{aligned} \qquad (9)$$

This new fractional derivative (9) may be viewed as interpolating the Hadamard fractional derivative and the Caputo–Hadamard fractional derivative. Indeed, for $\beta = 0$, this derivative reduces to the Hadamard fractional derivative, and when $\beta = 1$, we recover the Caputo–Hadamard fractional derivative.
$${}^H D_1^{\alpha,0} = {}^H D_1^\alpha, \text{ and } {}^H D_1^{\alpha,1} = {}^{Hc} D_1^\alpha.$$

From [31], we conclude the following lemma.

Lemma 2. *Let $g : [1,T] \times E \to E$ be such that $g(\cdot, u(\cdot)) \in C_{\gamma,\ln}([1,T])$ for any $u \in C_{\gamma,\ln}([1,T])$. Then, Problem (3) is equivalent to the following Volterra integral equation:*
$$u(t) = \frac{\phi_0}{\Gamma(\gamma)} (\ln t)^{\gamma-1} + ({}^H I_1^\alpha g(\cdot, u(\cdot)))(t).$$

Definition 16. *By a random solution of the coupled system (3) and (4), we mean a coupled measurable function $(u,v) \in C_{\gamma_1,\ln} \times C_{\gamma_2,\ln}$ that satisfies the conditions (4) and Equation (3) on $[1,T]$.*

Now, we give (without proof) similar existence and uniqueness results for the system (3) and (4). Let us introduce the following hypotheses:

(H_1') The functions g_i; $i = 1,2$ are Carathéodory.
(H_2') There exist measurable functions $p_i, q_i : \Omega \to (0,\infty)$; $i = 1,2$ such that:
$$\|g_i(t, u_1, v_1) - g_i(t, u_2, v_2)\| \le p_i(w)\|u_1 - u_2\| + q_i(w)\|v_1 - v_2\|;$$

for a.e. $t \in [1,T]$, and each $u_i, v_i \in \mathbb{R}^m$, $i = 1,2$.
(H_3') There exist measurable functions $a_i, b_i : \Omega \to (0,\infty)$; $i = 1,2$ such that:
$$\|g_i(t, u, v)\| \le a_i(w)\|u\| + b_i(w)\|v\|; \text{ for a.e. } t \in [1,T], \text{ and each } u, v \in \mathbb{R}^m.$$

Theorem 5. *Assume that the hypotheses* (H_1') *and* (H_2') *hold. If for every* $w \in \Omega$, *the matrix:*

$$\begin{pmatrix} \frac{(\ln T)^{\alpha_1}}{\Gamma(1+\alpha_1)} p_1(w) & \frac{(\ln T)^{\alpha_1}}{\Gamma(1+\alpha_1)} q_1(w) \\ \frac{(\ln T)^{\alpha_2}}{\Gamma(1+\alpha_2)} p_2(w) & \frac{(\ln T)^{\alpha_2}}{\Gamma(1+\alpha_2)} q_2(w) \end{pmatrix}$$

converges to zero, then the coupled system (3) *and* (4) *has a unique random solution.*

Theorem 6. *Assume that the hypotheses* (H_1') *and* (H_3') *hold. Then, the coupled system* (3) *and* (4) *has at at least a random solution.*

5. An Example

We equip the space $\mathbb{R}_-^* := (-\infty, 0)$ with the usual σ-algebra consisting of Lebesgue measurable subsets of $\mathbb{R}_-^*$. Consider the following random coupled Hilfer fractional differential system:

$$\begin{cases} (D_0^{\frac{1}{2},\frac{1}{2}} u)(t,w) = f(t, u(t,w), v(t,w), w); \\ (D_0^{\frac{1}{2},\frac{1}{2}} v)(t) = g(t, u(t,w), v(t,w), w); \\ (I_0^{\frac{1}{4}} u)(0,w) = \cos w, \\ (I_0^{\frac{1}{4}} v_n)(0,w) = \sin w, \end{cases} ; w \in \mathbb{R}_-^*, t \in [0,1], \qquad (10)$$

where:

$$f(t, u, v, w) = \frac{t^{\frac{-1}{4}} w^2 (u(t) + v(t)) \sin t}{64(1 + w^2 + \sqrt{t})(1 + |u| + |v|)}; t \in [0, 1],$$

$$g(t, u, v) = \frac{w^2 (u(t) + v(t)) \cos t}{64(1 + |u| + |v|)}; w \in \mathbb{R}_-^*, t \in [0, 1].$$

Set $\alpha_i = \beta_i = \frac{1}{2}$; $i = 1, 2$, then $\gamma_i = \frac{3}{4}$; $i = 1, 2$. The hypothesis (H_2) is satisfied with:

$$p_1(w) = p_2(w) = q_1(w) = q_2(w) = \frac{w^2}{64(1+w^2)}.$$

Furthermore, if for every $w \in \mathbb{R}_-^*$, the matrix:

$$\frac{w^2}{64(1+w^2)\Gamma(\frac{1}{2})} \begin{pmatrix} 1 & 1 \\ 1 & 1 \end{pmatrix}$$

converges to zero, hence, Theorem 3 implies that the system (10) has a unique random solution defined on $[0, 1]$.

Author Contributions: All the authors contributed in obtaining the results and writing the paper. All authors have read and approved the final manuscript.

Funding: This research was funded by National Natural Science Foundation of China (11671339).

Conflicts of Interest: The authors declare no conflict of interest.

References

1. Hilfer, R. *Applications of Fractional Calculus in Physics*; World Scientific: Singapore, 2000.
2. Tarasov, V.E. *Fractional Dynamics: Application of Fractional Calculus to Dynamics of Particles, Fields and Media*; Springer: Heidelberg, Germany; Higher Education Press: Beijing, China, 2010.
3. Abbas, S.; Benchohra, M.; Graef, J.R.; Henderson, J. *Implicit Fractional Differential and Integral Equations: Existence and Stability*; De Gruyter: Berlin, Germany, 2018.

4. Abbas, S.; Benchohra, M.; N'Guérékata, G.M. *Topics in Fractional Differential Equations*; Springer: New York, NY, USA, 2012.
5. Abbas, S.; Benchohra, M.; N'Guérékata, G.M. *Advanced Fractional Differential and Integral Equations*; Nova Science Publishers: New York, NY, USA, 2015.
6. Samko, S.G.; Kilbas, A.A.; Marichev, O.I. *Fractional Integrals and Derivatives. Theory and Applications*; Engl. Trans. from the Russian; Gordon and Breach: Amsterdam, The Netherlands, 1987.
7. Kilbas, A.A.; Srivastava, H.M.; Trujillo, J.J. *Theory and Applications of Fractional Differential Equations*; Elsevier Science B.V.: Amsterdam, The Netherlands, 2006.
8. Abbas, S.; Benchohra, M.; Lazreg, J.E.; Zhou, Y. A Survey on Hadamard and Hilfer fractional differential equations: Analysis and Stability. *Chaos Solitons Fractals* **2017**, *102*, 47–71. [CrossRef]
9. Benchohra, M.; Henderson, J.; Ntouyas, S.K.; Ouahab, A. Existence results for functional differential equations of fractional order. *J. Math. Anal. Appl.* **2008**, *338*, 1340–1350. [CrossRef]
10. Zhou, Y. Attractivity for fractional differential equations in Banach space. *Appl. Math. Lett.* **2018**, *75*, 1–6. [CrossRef]
11. Zhou, Y. Attractivity for fractional evolution equations with almost sectorial operators. *Fract. Calc. Appl. Anal.* **2018**, *21*, 786–800. [CrossRef]
12. Zhou, Y.; Shangerganesh, L.; Manimaran, J.; Debbouche, A. A class of time-fractional reaction-diffusion equation with nonlocal boundary condition. *Math. Meth. Appl. Sci.* **2018**, *41*, 2987–2999. [CrossRef]
13. Zhou, Y.; Peng, L.; Huang, Y.Q. Duhamel's formula for time-fractional Schrödinger equations. *Math. Meth. Appl. Sci.* **2018**, *41*, 8345–8349. [CrossRef]
14. Zhou, Y.; Peng, L.; Huang, Y.Q. Existence and Hölder continuity of solutions for time-fractional Navier-Stokes equations. *Math. Meth. Appl. Sci.* **2018**, *41*, 7830–7838. [CrossRef]
15. Furati, K.M.; Kassim, M.D. Non-existence of global solutions for a differential equation involving Hilfer fractional derivative. *Electron. J. Differ. Equ.* **2013**, *2013*, 1–10.
16. Furati, K.M.; Kassim, M.D.; Tatar, N.E. Existence and uniqueness for a problem involving Hilfer fractional derivative. *Comput. Math. Appl.* **2012**, *64*, 1616–1626. [CrossRef]
17. Kamocki, R.; Obczński, C. On fractional Cauchy-type problems containing Hilfer's derivative. *Electron. J. Qual. Theory Differ. Equ.* **2016**, *2016*, 1–12. [CrossRef]
18. Qassim, M.D.; Furati, K.M.; Tatar, N.-E. On a differential equation involving Hilfer-Hadamard fractional derivative. *Abstr. Appl. Anal.* **2012**, *2012*, 391062. [CrossRef]
19. Tomovski, Ž.; Hilfer, R.; Srivastava, H.M. Fractional and operational calculus with generalized fractional derivative operators and Mittag-Leffler type functions. *Integral Transform. Spec. Funct.* **2010**, *21*, 797–814. [CrossRef]
20. Wang, J.-R.; Zhang, Y. Nonlocal initial value problems for differential equations with Hilfer fractional derivative. *Appl. Math. Comput.* **2015**, *266*, 850–859. [CrossRef]
21. Abbas, S.; Albarakati, W.; Benchohra, M.; Zhou, Y. Weak solutions for partial pettis Hadamard fractional integral equations with random effects. *J. Integral Equ. Appl.* **2017**, *29*, 473–491. [CrossRef]
22. Abbas, S.; Benchohra, M.; Zhou, Y. Coupled Hilfer fractional differential systems with random effects. *Adv. Differ. Equ.* **2018**, *2018*, 369. [CrossRef]
23. Graef, J.R.; Henderson, J.; Ouahab, A. Some Krasnosel'skii type random fixed point theorems. *J. Nonlinear Funct. Anal.* **2017**, *2017*, 46.
24. Petre, I.R.; Petrusel, A. Krasnoselskii's theorem in generalized Banach spaces and applications. *Electron. J. Qual. Theory Differ. Equ.* **2012**, *85*, 1–20. [CrossRef]
25. Sinacer, M.L.; Nieto, J.J.; Ouahab, A. Random fixed point theorems in generalized Banach spaces and applications. *Random Oper. Stoch. Equ.* **2016**, *24*, 93–112. [CrossRef]
26. Itoh, S. Random fixed point theorems with applications to random differential equations in Banach spaces. *J. Math. Anal. Appl.* **1979**, *67*, 261–273. [CrossRef]
27. Engl, H.W. A general stochastic fixed-point theorem for continuous random operators on stochastic domains. *J. Math. Anal. Appl.* **1978**, *66*, 220–231. [CrossRef]
28. Allaire, G.; Kaber, S.M. *Numerical Linear Algebra*; Ser. Texts in Applied Mathematics; Springer: New York, NY, USA, 2008.
29. Varga, R.S. *Matrix Iterative Analysis*; Second Revised and Expanded Edition; Springer Series in Computational Mathematics 27; Springer: Berlin, Germany, 2000.

30. Kilbas, A.A. Hadamard-type fractional calculus. *J. Korean Math. Soc.* **2001**, *38*, 1191–1204.
31. Qassim, M.D.; Tatar, N.-E. Well-posedness and stability for a differential problem with Hilfer-Hadamard fractional derivative. *Abstr. Appl. Anal.* **2013**, *2013*, 605029.

© 2019 by the authors. Licensee MDPI, Basel, Switzerland. This article is an open access article distributed under the terms and conditions of the Creative Commons Attribution (CC BY) license (http://creativecommons.org/licenses/by/4.0/).

Article

On the Solvability of a Mixed Problem for a High-Order Partial Differential Equation with Fractional Derivatives with Respect to Time, with Laplace Operators with Spatial Variables and Nonlocal Boundary Conditions in Sobolev Classes

Onur Alp İlhan [1,*], Shakirbay G. Kasimov [2], Shonazar Q. Otaev [2] and Haci Mehmet Baskonus [3]

1. Department of Mathematics and Science Education, Faculty of Education, Erciyes University, Kayseri 38039, Turkey
2. Faculty of Mathematics, National University of Uzbekistan, Tashkent 100174, Uzbekistan; shokiraka@mail.ru (S.G.K.); otaev_sh@mail.ru (S.Q.O.)
3. Department of Mathematics and Science Education, Faculty of education, Harran University, Sanliurfa 63190, Turkey; hmbaskonus@gmail.com
* Correspondence: onuralp70@msn.com or oailhan@erciyes.edu.tr; Tel.: +90-54-3915-3652

Received: 16 November 2018; Accepted: 1 March 2019; Published: 5 March 2019

Abstract: In this paper, we study the solvability of a mixed problem for a high-order partial differential equation with fractional derivatives with respect to time, and with Laplace operators with spatial variables and nonlocal boundary conditions in Sobolev classes.

Keywords: Banach space; Sobolev space; Laplace operators; nonlocal boundary conditions.

MSC: 35M12; 46B25; 46E39

---

1. Introduction

The spectral theory of operators finds numerous uses in various fields of mathematics and their applications.

An important part of the spectral theory of differential operators is the distribution of their eigenvalues. This classical question was studied for a second-order operator on a finite interval by Liouville and Sturm. Later, G.D. Birkhoff [1–3] studied the distribution of eigenvalues for an ordinary differential operator of arbitrary order on a finite interval with regular boundary conditions.

For quantum mechanics, it is especially interesting to distribute the eigenvalues of operators defined throughout the space and having a discrete spectrum. E.C. Titchmarsh [4–9] was the first to rigorously establish the formula for the distribution of the number of eigenvalues for a one-dimensional Sturm-Liouville operator on the whole axis with potential growing at infinity. He also first strictly established the distribution formula for the Schrödinger operator. B.M. Levitan [10–12] deserves much credit for the improvement of E.C. Titchmarsh's method.

In solving many mathematical physics problems, the need arises for the expansion of an arbitrary function in a Fourier series with respect to Sturm-Liouville eigenvalues. The so-called regular case of the Sturm-Liouville problem corresponding to a finite interval and a continuous coefficient of the equation has been studied for a relatively long time and is usually described in detail in the manuals on the equations of mathematical physics and integral equations.

The Sturm-Liouville problem for the so-called singular case, as well as with nonlocal boundary conditions, is much less known.

As it is known, so-called fractal media are studied in solid-state physics and, in particular, diffusion phenomena in them. In one of the models studied in [13], diffusion in a strongly porous (fractal) medium is described by an equation of the type of heat-conduction equation, but with a fractional derivative with respect to time coordinate

$$D_t^{(\alpha)} u(x,t) = \frac{\partial^2 (u(x,t))}{\partial x^2}, \quad 0 < \alpha < 1. \tag{1}$$

The formulation of initial-boundary value problems for Equation (1), similar to the problems for parabolic differential equations, makes sense if by a regularized fractional derivative:

$$D^{(\alpha)} \varphi(t) = \frac{1}{\Gamma(1-\alpha)} \left[\frac{d}{dt} \int_0^1 (t-\tau)^{-\alpha} \varphi(\tau) d\tau - t^{-\alpha} \varphi(0) \right], \quad t \geq 0 \tag{2}$$

Study of the form equations

$$D_t^{(\alpha)} u = Au \tag{3}$$

where A is an elliptic operator (in [14–16]). In recent years, many authors studied fractional differential equations in [17–34].

2. Problem Formulation

In this work, we consider the equation of the form

$$D_{0t}^{\alpha} u(x,t) + (-\Delta)^{\nu} u(x,t) = f(x,t), \quad (x,t) \in \Pi \times (0,\infty), \ l-1 < \alpha \leq l, \ l, \nu \in N \tag{4}$$

with initial conditions

$$\lim_{t \to 0} D_{0t}^{\alpha-k} u(x,t) = \varphi_k(x), \quad k = 1, 2, \ldots, l \tag{5}$$

and boundary conditions

$$\begin{cases} \alpha_j \cdot (-\Delta)^i u(x_1,\ldots,x_j,\ldots,x_N,t) \mid_{x_j=0} + \beta_j \cdot (-\Delta)^i u(x_1,\ldots,x_j,\ldots,x_N,t) \mid_{x_j=\pi} = 0, \\ 1 \leq j \leq p, \\ \beta_j \cdot \frac{\partial (-\Delta)^i u(x_1,\ldots,x_j,\ldots,x_N,t)}{\partial x_j} \mid_{x_j=0} + \alpha_j \cdot \frac{\partial (-\Delta)^i u(x_1,\ldots,x_j,\ldots,x_N,t)}{\partial x_j} \mid_{x_j=\pi} = 0, \quad 1 \leq j \leq p, \\ (-\Delta)^i u(x_1,\ldots,x_j,\ldots,x_N,t) \mid_{x_j=0} = (-\Delta)^i u(x_1,\ldots,x_j,\ldots,x_N,t) \mid_{x_j=\pi}, \quad p+1 \leq j \leq q, \\ \frac{\partial (-\Delta)^i u(x_1,\ldots,x_j,\ldots,x_N,t)}{\partial x_j} \mid_{x_j=0} = \frac{\partial (-\Delta)^i u(x_1,\ldots,x_j,\ldots,x_N,t)}{\partial x_j} \mid_{x_j=\pi}, \quad p+1 \leq j \leq q, \\ (-\Delta)^i u(x_1,\ldots,x_j,\ldots,x_N,t) \mid_{x_j=0} = 0, \quad q+1 \leq j \leq N, \\ (-\Delta)^i u(x_1,\ldots,x_j,\ldots,x_N,t) \mid_{x_j=\pi} = 0, \quad q+1 \leq j \leq N, \\ 1 \leq p \leq q \leq N, \ i = 0, 1, \ldots, \nu - 1, \end{cases} \tag{6}$$

where $(x,t) = (x_1,\ldots,x_j,\ldots,x_N,t) \in \Pi \times (0,\infty)$, $\Pi = (0,\pi) \times \cdots \times (0,\pi)$, $\alpha_j = const$, $\beta_j = const$, and $f(x,t)$, $\varphi_k(x)$, $k = 1,2,\ldots,l$ are functions that can be expanded in terms of the system of eigenfunctions $\{v_n(x), n \in Z^N\}$ of the spectral problem:

$$(-\Delta)^{\nu} v(x) = \mu v(x), \tag{7}$$

$$\begin{cases} \alpha_j \cdot (-\Delta)^i v(x_1,\ldots,x_j,\ldots,x_N)\,|_{x_j=0} + \beta_j \cdot (-\Delta)^i v(x_1,\ldots,x_j,\ldots,x_N)\,|_{x_j=\pi} = 0, \\ 1 \le j \le p, \\ \beta_j \cdot \frac{\partial(-\Delta)^i v(x_1,\ldots,x_j,\ldots,x_N)}{\partial x_j}\,|_{x_j=0} + \alpha_j \cdot \frac{\partial(-\Delta)^i v(x_1,\ldots,x_j,\ldots,x_N)}{\partial x_j}\,|_{x_j=\pi} = 0, \quad 1 \le j \le p, \\ (-\Delta)^i v(x_1,\ldots,x_j,\ldots,x_N)\,|_{x_j=0} = (-\Delta)^i v(x_1,\ldots,x_j,\ldots,x_N)\,|_{x_j=\pi}, \quad p+1 \le j \le q, \\ \frac{\partial(-\Delta)^i v(x_1,\ldots,x_j,\ldots,x_N)}{\partial x_j}\,|_{x_j=0} = \frac{\partial(-\Delta)^i v(x_1,\ldots,x_j,\ldots,x_N)}{\partial x_j}\,|_{x_j=\pi}, \quad p+1 \le j \le q, \\ (-\Delta)^i v(x_1,\ldots,x_j,\ldots,x_N)\,|_{x_j=0} = 0, \quad q+1 \le j \le N, \\ (-\Delta)^i v(x_1,\ldots,x_j,\ldots,x_N)\,|_{x_j=\pi} = 0, \quad q+1 \le j \le N, \\ 1 \le p \le q \le N,\, i = 0,1,\ldots,\nu-1. \end{cases} \quad (8)$$

Here, for $\alpha < 0$, fractional integral D^α has the form

$$D^\alpha_{at} u(x,t) = \frac{\text{sign}(t-a)}{\Gamma(-\alpha)} \int_a^t \frac{u(x,\tau)\cdot d\tau}{|t-\tau|^{\alpha+1}},$$

$D^\alpha_{at} u(x,t) = u(x,t)$ for $\alpha = 0$, and for $l-1 < \alpha \le l, l \in N$, the fractional derivative has the form

$$D^\alpha_{at} u(x,t) = \text{sign}^l(t-a) \frac{d^l}{dt^l} D^{\alpha-l}_{at} u(x,t) =$$

$$= \frac{\text{sign}^{l+1}(t-a)}{\Gamma(l-\alpha)} \frac{d^l}{dt^l} \int_a^t \frac{u(x,\tau)\cdot d\tau}{|t-\tau|^{\alpha-l+1}}.$$

In [17], Problems (4)–(6) and, accordingly, spectral Problems (7) and (8) in the case $\nu = 1$, were considered.

3. Preliminaries

More detailed information for this section can be found in [17]. We look for eigenfunctions of spectral Problems (7) and (8) in the form of the product $v(x) = y_1(x_1) \cdot \ldots \cdot y_N(x_N)$. Then, we obtain, instead of spectral Problems (7) and (8), the following spectral problem:

$$-y''(x) = \mu y(x), \mu = \lambda^2 \quad (9)$$

$$\begin{cases} \alpha y(0) + \beta y(\pi) = 0, \\ \beta y'(0) + \alpha y'(\pi) = 0. \end{cases} \quad (10)$$

In the case of $|\alpha| = |\beta|$, i.e., with boundary conditions $y(0) = y(\pi)$, $y'(0) = y'(\pi)$ or $y(0) = -y(\pi)$, $y'(0) = -y'(\pi)$, spectral Problems (7) and (8) were investigated by many authors (see, for example, [35–41]). In order to simplify calculations, we confined ourselves to the case of $|\alpha| \ne |\beta|, \alpha \ne 0, \beta \ne 0$. It is not difficult to see that $\mu = 0$ is not an eigenvalue of Problems (9) and (10). In fact, if $\mu = 0$ is the eigenvalue, then $y'' = 0$, $y = ax + b$, $\alpha b + \beta(a\pi + b) = 0$, $\beta a + \alpha a = 0$. We obtained from here $a = 0$, $b = 0$, i.e., $y \equiv 0$. Similarly, for $\mu < 0$, Problems (9) and (10) have no nontrivial solutions.

For $\mu > 0$, the general solution of Problem (9) has the form

$$y(x) = A\cos\lambda x + B\sin\lambda x.$$

From boundary conditions, we have:

$$\alpha y(0) + \beta y(\pi) = \alpha A + \beta(A\cos\lambda\pi + B\sin\lambda\pi) = 0,$$

i.e.,
$$\beta y'(0) + \alpha y'(\pi) = \beta(\lambda B) + \alpha(\lambda B \cos\lambda\pi - \lambda A \sin\lambda\pi) = 0,$$

$$\begin{cases} (\alpha + \beta\cos\lambda\pi)A + \beta\sin\lambda\pi B = 0, \\ \alpha\sin\lambda\pi A - (\beta + \alpha\cos\lambda\pi)B = 0. \end{cases}$$

Hence, the nontrivial solutions of Problems (9) and (10) are only possible in the case of

$$(\alpha + \beta\cos\lambda\pi)(-\beta - \alpha\cos\lambda\pi) - \alpha\beta\sin^2(\lambda\pi) = 0.$$

Furthermore,

$$-\alpha\beta - \alpha^2\cos\lambda\pi - \beta^2\cos\lambda\pi - \alpha\beta\cos^2\lambda\pi - \alpha\beta\sin^2\lambda\pi = 0,$$

i.e., $-(\alpha^2 + \beta^2)\cos\lambda\pi = 2\alpha\beta$ or $\cos\lambda\pi = \dfrac{-2\alpha\beta}{\alpha^2 + \beta^2}$.

Therefore, $\lambda\pi = \arccos\dfrac{-2\alpha\beta}{\alpha^2 + \beta^2}$ or

$$\lambda\pi = \pm \arccos\dfrac{-2\alpha\beta}{\alpha^2 + \beta^2} + 2n\pi, \quad n \in Z.$$

Further,

$$\mu_n^\pm = (2n + \varepsilon_n\varphi)^2 = (-2n - \varepsilon_n\varphi)^2 = \mu_{-n}^\mp, \quad \varepsilon_n = \pm 1, \quad \varphi = \dfrac{1}{\pi}\arccos\dfrac{-2\alpha\beta}{\alpha^2 + \beta^2}, \quad n \in Z.$$

That's why $\mu_n^\pm \neq \mu_{-n}^\pm$ means that $\varepsilon_{-n} \neq -\varepsilon_n$, i.e., $\varepsilon_{-n} = \varepsilon_n$, $n \in Z$. Thus, the eigenvalues and eigenfunctions of Problems (9) and (10) are

$$\mu_n = \lambda_n^2 = (2n + \varepsilon_n\varphi)^2, \quad \varphi = \dfrac{1}{\pi}\arccos\dfrac{-2\alpha\beta}{\alpha^2 + \beta^2}, \quad \varepsilon_n = \pm 1, \varepsilon_{-n} = \varepsilon_n, \quad n \in Z$$

and

$$y_n(x) = B_n \left(\dfrac{\beta + \alpha\cos\lambda_n\pi}{\alpha\sin\lambda_n\pi} \cos\lambda_n x + \sin\lambda_n x \right),$$

respectively, where

$$\dfrac{\beta + \alpha\cos\lambda_n\pi}{\alpha\sin\lambda_n\pi} = \dfrac{\beta - \dfrac{2\alpha^2\beta}{\alpha^2+\beta^2}}{\varepsilon_n\alpha\sqrt{1 - \dfrac{4\alpha^2\beta^2}{(\alpha^2+\beta^2)^2}}} = \dfrac{\beta(\beta^2 - \alpha^2)}{\varepsilon_n\alpha \mid \beta^2 - \alpha^2 \mid} = \varepsilon_n \mathrm{sign}(\beta^2 - \alpha^2)\dfrac{\beta}{\alpha},$$

hence, $y_n(x) = B_n \left(\varepsilon_n \mathrm{sign}(\beta^2 - \alpha^2)\dfrac{\beta}{\alpha}\cos\lambda_n x + \sin\lambda_n x \right)$. Choosing

$$B_n = \varepsilon_n \mathrm{sign}(\beta^2 - \alpha^2)\dfrac{\alpha}{\sqrt{\alpha^2 + \beta^2}}\sqrt{\dfrac{2}{\pi}}\dfrac{1}{\sqrt{1 + (2n)^{2s}}}$$

we obtain

$$y_n(x) = \sqrt{\dfrac{2}{\pi}}\dfrac{1}{\sqrt{\alpha^2 + \beta^2}}\dfrac{1}{\sqrt{1 + (2n)^{2s}}}\left(\beta\cos\lambda_n x + \varepsilon_n \mathrm{sign}(\beta^2 - \alpha^2)\alpha\sin\lambda_n x \right).$$

Denote $\omega_n = \sqrt{\frac{2}{\pi}} \frac{1}{\sqrt{\alpha^2+\beta^2}} \frac{1}{\sqrt{1+(2n)^{2s}}}$. Then,

$$y_n(x) = \omega_n \left(\beta \cos \lambda_n x + \varepsilon_n \text{sign}(\beta^2 - \alpha^2) \alpha \sin \lambda_n x\right).$$

The norm in space $W_2^s(0, \pi)$ is introduced as follows:

$$\|f\|^2_{W_2^s(0,\pi)} = \|f\|^2_{L_2(0,\pi)} + \|D^s f\|^2_{L_2(0,\pi)}.$$

Let $\varepsilon_n = \varepsilon_{-n}$. Then, system of vectors

$$z_n(x) = \omega_n \left(\beta \cos 2nx + \varepsilon_n \text{sign}(\beta^2 - \alpha^2) \alpha \sin 2nx\right)$$

forms the complete orthonormal system in $W_2^s(0, \pi)$. The following lemma holds.

Lemma 1. *Let $\{a_n\}$ be a finite system of complex numbers. Then, inequalities*

$$\left\|\sum_{-N}^{N} a_n(y_n(x) - z_n(x))\right\|_{L_2(0,\pi)} \leq \sqrt{2} \cdot \max_{x \in [0,\pi]} \left|e^{i\varphi x} - 1\right| \cdot \sqrt{\sum_{-N}^{N} |a_n \cdot c_n|^2}$$

are valid where

$$c_n = \frac{1}{\sqrt{1+(2n)^{2s}}}, \quad s = 1,2,3,\dots.$$

Proof. Calculating the difference of $y_n(x) - z_n(x)$, we obtain

$$y_n(x) - z_n(x) =$$

$$\omega_n[\beta(\cos \lambda_n x - \cos 2nx) + \varepsilon_n \text{sign}(\beta^2 - \alpha^2)\alpha(\sin \lambda_n x - \sin 2nx)] =$$

$$= \omega_n[(\varepsilon_n \text{sign}(\beta^2 - \alpha^2)\alpha + \beta i)\frac{e^{i\varepsilon_n \varphi x} - 1}{2i} e^{2nix} +$$

$$+(\varepsilon_n \text{sign}(\beta^2 - \alpha^2)\alpha - \beta i)\frac{1 - e^{-i\varepsilon_n \varphi x}}{2i} e^{-2nix}].$$

Then,

$$\sum_{-N}^{N} a_n(y_n - z_n) = \sum_{-N}^{N} a_n \omega_n \left[(\varepsilon_n \text{sign}(\beta^2 - \alpha^2)\alpha + \beta i)\frac{e^{i\varepsilon_n \varphi x} - 1}{2i} e^{2nix} +\right.$$

$$\left.+(\varepsilon_n \text{sign}(\beta^2 - \alpha^2)\alpha - \beta i)\frac{1 - e^{-i\varepsilon_n \varphi x}}{2i} e^{-2nix}\right].$$

Using properties of the norm, we have

$$\left\|\sum_{-N}^{N} a_n(y_n - z_n)\right\|_{L_2(0,\pi)} =$$

$$= \left\|\frac{\text{sign}(\beta^2 - \alpha^2)\alpha + \beta i}{2i}(e^{i\varphi x} - 1) \sum_{-N, \varepsilon_n=1}^{N} a_n \omega_n e^{2nix} +\right.$$

$$\left.+\frac{-\text{sign}(\beta^2 - \alpha^2)\alpha + \beta i}{2i}(e^{-i\varphi x} - 1) \sum_{-N, \varepsilon_n=-1}^{N} a_n \omega_n e^{2nix} +\right.$$

$$+\frac{sign(\beta^2-\alpha^2)\alpha-\beta i}{2i}(1-e^{-i\varphi x})\sum_{-N,\varepsilon_n=1}^{N}a_n\omega_n e^{-2nix}+$$

$$+\frac{-sign(\beta^2-\alpha^2)\alpha-\beta i}{2i}(1-e^{i\varphi x})\sum_{-N,\varepsilon_n=-1}^{N}a_n\omega_n e^{-2nix}\bigg\|_{L_2(0,\pi)}=$$

$$=\bigg\|\frac{sign(\beta^2-\alpha^2)\alpha+\beta i}{2i}(e^{i\varphi x}-1)\times$$

$$\times\left(\sum_{-N,\varepsilon_n=1}^{N}a_n\omega_n e^{2nix}+\sum_{-N,\varepsilon_n=-1}^{N}a_n\omega_n e^{-2nix}\right)+$$

$$+\frac{-sign(\beta^2-\alpha^2)\alpha+\beta i}{2i}(e^{-i\varphi x}-1)\times$$

$$\times\left(\sum_{-N,\varepsilon_n=-1}^{N}a_n\omega_n e^{2nix}+\sum_{-N,\varepsilon_n=1}^{N}a_n\omega_n e^{-2nix}\right)\bigg\|_{L_2(0,\pi)}\leq$$

$$\leq\frac{\sqrt{\alpha^2+\beta^2}}{2}\cdot\max_{x\in[0,\pi]}\left|e^{i\varphi x}-1\right|\times$$

$$\times\left(\bigg\|\sum_{-N,\varepsilon_n=1}^{N}a_n\omega_n e^{2nix}+\sum_{-N,\varepsilon_n=-1}^{N}a_n\omega_n e^{-2nix}\bigg\|_{L_2(0,\pi)}+$$

$$+\bigg\|\sum_{-N,\varepsilon_n=-1}^{N}a_n\omega_n e^{2nix}+\sum_{-N,\varepsilon_n=1}^{N}a_n\omega_n e^{-2nix}\bigg\|_{L_2(0,\pi)}\right)=$$

$$=\frac{\sqrt{\alpha^2+\beta^2}}{2}\max_{x\in[0,\pi]}\left|e^{i\varphi x}-1\right|\left(\sqrt{\sum_{-N,\varepsilon_n=1}^{N}|a_n\omega_n|^2+\sum_{-N,\varepsilon_n=-1}^{N}|a_n\omega_n|^2}+\right.$$

$$\left.+\sqrt{\sum_{-N,\varepsilon_n=-1}^{N}|a_n\omega_n|^2+\sum_{-N,\varepsilon_n=1}^{N}|a_n\omega_n|^2}\right)\cdot\sqrt{\pi}=$$

$$=\sqrt{\alpha^2+\beta^2}\cdot\max_{x\in[0,\pi]}\left|e^{i\varphi x}-1\right|\cdot\sqrt{\pi}\cdot\sqrt{\sum_{-N}^{N}|a_n\omega_n|^2}.$$

Thus, denoting $c_n=\frac{1}{\sqrt{1+(2n)^{2s}}}$, we obtain

$$\bigg\|\sum_{-N}^{N}a_n(y_n(x)-z_n(x))\bigg\|_{L_2(0,\pi)}\leq\sqrt{2}\cdot\max_{x\in[0,\pi]}\left|e^{i\varphi x}-1\right|\cdot\sqrt{\sum_{-N}^{N}|a_n\cdot c_n|^2}.$$

□

Lemma 2. *Let $\{a_n\}$ be a finite system of complex numbers. Then, inequalities*

$$\bigg\|D^s\sum_{-N}^{N}a_n(y_n(x)-z_n(x))\bigg\|_{L_2(0,\pi)}\leq$$

$$\leq\sqrt{2}\Big[\max_{x\in[0,\pi]}\left|e^{i\varphi x}-1\right|+(\varphi+1)^s-1\Big]\cdot\sqrt{\sum_{-N}^{N}|a_n\cdot c_n\cdot(2n)^s|^2}$$

are valid at $s = 1, 2, 3, \ldots$.

Proof. Denote
$$\theta = \sqrt{2} \cdot \max_{x \in [0,\pi]} \left| e^{i\varphi x} - 1 \right|,$$

since
$$\sum_{-N}^{N} a_n(y_n - z_n) = \frac{\text{sign}(\beta^2 - \alpha^2)\alpha + \beta i}{2i} \cdot (e^{i\varphi x} - 1) \cdot$$
$$\cdot \left(\sum_{-N,\varepsilon_n=1}^{N} a_n \cdot \omega_n \cdot e^{2nix} + \sum_{-N,\varepsilon_n=-1}^{N} a_n \cdot \omega_n \cdot e^{-2nix} \right) +$$
$$+ \frac{-\text{sign}(\beta^2 - \alpha^2)\alpha + \beta i}{2i} \cdot \left(e^{-i\varphi x} - 1 \right) \cdot$$
$$\cdot \left(\sum_{-N,\varepsilon_n=-1}^{N} a_n \cdot \omega_n \cdot e^{2nix} + \sum_{-N,\varepsilon_n=1}^{N} a_n \cdot \omega_n \cdot e^{-2nix} \right),$$

using properties of the norm, we have

$$\left\| D^s \sum_{-N}^{N} a_n(y_n - z_n) \right\|_{L_2(0,\pi)} \leq \frac{\sqrt{\alpha^2 + \beta^2}}{2} \cdot \left(\left\| \sum_{k=0}^{s} C_s^k \cdot D^k(e^{i\varphi x} - 1) \cdot \right. \right.$$
$$\left. \cdot D^{s-k} \left(\sum_{-N,\varepsilon_n=1}^{N} a_n \cdot \omega_n \cdot e^{2nix} + \sum_{-N,\varepsilon_n=-1}^{N} a_n \cdot \omega_n \cdot e^{-2nix} \right) \right\|_{L_2(0,\pi)} +$$
$$+ \left\| \sum_{k=0}^{s} C_s^k \cdot D^k(e^{-i\varphi x} - 1) \cdot \right.$$
$$\left. \left. \cdot D^{s-k} \left(\sum_{-N,\varepsilon_n=-1}^{N} a_n \cdot \omega_n \cdot e^{2nix} + \sum_{-N,\varepsilon_n=1}^{N} a_n \cdot \omega_n \cdot e^{-2nix} \right) \right\|_{L_2(0,\pi)} \right) \leq$$
$$\leq \frac{\sqrt{\alpha^2 + \beta^2}}{2} \cdot \left(\max_{x \in [0,\pi]} \left| e^{i\varphi x} - 1 \right| \times \right.$$
$$\times \left\| \sum_{-N,\varepsilon_n=1}^{N} a_n \cdot \omega_n \cdot (2n)^s e^{2nix} + \sum_{-N,\varepsilon_n=-1}^{N} a_n \cdot \omega_n \cdot (-2n)^s e^{-2nix} \right\|_{L_2(0,\pi)} +$$
$$+ \sum_{k=1}^{s} C_s^k \cdot \varphi^k \cdot \left\| \sum_{-N,\varepsilon_n=1}^{N} a_n \cdot \omega_n \cdot (2n)^{s-k} e^{2nix} + \right.$$
$$\left. + \sum_{-N,\varepsilon_n=-1}^{N} a_n \cdot \omega_n \cdot (-2n)^{s-k} e^{-2nix} \right\|_{L_2(0,\pi)} + \max_{x \in [0,\pi]} \left| e^{i\varphi x} - 1 \right| \times$$
$$\times \left\| \sum_{-N,\varepsilon_n=-1}^{N} a_n \cdot \omega_n \cdot (2n)^s \cdot e^{2nix} + \sum_{-N,\varepsilon_n=1}^{N} a_n \cdot \omega_n \cdot (-2n)^s \cdot e^{-2nix} \right\|_{L_2(0,\pi)} +$$
$$+ \sum_{k=1}^{s} C_s^k \cdot \varphi^k \cdot \left\| \sum_{-N,\varepsilon_n=-1}^{N} a_n \cdot \omega_n \cdot (2n)^{s-k} \cdot e^{2nix} + \right.$$
$$\left. \left. + \sum_{-N,\varepsilon_n=1}^{N} a_n \cdot \omega_n \cdot (-2n)^{s-k} \cdot e^{-2nix} \right\|_{L_2(0,\pi)} \right) \leq \sqrt{\alpha^2 + \beta^2} \times$$

$$\times \left(\max_{x \in [0,\pi]} \left| e^{i\varphi x} - 1 \right| \sqrt{\sum_{-N}^{N} |a_n \omega_n (2n)^s|^2} + \sum_{k=1}^{s} C_s^k \varphi^k \sqrt{\sum_{-N}^{N} |a_n \omega_n (2n)^{s-k}|^2} \right) \times$$

$$\times \sqrt{\pi} = \sqrt{2} \left(\max_{x \in [0,\pi]} \left| e^{i\varphi x} - 1 \right| \cdot \sqrt{\sum_{-N}^{N} |a_n \cdot c_n \cdot (2n)^s|^2} + \right.$$

$$\left. + \sum_{k=1}^{s} C_s^k \cdot \varphi^k \cdot \sqrt{\sum_{-N}^{N} |a_n \cdot c_n \cdot (2n)^{s-k}|^2} \right) \leq$$

$$\leq \sqrt{2} \left[\max_{x \in [0,\pi]} \left| e^{i\varphi x} - 1 \right| + (\varphi + 1)^s - 1 \right] \cdot \sqrt{\sum_{-N}^{N} |a_n \cdot c_n \cdot (2n)^s|^2}.$$

Thus, inequalities

$$\left\| D^s \sum_{-N}^{N} a_n (y_n(x) - z_n(x)) \right\|_{L_2(0,\pi)} \leq$$

$$\leq \sqrt{2} \left[\max_{x \in [0,\pi]} \left| e^{i\varphi x} - 1 \right| + (\varphi + 1)^s - 1 \right] \cdot \sqrt{\sum_{-N}^{N} |a_n \cdot c_n \cdot (2n)^s|^2}$$

hold at $s = 1, 2, 3, \ldots$. □

Using Lemmas 1 and 2, we obtain

Lemma 3. *Let $\{a_n\}$ be a finite system of complex numbers. Then the following inequality*

$$\left\| \sum_{-N}^{N} a_n (y_n(x) - z_n(x)) \right\|_{W_2^s(0,\pi)} \leq$$

$$\leq \sqrt{\theta^2 + 2 \left(\frac{\theta}{\sqrt{2}} + (\varphi + 1)^s - 1 \right)^2 \cdot \sigma(s)} \cdot \sqrt{\sum_{-N}^{N} |a_n|^2}$$

is valid where $\sigma(0) = \frac{1}{\sqrt{2}}$, $\sigma(s) = 1$ *at* $s > 0$.

Lemma 4. *Let $\alpha \neq 0$, $\beta \neq 0$, $|\alpha| \neq |\beta|$ be real numbers, and*

$$\rho = \sqrt{\theta^2 + 2 \left(\frac{\theta}{\sqrt{2}} + (\varphi + 1)^s - 1 \right)^2 \cdot \sigma(s)} < 1$$

where $\sigma(0) = \frac{1}{\sqrt{2}}$, $\sigma(s) = 1$ *at* $s > 0$, $\theta = \sqrt{2} \cdot \max_{x \in [0,\pi]} \left| e^{i\varphi x} - 1 \right|$, $\lambda_n = 2n + \varepsilon_n \cdot \varphi$, $\varphi = \frac{1}{\pi} \arccos \frac{-2\alpha\beta}{\alpha^2 + \beta^2}$, $\varepsilon_n = \varepsilon_{-n} = \pm 1$ *at* $n \in Z$.

Then, eigenfunction system

$$y_n(x) = \sqrt{\frac{2}{\pi}} \cdot \frac{\beta \cos \lambda_n x + \varepsilon_n \cdot \text{sign}(\beta^2 - \alpha^2) \cdot \alpha \sin \lambda_n x}{\sqrt{\alpha^2 + \beta^2} \cdot \sqrt{1 + (2n)^{2s}}}, \quad n \in Z,$$

of spectral Problems (9) and (10) forms the Riesz basis in the space $W_2^s(0, \pi)$.

Proof. Vector system

$$z_n(x) = \sqrt{\frac{2}{\pi}} \cdot \frac{\beta \cos 2nx + \varepsilon_n \cdot sign(\beta^2 - \alpha^2) \cdot \alpha \sin 2nx}{\sqrt{\alpha^2 + \beta^2} \cdot \sqrt{1 + (2n)^{2s}}}, n \in Z$$

forms the complete orthonormal system in Hilbert space $W_2^s(0, \pi)$,, and vector system

$$y_n(x) = \sqrt{\frac{2}{\pi}} \cdot \frac{\beta \cos \lambda_n x + \varepsilon_n \cdot sign(\beta^2 - \alpha^2) \cdot \alpha \sin \lambda_n x}{\sqrt{\alpha^2 + \beta^2} \cdot \sqrt{1 + (2n)^{2s}}}, n \in Z$$

by virtue of Lemma 3 satisfying the theorem conditions by R. Paley and N. Wiener (see p. 224, [39]). This theorem implies that system of vectors $\{y_n(x)\}_{n\in Z}$ forms the Riesz basis in space $W_2^s(0, \pi)$. □

Lemma 5. *Operator*

$$Ly = -y''$$

with domain

$$D(L) = \{y(x) : y(x) \in C^2(0, \pi) \cap C^1[0, \pi], y'' \in L_2(0, \pi),$$
$$\alpha y(0) + \beta y(\pi) = 0, \beta y'(0) + \alpha y'(0) = 0\}$$

is a symmetric operator in class $L_2(0, \pi)$.

Proof. Indeed, since functions f and $\overline{g}$ belong to domain $D(L)$, we have $Lf \in L_2(0, \pi)$, $L\overline{g} = \overline{Lg} \in L_2(0, \pi)$, and the second Green formula

$$\int_G (Lu \cdot v - u \cdot Lv) dx = -\int_{\partial G} \left(\frac{\partial u}{\partial n} \cdot v - u \cdot \frac{\partial v}{\partial n} \right) ds$$

at $u = f$ and $v = \overline{g}$ takes the form

$$\int_0^\pi (Lf \cdot \overline{g} - f \cdot \overline{Lg}) dx = - \left(f'(x)\overline{g(x)} - f(x)\overline{g'(x)} \right) \Big|_0^\pi.$$

Further, functions f and $\overline{g}$ satisfy the boundary conditions:

$$\alpha f(0) + \beta f(\pi) = 0, \ \beta f'(0) + \alpha f'(\pi) = 0, \ \alpha \overline{g(0)} + \beta \overline{g(\pi)} = 0, \ \beta \overline{g'(0)} + \alpha \overline{g'(\pi)} = 0.$$

By assumption, $\alpha \neq 0, \beta \neq 0$. Therefore,

$$f(0) \cdot \overline{g(\pi)} - f(\pi) \cdot \overline{g(0)} = 0$$

and

$$f'(0) \cdot \overline{g'(\pi)} - f'(\pi) \cdot \overline{g'(0)} = 0,$$

i.e., $f(0) \cdot \overline{g(\pi)} = f(\pi) \cdot \overline{g(0)}$ and $f'(0) \cdot \overline{g'(\pi)} = f'(\pi) \cdot \overline{g'(0)}$. For here, we obtain

$$\frac{f(\pi)}{f(0)} = \frac{\overline{g(\pi)}}{\overline{g(0)}} = k_0 = -\frac{\alpha}{\beta}$$

and

$$\frac{f'(\pi)}{f'(0)} = \frac{\overline{g'(\pi)}}{\overline{g'(0)}} = k_1 = -\frac{\beta}{\alpha}, \ k_0 \cdot k_1 = 1.$$

So, $f(\pi) = k_0 f(0), \overline{g(\pi)} = k_0 \overline{g(0)}$ è $f'(\pi) = k_1 f'(0), \overline{g'(\pi)} = k_1 \overline{g'(0)}$. Thus,

$$\int_0^\pi (Lf \cdot \overline{g} - f \cdot \overline{Lg})\, dx = -\left(f'(x) \cdot \overline{g(x)} - f(x) \cdot \overline{g'(x)}\right)\Big|_0^\pi =$$

$$= -\left(f'(\pi) \cdot \overline{g(\pi)} - f(\pi) \cdot \overline{g'(\pi)}\right) + \left(f'(0) \cdot \overline{g(0)} - f(0) \cdot \overline{g'(0)}\right) =$$

$$= -\left(f'(0) \cdot \overline{g(0)} - f(0) \cdot \overline{g'(0)}\right) + \left(f'(0) \cdot \overline{g(0)} - f(0) \cdot \overline{g'(0)}\right) = 0.$$

Thereby, $(Lf, g) = (f, Lg), \forall f, g \in D(L)$. □

Theorem 1. *Let $\alpha \neq 0, \beta \neq 0, |\alpha| \neq |\beta|$ be real number, and*

$$\rho = \sqrt{\theta^2 + 2(\frac{\theta}{\sqrt{2}} + (\varphi + 1)^s - 1)^2} \cdot \sigma(s) < 1$$

where $\sigma(0) = \dfrac{1}{\sqrt{2}}, \sigma(s) = 1$ at $s > 0, \theta = \sqrt{2} \cdot \max\limits_{x \in [0, \pi]} |e^{i\varphi x} - 1|, \lambda_n = 2n + \varepsilon_n \cdot \varphi, \varphi = \dfrac{1}{\pi} \arccos \dfrac{-2\alpha\beta}{\alpha^2 + \beta^2}$, $\varepsilon_n = \varepsilon_{-n} = \pm 1$ at $n \in Z$. Then the system of eigenfunctions

$$\bar{y}_n(x) = \sqrt{\frac{2}{\pi}} \cdot \frac{\beta \cos \lambda_n x + \varepsilon_n \cdot \text{sign}(\beta^2 - \alpha^2) \cdot \alpha \sin \lambda_n x}{\sqrt{\alpha^2 + \beta^2} \cdot \sqrt{1 + |\lambda_n|^{2s}}}, \quad n \in Z,$$

of spectral Problems (9) and (10) form the complete orthonormal system in Sobolev classes $W_2^s(0, \pi)$.

Proof. Symmetry of operator L implies that eigenfunctions $\{\bar{y}_n(x)\}_{n \in Z}$ of operator L, corresponding to the different eigenvalues, are orthogonal in classes $L_2(0, \pi)$.

System of functions $\{D^\alpha \bar{y}_n(x)\}_{n \in Z}$ is also the system of eigenfunctions of a similar operator corresponding to different eigenvalues, which implies that functions of system $\{D^\alpha \bar{y}_n(x)\}_{n \in Z}$ are orthogonal in classes $L_2(0, \pi)$.

As a result, we see that system of eigenfunctions $\{\bar{y}_n(x)\}_{n \in Z}$ of operator L, corresponding to different eigenvalues, are orthogonal in the Sobolev classes $W_2^s(0, \pi)$. It is known that, if a sequence of vectors $\{\psi_n(x)\}_{n \in Z}$ forms the Riesz basis in a Hilbert space H, then system of vectors

$$\{\widehat{\psi}_n(x)\}_{n \in Z} \left(\widehat{\psi}_n(x) = \frac{\psi_n(x)}{\|\psi_n(x)\|}, n \in Z\right)$$

also forms the Riesz basis in H (see p. 374, [42]).

By virtue of Lemma 4, system of eigenvectors $\{y_n(x)\}_{n \in Z}$ forms the Riesz basis in space $W_2^s(0, \pi)$. The orthogonality of this system implies that $\{\bar{y}_n(x)\}_{n \in Z}$ is a complete orthonormal system in the Sobolev classes $W_2^s(0, \pi)$. □

Theorem 1 and the Sobolev embedding theorem imply the following corollaries.

Corollary 1. *Let $\alpha \neq 0, \beta \neq 0, |\alpha| \neq |\beta|$ be real numbers, and*

$$\rho = \sqrt{\theta^2 + 2\left(\frac{\theta}{\sqrt{2}} + \varphi\right)^2} < 1$$

where $\theta = \sqrt{2} \cdot \max_{x \in [0,\pi]} |e^{i\varphi x} - 1|$, $\lambda_n = 2n + \varepsilon_n \cdot \varphi$, $\varphi = \frac{1}{\pi} \arccos \frac{-2\alpha\beta}{\alpha^2 + \beta^2}$, $\varepsilon_n = \varepsilon_{-n} = \pm 1$ at $n \in Z$. Then, the Fourier series for function $f(x) \in W_2^1(0, \pi) \cap C[0, \pi]$ in orthonormal eigenfunctions

$$\bar{y}_n(x) = \sqrt{\frac{2}{\pi}} \cdot \frac{\beta \cos \lambda_n x + \varepsilon_n \cdot \operatorname{sign}(\beta^2 - \alpha^2) \cdot \alpha \sin \lambda_n x}{\sqrt{\alpha^2 + \beta^2} \cdot \sqrt{1 + |\lambda_n|^2}}, \quad n \in Z$$

of spectral Problems (9) and (10) uniformly converges on segment $[0, \pi]$ to function $f(x)$.

Corollary 2. Let $\alpha \neq 0$, $\beta \neq 0$, $|\alpha| \neq |\beta|$ be real numbers, and

$$\rho = \sqrt{\theta^2 + 2\left(\frac{\theta}{\sqrt{2}} + (\varphi + 1)^s - 1\right)^2} < 1$$

where $s > k$, $\theta = \sqrt{2} \cdot \max_{x \in [0,\pi]} |e^{i\varphi x} - 1|$, $\lambda_n = 2n + \varepsilon_n \cdot \varphi$, $\varphi = \frac{1}{\pi} \arccos \frac{-2\alpha\beta}{\alpha^2 + \beta^2}$, $\varepsilon_n = \varepsilon_{-n} = \pm 1$ at $n \in Z$. Then the Fourier series for function $f(x) \in W_2^s(0, \pi) \cap C^k[0, \pi]$ in orthonormal eigenfunctions

$$\bar{y}_n(x) = \sqrt{\frac{2}{\pi}} \cdot \frac{\beta \cos \lambda_n x + \varepsilon_n \cdot \operatorname{sign}(\beta^2 - \alpha^2) \cdot \alpha \sin \lambda_n x}{\sqrt{\alpha^2 + \beta^2} \cdot \sqrt{1 + |\lambda_n|^2}}, \quad n \in Z,$$

of spectral Problems (9) and (10) converges in the norm of space $C^k[0, \pi]$ to function $f(x)$.

The scalar product in space $W_2^{s_1,s_2}((0, \pi) \times (0, \pi))$ is introduced in the following way:

$$(f(x,y), g(x,y))_{W_2^{s_1,s_2}((0,\pi)\times(0,\pi))} = (f(x,y), g(x,y))_{L_2((0,\pi)\times(0,\pi))} +$$

$$+ (D_x^{s_1} f(x,y), D_x^{s_1} g(x,y))_{L_2((0,\pi)\times(0,\pi))} + (D_y^{s_2} f(x,y), D_y^{s_2} g(x,y))_{L_2((0,\pi)\times(0,\pi))} +$$

$$+ (D_{x,y}^{s_1,s_2} f(x,y), D_{x,y}^{s_1,s_2} g(x,y))_{L_2((0,\pi)\times(0,\pi))}.$$

Respectively, the norm in this space is introduced as follows:

$$\|f(x,y)\|^2_{W_2^{s_1,s_2}((0,\pi)\times(0,\pi))} =$$

$$= \|f(x,y)\|^2_{L_2((0,\pi)\times(0,\pi))} + \|D_x^{s_1} f(x,y)\|^2_{L_2((0,\pi)\times(0,\pi))} +$$

$$+ \|D_y^{s_2} f(x,y)\|^2_{L_2((0,\pi)\times(0,\pi))} + \|D_{x,y}^{s_1,s_2} f(x,y)\|^2_{L_2((0,\pi)\times(0,\pi))}.$$

Lemma 6. If $\{\psi_m^{(1)}(x)\}$ and $\{\psi_n^{(2)}(y)\}$ are complete orthonormal systems in $W_2^{s_1}(0, \pi)$ and $W_2^{s_2}(0, \pi)$, respectively, then the system of all products

$$f_{mn}(x,y) = \psi_m^{(1)}(x) \cdot \psi_n^{(2)}(y)$$

is a complete orthonormal system in $W_2^{s_1,s_2}((0, \pi) \times (0, \pi))$, where $s_1, s_2 = 1, 2, 3, \ldots$ and $x, y \in (0, \pi)$

Proof. By virtue of the Fubini theorem,

$$\|f_{mn}(x,y)\|^2_{W_2^{s_1,s_2}((0,\pi)\times(0,\pi))} = \|\psi_m^{(1)}(x)\|^2_{L_2(0,\pi)} \cdot \|\psi_n^{(2)}(y)\|^2_{L_2(0,\pi)} +$$

$$+\left\|D_x^{s_1}\psi_m^{(1)}(x)\right\|_{L_2(0,\pi)}^2 \cdot \left\|\psi_n^{(2)}(y)\right\|_{L_2(0,\pi)}^2 + \left\|\psi_m^{(1)}(x)\right\|_{L_2(0,\pi)}^2 \cdot \left\|D_y^{s_2}\psi_n^{(2)}(y)\right\|_{L_2(0,\pi)}^2 +$$

$$+\left\|D_x^{s_1}\psi_m^{(1)}(x)\right\|_{L_2(0,\pi)}^2 \cdot \left\|D_y^{s_2}\psi_n^{(2)}(y)\right\|_{L_2(0,\pi)}^2 =$$

$$= \left(\left\|\psi_m^{(1)}(x)\right\|_{L_2(0,\pi)}^2 + \left\|D_x^{s_1}\psi_m^{(1)}(x)\right\|_{L_2(0,\pi)}^2\right) \cdot \left\|\psi_n^{(2)}(y)\right\|_{L_2(0,\pi)}^2 +$$

$$+ \left(\left\|\psi_m^{(1)}(x)\right\|_{L_2(0,\pi)}^2 + \left\|D_x^{s_1}\psi_m^{(1)}(x)\right\|_{L_2(0,\pi)}^2\right) \cdot \left\|D_y^{s_2}\psi_n^{(2)}(y)\right\|_{L_2(0,\pi)}^2 =$$

$$= \left(\left\|\psi_m^{(1)}(x)\right\|_{L_2(0,\pi)}^2 + \left\|D_x^{s_1}\psi_m^{(1)}(x)\right\|_{L_2(0,\pi)}^2\right) \cdot$$

$$\cdot \left(\left\|\psi_n^{(2)}(y)\right\|_{L_2(0,\pi)}^2 + \left\|D_y^{s_2}\psi_n^{(2)}(y)\right\|_{L_2(0,\pi)}^2\right) = 1.$$

If $m \neq m_1$ or $n \neq n_1$, by the same theorem

$$(f_{mn}(x,y), f_{m_1 n_1}(x,y))_{W_2^{s_1,s_2}((0,\pi)\times(0,\pi))} =$$

$$= (f_{mn}(x,y), f_{m_1 n_1}(x,y))_{L_2((0,\pi)\times(0,\pi))} +$$

$$+ (D_x^{s_1} f_{mn}(x,y), D_x^{s_1} f_{m_1 n_1}(x,y))_{L_2((0,\pi)\times(0,\pi))} +$$

$$+ (D_y^{s_2} f_{mn}(x,y), D_y^{s_2} f_{m_1 n_1}(x,y))_{L_2((0,\pi)\times(0,\pi))} +$$

$$+ (D_{x,y}^{s_1,s_2} f_{mn}(x,y), D_{x,y}^{s_1,s_2} f_{m_1 n_1}(x,y))_{L_2((0,\pi)\times(0,\pi))} =$$

$$= (\psi_m^{(1)}(x), \psi_{m_1}^{(1)}(x))_{L_2(0,\pi)} \cdot (\psi_n^{(2)}(y), \psi_{n_1}^{(2)}(y))_{L_2(0,\pi)} +$$

$$+ (D_x^{s_1}\psi_m^{(1)}(x), D_x^{s_1}\psi_{m_1}^{(1)}(x))_{L_2(0,\pi)} \cdot (\psi_n^{(2)}(y), \psi_{n_1}^{(2)}(y))_{L_2(0,\pi)} +$$

$$+ (\psi_m^{(1)}(x), \psi_{m_1}^{(1)}(x))_{L_2(0,\pi)} \cdot (D_y^{s_2}\psi_n^{(2)}(y), D_y^{s_2}\psi_{n_1}^{(2)}(y))_{L_2(0,\pi)} +$$

$$+ (D_x^{s_1}\psi_m^{(1)}(x), D_x^{s_1}\psi_{m_1}^{(1)}(x))_{L_2(0,\pi)} \cdot (D_y^{s_2}\psi_n^{(2)}(y), D_y^{s_2}\psi_{n_1}^{(2)}(y))_{L_2(0,\pi)} =$$

$$= ((\psi_m^{(1)}(x), \psi_{m_1}^{(1)}(x))_{L_2(0,\pi)} + (D_x^{s_1}\psi_m^{(1)}(x), D_x^{s_1}\psi_{m_1}^{(1)}(x))_{L_2(0,\pi)}) \cdot$$

$$\cdot (\psi_n^{(2)}(y), \psi_{n_1}^{(2)}(y))_{L_2(0,\pi)} +$$

$$+ ((\psi_m^{(1)}(x), \psi_{m_1}^{(1)}(x))_{L_2(0,\pi)} + (D_x^{s_1}\psi_m^{(1)}(x), D_x^{s_1}\psi_{m_1}^{(1)}(x))_{L_2(0,\pi)}) \cdot$$

$$\cdot (D_y^{s_2}\psi_n^{(2)}(y), D_y^{s_2}\psi_{n_1}^{(2)}(y))_{L_2(0,\pi)} =$$

$$= ((\psi_m^{(1)}(x), \psi_{m_1}^{(1)}(x))_{L_2(0,\pi)} + (D_x^{s_1}\psi_m^{(1)}(x), D_x^{s_1}\psi_{m_1}^{(1)}(x))_{L_2(0,\pi)}) \cdot$$

$$\cdot ((\psi_n^{(2)}(y), \psi_{n_1}^{(2)}(y))_{L_2(0,\pi)} + (D_y^{s_2}\psi_n^{(2)}(y), D_y^{s_2}\psi_{n_1}^{(2)}(y))_{L_2(0,\pi)}) = 0$$

since scalar product $(f_{mn}(x,y), f_{m_1 n_1}(x,y))_{W_2^{s_1,s_2}((0,\pi)\times(0,\pi))}$ of two variables exist on $\Pi = (0,\pi) \times (0,\pi)$. Let us prove the completeness of system $\{f_{mn}(x,y)\}$. Assume that there exists a function $f(x,y)$ in $W_2^{s_1,s_2}((0,\pi) \times (0,\pi))$ that is orthogonal to all functions $f_{mn}(x,y)$. Set

$$F_m(y) = (f(x,y), \psi_m^{(1)}(x))_{W_2^{s_1}(0,\pi)}.$$

It is easy to see, that function $F_m(y)$ belongs to class $W_2^{s_2}(0, \pi)$. That's why for any n, m again applying the Fubini theorem, we obtain

$$(F_m(y), \psi_n^{(2)}(y))_{W_2^{s_2}(0,\pi)} = (f(x,y), f_{mn}(x,y))_{W_2^{s_1,s_2}((0,\pi)\times(0,\pi))} = 0.$$

By completeness of system $\psi_n^{(2)}(y)$, for almost all y

$$F_m(y) = 0.$$

But then, for almost every y,, equalities

$$(f(x,y), \psi_m^{(1)}(x))_{W_2^{s_1}(0,\pi)} = 0$$

hold for all m. Completeness of system $\psi_m^{(1)}(x)$ implies that, for almost all y, the set of those x, for which

$$f(x,y) \neq 0,$$

has the measure zero. By virtue of the Fubini theorem, this means that, on $\Pi = (0, \pi) \times (0, \pi)$, function $f(x, y)$ is zero almost everywhere. □

The scalar product in space $W_2^{s_1, s_2, \ldots, s_N}(\Pi)$ is introduced in the following way:

$$(f(x), g(x))_{W_2^{s_1,s_2,\ldots,s_N}(\Pi)} = (f(x), g(x))_{L_2(\Pi)} +$$

$$+ \sum_{j_1=1}^{N} (D_{x_{j_1}}^{s_{j_1}} f(x), D_{x_{j_1}}^{s_{j_1}} g(x))_{L_2(\Pi)} +$$

$$+ \sum_{1 \leq j_1 < j_2 \leq N} (D_{x_{j_1}}^{s_{j_1}} D_{x_{j_2}}^{s_{j_2}} f(x), D_{x_{j_1}}^{s_{j_1}} D_{x_{j_2}}^{s_{j_2}} g(x))_{L_2(\Pi)} + \cdots +$$

$$+ \sum_{1 \leq j_1 < j_2 < \cdots < j_N \leq N} (D_{x_{j_1}}^{s_{j_1}} D_{x_{j_2}}^{s_{j_2}} \ldots D_{x_{j_N}}^{s_{j_N}} f(x), D_{x_{j_1}}^{s_{j_1}} D_{x_{j_2}}^{s_{j_2}} \ldots D_{x_{j_N}}^{s_{j_N}} g(x))_{L_2(\Pi)}.$$

Respectively, the norm in this space is introduced as follows:

$$\|f(x)\|^2_{W_2^{s_1,s_2,\ldots,s_N}(\Pi)} = \|f(x)\|^2_{L_2(\Pi)} + \sum_{j_1=1}^{N} \left\|D_{x_{j_1}}^{s_{j_1}} f(x)\right\|^2_{L_2(\Pi)} +$$

$$+ \sum_{1 \leq j_1 < j_2 \leq N} \left\|D_{x_{j_1}}^{s_{j_1}} D_{x_{j_2}}^{s_{j_2}} f(x)\right\|^2_{L_2(\Pi)} +$$

$$+ \cdots + \sum_{1 \leq j_1 < j_2 < \cdots < j_N \leq N} \left\|D_{x_{j_1}}^{s_{j_1}} D_{x_{j_2}}^{s_{j_2}} \ldots D_{x_{j_N}}^{s_{j_N}} f(x)\right\|^2_{L_2(\Pi)}.$$

Using the method of mathematical induction and Lemma 6, we obtain the following:

Lemma 7. *If $\{\psi_{m_1}^{(1)}(x_1)\}, \ldots, \{\psi_{m_N}^{(N)}(x_N)\}$ are complete orthonormal systems in spaces $W_2^{s_1}(0, \pi), \ldots, W_2^{s_N}(0, \pi)$, respectively, then system of all products*

$$f_m(x) = f_{m_1 \ldots m_N}(x_1, \ldots, x_N) = \psi_{m_1}^{(1)}(x_1) \cdot \ldots \cdot \psi_{m_N}^{(N)}(x_N)$$

is a complete orthonormal system in $W_2^{s_1, s_2, \ldots, s_N}(\Pi)$.

Let us apply Lemma 7 to our orthonormal systems. In space $W_2^{s_1,s_2,\ldots,s_N}(\Pi)$ of functions of N variables $f(x) = f(x_1,\ldots,x_N)$ all products

$$v_{m_1\ldots m_N}(x_1,\ldots,x_N) = \overline{y}_{m_1}^{(1)}(x_1) \cdot \ldots \cdot \overline{y}_{m_N}^{(N)}(x_N)$$

form the complete orthonormal system. Here,

$$\overline{y}_{m_j}^{(j)}(x_j) = \sqrt{\frac{2}{\pi}} \cdot \frac{\beta_j \cos \lambda_{m_j} x_j + \varepsilon_{m_j} \cdot \operatorname{sign}(\beta_j^2 - \alpha_j^2) \cdot \alpha_j \sin \lambda_{m_j} x_j}{\sqrt{\alpha_j^2 + \beta_j^2} \cdot \sqrt{1 + |\lambda_{m_j}|^{2s_j}}}, \quad m_j \in Z$$

at $1 \leq j \leq p$,

$$\overline{y}_{m_j}^{(j)}(x_j) = \frac{1}{\sqrt{\pi}} \frac{1}{\sqrt{1 + |2m_j|^{2s_j}}} \exp(i 2 m_j x_j), \quad m_j \in Z$$

at $p+1 \leq j \leq q$,

$$\overline{y}_{m_j}^{(j)}(x_j) = \sqrt{\frac{2}{\pi}} \frac{1}{\sqrt{1 + |m_j|^{2s_j}}} \sin(m_j x_j), \quad m_j \in N$$

at $q+1 \leq j \leq N$.

Thus, the following statement is valid:

Theorem 2. *Let $\alpha_j \neq 0$, $\beta_j \neq 0$, $|\alpha_j| \neq |\beta_j|$ be real numbers at every $1 \leq j \leq p$, and*

$$\rho = \max_{1 \leq j \leq p} \sqrt{\theta_j^2 + 2\left(\frac{\theta_j}{\sqrt{2}} + (\varphi_j + 1)^{s_j} - 1\right)^2} \cdot \sigma(s_j) < 1$$

where $\sigma(0) = \frac{1}{\sqrt{2}}$, $\sigma(s_j) = 1$, at $s_j > 0$, $\theta_j = \sqrt{2} \cdot \max_{x \in [0,\pi]} |e^{i\varphi_j x} - 1|$, $\lambda_{m_j} = 2m_j + \varepsilon_{m_j} \cdot \varphi_j$, $\varphi_j = \frac{1}{\pi} \arccos \frac{-2\alpha_j \beta_j}{\alpha_j^2 + \beta_j^2}$, $\varepsilon_{m_j} = \varepsilon_{-m_j} = \pm 1$ at $m_j \in Z$. Then, system of eigenfunctions

$$\{v_{m_1\ldots m_N}(x_1,\ldots,x_N)\}_{(m_1,\ldots,m_p) \in Z^p, (m_{p+1},\ldots,m_q) \in Z^{q-p}, (m_{q+1},\ldots,m_N) \in N^{N-q}} =$$

$$= \left\{\prod_{j=1}^{p} \sqrt{\frac{2}{\pi}} \frac{\beta_j \cos \lambda_{m_j} x_j + \varepsilon_{m_j} \operatorname{sign}(\beta_j^2 - \alpha_j^2) \cdot \alpha_j \sin \lambda_{m_j} x_j}{\sqrt{\alpha_j^2 + \beta_j^2} \cdot \sqrt{1 + |\lambda_{m_j}|^{2s_j}}}\right\}_{(m_1,\ldots,m_p) \in Z^p} \times$$

$$\times \left\{\prod_{j=p+1}^{q} \frac{1}{\sqrt{\pi}} \frac{1}{\sqrt{1 + |2m_j|^{2s_j}}} \exp(i 2 m_j x_j)\right\}_{(m_{p+1},\ldots,m_q) \in Z^{q-p}} \times$$

$$\times \left\{\prod_{j=q+1}^{N} \sqrt{\frac{2}{\pi}} \frac{1}{\sqrt{1 + |m_j|^{2s_j}}} \sin(m_j x_j)\right\}_{(m_{q+1},\ldots,m_N) \in N^{N-q}}$$

of spectral Problems (7) and (8) forms the complete orthonormal system in Sobolev classes $W_2^{s_1,s_2,\ldots,s_N}(\Pi)$.

Corollary 3. *Let $\alpha_j \neq 0$, $\beta_j \neq 0$, $|\alpha_j| \neq |\beta_j|$ be real numbers at every $1 \leq j \leq p$, and*

$$\rho = \max_{1 \leq j \leq p} \sqrt{\theta_j^2 + 2\left(\frac{\theta_j}{\sqrt{2}} + (\varphi_j + 1)^{s_j} - 1\right)^2} \cdot \sigma(s_j) < 1$$

where $\sigma(0) = \frac{1}{\sqrt{2}}$, $\sigma(s_j) = 1$ at $s_j > 0$, $\theta_j = \sqrt{2} \cdot \max_{x \in [0,\pi]} |e^{i\varphi_j x} - 1|$, $\lambda_{m_j} = 2m_j + \varepsilon_{m_j} \cdot \varphi_j$, $\varphi_j = \frac{1}{\pi} \arccos \frac{-2\alpha_j \beta_j}{\alpha_j^2 + \beta_j^2}$, $\varepsilon_{m_j} = \varepsilon_{-m_j} = \pm 1$ at $m_j \in Z$, $s_j > k + \frac{N}{2}$, $k \geq 0$, $k \in Z$. Then, the Fourier series for function $f(x) \in W_2^{s_1,s_2,\ldots,s_N}(\Pi) \cap C^k(\Pi)$ in orthonormal eigenfunctions

$$\{v_{m_1 \ldots m_N}(x_1, \ldots, x_N)\}_{(m_1,\ldots,m_p) \in Z^p, (m_{p+1},\ldots,m_q) \in Z^{q-p}, (m_{q+1},\ldots,m_N) \in N^{N-q}} =$$

$$= \left\{ \prod_{j=1}^{p} \sqrt{\frac{2}{\pi}} \frac{\beta_j \cos \lambda_{m_j} x_j + \varepsilon_{m_j} \operatorname{sign}(\beta_j^2 - \alpha_j^2) \cdot \alpha_j \sin \lambda_{m_j} x_j}{\sqrt{\alpha_j^2 + \beta_j^2} \cdot \sqrt{1 + |\lambda_{m_j}|^{2s_j}}} \right\}_{(m_1,\ldots,m_p) \in Z^p} \times$$

$$\times \left\{ \prod_{j=p+1}^{q} \frac{1}{\sqrt{\pi}} \frac{1}{\sqrt{1 + |2m_j|^{2s_j}}} \exp(i 2 m_j x_j) \right\}_{(m_{p+1},\ldots,m_q) \in Z^{q-p}} \times$$

$$\times \left\{ \prod_{j=q+1}^{N} \sqrt{\frac{2}{\pi}} \frac{1}{\sqrt{1 + |m_j|^{2s_j}}} \sin(m_j x_j) \right\}_{(m_{q+1},\ldots,m_N) \in N^{N-q}}$$

of spectral Problems (7) and (8) converges in the norm of space $C^k(\Pi)$ to function $f(x)$.

The proof of Corollary 3 is carried out using Theorem 2 and the Sobolev embedding theorem. The following are true:

4. Main Results

In this section, we give the most general case of the works done in [17].

Theorem 3. Let $\alpha_j \neq 0$, $\beta_j \neq 0$, $|\alpha_j| \neq |\beta_j|$ be real numbers at every $1 \leq j \leq p$, and

$$\rho = \max_{1 \leq j \leq p} \sqrt{\theta_j^2 + 2(\frac{\theta_j}{\sqrt{2}} + (\varphi_j + 1)^{s_j} - 1)^2} \cdot \sigma(s_j) < 1$$

where $\sigma(0) = \frac{1}{\sqrt{2}}$, $\sigma(s_j) = 1$ at $s_j > 0$, $\theta_j = \sqrt{2} \cdot \max_{x \in [0,\pi]} |e^{i\varphi_j x} - 1|$, $\lambda_{m_j} = 2m_j + \varepsilon_{m_j} \cdot \varphi_j$, $\varphi_j = \frac{1}{\pi} \arccos \frac{-2\alpha_j \beta_j}{\alpha_j^2 + \beta_j^2}$, $\varepsilon_{m_j} = \varepsilon_{-m_j} = \pm 1$ at $m_j \in Z$, $s_j > k + \frac{N}{2}$, $k \geq 0$, $k \in Z$ and $\varphi_j(x) \in W_2^{s_1+j-\frac{N}{2}, s_2+j-\frac{N}{2},\ldots,s_N+j-\frac{N}{2}}(\Pi)$, $f(x,t) \in W_2^{s_1,s_2,\ldots,s_N,s_{N+1}}(\Pi \times (0,+\infty))$. Then, the solution of problems (4)–(6) exists, it is unique, and is represented in the form of series

$$u(x,t) = \sum_{m_1=-\infty}^{\infty} \cdots \sum_{m_q=-\infty}^{\infty} \sum_{m_{q+1}=1}^{\infty} \cdots \sum_{m_N=1}^{\infty} \sum_{j=1}^{n} \varphi_{j,(m_1\ldots m_N)} t^{\alpha-j} E_{\alpha,\alpha-j+1}(-\mu_{m_1\ldots m_N} \cdot t^\alpha) +$$

$$+ \int_0^t (t-\tau)^{\alpha-1} \cdot E_{\alpha,\alpha}[-\mu_{m_1\ldots m_N}(t-\tau)^\alpha] f_{m_1\ldots m_N}(\tau) d\tau \cdot v_{m_1\ldots m_N}(x_1,\ldots,x_N) \quad (11)$$

where coefficients are determined in the following way:

$$E_{\alpha,\alpha-j+1}(-\mu_{m_1\ldots m_N} \cdot t^\alpha) = \sum_{i=0}^{\infty} \frac{(-\mu_{m_1\ldots m_N} \cdot t^\alpha)^i}{\Gamma(\alpha i + \alpha - j + 1)},$$

$$E_{\alpha,\alpha}\left(-\mu_{m_1...m_N}\cdot(t-\tau)^\alpha\right) = \sum_{i=1}^{\infty}\frac{(-\mu_{m_1...m_N})^{i-1}\cdot(t-\tau)^{\alpha(i-1)}}{\Gamma(\alpha\cdot i)},$$

$$f(x,t) = \sum_{m_1=-\infty}^{\infty}\cdots\sum_{m_q=-\infty}^{\infty}\sum_{m_{q+1}=1}^{\infty}\cdots\sum_{m_N=1}^{\infty} f_{m_1...m_N}(t)\cdot v_{m_1...m_N}(x_1,\ldots,x_N),$$

$$\varphi_j(x) = \sum_{m_1=-\infty}^{\infty}\cdots\sum_{m_q=-\infty}^{\infty}\sum_{m_{q+1}=1}^{\infty}\cdots\sum_{m_N=1}^{\infty}\varphi_{j,(m_1...m_N)}\cdot v_{m_1...m_N}(x_1,\ldots,x_N),$$

$$j=1,2,\ldots,n,\ \mu_{m_1...m_N} = \lambda_{m_1}^2+\cdots+\lambda_{m_N}^2.$$

Proof. Since system of eigenfunctions

$$\{v_{m_1...m_N}(x_1,\ldots,x_N)\}_{(m_1,\ldots,m_p)\in Z^p,(m_{p+1},\ldots,m_q)\in Z^{q-p},(m_{q+1},\ldots,m_N)\in N^{N-q}}$$

of spectral Problems (7) and (8) forms the complete orthonormal system in Sobolev classes $W_2^{s_1,s_2,\ldots,s_N}(\Pi)$, any function from class $W_2^{s_1,s_2,\ldots,s_N}(\Pi)$ can be represented as a convergent Fourier series in this system. For any $t>0$, expand solution $u(x,t)$ of Problems (4)–(6) into the Fourier series in eigenfunctions

$$\{v_{m_1...m_N}(x_1,\ldots,x_N)\}_{(m_1,\ldots,m_p)\in Z^p,(m_{p+1},\ldots,m_q)\in Z^{q-p},(m_{q+1},\ldots,m_N)\in N^{N-q}}$$

of spectral Problems (4) and (5):

$$u(x,t) = \sum_{m_1=-\infty}^{\infty}\cdots\sum_{m_q=-\infty}^{\infty}\sum_{m_{q+1}=1}^{\infty}\cdots\sum_{m_N=1}^{\infty} T_{m_1...m_N}(t)\cdot v_{m_1...m_N}(x), \quad (12)$$

$$T_{m_1...m_N}(t) = (u(x,t),v_{m_1...m_N}(x)).$$

By virtue of Problems (4) and (5), unknown functions $T_{m_1...m_N}(t)$ must satisfy equation

$$D_{0t}^\alpha T_{m_1...m_N}(t) + \mu_{m_1...m_N} T_{m_1...m_N}(t) = f_{m_1...m_N}(t),\ l-1<\alpha\leq l,\ l\in N \quad (13)$$

with initial conditions

$$\lim_{t\to 0} D_{0t}^{\alpha-k} T_{m_1...m_N}(t) = \varphi_{k,m_1...m_N},\ k=1,2,\ldots,l,\ \mu_{m_1...m_N} = \lambda_{m_1}^2+\cdots+\lambda_{m_N}^2. \quad (14)$$

The solution of Cauchy Problems (13) and (14) has the form

$$T_{m_1...m_N}(t) = \sum_{j=1}^{n}\varphi_{j,(m_1...m_N)} t^{\alpha-j} E_{\alpha,\alpha-j+1}(-\mu_{m_1...m_N}\cdot t^\alpha) +$$

$$+ \int_0^t (t-\tau)^{\alpha-1}\cdot E_{\alpha,\alpha}[-\mu_{m_1...m_N}(t-\tau)^\alpha] f_{m_1...m_N}(\tau)d\tau \quad (15)$$

where coefficients are determined as follows:

$$E_{\alpha,\alpha-j+1}(-\mu_{m_1...m_N}\cdot t^\alpha) = \sum_{i=0}^{\infty}\frac{(-\mu_{m_1...m_N}\cdot t^\alpha)^i}{\Gamma(\alpha i+\alpha-j+1)},$$

$$E_{\alpha,\alpha}\left(-\mu_{m_1...m_N}\cdot(t-\tau)^\alpha\right) = \sum_{i=1}^{\infty} \frac{(-\mu_{m_1...m_N})^{i-1}\cdot(t-\tau)^{\alpha(i-1)}}{\Gamma(\alpha\cdot i)},$$

$$f(x,t) = \sum_{m_1=-\infty}^{\infty}\cdots\sum_{m_q=-\infty}^{\infty}\sum_{m_{q+1}=1}^{\infty}\cdots\sum_{m_N=1}^{\infty} f_{m_1...m_N}(t)\cdot v_{m_1...m_N}(x_1,\ldots,x_N),$$

$$\varphi_j(x) = \sum_{m_1=-\infty}^{\infty}\cdots\sum_{m_q=-\infty}^{\infty}\sum_{m_{q+1}=1}^{\infty}\cdots\sum_{m_N=1}^{\infty} \varphi_{j,(m_1...m_N)}\cdot v_{m_1...m_N}(x_1,\ldots,x_N), j=1,2,\ldots,n.$$

After substituting Problem (15) into Problem (12), we obtain the unique solution of Problems (4)–(6) in the form of Series (8).

Let $\nu > 1$. Consider mixed Problems (4)–(6). If we look for a solution $u(x,t)$ to Problems (4)–(6) in the form of Fourier series expansion

$$u(x,t) = \sum_{m_1=-\infty}^{\infty}\cdots\sum_{m_q=-\infty}^{\infty}\sum_{m_{q+1}=1}^{\infty}\cdots\sum_{m_N=1}^{\infty} T_{m_1...m_N}(t)\cdot v_{m_1...m_N}(x),$$

where are $T_{m_1...m_N}(t) = (u(x,t), v_{m_1...m_N}(x))$ are the coefficients of the series, $\{v_{m_1...m_N}\}$ is the system of eigenfunctions of spectral Problems (7) and (8).

Differential operator $(-\Delta)^\nu$, generated by a differential expression $l^{(\nu)}(v(x)) = (-\Delta)^\nu v(x)$ with domain definition

$$D\left((-\Delta)^\nu\right) = \{v(x) : v(x) \in C^{2\nu}(\Pi) \cap C^{2\nu-1}(\overline{\Pi}), l^{(\nu)}(v(x)) \in L_2(\Pi)\}$$

satisfies Condition (8).

Similarly, as Lemma 5, it can be shown that operator $(-\Delta)^\nu$, is a symmetric and positive operator in space $L_2(\Pi)$. The eigenvalues of Problems (7) and (8) $\mu_{m_1...m_N} \geq 0$, and each $\mu_{\mu_{m_1...m_N}} = \left(\lambda_{m_1}^2 + \cdots \lambda_{m_1}^2\right)^\nu$ corresponds to an eigenvalue of Problems (9) and (10), and the eigenfunctions $\{v_{m_1...m_N}(x)\}$ of Problems (7) and (8) and eigenfunctions $\{y_{m_1...m_N}(x)\}$ of Problems (9) and (10) coincide, i.e.,

$$v_{m_1...m_N}(x) \equiv y_{m_1...m_N}(x).$$

□

Therefore, the following theorem is valid:

Theorem 4. Let $\alpha_j \neq 0, \beta_j \neq 0, |\alpha_j| \neq |\beta_j|$ be real numbers at every $1 \leq j \leq p$, and

$$\rho = \max_{1\leq j \leq p}\sqrt{\theta_j^2 + 2(\frac{\theta_j}{\sqrt{2}} + (\varphi_j+1)^{s_j} - 1)^2\cdot \sigma(s_j)} < 1$$

where $\sigma(0) = \frac{1}{\sqrt{2}}$, $\sigma(s_j) = 1$ at $s_j > 0$, $\theta_j = \sqrt{2}\cdot\max_{x\in[0,\pi]}|e^{i\varphi_j x}-1|$, $\lambda_{m_j} = 2m_j + \varepsilon_{m_j}\cdot\varphi_j$, $\varphi_j = \frac{1}{\pi}\arccos\frac{-2\alpha_j\beta_j}{\alpha_j^2+\beta_j^2}$, $\varepsilon_{m_j} = \varepsilon_{-m_j} = \pm 1$ at $m_j \in Z$, $s_j > (k+\frac{N}{2})\nu$, $k\geq 0$, $k\in Z$ and

$\varphi_j(x) \in W_2^{(s_1+j-\frac{N}{2})\nu,(s_2+j-\frac{N}{2})\nu,\ldots,(s_N+j-\frac{N}{2})\nu}(\Pi)$, $f(x,t) \in W_2^{s_1,s_2,\ldots,s_N,s_{N+1}}(\Pi \times (0,+\infty))$. Then the solution of Problems (4)–(6) exists, it is unique, and is represented in the form of series

$$u(x,t) = \sum_{m_1=-\infty}^{\infty} \cdots \sum_{m_q=-\infty}^{\infty} \sum_{m_{q+1}=1}^{\infty} \cdots \sum_{m_N=1}^{\infty} \sum_{j=1}^{n} \varphi_{j,(m_1\ldots m_N)} t^{\alpha-j} E_{\alpha,\alpha-j+1}(-\mu_{m_1\ldots m_N} \cdot t^\alpha) +$$

$$+ \int_0^t (t-\tau)^{\alpha-1} \cdot E_{\alpha,\alpha}[-\mu_{m_1\ldots m_N}(t-\tau)^\alpha] f_{m_1\ldots m_N}(\tau) d\tau \cdot v_{m_1\ldots m_N}(x_1,\ldots,x_N)$$

where coefficients are determined in the following way:

$$E_{\alpha,\alpha-j+1}(-\mu_{m_1\ldots m_N} \cdot t^\alpha) = \sum_{i=0}^{\infty} \frac{(-\mu_{m_1\ldots m_N} \cdot t^\alpha)^i}{\Gamma(\alpha i + \alpha - j + 1)},$$

$$E_{\alpha,\alpha}\left(-\mu_{m_1\ldots m_N} \cdot (t-\tau)^\alpha\right) = \sum_{i=1}^{\infty} \frac{(-\mu_{m_1\ldots m_N})^{i-1} \cdot (t-\tau)^{\alpha(i-1)}}{\Gamma(\alpha \cdot i)},$$

$$f(x,t) = \sum_{m_1=-\infty}^{\infty} \cdots \sum_{m_q=-\infty}^{\infty} \sum_{m_{q+1}=1}^{\infty} \cdots \sum_{m_N=1}^{\infty} f_{m_1\ldots m_N}(t) \cdot v_{m_1\ldots m_N}(x_1,\ldots,x_N),$$

$$\varphi_j(x) = \sum_{m_1=-\infty}^{\infty} \cdots \sum_{m_q=-\infty}^{\infty} \sum_{m_{q+1}=1}^{\infty} \cdots \sum_{m_N=1}^{\infty} \varphi_{j,(m_1\ldots m_N)} \cdot v_{m_1\ldots m_N}(x_1,\ldots,x_N), \quad j=1,2,\ldots,n,$$

$$\mu_{m_1\ldots m_N} = \left(\lambda_{m_1}^2 + \cdots + \lambda_{m_N}^2\right)^\nu.$$

5. Conclusions

In this paper, we considered questions on the unique solvability of a mixed problem for a partial differential equation of high order with fractional Riemann-Liouville derivatives with respect to time, and with Laplace operators with spatial variables and with nonlocal boundary conditions in Sobolev classes. The solution was found in the form of a series of expansions in eigenfunctions of the Laplace operator with nonlocal boundary conditions. Initial and boundary problems with fractional Riemann-Liouville derivatives with respect to time have many applications [13]. In connection to this, we chose the fractional Riemann-Liouville derivative, although we could consider other types of fractional derivatives.

Author Contributions: Methodology, O.A.İ.; Resources, S.G.K.; Writing—original draft, S.Q.O.; Writing—review editing, H.M.B.

Funding: This research received no external funding.

Acknowledgments: The authors gratefully thank the referees for their several suggestions and comments.

Conflicts of Interest: The authors declare no conflict of interest.

References

1. Birkhoff, G.D. On the asymptotic character of the solutions of certain linear differential equations containing a parameter. *Trans. Am. Math. Soc.* **1908**, *9*, 219–231. [CrossRef]
2. Birkhoff, G.D. Boundary value and expansion problems of ordinary linear differential equations. *Trans. Am. Math. Soc.* **1908**, *9*, 373–395. [CrossRef]
3. Birkhoff, G.D. Existence and oscillation theorem for a certain boundary value problem. *Trans. Am. Math. Soc.* **1909**, *10*, 259–270. [CrossRef]
4. Titchmarsh, E.C. *Eigenfunction Expansions*; Oxford University Press: Oxford, UK, 1953 and 1958; Volume I.
5. Titchmarsh, E.C. *Eigenfunction Expansions*; Oxford University Press: Oxford, UK, 1953 and 1958; Volume II.

6. Titchmarsh, E.C. On the asymptotic distribution of eigenvalues. *Q. J. Math.* **1954**, *5*, 228–240. [CrossRef]
7. Titchmarsh, E.C. On the eigenvalues in problems with spherical symmetry. *Proc. R. Soc. A* **1958**, *245*, 147–155.
8. Titchmarsh, E.C. On the eigenvalues in problems with spherical symmetry. II. *Proc. R. Soc. A* **1959**, *251*, 46–54.
9. Titchmarsh, E.C. On the eigenvalues in problems with spherical symmetry. III. *Proc. R. Soc. A* **1959**, *252*, 436–444.
10. Levitan, B.M. *Razlozenie po Sobstvennym Funkciyam Differencialnyh Uravnenii Vtorogo Poryadka*; Expansion in Characteristic Functions of Differential Equations of the Second Order; Gosudarstv. Izdat. Tehn.-Teor. Lit.: Leningrad, Moscow, Russia, 1950; p. 159. (In Russian)
11. Levitan, B.M. On expansion in eigenfunctions of the Laplace operator. *Doklady Akad. Nauk SSSR* **1954**, *35*, 267–316.
12. Levitan, B.M. On expansion in eigenfunctions of the Schrödinger operator in the case of a potential increasing without bound. *Dokl. Akad. Nauk SSSR* **1955**, *103*, 191–194
13. Nigmatullin, R.R. The realization of the generalized transfer equation in a medium with fractal geometry. *Phys. Status Solidi B* **1986**, *133*, 425–430. [CrossRef]
14. Kochubey, A.N. Cauchy problem for evolutionary equations of fractional order. *Differ. Equ.* **1989**, *25*, 1359–1368.
15. Kochubey, A.N. Fractional diffusion. *Differ. Equ.* **1990**, *26*, 660–670.
16. Samko, S.G.; Kilbas, A.A.; Marichev, O.I. *Integrals and Derivatives of Fractional Order and Some of Their Applications*; Science and Technology Publishing: Minsk, Belarus, 1987; p. 688
17. Kasimov, S.G.; Ataev, S.K. On solvability of the mixed problem for a partial equation of a fractional order with Laplace operators and nonlocal boundary conditions in the Sobolev classes. *Uzb. Math. J.* **2018**, *1*, 73–89. [CrossRef]
18. Srivastava, H.M.; El-Sayed, A.M.A.; Gaafar, F.M. A class of nonlinear boundary value problems for an arbitrary fractional-order differential equation with the Riemann-Stieltjes functional integral and infinite-point boundary conditions. *Symmetry* **2018**, *10*, 508. [CrossRef]
19. Jiang, J.; Feng, Y.; Li, S. Exact solutions to the fractional differential equations with mixed partial derivatives. *Axioms* **2018**, *7*, 10. [CrossRef]
20. Asawasamrit, S.; Ntouyas, S.K.; Tariboon, J.; Nithiarayaphaks, W. Coupled systems of sequential caputo and hadamard fractional differential equations with coupled separated boundary conditions. *Symmetry* **2018**, *10*, 701. [CrossRef]
21. Bazhlekova, E. Subordination principle for a class of fractional order differential equations. *Mathematics* **2015**, *3*, 412–427. [CrossRef]
22. Bulut, H.; Yel, G.; Baskonus, H.M. An application of improved bernoulli sub-equation function method to the nonlinear time-fractional burgers equation. *Turk. J. Math. Comput. Sci.* **2016**, *5*, 1–17.
23. Baskonus, H.M.; Bulut, H. On the Numerical solutions of some fractional ordinary differential equations by fractional Adams-Bashforth-Moulton method. *Open Math.* **2015**, *13*, 547–556. [CrossRef]
24. Kumar, D.; Singh, J.; Baskonus, H.M.; Bulut, H. An effective computational approach for solving local fractional telegraph equations. *Nonlinear Sci. Lett. A Math. Phys. Mech.* **2017**, *8*, 200–206.
25. Gencoglu, M.T.; Baskonus, H.M; Bulut, H. Numerical simulations to the nonlinear model of interpersonal Relationships with time fractional derivative. *AIP Conf. Proc.* **2017**, *1798*, 1–9.
26. Ravichandran, C.; Jothimani, K.; Baskonus, H.M.; Valliammal, N. New results on nondensely characterized integrodifferential equations with fractional order. *Eur. Phys. J. Plus* **2018**, *133*, 1–10. [CrossRef]
27. Bulut, H.; Sulaiman, T.A.; Baskonus, H.M. Dark, bright optical and other solitons with conformable space-time fractional second-order spatiotemporal dispersion. *Opt.-Int. J. Light Electron Opt.* **2018**, *163*, 1–7. [CrossRef]
28. Dokuyucu, M.A.; Celik, E.; Bulut, H.;t Baskonus, H.M. Cancer treatment model with the Caputo-Fabrizio fractional derivative. *Eur. Phys. J. Plus* **2018**, *133*, 1–7. [CrossRef]
29. Esen, A.; Sulaiman, T.A.; Bulut, H.; Baskonus, H.M. Optical solitons to the space-time fractional (1+1)-dimensional coupled nonlinear Schrödinger equation. *Opt.-Int. J. Light Electron Opt.* **2018**, *167*, 150–156. [CrossRef]
30. Yavuz, M.; Ozdemir, N.; Baskonus, H.M. Solutions of partial differential equations using the fractional operator involving Mittag-Leffler kernel. *Eur. Phys. J. Plus* **2018**, *133*, 1–12. [CrossRef]
31. Bulut, H.; Kumar, D.; Singh, J.; Swroop, R.; Baskonus, H.M. Analytic study for a fractional model of HIV infection of CD4+TCD4+T lymphocyte cells. *Math. Nat. Sci.* **2018**, *2*, 33–43. [CrossRef]

32. Bulut, H.; Sulaiman, T.A.; Baskonus, H.M.; Rezazadeh, H.; Eslami, M.; Mirzazadeh, M. Optical solitons and other solutions to the conformable space-time fractional Fokas-Lenells equation. *Opt.-Int. J. Light Electron Opt.* **2018**, *172*, 20–27. [CrossRef]
33. Veeresha, P.; Prakasha, D.G.; Baskonus, H.M. New numerical surfaces to the mathematical model of cancer chemotherapy effect in caputo fractional derivatives. *AIP Chaos Interdiscip. J. Nonlinear Sci.* **2019**, *29*, 1–14. [CrossRef] [PubMed]
34. Kilbas, A.A., Srivastava, H.M. and Trujillo, J.J. *Theory and Applications of Fractional Differential Equations*; North-Holland Mathematical Studies Volume 204; Elsevier (North-Holland) Science Publishers: Amsterdam, The Netherland, 2006;
35. Naimark, M.A. *Linear Differential Operators*; Nauka: Moscow, Russia, 1969.
36. Levitan, B.M.; Sargsyan, I.S. *Introduction to Spectral Theory: Selfadjoint Ordinary Differential Operators*; English translation, Translation of Mathematical Monographs; Nauka: Moscow, Russia, 1979; American Mathematical Society: Providence, RI, USA, 1975; Volume 39.
37. Levitan, B.M.; Sargsyan I.S., *Sturm-Liouville and Dirac Operators*; Nauka: Moscow, Russia, 1988.
38. Kostychenko, A.G.; Sargsyan, I.S. *Distribution of Eigenvalues: Selfadjoint Ordinary Differential Operators*; Nauka: Moscow, Russia, 1979.
39. Riesz, F.; Szökefalvi-Nagy, B. *Functional Analysis*; Frederick Ungar Publishing Co.: New York, NY, USA, 1955.
40. Sadovnichiy, V.A. Theory of Operators; MSU Press: Moscow, Russia, 1986.
41. Kasimov, S.G.; Ataev, S.K. On completeness of the system of orthonormal vectors of a generalized spectral problem. *Uzb. Math. J.* **2009**, *2*, 101–111.
42. Gokhberg, I.T.; Krein, M.G. *Introduction to the Theory of Linear Nonselfadjoint Operators in Hilbert Space*; English translation, Translation of Mathematical Monographs; Nauka: Moscow, Russia, 1965; American Mathematical Society: Providence, RI, USA, 1969.

 © 2019 by the authors. Licensee MDPI, Basel, Switzerland. This article is an open access article distributed under the terms and conditions of the Creative Commons Attribution (CC BY) license (http://creativecommons.org/licenses/by/4.0/).

Article

Solving Non-Linear Fractional Variational Problems Using Jacobi Polynomials

Harendra Singh [1], Rajesh K. Pandey [2,3,*] and Hari Mohan Srivastava [4,5]

1. Department of Mathematics, Post Graduate College, Ghazipur 233001, India; harendra059@gmail.com
2. Department of Mathematical Sciences, Indian Institute of Technology (BHU) Varanasi, Varanasi 221005, India
3. Centre for Advanced Biomaterials and Tissue Engineering, Indian Institute of Technology (BHU) Varanasi, Varanasi 221005, India
4. Department of Mathematics and Statistics, University of Victoria, Victoria, BC V8W 3R4, Canada; harimsri@math.uvic.ca
5. Department of Medical Research, China Medical University Hospital, China Medical University, Taichung 40402, Taiwan
* Correspondence: rkpandey.mat@iitbhu.ac.in

Received: 15 January 2019; Accepted: 22 February 2019; Published: 27 February 2019

Abstract: The aim of this paper is to solve a class of non-linear fractional variational problems (NLFVPs) using the Ritz method and to perform a comparative study on the choice of different polynomials in the method. The Ritz method has allowed many researchers to solve different forms of fractional variational problems in recent years. The NLFVP is solved by applying the Ritz method using different orthogonal polynomials. Further, the approximate solution is obtained by solving a system of nonlinear algebraic equations. Error and convergence analysis of the discussed method is also provided. Numerical simulations are performed on illustrative examples to test the accuracy and applicability of the method. For comparison purposes, different polynomials such as 1) Shifted Legendre polynomials, 2) Shifted Chebyshev polynomials of the first kind, 3) Shifted Chebyshev polynomials of the third kind, 4) Shifted Chebyshev polynomials of the fourth kind, and 5) Gegenbauer polynomials are considered to perform the numerical investigations in the test examples. Further, the obtained results are presented in the form of tables and figures. The numerical results are also compared with some known methods from the literature.

Keywords: non-linear fractional variational problems; orthogonal polynomials; Rayleigh-Ritz method; error analysis; convergence analysis

1. Introduction

It is necessary to determine the maxima and minima of certain functionals in study problems in analysis, mechanics, and geometry. These problems are known as variational problems in calculus of variations. Variational problems have many applications in various fields like physics [1], engineering [2], and areas in which energy principles are applicable [3–5].

Nowadays, fractional calculus is a very interesting branch of mathematics. Fractional calculus has many real applications in science and engineering, such as fluid dynamics [6], biology [7], chemistry [8], viscoelasticity [9,10], signal processing [11], bioengineering [12], control theory [13], and physics [14]. Due to the importance of the fractional derivatives established through real-life applications, several authors have considered problems in calculus of variations by replacing the integer-order derivative with fractional orders in objective functionals, and this is thus known as fractional calculus of variations. Some of these studies are of a fractionally damped system [15], energy control for a fractional linear control system [16], a fractional model of a vibrating string [17], and an optimal control problem [18]. In this paper, our aim is to minimize non-linear fractional variational problems (NLFVPs) [19] of the following form:

$$J(y) = \int_0^1 \left(g(x) D^\alpha y(x) + g'(x) I^{1-\alpha} y(x) + h'(x) \right)^2 dx \tag{1}$$

under the constraints

$$y(0) = a, \qquad I^{1-\alpha} y(1) = \epsilon, \tag{2}$$

where g and h are two functions of class C^1 with $g(x) \neq 0$ on $[0, 1]$, α and ϵ are real numbers with $\alpha \in (0, 1)$, and a is a constant.

The pioneer approach for solving the fractional variational problems originates in reference [20] where Agrawal derived the formulation of the Euler-Langrage equation for fractional variational problems. Further, in reference [4], he gave a general formulation for fractional variational problems. In reference [5], the authors used an analytical algorithm based on the Adomian decomposition method (ADM) for solving problems in calculus of variations. In [21,22], Legendre orthonormal polynomials and Jacobi orthonormal polynomials, respectively, were used to obtain an approximate numerical solution of fractional optimum control problems. In [23], the Haar wavelet method was used to obtain numerical solution of these problems. Some other numerical methods for the approximate solution of fractional variational problems are given in [24–34]. Recently, in [19], the authors gave a new class of fractional variational problems and solved this using a decomposition formula based on Jacobi polynomials. The operational matrix methods (see [35–41]) have been found to be useful for solving problems in fractional calculus.

In present paper, we extend the Rayleigh-Ritz method together with operational matrices of different orthogonal polynomials such as Shifted Legendre polynomials, Shifted Chebyshev polynomials of the first kind, Shifted Chebyshev polynomials of the third kind, Shifted Chebyshev polynomials of the fourth kind, and Gegenbauer polynomials to solve a special class of NLFVPs. The Rayleigh-Ritz methods have been discussed by many researchers in the literature for different kinds of variational problems, i.e., fractional optimal control problems [18,21,22,32,33]; here we cite only few, and many more can be found in the literature. In this method, first we take a finite-dimensional approximation of the unknown function. Further, using an operational matrix of integration and the Rayleigh-Ritz method in the variational problem, we obtain a system of non-linear algebraic equations whose solution gives an approximate solution for the non-linear variational problem. Error analysis of the method for different orthogonal polynomials is given, and convergence of the approximate numerical solution to the exact solution is shown. A comparative study using absolute error and root-mean-square error tables for all five kinds of polynomials is analyzed. Numerical results are discussed in terms of the different values of fractional order involved in the problem and are shown through tables and figures.

2. Basic Preliminaries

The definition of fractional order integration in the Riemann-Liouville sense is defined as follows.

Definition 1. *The Riemann-Liouville fractional order integral operator is given by*

$$I^\alpha f(x) = \begin{cases} \frac{1}{\Gamma(\alpha)} \int_0^x (x-t)^{\alpha-1} f(t) dt, & \alpha > 0, x > 0, \\ f(x), & \alpha = 0. \end{cases}$$

The analytical form of the shifted Jacobi polynomial of degree i on $[0, 1]$ is given as

$$\Psi_i(x) = \sum_{k=0}^{i} (-1)^{i-k} \frac{\Gamma(i+b+1)\Gamma(i+k+a+b+1)}{\Gamma(k+b+1)\Gamma(i+a+b+1)(i-k)!k!} x^k \qquad (3)$$

where a and b are certain constants. Jacobi polynomials are orthogonal in the interval $[0, 1]$ with respect to the weight function $w^{(a,b)}(x) = (1-x)^a x^b$ and have the orthogonality property

$$\int_0^1 \Psi_n(x)\Psi_m(x) w^{(a,b)}(x) dx = v_n^{a,b} \delta_{mn} \qquad (4)$$

where δ_{mn} is the Kronecker delta function and

$$v_n^{a,b} = \frac{\Gamma(n+a+1)\Gamma(n+b+1)}{(2n+a+b+1)n!\Gamma(n+a+b+1)}. \qquad (5)$$

For certain values of the constants a and b, the Jacobi polynomials take the form of some well-known polynomials, defined as follows.

Case 1: Legendre polynomials (S1) For $a = 0$, $b = 0$ in Equation (3), we get Legendre polynomials.

$$\Psi_i(x) = \sum_{k=0}^{i} (-1)^{i-k} \frac{\Gamma(i+1)\Gamma(i+k+1)}{\Gamma(k+1)\Gamma(i+1)(i-k)!k!} x^k \qquad (6)$$

Case 2: Chebyshev polynomials of the first kind (S2) For $a = \frac{1}{2}$, $b = \frac{1}{2}$ in Equation (3), we get Chebyshev polynomials of the first kind.

$$\Psi_i(x) = \sum_{k=0}^{i} (-1)^{i-k} \frac{\Gamma(i+\frac{3}{2})\Gamma(i+k+2)}{\Gamma(k+\frac{3}{2})\Gamma(i+2)(i-k)!k!} x^k \qquad (7)$$

Case 3: Chebyshev polynomials of the third kind (S3) For $a = \frac{1}{2}$, $b = -\frac{1}{2}$ in Equation (3), we get Chebyshev polynomials of the third kind.

$$\Psi_i(x) = \sum_{k=0}^{i} (-1)^{i-k} \frac{\Gamma\left(i+\frac{1}{2}\right)\Gamma(i+k+1)}{\Gamma\left(k+\frac{1}{2}\right)\Gamma(i+1)(i-k)!k!} x^k \qquad (8)$$

Case 4: Chebyshev polynomials of the fourth kind (S4) For $a = -\frac{1}{2}$, $b = \frac{1}{2}$ in Equation (3), we get Chebyshev polynomials of the fourth kind.

$$\Psi_i(x) = \sum_{k=0}^{i} (-1)^{i-k} \frac{\Gamma(i+\frac{3}{2})\Gamma(i+k+1)}{\Gamma(k+\frac{3}{2})\Gamma(i+1)(i-k)!k!} x^k \qquad (9)$$

Case 5: Gegenbauer polynomials (S5) For $a = b = a - \frac{1}{2}$ in Equation (3), we get Gegenbauer polynomials.

$$\Psi_i(x) = \sum_{k=0}^{i} (-1)^{i-k} \frac{\Gamma\left(i+a+\frac{1}{2}\right)\Gamma(i+k+2a)}{\Gamma\left(k+a+\frac{1}{2}\right)\Gamma(i+2a)(i-k)!k!} x^k \qquad (10)$$

A function $f \in L^2[0,1]$ with $|f''(t)| \le K$ can be expanded as

$$f(t) = \lim_{n \to \infty} \sum_{i=0}^{n} c_i \Psi_i(t), \qquad (11)$$

where $c_i = \langle f(t), \Psi_i(t) \rangle$ and $\langle -, - \rangle$ is the usual inner product space.

Equation (11) for finite-dimensional approximation is written as

$$f \cong \sum_{i=0}^{m} c_i \Psi_i(t) = C^T \phi_m(t), \tag{12}$$

where C and $\phi_m(t)$ are $(m+1) \times 1$ matrices given by $C = [c_0, c_1, \ldots, c_m]^T$ and $\phi_m(t) = [\Psi_0, \Psi_1, \ldots, \Psi_m]^T$.

Theorem 1. *Let H be a Hilbert space and Z be a closed subspace of H with dim $Z < \infty$; let $\{z_1, z_2, \ldots, z_N\}$ be any basis for Z. Suppose that y is an arbitrary element in H and z_0 is the unique best approximation to y out of Z. Then*

$$\|y - z_0\|_2^2 = \frac{T(y; z_1, z_2, \ldots, z_N)}{T(z_1, z_2, \ldots, z_N)},$$

where

$$T(y; z_1, z_2, \ldots, z_N) = \begin{vmatrix} \langle y, y \rangle & \langle y, Z_1 \rangle & \cdots & \langle y, Z_N \rangle \\ \langle Z_1, Z_N \rangle & \langle Z_1, Z_1 \rangle & \cdots & \langle Z_1, Z_N \rangle \\ \cdot & \cdot & \vdots & \cdot \\ \cdot & \cdot & \vdots & \cdot \\ \cdot & \cdot & \vdots & \cdot \\ \langle Z_N, y \rangle & \langle Z_N, Z_1 \rangle & \cdots & \langle Z_N, Z_N \rangle \end{vmatrix}$$

Proof. Please see references [42,43]. □

Theorem 2. *Suppose that $f_N(x)$ is the Nth approximation of the function $f \in L^2_{w^{(a,b)}}[0,1]$, and suppose*

$$S_N(f) = \int_0^1 [f(x) - f_N(x)]^2 w^{(a,b)}(x) dx;$$

then we have

$$\lim_{N \to \infty} S_N(f) = 0.$$

Proof. Please see Appendix A. □

3. Operational Matrices

Theorem 3. *Let $\phi_n = [\Psi_0(x), \Psi_1(x), \ldots, \Psi_n(x)]^T$ be a Shifted Jacobi vector and suppose $v > 0$; then*

$$I^v \Psi_i(x) = I^{(v)} \phi_n(x)$$

where $I^{(v)} = (\mu(i,j))$ is an $(n+1) \times (n+1)$ operational matrix of the fractional integral of order v and its (i,j)th entry is given by

$$\mu(i,j) = \sum_{k=0}^{i} \sum_{l=0}^{j} (-1)^{i+j-k-l} \frac{\Gamma(a+1)\Gamma(i+b+1)\Gamma(i+k+a+b+1)\Gamma(j+l+a+b+1)\Gamma(v+k+l+a+b+1)(2j+a+b+1)j!}{(i-k)!(j-l)!(l)!\,\Gamma(k+b+1)\Gamma(i+a+b+1)\Gamma(v+k+1)\Gamma(j+a+1)\Gamma(l+b+1)\Gamma(k+l+v+a+b+1)}. \tag{13}$$

Proof. We refer to reference [44] for the proof. □

Now, in particular cases, the operational matrix of integration for various polynomials is given as follows.

For Shifted Legendre polynomials (S1), the (i,j)th entry of the operational matrix of integration is given as

$$\mu(i,j) = \sum_{k=0}^{i}\sum_{l=0}^{j}(-1)^{i+j+k+l}\frac{(i+k)!(j+l)!}{(i-k)!(j-l)!(k)!(l!)^2(\alpha+k+l+1)\Gamma(\alpha+k+l)}. \tag{14}$$

For Shifted Chebyshev polynomials of the first kind (S2), the (i,j)th entry of the operational matrix of integration is given as:

$$\mu(i,j) = \sum_{k=0}^{i}\sum_{l=0}^{j}(-1)^{i+j-k-l}\frac{\Gamma(\frac{3}{2})\Gamma(i+\frac{3}{2})\Gamma(i+k+2)\Gamma(j+l+2)\Gamma(\alpha+k+l+\frac{3}{2})(2j+2)j!}{(i-k)!(j-l)!(l)!\,\Gamma(k+\frac{3}{2})\Gamma(i+2)\Gamma(\alpha+k+1)\Gamma(j+\frac{3}{2})\Gamma(l+\frac{3}{2})\Gamma(k+l+\alpha+3)}. \tag{15}$$

For Shifted Chebyshev polynomials of the third kind (S3), the (i,j)th entry of the operational matrix of integration is given as

$$\mu(i,j) = \sum_{k=0}^{i}\sum_{l=0}^{j}(-1)^{i+j-k-l}\frac{\Gamma(\frac{3}{2})\Gamma(i+\frac{1}{2})\Gamma(i+k+1)\Gamma(j+l+1)\Gamma(\alpha+k+l+\frac{1}{2})(2j+1)j!}{(i-k)!(j-l)!(l)!\,\Gamma(k+\frac{1}{2})\Gamma(i+1)\Gamma(\alpha+k+1)\Gamma(j+\frac{3}{2})\Gamma(l+\frac{1}{2})\Gamma(k+l+\alpha+2)}. \tag{16}$$

For Shifted Chebyshev polynomials of the fourth kind (S4), the (i,j)th entry of the operational matrix of integration is given as

$$\mu(i,j) = \sum_{k=0}^{i}\sum_{l=0}^{j}(-1)^{i+j-k-l}\frac{\Gamma(\frac{1}{2})\Gamma(i+\frac{3}{2})\Gamma(i+k+1)\Gamma(j+l+1)\Gamma(\alpha+k+l+\frac{3}{2})(2j+1)j!}{(i-k)!(j-l)!(l)!\,\Gamma(k+\frac{3}{2})\Gamma(i+1)\Gamma(\alpha+k+1)\Gamma(j+\frac{1}{2})\Gamma(l+\frac{3}{2})\Gamma(k+l+\alpha+2)}. \tag{17}$$

For Shifted Gegenbauer polynomials (S5), the (i,j)th entry of the operational matrix of integration is given as

$$\mu(i,j) = \sum_{k=0}^{i}\sum_{l=0}^{j}(-1)^{i+j-k-l}\frac{\Gamma(i+a+\frac{1}{2})\Gamma(i+k+2a)\Gamma(j+l+2a)\Gamma(a+\frac{1}{2})\Gamma(\alpha+k+l+a+\frac{1}{2})(2j+2a)j!}{(i-k)!(j-l)!(l)!\,\Gamma(k+a+\frac{1}{2})\Gamma(i+2a)\Gamma(\alpha+k+1)\Gamma(j+a+\frac{1}{2})\Gamma(l+a+\frac{1}{2})\Gamma(2a+k+l+\alpha+1)}. \tag{18}$$

4. Method of Solution

Approximating the unknown function in terms of orthogonal polynomials has been practiced in several papers in recent years [18,21,22,32,33] for different types of problems. Here, for solving the problem in Equation (1), we approximate

$$D^\alpha y(x) = C^T \Phi_n(x). \tag{19}$$

We are approximating the derivative first because we want to use the initial condition. Taking the integral of order α on both sides of Equation (19), we get

$$y(x) = C^T I^\alpha \Phi_n(x) + y(0). \tag{20}$$

Using the operational matrix of integration, Equation (20) can be written as

$$y(x) \cong C^T I^{(\alpha)} \Phi_n(x) + A^T \Phi_n(x) \tag{21}$$

where $y(0) = a \cong A^T \Phi_n(x)$ and $I^{(\alpha)}$ is the operational matrix of integration of order α.

Using Equation (19), we can write

$$I^{1-\alpha} y(x) = I D^\alpha y(x) = C^T I \Phi_n(x) \cong C^T I^{(1)} \Phi_n(x). \tag{22}$$

Using Equations (19) and (22) in Equation (1), we obtain

$$J(c_0, c_1, \ldots, c_n) = \int_0^1 \left(g(x) C^T \Phi_n(x) + g'(x) C^T I \Phi_n(x) + h'(x) \right)^2 dx. \tag{23}$$

Equation (23) can then be written as

$$J(c_0, c_1, \ldots, c_n) = \int_0^1 \left(C^T g(x) \Phi_n(x) + C^T I^{(1)} g'(x) \Phi_n(x) + h'(x) \right)^2 dx. \tag{24}$$

We further take the following approximations:

$$g(x) \Phi_i(x) \cong E_1^{i,T} \Phi_n(x) \tag{25}$$

$$g'(x) \Phi_i(x) \cong E_2^{i,T} \Phi_n(x) \tag{26}$$

$$h'(x) \cong E_3^T \Phi_n(x) \tag{27}$$

where $E_1^{i,T} = [e_{1,0}^i, e_{1,1}^i, \ldots, e_{1,n}^i]$, $E_2^{i,T} = [e_{2,0}^i, e_{2,1}^i, \ldots, e_{2,n}^i]$, $E_3^T = [e_{3,0}, e_{3,1}, \ldots, e_{3,n}]$, and $e_{1,j}^i = \langle g(x) \Phi_i(x), \Psi_j(x) \rangle$, $e_{2,j}^i = \langle g'(x) \Phi_i(x), \Psi_j(x) \rangle$, $e_{3,j} = \langle h'(x), \Psi_j(x) \rangle$, $0 \leq i, j \leq n$, and $\langle -, - \rangle$ is the usual inner product space.

Using Equations (25) and (26) we can write

$$g(x) \Phi_n(x) \cong E_1^T \Phi_n(x) \tag{28}$$

$$g'(x) \Phi_n(x) \cong E_2^T \Phi_n(x) \tag{29}$$

where

$$E_1^T = \left(E_1^{i,T} \right)_{0 \leq i \leq n} \text{ and } E_2^T = \left(E_2^{i,T} \right)_{0 \leq i \leq n}. \tag{30}$$

From Equations (24) and (27)–(29), we get

$$J(c_0, c_1, \ldots, c_n) = \int_0^1 \left(C^T E_1^T \Phi_n(x) + C^T I^{(1)} E_2^T \Phi_n(x) + E_3^T \Phi_n(x) \right)^2 dx. \tag{31}$$

Let

$$E^T = C^T \left(E_1^T + I^{(1)} E_2^T \right) + E_3^T. \tag{32}$$

From Equations (31) and (32), we get

$$\begin{aligned} J(c_0, c_1, \ldots, c_n) &= \int_0^1 \left(E^T \Phi_n(x) \right)^2 dx \\ &= \int_0^1 E^T \Phi_n(x) \Phi_n(x)^T E \, dx, \\ &= E^T P E \end{aligned} \tag{33}$$

where P is a square matrix given by $P = \int_0^1 \Phi_n(x) \Phi_n(x)^T dx$.

Using Equation (22), the boundary condition can be written as

$$I^{1-\alpha} y(1) \cong C^T I^{(1)} \Phi_n(1) = \epsilon. \tag{34}$$

Using the Lagrange multiplier method [18,20–22,32,33], the necessary extremal condition for the functional in Equation (33) becomes

$$\frac{\partial J}{\partial c_0} = 0, \ \frac{\partial J}{\partial c_1} = 0, \ \ldots, \ \frac{\partial J}{\partial c_{n-1}} = 0. \tag{35}$$

From Equations (34) and (35), we get a set of $n+1$ equations. Solving these $n+1$ equations, we get unknown parameters $c_0, c_1, \ldots, c_n$. Using these unknown parameters in Equation (21), we get the unknown function's extreme values of the non-linear fractional functional.

5. Error Analysis

The upper bound of error for the operational matrix of fractional integration of a Jacobi polynomial of the ith degree is given as

$$e_i^\alpha = I^{(\alpha)} \Psi_i(x) - I^\alpha \Psi_i(x). \tag{36}$$

From Equation (36), we can write

$$\|e_i^\alpha\|_2 = \left\| I^\alpha \Psi_i(x) - \sum_{j=0}^{n} \mu(i,j) \Psi_j(x) \right\|_2. \tag{37}$$

Taking the integral operator of order α on both sides of Equation (3), we get

$$I^\alpha \Psi_i(x) = \sum_{k=0}^{i} (-1)^{i-k} \frac{\Gamma(i+b+1)\Gamma(i+k+a+b+1)}{(i-k)!\,\Gamma(k+b+1)\Gamma(i+a+b+1)\Gamma(\alpha+k+1)} x^{\alpha+k}. \tag{38}$$

From the construction of the operational matrix we can write

$$\mu(i,j) = \sum_{k=0}^{i} (-1)^{i-k} \frac{\Gamma(i+b+1)\Gamma(i+k+a+b+1)}{(i-k)!\,\Gamma(k+b+1)\Gamma(i+a+b+1)\Gamma(\alpha+k+1)} c_{j,k}, \ j = 0, 1, \ldots, n. \tag{39}$$

Using Theorem 1 we can write

$$\left\| x^{\alpha+k} - \sum_{j=0}^{n} c_{j,k} \Psi_j(x) \right\|_2 = \left(\frac{T\left(x^{\alpha+k}; \Psi_0(x), \Psi_1(x), \ldots, \Psi_n(x)\right)}{T(\Psi_0(x), \Psi_1(x), \ldots, \Psi_n(x))} \right)^2. \tag{40}$$

From Equations (37)–(39), we get

$$\|e_i^\alpha\|_2 = \left\| \begin{array}{l} \sum_{k=0}^{i} (-1)^{i-k} \frac{\Gamma(i+b+1)\Gamma(i+k+a+b+1)}{(i-k)!\,\Gamma(k+b+1)\Gamma(i+a+b+1)\Gamma(\alpha+k+1)} x^{\alpha+k} \\ - \sum_{j=0}^{n} \sum_{k=0}^{i} (-1)^{i-k} \frac{\Gamma(i+b+1)\Gamma(i+k+a+b+1)}{(i-k)!\,\Gamma(k+b+1)\Gamma(i+a+b+1)\Gamma(\alpha+k+1)} c_{j,k} \Psi_j(x) \end{array} \right\|$$

$$\leq \sum_{k=0}^{i} \left| \frac{\Gamma(i+b+1)\Gamma(i+k+a+b+1)}{(i-k)!\,\Gamma(k+b+1)\Gamma(i+a+b+1)\Gamma(\alpha+k+1)} \right| \left\| x^{\alpha+k} - \sum_{j=0}^{n} c_{j,k} \Psi_j(x) \right\|_2. \tag{41}$$

Using Equation (40) in Equation (41), we obtain the error bound for the operational matrix of integration of an ith-degree polynomial, which is given as

$$\|e_i^\alpha\|_2 \leq \sum_{k=0}^{i} \left| \frac{\Gamma(i+b+1)\Gamma(i+k+a+b+1)}{(i-k)!\,\Gamma(k+b+1)\Gamma(i+a+b+1)\Gamma(\alpha+k+1)} \right| \left(\frac{T\left(x^{\alpha+k}; \Psi_0(x), \Psi_1(x), \ldots, \Psi_n(x)\right)}{T(\Psi_0(x), \Psi_1(x), \ldots, \Psi_n(x))} \right)^2, \ i = 0, 1, 2, \ldots, n. \tag{42}$$

Now, in particular cases, the error bounds for different orthogonal polynomials are given as follows.

Case 1: For Legendre polynomials (S1) the error bound is given as

$$\|e_i^\alpha\|_2 \leq \sum_{k=0}^{i} \left| \frac{\Gamma(i+1)\Gamma(i+k+1)}{(i-k)!\,\Gamma(k+1)\Gamma(i+1)\Gamma(\alpha+k+1)} \right| \left(\frac{T(x^{\alpha+k};\Psi_0(x),\Psi_1(x),\ldots,\Psi_n(x))}{T(\Psi_0(x),\Psi_1(x),\ldots,\Psi_n(x))} \right)^2, i=0,1,2,\ldots,n. \quad (43)$$

Case 2: For Chebyshev polynomials of the first kind (S2) the error bound is given as

$$\|e_i^\alpha\|_2 \leq \sum_{k=0}^{i} \left| \frac{\Gamma\left(i+\frac{3}{2}\right)\Gamma(i+k+2)}{(i-k)!\,\Gamma\left(k+\frac{3}{2}\right)\Gamma(i+2)\Gamma(\alpha+k+1)} \right| \left(\frac{T(x^{\alpha+k};\Psi_0(x),\Psi_1(x),\ldots,\Psi_n(x))}{T(\Psi_0(x),\Psi_1(x),\ldots,\Psi_n(x))} \right)^2, i=0,1,2,\ldots,n. \quad (44)$$

Case 3: For Chebyshev polynomials of the third kind (S3) the error bound is given as

$$\|e_i^\alpha\|_2 \leq \sum_{k=0}^{i} \left| \frac{\Gamma\left(i+\frac{1}{2}\right)\Gamma(i+k+1)}{(i-k)!\,\Gamma\left(k+\frac{1}{2}\right)\Gamma(i+1)\Gamma(\alpha+k+1)} \right| \left(\frac{T(x^{\alpha+k};\Psi_0(x),\Psi_1(x),\ldots,\Psi_n(x))}{T(\Psi_0(x),\Psi_1(x),\ldots,\Psi_n(x))} \right)^2, i=0,1,2,\ldots,n. \quad (45)$$

Case 4: For Chebyshev polynomials (S4) the error bound is given as

$$\|e_i^\alpha\|_2 \leq \sum_{k=0}^{i} \left| \frac{\Gamma\left(i+\frac{3}{2}\right)\Gamma(i+k1)}{(i-k)!\,\Gamma\left(k+\frac{3}{2}\right)\Gamma(i+1)\Gamma(\alpha+k+1)} \right| \left(\frac{T(x^{\alpha+k};\Psi_0(x),\Psi_1(x),\ldots,\Psi_n(x))}{T(\Psi_0(x),\Psi_1(x),\ldots,\Psi_n(x))} \right)^2, i=0,1,2,\ldots,n. \quad (46)$$

Case 5: For Gegenbauer polynomials (S5) the error bound is given as

$$\|e_i^\alpha\|_2 \leq \sum_{k=0}^{i} \left| \frac{\Gamma(i+2)\Gamma(i+k+3)}{(i-k)!\,\Gamma(k+2)\Gamma(i+3)\Gamma(\alpha+k+1)} \right| \left(\frac{T(x^{\alpha+k};\Psi_0(x),\Psi_1(x),\ldots,\Psi_n(x))}{T(\Psi_0(x),\Psi_1(x),\ldots,\Psi_n(x))} \right)^2, i=0,1,2,\ldots,n. \quad (47)$$

Let $e_{I,n}^{\alpha,w}$ denote the error vector for the operational matrix of integration of order α obtained by using $(n+1)$ orthogonal polynomials in $L_w^2[0,1]$; then

$$e_{I,n}^{\alpha,w} = I^{(\alpha)}\Phi_n(x) - I^\alpha \Phi_n(x). \quad (48)$$

From Theorems 1 and 2 and from Equations (43)–(47), it is clear that as $n \to \infty$ the error vector in Equation (48) tends to zero.

6. Convergence Analysis

A set of orthogonal polynomials on $[0,1]$ forms a basis for $L_w^2[0,1]$. Let S_n be the n-dimensional subspace of $L_w^2[0,1]$ generated by $(\Phi_i)_{0 \leq i \leq n}$. Thus, every functional on S_n can be written as a linear combination of orthogonal polynomials $(\Phi_i)_{0 \leq i \leq n}$. The scalars in the linear combinations can be chosen in such a way that the functional minimizes. Let the minimum value of a functional on space S_n be denoted by m_n. From the construction of S_n and m_n, it is clear that $S_n \subset S_{n+1}$ and $m_{n+1} \geq m_n$.

Theorem 4. *Consider the functional J, then*

$$\lim_{n \to \infty} m_n = m = \inf_{x \in L_w^2[0,1]} J[x].$$

Proof. Using Equation (48) in Equation (23), we have

$$J(c_0, c_1, \ldots, c_n) = \int_0^1 \left(C^T g(x) \Phi_n(x) + C^T I^{(1)} g'(x) \Phi_n(x) + C^T e_{I,n}^1 g'(x) + h'(x) \right)^2 dx. \quad (49)$$

Taking $n \to \infty$ and using Equations (25)–(27) and (48) in Equation (49), we get

$$J^e(c_0, c_1, \ldots, c_n) = \int_0^1 \left(C^T \left(\sum_{i=0}^n (E_1^{i,T} \Phi_n(x) + e_{E_1^i,n}^w) \right) \right.$$
$$\left. + C^T I^{(1)} \left(\sum_{i=0}^n \left(E_2^{i,T} \Phi_n(x) + e_{E_2^i,n}^w \right) \right) + E_3^T \Phi_n(x) + e_{E_3,n}^w \right)^2 dx \quad (50)$$

where
$$e_{E_1^i,n}^w = E_1^{i,T} \Phi(x) - E_1^{i,T} \Phi_n(x),$$
$$e_{E_2^i,n}^w = E_2^{i,T} \Phi(x) - E_2^{i,T} \Phi_n(x),$$
$$e_{E_3,n}^w = E_3^T \Phi(x) - E_3^T \Phi_n(x),$$

and J^e is the error term of the functional.

Using Equations (30) and (32) in Equation (50), we get

$$J^e(c_0, c_1, \ldots, c_n) = \int_0^1 \left(E^T \Phi_n(x) + e_n^w \right)^2 dx \quad (51)$$

where

$$e_n^w = C^T \sum_{i=0}^n e_{E_1^i,n}^w + C^T I^{(1)} \sum_{i=0}^n e_{E_2^i,n}^w. \quad (52)$$

Solving Equation (51) similarly to the original functional, Equation (51) reduces to the following form:

$$J^e(c_0, c_1, \ldots, c_n) = E^T P E + e_n^w(J^e). \quad (53)$$

Using Equation (48) in Equation (34), we get

$$C^T I^{(1)} \Phi_n(1) + C^T e_{I,n}^{1,w} = \epsilon. \quad (54)$$

Similar to above, by using the Rayleigh-Ritz method on Equation (53) with the boundary condition in Equation (54) we obtain the extreme value of the functional defined in Equation (53). Let this extreme value be denoted by $m_n^*(t)$.

Now, from Equation (48), it is obvious that $e_{E_1^i,n}^w, e_{E_2^i,n}^w, e_{E_3,n}^w \to 0$ as $n \to \infty$, which implies that $e_n^w(J^e) \to 0$ as $n \to \infty$. So, it is clear that as $n \to \infty$, the functional J^e in Equation (53) comes close to the functional J in Equation (23) and the boundary condition in Equation (54) comes close to Equation (34).

So, for large values of n,

$$m_n^*(t) \to m_n(t). \quad (55)$$

From Theorem 4 and Equation (55), we conclude that

$$\lim_{n \to \infty} m_n^*(t) = m(t).$$

Proof completed. □

7. Numerical Results and Discussions

In this section, we investigate the accuracy of the method by testing it on some numerical examples. We apply the numerical algorithm to two test problems using different orthogonal polynomials as a basis. The results for the test problems are shown through the figures and tables.

Example 1. *Consider a non-linear fractional variational problem as in Equation (1) with $g(x) = h(x) = \frac{1}{1+x^\beta}$; we then have the following non-linear fractional variational problem [19]:*

$$J(y) = \int_0^1 \left(\frac{1}{1+x^\beta} D^\alpha y(x) - \left(I^{1-\alpha}y(x) + 1\right) \frac{\beta x^{\beta-1}}{(1+x^\beta)^2} \right)^2 dx \tag{56}$$

under the constraints

$$y(0) = 0, \ I^{1-\alpha}y(1) = \epsilon.$$

The exact solution of the above equation is given as

$$y_{exact}(x) = \left(\frac{1}{2}(1+\epsilon) - 1\right) \left(\frac{\Gamma(\beta+2)}{\Gamma(\beta+\alpha+1)} x^{\beta+\alpha} + \frac{1}{\Gamma(\alpha+1)} x^\alpha \right) + \frac{\Gamma(\beta+1)}{\Gamma(\alpha+\beta)} x^{\beta+\alpha-1}.$$

We discuss this example for different values of $\alpha = 0.5, 0.6, 0.7, 0.8, 0.9,$ or $1, \beta = 5,$ and $\epsilon = 1$.

In Figures 1–5, it is shown that the solutions for the two different values of $\alpha = 0.8$ and $\alpha = 1$ coincide with the exact solutions for different orthogonal polynomials at $n = 5$.

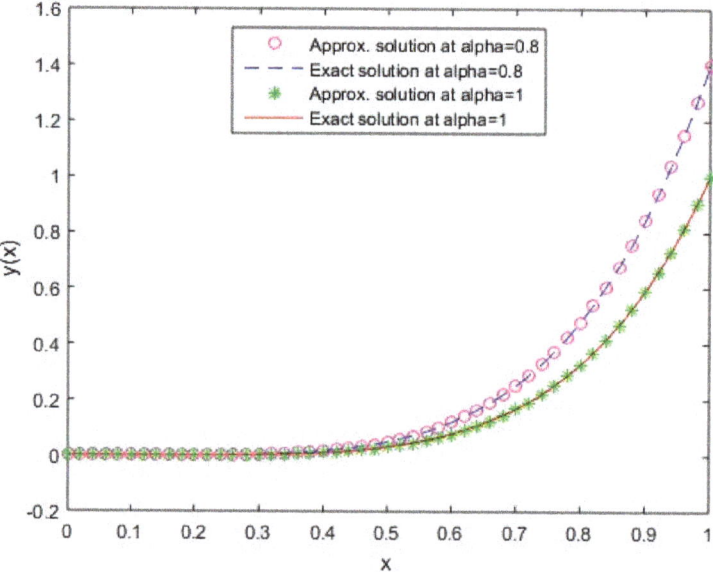

Figure 1. Comparison of exact and numerical solutions using S1 for $\alpha = 0.8$ and $\alpha = 1$, Example 1.

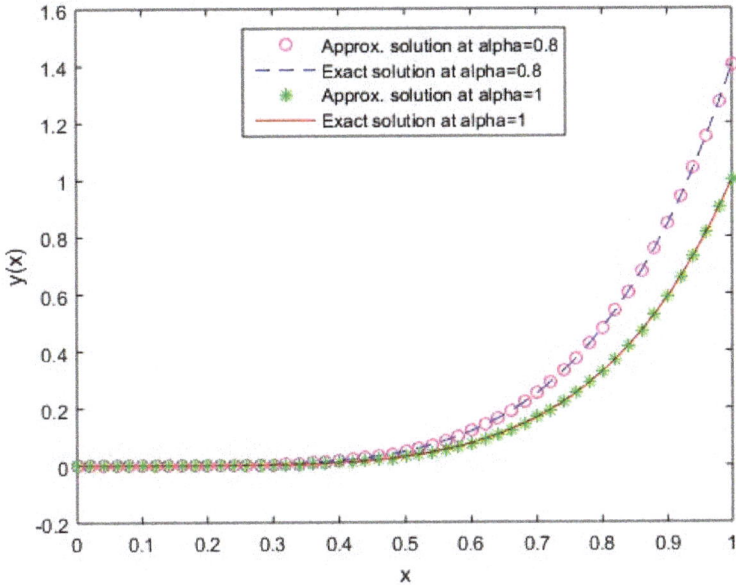

Figure 2. Comparison of exact and numerical solutions using S2 for $\alpha = 0.8$ and $\alpha = 1$, Example 1.

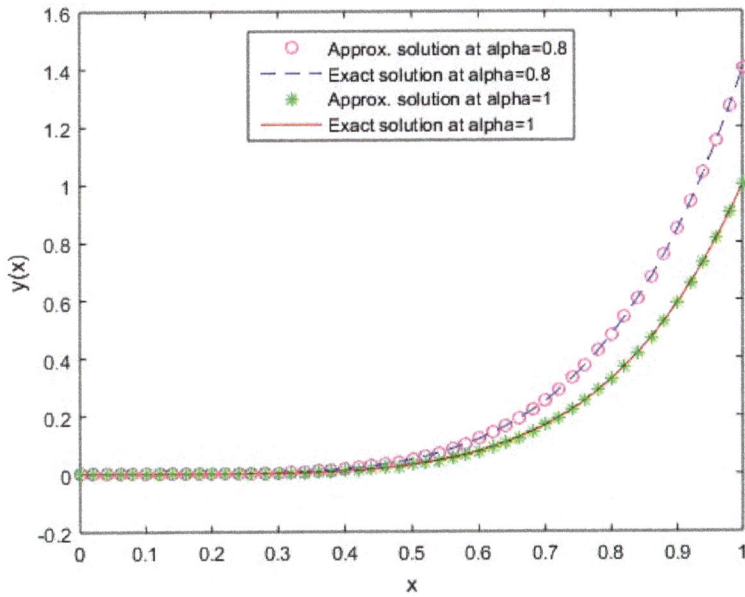

Figure 3. Comparison of exact and numerical solutions using S3 for $\alpha = 0.8$ and $\alpha = 1$, Example 1.

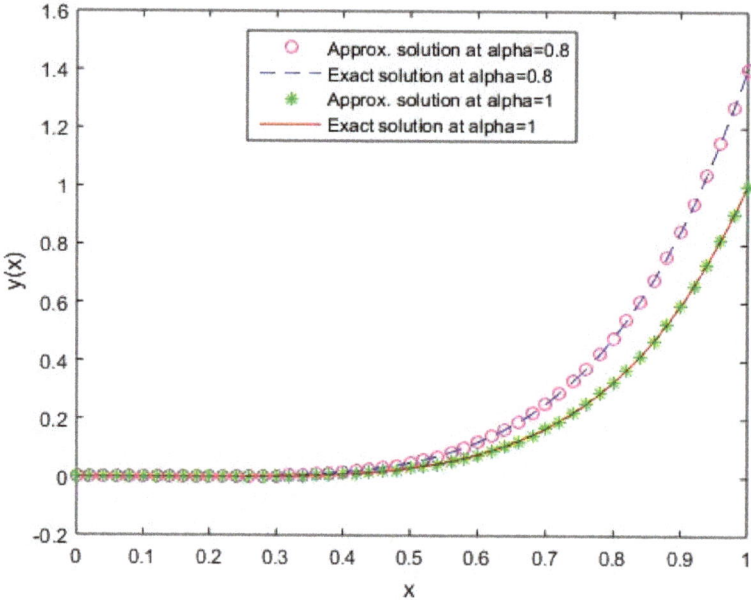

Figure 4. Comparison of exact and numerical solutions using S4 for $\alpha = 0.8$ and $\alpha = 1$, Example 1.

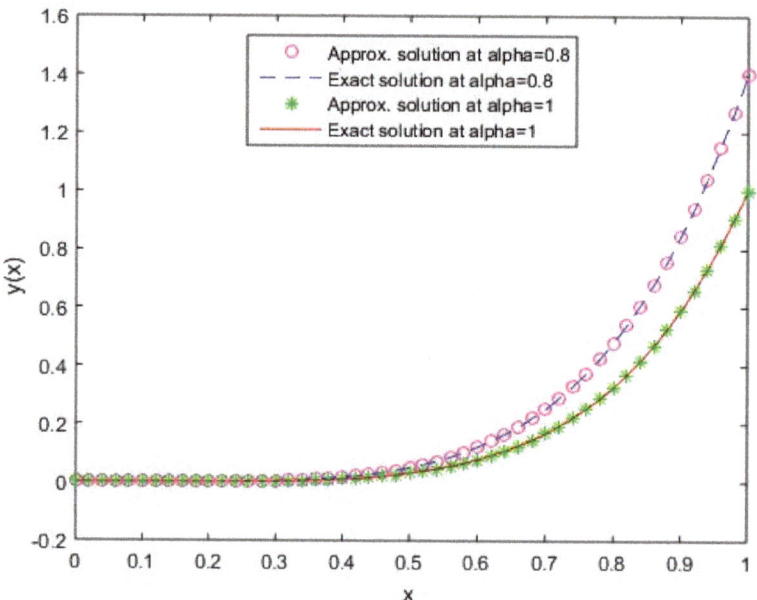

Figure 5. Comparison of exact and numerical solutions using S5 for $\alpha = 0.8$ and $\alpha = 1$, Example 1.

In Figures 6–10, it is shown that the solution varies continuously for Shifted Legendre polynomials, Shifted Chebyshev polynomials of the second kind, Shifted Chebyshev polynomials of the third kind, Shifted Chebyshev polynomials of the fourth kind, and Gegenbauer polynomials, respectively, with different values of fractional order.

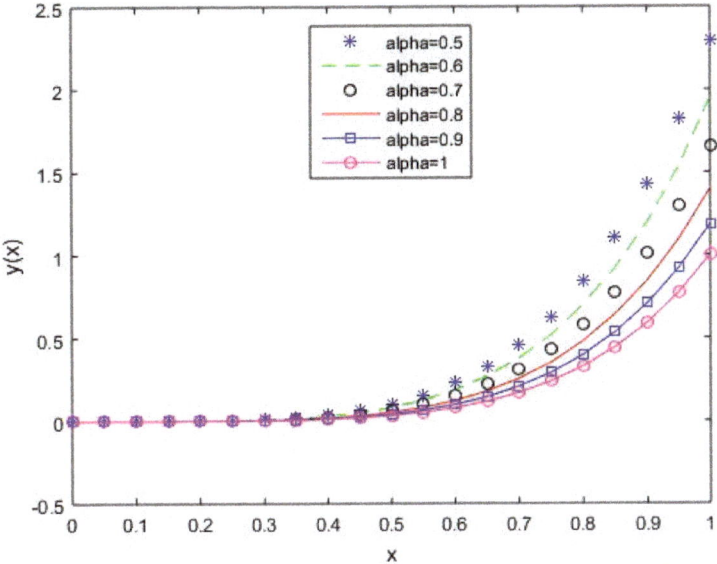

Figure 6. The behavior of solutions using S1 for α values of 0.5, 0.6, 0.7, 0.8, 0.9, and 1, Example 1.

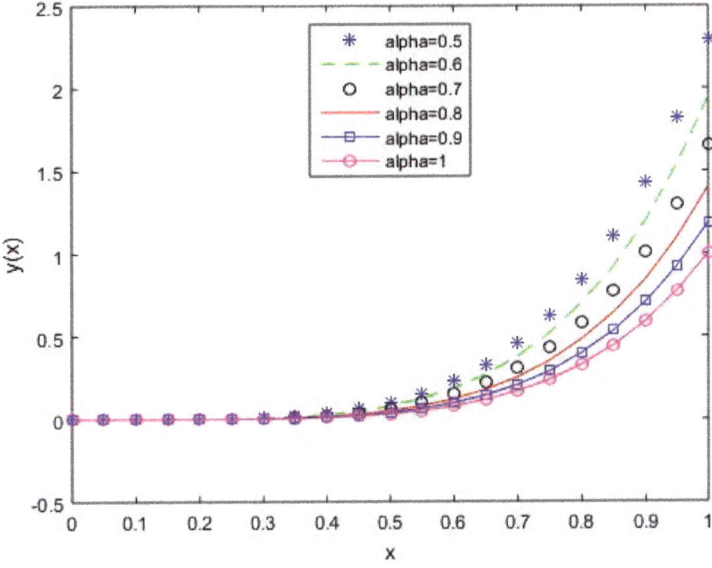

Figure 7. The behavior of solutions using S2 for α values of 0.5, 0.6, 0.7, 0.8, 0.9, and 1, Example 1.

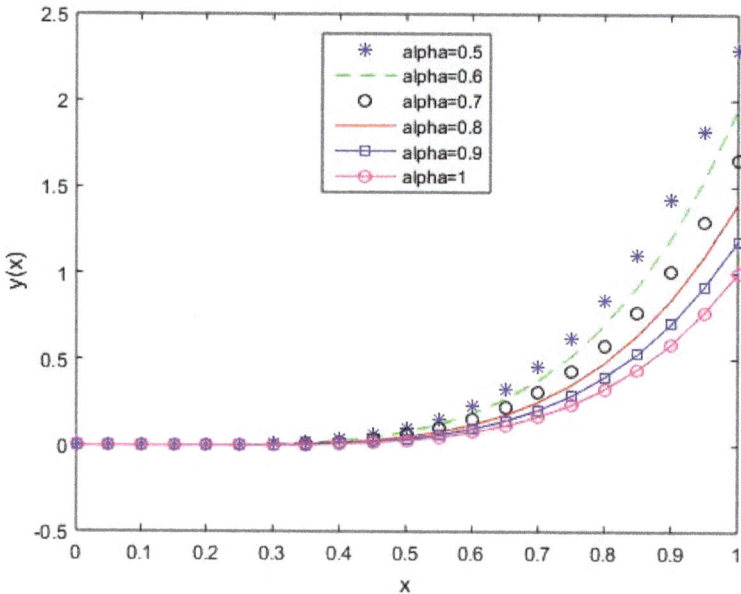

Figure 8. The behavior of solutions using S3 for α values of 0.5, 0.6, 0.7, 0.8, 0.9, and 1, Example 1.

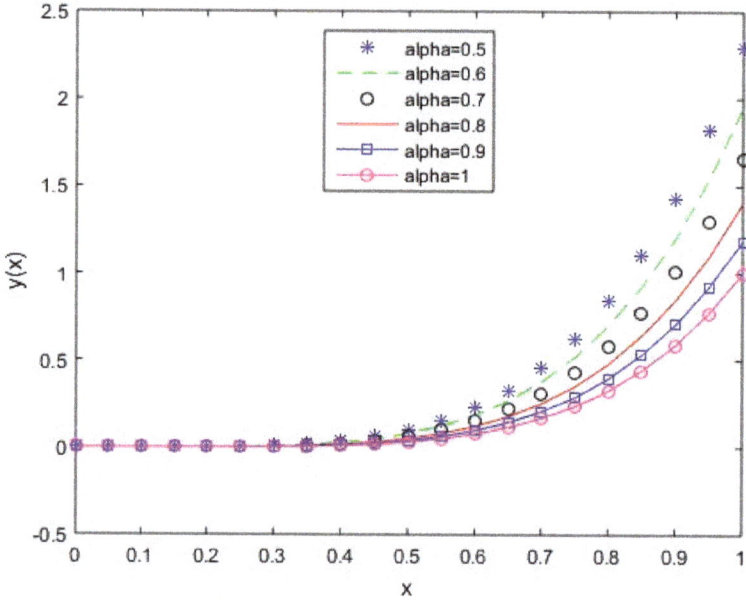

Figure 9. The behavior of solutions using S4 for α values of 0.5, 0.6, 0.7, 0.8, 0.9, and 1, Example 1.

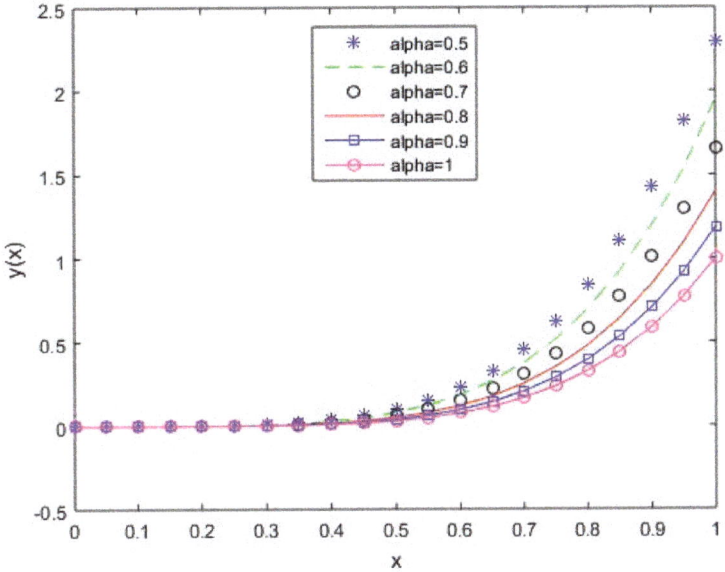

Figure 10. The behavior of solutions using S5 for α values of 0.5, 0.6, 0.7, 0.8, 0.9, and 1, Example 1.

In Table 1, we have listed the maximum absolute errors (MAE) and root-mean-square errors (RMSE) for Example 1 for the two different n values of 2 and 6.

Table 1. Result comparison of Example 1 for different orthogonal polynomials at different values of n.

Polynomials	Maximum Absolute Errors		Root-Mean-Square Errors	
	$n = 2$	$n = 6$	$n = 2$	$n = 6$
S1	1.4584×10^{-1}	1.8326×10^{-7}	1.3923×10^{-2}	2.4900×10^{-8}
S2	1.6154×10^{-1}	4.2127×10^{-7}	2.0960×10^{-2}	7.1127×10^{-8}
S3	2.2296×10^{-1}	3.6897×10^{-7}	1.3179×10^{-2}	3.3726×10^{-8}
S4	4.1764×10^{-1}	1.0973×10^{-6}	3.2039×10^{-2}	1.7307×10^{-7}
S5	1.9055×10^{-1}	3.1593×10^{-7}	2.2138×10^{-2}	4.2368×10^{-8}

In Table 1, we have compared results for different polynomials, and it is observed that the results for Shifted Legendre polynomials and Gegenbauer polynomials are better than those for the other polynomials. It is also observed that the MAE and RMSE decrease with increasing n.

Example 2. Consider a non-linear fractional variational problem as in Equation (1) with $g(x) = h(x) = e^{-vx}$; we then have the following non-linear fractional variational problem [19]:

$$J(y) = \int_0^1 \left(e^{-vx} D^\alpha y(x) - v \left(I^{1-\alpha} y(x) + 1 \right) e^{-vx} \right)^2 dx \qquad (57)$$

under the constraints

$$y(0) = 0, \qquad I^{1-\alpha} y(1) = \epsilon.$$

The exact solution of the above equation is given as

$$y_{exact}(x) = \left(e^{-1}(1+\epsilon) - 1 \right) v^{-\alpha} \left(\sum_{k=0}^{\infty} \frac{(k+1)}{\Gamma(k+\alpha+1)} (vx)^{k+\alpha} \right) + x^{\alpha-1} E_{1,\alpha}(vx) - \frac{x^{\alpha-1}}{\Gamma(\alpha)}$$

where $E_{a,b}(x)$ is the Mittag-Leffler function of order a and b and is defined as

$$E_{a,b}(x) = \sum_{k=0}^{\infty} \frac{x^k}{\Gamma(ak+b)}.$$

We discuss Example 2 for different α values of 0.5, 0.6, 0.7, 0.8, 0.9, and 1 and $\epsilon = 2$.

In Figures 11–15, it is shown that the solutions for the two different values of $\alpha = 0.8$ and $\alpha = 1$ coincide with the exact solutions for different orthogonal polynomials at $n = 5$.

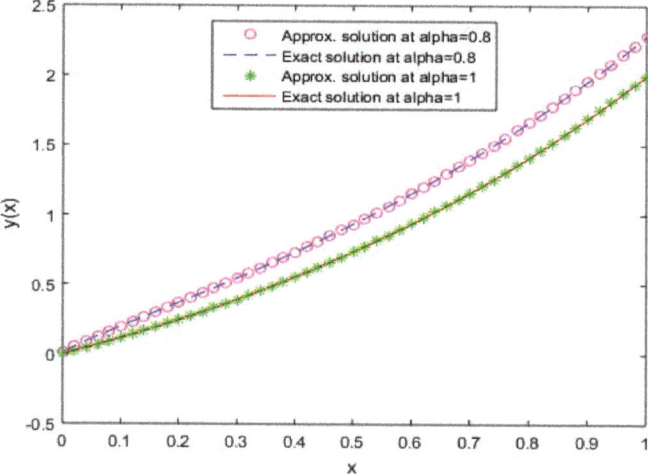

Figure 11. Comparison of exact and numerical solutions using S1 for $\alpha = 0.8$ and $\alpha = 1$, Example 2.

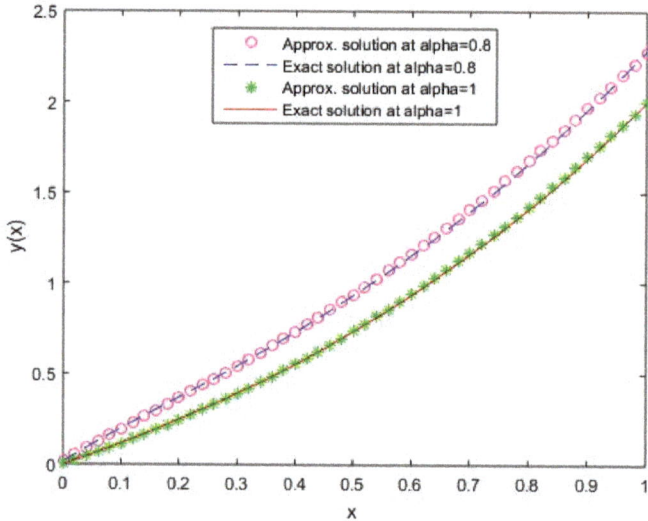

Figure 12. Comparison of exact and numerical solutions using S2 for $\alpha = 0.8$ and $\alpha = 1$, Example 2.

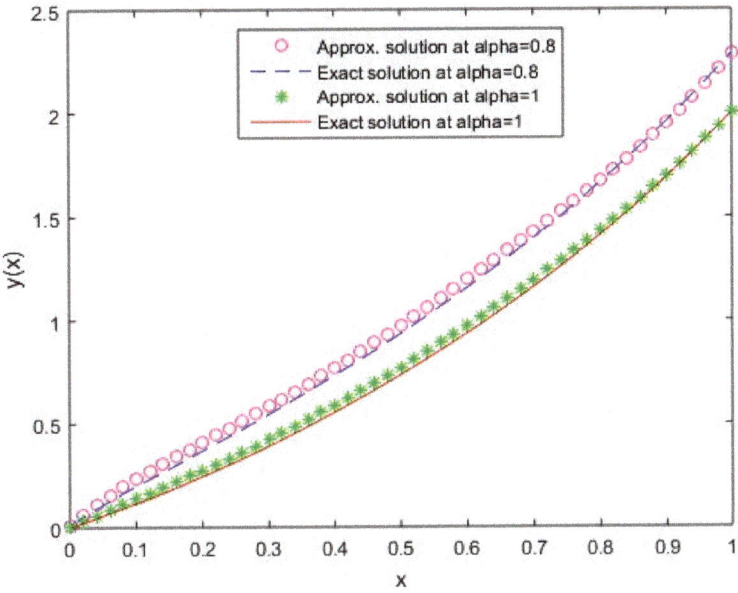

Figure 13. Comparison of exact and numerical solutions using S3 for $\alpha = 0.8$ and $\alpha = 1$, Example 2.

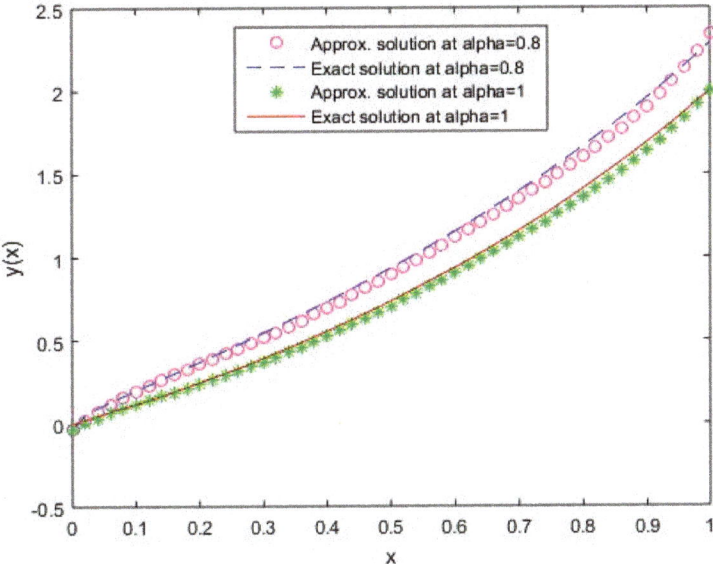

Figure 14. Comparison of exact and numerical solutions using S4 for $\alpha = 0.8$ and $\alpha = 1$, Example 2.

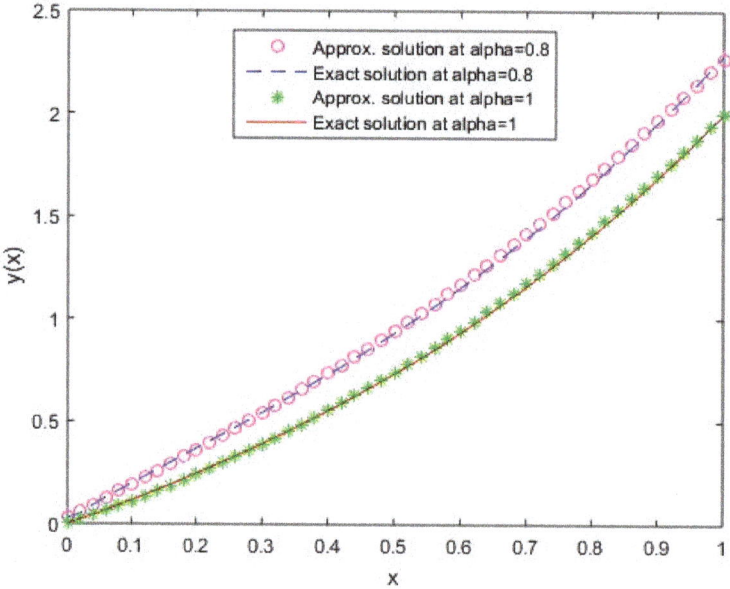

Figure 15. Comparison of exact and numerical solutions using S5 for $\alpha = 0.8$ and $\alpha = 1$, Example 2.

Figures 16–20 reflect that the approximate solution varies continuously for Shifted Legendre polynomials, Shifted Chebyshev polynomials of the second kind, Shifted Chebyshev polynomials of the third kind, Shifted Chebyshev polynomials of the fourth kind, and Gegenbauer polynomials, respectively, with different values of fractional order.

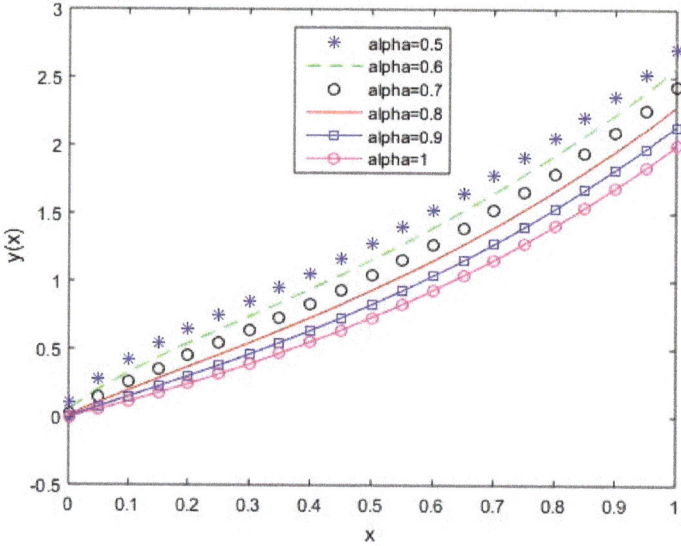

Figure 16. The behavior of solutions using S1 for α values of 0.5, 0.6, 0.7, 0.8, 0.9, and 1, Example 2.

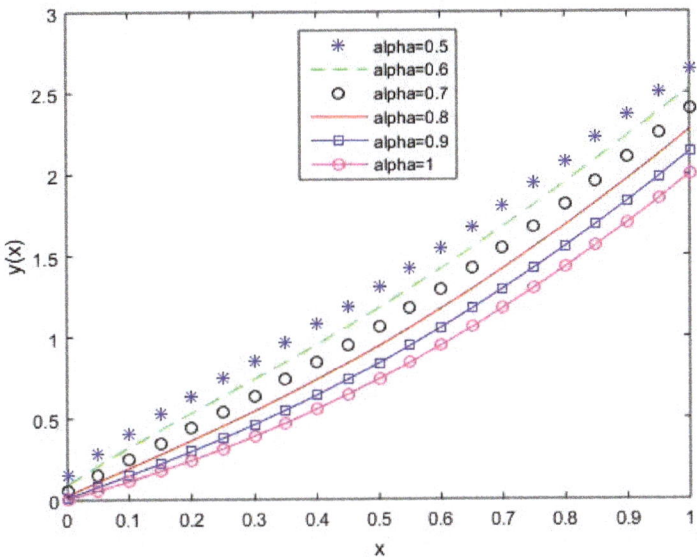

Figure 17. The behavior of solutions using S2 for α values of 0.5, 0.6, 0.7, 0.8, 0.9, and 1, Example 2.

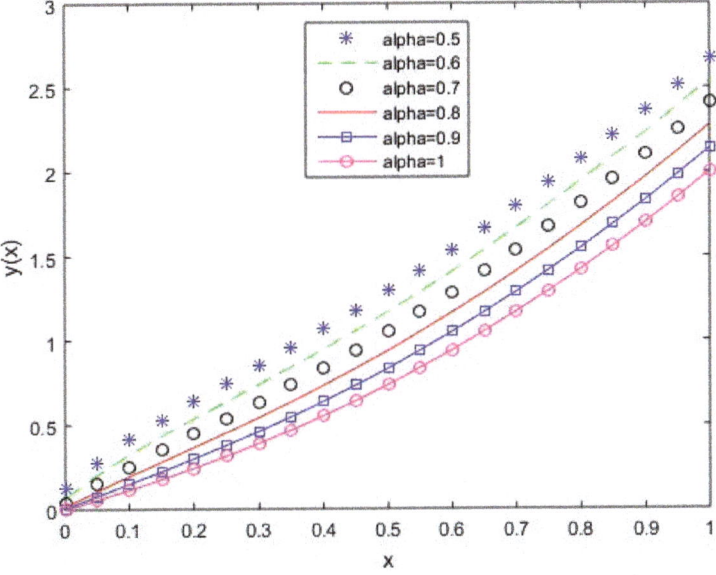

Figure 18. The behavior of solutions using S3 for α values of 0.5, 0.6, 0.7, 0.8, 0.9, and 1, Example 2.

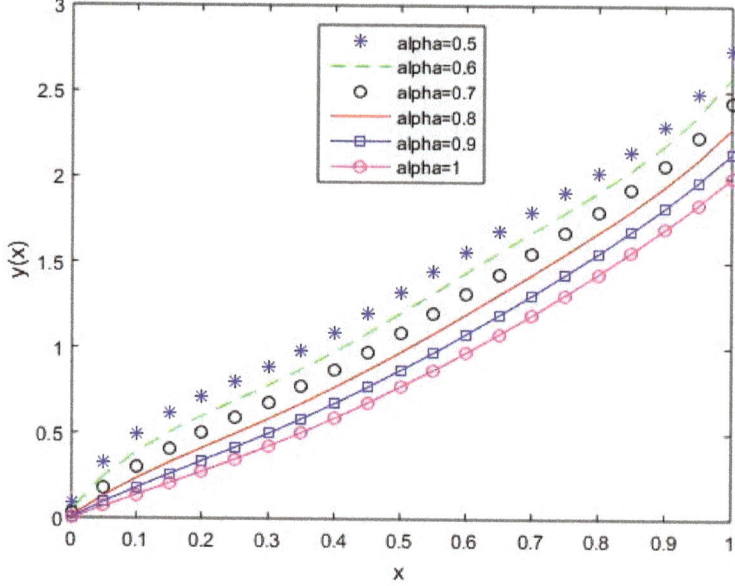

Figure 19. The behavior of solutions using S4 for α values of 0.5, 0.6, 0.7, 0.8, 0.9, and 1, Example 2.

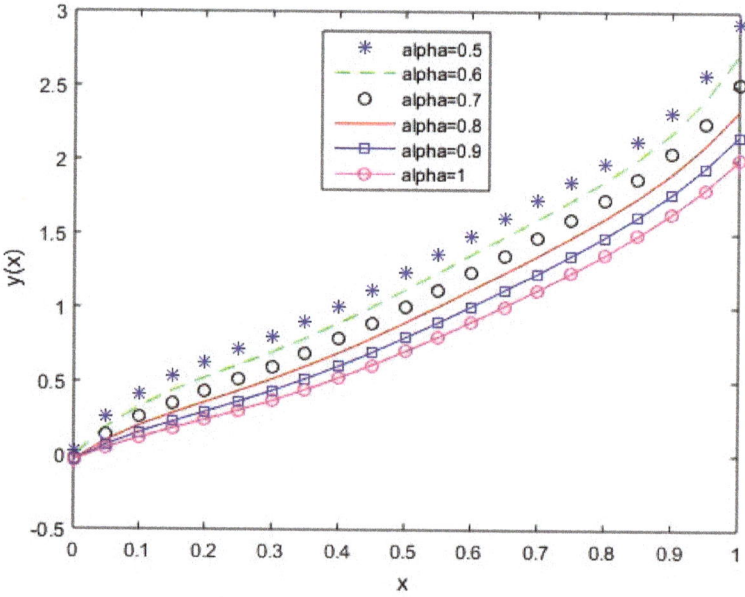

Figure 20. The behavior of solutions using S5 for α values of 0.5, 0.6, 0.7, 0.8, 0.9, and 1, Example 2.

In Table 2, we have listed the maximum absolute errors (MAE) and root-mean-square errors (RMSE) for Example 2 for the two n values 2 and 6.

Table 2. Result comparison of Example 2 for different orthogonal polynomials at different values of n.

Polynomials	Maximum Absolute Errors		Root-Mean-Square Errors	
	$n = 2$	$n = 6$	$n = 2$	$n = 6$
S1	2.0407×10^{-2}	1.4819×10^{-7}	2.7038×10^{-3}	1.5021×10^{-8}
S2	2.4295×10^{-2}	1.3713×10^{-2}	3.1356×10^{-3}	1.6490×10^{-3}
S3	8.0010×10^{-2}	4.8371×10^{-2}	1.2590×10^{-2}	8.3236×10^{-3}
S4	1.1193×10^{-1}	5.0558×10^{-2}	9.8251×10^{-3}	6.8594×10^{-3}
S5	2.5349×10^{-2}	1.9316×10^{-2}	3.2343×10^{-3}	2.4746×10^{-3}

In Table 2, we have compared results for different polynomials, and it is observed that the results for the Shifted Legendre polynomial are better than those for the other polynomials. It is also observed that the MAE and RMSE decrease as n increases.

8. Conclusions

We extended the Ritz method [18,20–22,32,33] for solving a class of NLFVPs using different orthogonal polynomials such as shifted Legendre polynomials, shifted Chebyshev polynomials of the first kind, shifted Chebyshev polynomials of the third kind, shifted Chebyshev polynomials of the fourth kind, and Gegenbauer polynomials. These polynomials were used to approximate the unknown function in the NLFVP. The advantage of the method is that it converts the given NLFVPs into a set of non-linear algebraic equations which are then solved numerically. The error bound of the approximation method for NLFVP was established. It was also shown that the approximate numerical solution converges to the exact solution as we increase the number of basis functions in the approximation. At the end, numerical results were provided by applying the method to two test examples, and it was observed that the results showed good agreement with the exact solution. Numerical results obtained using different orthogonal polynomials were compared. A comparative study showed that the shifted Legendre polynomials were more accurate in approximating the numerical solution.

Author Contributions: All authors contributed equally to the writing of this paper. All authors read and approved the final manuscript.

Funding: This research received no external funding.

Acknowledgments: The authors are very grateful to the referees for their constructive comments and suggestions for the improvement of the paper.

Conflicts of Interest: The authors declare no conflict of interest.

Appendix A

Theorem A1. *Let* $f : [0, 1] \to R$ *be a function such that* $f \in C^{(N+1)}[0, 1]$ *and let* $f_N(x)$ *be the Nth approximation of the function from* $P_N^{(a, b)}(x) = \text{span} \{\Psi_0(x), \Psi_1(x), \ldots, \Psi_N(x)\}$; *then* [45]

$$\|f(x) - f_N(x)\|^2_{w^{(a, b)}} \leq \frac{K}{(N+1)!} \sqrt{\frac{\Gamma(1+a)\Gamma(3+2N+b)}{\Gamma(4+2N+a+b)}},$$

where $K = \max_{x \in [0,1]} \left| f^{(N+1)}(x) \right|$.

Proof. Since $f \in C^{(N+1)}[0, 1]$, the Taylor polynomial of f at $x = 0$, is given as

$$g_1(x) = f(0) + f'(0)x + \cdots + f^N(0)\frac{x^N}{N!}.$$

The upper bound of the error of the Taylor polynomial is given as

$$|f(x) - g_1(x)| \leq \frac{Kx^{N+1}}{(N+1)!},$$

where $K = \max\limits_{x \in [0,1]} \left| f^{(N+1)}(x) \right|$.

Since $f_N(x)$ and $g_1(x) \in P_N^{(a,\,b)}(x)$, we have

$$\|f(x) - f_N(x)\|^2_{w^{(a,\,b)}} \leq \|f(x) - g_1(x)\|^2_{w^{(a,\,b)}} \leq \left(\frac{K}{(N+1)!}\right)^2 \int_0^1 x^{2N+2+b}(1-x)^a dx$$

$$= \left(\frac{K}{(N+1)!}\right)^2 \frac{\Gamma(1+a)\Gamma(3+2N+b)}{\Gamma(4+2N+a+b)},$$

$$\|f(x) - f_N(x)\|^2_{w^{(a,\,b)}} \leq \frac{K}{(N+1)!} \sqrt{\frac{\Gamma(1+a)\Gamma(3+2N+b)}{\Gamma(4+2N+a+b)}},$$

which shows that $\lim\limits_{N \to \infty} \|f(x) - f_N(x)\|^2_{w^{(a,\,b)}} = 0$. □

References

1. Dym, C.L.; Shames, I.H. *Solid Mechanics: A Variational Approach*; McGraw-Hill: New York, NY, USA, 1973.
2. Frederico, G.S.F.; Torres, D.F.M. Fractional conservation laws in optimal control theory. *Nonlinear Dyn.* **2008**, *53*, 215–222. [CrossRef]
3. Pirvan, M.; Udriste, C. Optimal control of electromagnetic energy. *Balk. J. Geom. Appl.* **2010**, *15*, 131–141.
4. Agrawal, O.P. A general finite element formulation for fractional variational problems. *J. Math. Anal. Appl.* **2008**, *337*, 1–12. [CrossRef]
5. Dehghan, M.; Tatari, M. The use of Adomian decomposition method for solving problems in calculus of variations. *Math. Probl. Eng.* **2006**, *2006*, 1–12. [CrossRef]
6. Singh, H. A new stable algorithm for fractional Navier-Stokes equation in polar coordinate. *Int. J. Appl. Comp. Math.* **2017**, *3*, 3705–3722. [CrossRef]
7. Robinson, A.D. The use of control systems analysis in neurophysiology of eye movements. *Ann. Rev. Neurosci.* **1981**, *4*, 462–503. [CrossRef] [PubMed]
8. Singh, H. Operational matrix approach for approximate solution of fractional model of Bloch equation. *J. King Saud Univ.-Sci.* **2017**, *29*, 235–240. [CrossRef]
9. Bagley, R.L.; Torvik, P.J. Fractional calculus a differential approach to the analysis of viscoelasticity damped structures. *AIAA J.* **1983**, *21*, 741–748. [CrossRef]
10. Bagley, R.L.; Torvik, P.J. Fractional calculus in the transient analysis of viscoelasticity damped structures. *AIAA J.* **1985**, *23*, 918–925. [CrossRef]
11. Panda, R.; Dash, M. Fractional generalized splines and signal processing. *Signal Process.* **2006**, *86*, 2340–2350. [CrossRef]
12. Magin, R.L. Fractional calculus in bioengineering. *Crit. Rev. Biomed. Eng.* **2004**, *32*, 1–104. [CrossRef] [PubMed]
13. Bohannan, G.W. Analog fractional order controller in temperature and motor control applications. *J. Vib. Control.* **2008**, *14*, 1487–1498. [CrossRef]
14. Novikov, V.V.; Wojciechowski, K.W.; Komkova, O.A.; Thiel, T. Anomalous relaxation in dielectrics. Equations with fractional derivatives. *Mater. Sci.* **2005**, *23*, 977–984.
15. Agrawal, O.P. A new Lagrangian and a new Lagrange equation of motion for fractionally damped systems. *J. Appl. Mech.* **2013**, *237*, 339–341. [CrossRef]
16. Mozyrska, D.; Torres, D.F.M. Minimal modified energy control for fractional linear control systems with the Caputo derivative. *Carpath. J. Math.* **2010**, *26*, 210–221.
17. Almeida, R.; Malinowska, R.; Torres, D.F.M. A fractional calculus of variations for multiple integrals with application to vibrating string. *J. Math. Phys.* **2010**, *51*, 033503. [CrossRef]

18. Lotfi, A.; Dehghan, M.; Yousefi, S.A. A numerical technique for solving fractional optimal control problems. *Comput. Math. Appl.* **2011**, *62*, 1055–1067. [CrossRef]
19. Khosravian-Arab, H.; Almeida, R. Numerical solution for fractional variational problems using the Jacobi polynomials. *Appl. Math. Modell.* **2015**. [CrossRef]
20. Agrawal, O.P. Formulation of Euler-Lagrange equations for fractional variational problems. *J. Math. Anal. Appl.* **2002**, *272*, 368–379. [CrossRef]
21. Doha, E.H.; Bhrawy, A.H.; Baleanu, D.; Ezz-Eldien, S.S.; Hafez, R.M. An efficient numerical scheme based on the shifted orthonormal Jacobi polynomials for solving fractional optimal control problems. *Adv. Differ. Equ.* **2015**. [CrossRef]
22. Ezz-Eldie, S.S.; Doha, E.H.; Baleanu, D.; Bhrawy, A.H. A numerical approach based on Legendre orthonormal polynomials for numerical solutions of fractional optimal control problems. *J. Vib. Ctrl.* **2015**. [CrossRef]
23. Osama, H.M.; Fadhel, S.F.; Zaid, A.M. Numerical solution of fractional variational problems using direct Haar wavelet method. *Int. J. Innov. Res. Sci. Eng. Technol.* **2014**, *3*, 12742–12750.
24. Ezz-Eldien, S.S. New quadrature approach based on operational matrix for solving a class of fractional variational problems. *J. Comp. Phys.* **2016**, *317*, 362–381. [CrossRef]
25. Bastos, N.; Ferreira, R.; Torres, D.F.M. Discrete-time fractional variational problems. *Signal Process.* **2011**, *91*, 513–524. [CrossRef]
26. Wang, D.; Xiao, A. Fractional variational integrators for fractional variational problems. *Commun. Nonlinear Sci. Numer. Simul.* **2012**, *17*, 602–610. [CrossRef]
27. Odzijewicz, T.; Malinowska, A.B.; Torres, D.F.M. Fractional variational calculus with classical and combined Caputo derivatives. *Nonlinear Anal.* **2012**, *75*, 1507–1515. [CrossRef]
28. Bhrawy, A.H.; Ezz-Eldien, S.S. A new Legendre operational technique for delay fractional optimal control problems. *Calcolo* **2015**. [CrossRef]
29. Tavares, D.; Almeida, R.; Torres, D.F.M. Optimality conditions for fractional variational problems with dependence on a combined Caputo derivative of variable order. *Optimization* **2015**, *64*, 1381–1391. [CrossRef]
30. Ezz-Eldien, S.S.; Hafez, R.M.; Bhrawy, A.H.; Baleanu, D.; El-Kalaawy, A.A. New numerical approach for fractional variational problems using shifted Legendre orthonormal polynomials. *J. Optim. Theory Appl.* **2017**, *174*, 295–320. [CrossRef]
31. Almeida, R. Variational problems involving a Caputo-type fractional derivative. *J. Optim. Theory Appl.* **2017**, *174*, 276–294. [CrossRef]
32. Pandey, R.K.; Agrawal, O.P. Numerical Scheme for Generalized Isoparametric Constraint Variational Problems with A-Operator. In Proceedings of the ASME International Design Engineering Technical Conferences and Computers and Information in Engineering Conference, Portland, OR, USA, 4–7 August 2013. [CrossRef]
33. Pandey, R.K.; Agrawal, O.P. Numerical scheme for a quadratic type generalized isoperimetric constraint variational problems with A.-operator. *J. Comput. Nonlinear Dyn.* **2015**, *10*, 021003. [CrossRef]
34. Pandey, R.K.; Agrawal, O.P. Comparison of four numerical schemes for isoperimetric constraint fractional variational problems with A-operator. In Proceedings of the ASME 2015 International Design Engineering Technical Conferences and Computers and Information in Engineering Conference, Boston, MA, USA, 2–5 August 2015. [CrossRef]
35. Singh, H.; Srivastava, H.M.D.; Kumar, D. A reliable numerical algorithm for the fractional vibration equation. *Chaos Solitons Fractals* **2017**, *103*, 131–138. [CrossRef]
36. Singh, O.P.; Singh, V.K.; Pandey, R.K. A stable numerical inversion of Abel's integral equation using almost Bernstein operational matrix. *J. Quant. Spec. Rad. Trans.* **2012**, *111*, 567–579. [CrossRef]
37. Zhou, F.; Xu, X. Numerical solution of convection diffusions equations by the second kind Chebyshev wavelets. *Appl. Math. Comput.* **2014**, *247*, 353–367. [CrossRef]
38. Yousefi, S.A.; Behroozifar, M.; Dehghan, M. The operational matrices of Bernstein polynomials for solving the parabolic equation subject to the specification of the mass. *J. Comput. Appl. Math.* **2011**, *235*, 5272–5283. [CrossRef]
39. Singh, H. A New Numerical Algorithm for Fractional Model of Bloch Equation in Nuclear Magnetic Resonance. *Alex. Eng. J.* **2016**, *55*, 2863–2869. [CrossRef]

40. Khalil, H.; Khan, R.A. A new method based on Legendre polynomials for solutions of the fractional two dimensional heat conduction equations. *Comput. Math. Appl.* **2014**, *67*, 1938–1953. [CrossRef]
41. Singh, C.S.; Singh, H.; Singh, V.K.; Singh, O.P. Fractional order operational matrix methods for fractional singular integro-differential equation. *Appl. Math. Modell.* **2016**, *40*, 10705–10718. [CrossRef]
42. Rivlin, T.J. *An Introduction to the Approximation of Functions*; Dover Publication: New York, NY, USA, 1981.
43. Kreyszig, E. *Introductory Functional Analysis with Applications*; John Wiley and Sons, Inc.: Hoboken, NJ, USA, 1978.
44. Bhrawy, A.H.; Tharwat, M.M.; Alghamdi, M.A. A new operational matrix of fractional integration for shifted Jacobi polynomials. *Bull. Malays. Math. Sci. Soc.* **2014**, *37*, 983–995.
45. Behroozifar, M.; Sazmand, A. An approximate solution based on Jacobi polynomials for time-fractional convection–diffusion equation. *Appl. Math. Comput.* **2017**, *296*, 1–17. [CrossRef]

© 2019 by the authors. Licensee MDPI, Basel, Switzerland. This article is an open access article distributed under the terms and conditions of the Creative Commons Attribution (CC BY) license (http://creativecommons.org/licenses/by/4.0/).

Article

Approximate Controllability of Sub-Diffusion Equation with Impulsive Condition

Lakshman Mahto [1], Syed Abbas [2,*], Mokhtar Hafayed [3] and Hari M. Srivastava [4,5]

[1] Department of Science and Humanities, Indian Institute of Information Technology Dharwad, Hubli 580029, India; lm.iitmandi@gmail.com
[2] School of Basic Sciences, Indian Institute of Technology Mandi, Mandi 175001, H.P., India
[3] Laboratory of Applied Mathematics, Biskra University, Biskra 07000, Algeria; hafa.mokh@yahoo.com
[4] Department of Mathematics and Statistics, University of Victoria, Victoria, BC V8W 3R4, Canada; harimsri@math.uvic.ca
[5] Department of Medical Research, China Medical University Hospital, China Medical University, Taichung 40402, Taiwan
* Correspondence: sabbas.iitk@gmail.com

Received: 7 January 2019; Accepted: 14 February 2019; Published: 17 February 2019

Abstract: In this work, we study an impulsive sub-diffusion equation as a fractional diffusion equation of order $\alpha \in (0,1)$. Existence, uniqueness and regularity of solution of the problem is established via eigenfunction expansion. Moreover, we establish the approximate controllability of the problem by applying a unique continuation property via internal control which acts on a sub-domain.

Keywords: fractional diffusion equation; controllability; impulsive system; unique continuation property

MSC: 93B05; 34A08; 47H10; 49J15; 26A33; 45K05

---

1. Introduction

1.1. Fractional Diffusion Equations

A fractional diffusion equation of order $\alpha \in (0,1)$ is obtained by rewriting a normal diffusion equation in integral form as

$$u(x,t) + \int_0^t Au(x,t)dt = u_0 + \int_0^t f(x,t)dt, \quad (x,t) \in \Omega \times (0,T). \tag{1}$$

Then, replacing the first of right-hand side (RHS) integral of Equation (1) by a Riemann-Liouville fractional integral, I^α of order $0 < \alpha < 1$, we get

$$u(x,t) + \int_0^t \frac{(t-s)^{\alpha-1}}{\Gamma(\alpha)} Au(x,t)dt = u_0 + \int_0^t f(x,t)dt, \quad (x,t) \in \Omega \times (0,T).$$

Now, differentiating the above equation on both sides with respect to t, we get the following fractional diffusion equation:

$$\begin{aligned}
\partial_t u + \partial_t^{1-\alpha} Au &= f(x,t) & &\text{in} \quad \Omega \times (0,T), \\
u &= 0 & &\text{on} \quad (\Gamma = \partial\Omega) \times (0,T), \\
u(\cdot, 0) &= u_0 & &\text{in} \quad \Omega.
\end{aligned} \tag{2}$$

If $\alpha = 1$, then Equation (2) is a classical diffusion equation. Equation (2) with $0 < \alpha < 1$ is called the fractional diffusion equation. These equations appear in the model of anomalous diffusion in heterogeneous media. Anomalous diffusion is one of the most ubiquitous phenomena in nature; it has been observed in various fields of physical sciences, for example, surface growth, transport of fluid in porous media, two-dimensional rotating flow and diffusion of plasma. Because of such anomalies, the classical diffusion models can not be used to study the dynamics of such systems. In this situation, fractional derivatives extend the help and play a crucial role in characterizing such diffusion. The model corresponding to such derivative is called a fractional partial differential equation. From the continuous time random walk (CTRW) model, Metzler and Klafter [1] derived Equation (3) with $0 < \alpha < 1$ as a macroscopic model.

1.2. Impulsive Partial Differential Equations

Impulsive partial differential equations are a very important class of differential equations. These equations arise from the modelling of various real world processes having memory and are subject to short time fluctuations. The theory of impulsive differential equation is very rich and wide. It is mainly due to the fact that the it inherit intrinsic difficulties of the problems. These kinds of equations have lots of applications in different branches of Science and Engineering. These kinds of equations arise naturally from several physical and natural processes like earthquakes and pulse vaccination strategy. For more information, we refer to [2–4] and references therein. For more theoretical work, one can see the interesting book by Bainov and Simeonov [5]. The authors Shun et al. in [6] consider second-order impulsive Hamiltonian systems and established the existence of infinitely many solutions.

1.3. Controllability

In mathematical control theory, controllability and optimal control are two important concepts. In controllability, one studies the steering of a dynamical system from a given initial state to any other state or in the neighborhood of the state under some admissible control input. The cases where target states are defined in a given subregion are particularly very important; this situation arises in many real world applications. The last few decades have seen tremendous work in the controllability problems for integer order systems. Several techniques have been developed for solving such problems [7,8]. It has been seen that mostly authors worked on the problems with hard constraints on the state or control. This is mainly due to its applicability and importance in various applications in optimal control. Moreover, many authors have studied controllability of the semilinear, partial evolution equations, we refer to [9–17] and references therein. In a very interesting paper [14], Kenichi Fujishiro and Masahiro Yamamoto consider a partial differential equations with fractional order time derivatives and established approximate controllability by interior control.

1.4. The Problem under Consideration

Let Ω be a bounded domain of $\mathbb{R}^d$ with C^2 boundary $\Gamma = \partial\Omega$. We consider the following initial value/boundary value problem of an impulsive sub-diffusion equation of order $\alpha \in (0,1)$:

$$\begin{cases} \partial_t u + \partial_t^{1-\alpha} Au = f(x,t) & \text{in } \Omega \times (0,T), \\ \Delta u(\cdot, t_i) = I_i(u(\cdot, t_i)), i = 1,2,3,\cdots,P, \\ u = 0 & \text{on } \Gamma \times (0,T), \\ u(\cdot, 0) = u_0 & \text{in } \Omega. \end{cases} \quad (3)$$

In Equation (3), $u = u(x,t)$ is the state to be controlled and $f = f(x,t)$ is the control which is localized in a subdomain ω of Ω. We will act by f to drive the initial state $u_0 = u_0(x)$ to some target function $u_1 = u_1(x)$. The operator A is a symmetric and uniformly elliptic operator. The details will be specified later; $T > 0$ is also a constant. Several problems in applications can be modeled by the above equation. Some of them are: thermal diffusion in media with fractional geometry, underground environmental problems, highly heterogeneous aquifer, etc. [18]. In this paper, we study approximate controllability for fractional partial differential equations with impulses. We say that Equation (3) is approximately controllable if, for any $u_1 \in L^2(\Omega)$ and $\varepsilon > 0$, there exists a control f such that the solution u of (3) satisfies

$$\|u(\cdot,T) - u_1\|_{L^2(\Omega)} \leq \varepsilon. \tag{4}$$

This paper is divided into four sections. In Section 2, we study requisite function spaces and some important basic results. In Section 3, we analyse the mild solutions of the Equation (3) by eigenfunction expansion. Section 4 is devoted to the study of a dual system of (3) and to establish a unique continuation property. In the last section, we establish the proof of approximate controllability.

2. Preliminaries

In this section, we state a few function spaces, notations and results in order to establish our main results. For the smooth reading of the manuscript, we first define the following class of spaces (for more details, we refer to Adams [19], Mahto [12]):

$$L^p[a,b] = \left\{ f : [a,b] \to \mathbb{R} \mid f \text{ is Lebesgue measurable and } \int_a^b |f(t)|^p dt < \infty \right\},$$

$$AC[a,b] = \left\{ f : [0,T] \to \mathbb{R} \mid f \text{ is absolutely continuous on } [a,b] \right\},$$

$$C[a,b] = \left\{ f : [0,T] \to \mathbb{R} \mid f \text{ is continuous on } [a,b] \right\},$$

$$L^p(\Omega) = \left\{ f : \Omega \to \mathbb{R} \mid f \text{ is Lebesgue measurable and } \int_\Omega |f(x)|^p dx < \infty \right\},$$

$$H^1(\Omega) = \left\{ f : f, \frac{\partial f}{\partial x_1}, \cdots, \frac{\partial f}{\partial x_d} \in L^2(\Omega) \right\},$$

$$H_0^1(\Omega) = \left\{ f : f \in H^1(\Omega) \text{ and } f = 0 \text{ on } \Gamma \right\},$$

$$AC(0,T; L^2(\Omega)) = \left\{ f : [0,T] \to L^2(\Omega) \mid f \in AC([t_0,t_1]; L^2(\Omega)) \cup AC((t_i, t_{i+1}], L^2(\Omega), i = 1, \right.$$
$$\left. 2, \cdots, P, x(t_i^+), x(t_i^-) \text{ exist and } x(t_i) = x(t_i^-) \right\},$$

$$PC(0,T; L^2(\Omega)) = \left\{ f : [0,T] \to L^2(\Omega) \mid f \in C([t_0,t_1]; L^2(\Omega)) \cup C((t_i, t_{i+1}], L^2(\Omega), i = 1, 2, \right.$$
$$\left. \cdots, P, x(t_i^+), x(t_i^-) \text{ exist and } x(t_i) = x(t_i^-) \right\}.$$

The functions and operators defined below are very standard in the fractional calculus. For more details, we refer to [20]:

1. Mittag-Leffler function by

$$E_{\alpha,\beta}(z) := \sum_{k=0}^{\infty} \frac{z^k}{\Gamma(\alpha k + \beta)}, \quad z \in \mathbb{C},$$

where $\alpha > 0$ and $\beta \in \mathbb{R}$ are arbitrary constants. We can directly verify that $E_{\alpha,\beta}(z)$ is an entire function of $z \in \mathbb{C}$. As for the Mittag–Leffler functions, we have the following lemma.

Lemma 1. Let $0 < \alpha < 2$ and $\beta \in \mathbb{R}$ be arbitrary and μ satisfy $\pi\alpha/2 < \mu < \min\{\pi, \pi\alpha\}$. Then, there exists a constant $C = C(\alpha, \beta, \mu) > 0$ such that

$$|E_{\alpha,\beta}(z)| \leq \frac{C}{1+|z|}, \quad \mu \leq |\arg(z)| \leq \pi. \tag{5}$$

2. Reimann-Liouville integrals: For $\alpha > 0$ and $f \in L^1(0,T)$, we define α-th order forward and backward integrals of f by

$$I_{0+}^{\alpha} f(t) := \frac{1}{\Gamma(\alpha)} \int_0^t (t-\tau)^{\alpha-1} f(\tau) d\tau,$$

$$I_{T-}^{\alpha} f(t) := \frac{1}{\Gamma(\alpha)} \int_t^T (\tau-t)^{\alpha-1} f(\tau) d\tau.$$

In other words, the forward integral operators of α-th order is the convolution with $t^{\alpha-1}/\Gamma(\alpha)$ and consequently $I_{0+}^{\alpha} f$ also belongs to $L^1(0,T)$. The same argument is also valid for the backward integrals.

3. The Riemann-Liouvill fractional derivatives: For $\alpha \in (0,1)$, we define the forward and backward fractional derivatives of $f \in AC[0,T]$ by

$$\partial_t^{\alpha} f(t) := \frac{d}{dt} I^{1-\alpha} h(t) = \frac{1}{\Gamma(1-\alpha)} \frac{d}{dt} \int_0^t (t-\tau)^{-\alpha} h(\tau) d\tau, \tag{6}$$

$$D_t^{\alpha} f(t) := \frac{1}{\Gamma(1-\alpha)} \left(-\frac{d}{dt}\right) \int_t^T (\tau-t)^{-\alpha} h(\tau) d\tau. \tag{7}$$

We also have the following lemmas for fractional integration by parts.

Lemma 2. Let $\alpha > 0$. If $f, g \in PC([0,T], L^2(\Omega))$, then

$$\int_0^T I_{0+}^{\alpha} f(t) g(t) dt = \int_0^T f(t) I_{T-}^{\alpha} g(t) dt.$$

Proof.

$$\int_0^T g(t) I_{0+}^{\alpha} f(t) dt = \int_0^T g(t) \int_0^t \frac{(t-s)^{\alpha-1}}{\Gamma(\alpha)} f(s) ds dt$$

$$= \int_0^T f(t) \int_t^T \frac{(s-t)^{\alpha-1}}{\Gamma(\alpha)} g(s) ds dt$$

(using Fubini theorem for change of order of integration.)

$$= \int_0^T f(t) I_{T-}^{\alpha} g(t) dt.$$

□

Lemma 3. Let $f \in PC(0,T), g \in C_0^{\infty}(0,T)$. Then, we have the following identity:

$$\int_0^T g(t) \partial_t^{\alpha} f(t) dt = \int_0^T f(t) D_t^{\alpha} g(t) dt. \tag{8}$$

Proof. By substituing the value of R-L fractional derivative, we obtain

$$\int_0^T g(t)\partial_t^\alpha f(t)dt$$

$$= \int_0^T g(t)\frac{d}{dt}\int_0^t \frac{(t-s)^{-\alpha}}{\Gamma(1-\alpha)}f(s)dsdt$$

$$= \left(\frac{g(t)}{\Gamma(1-\alpha)}\int_0^t (t-s)^{-\alpha}f(s)ds\right)_{t=0}^{t=T} - \int_0^T g'(t)\int_0^t \frac{(t-s)^{-\alpha}}{\Gamma(1-\alpha)}f(s)dsdt$$

(using integration by parts.)

$$= \frac{g(T)}{\Gamma(1-\alpha)}\int_0^T (t-s)^{-\alpha}f(s)ds - \int_0^T g'(t)\int_0^t \frac{(t-s)^{-\alpha}}{\Gamma(1-\alpha)}f(s)dsdt$$

$$= -\int_0^T g'(t)\int_0^t \frac{(t-s)^{-\alpha}}{\Gamma(1-\alpha)}f(s)dsdt \quad (\because g(T)=0.)$$

$$= -\int_0^T f(t)\int_t^T \frac{(s-t)^{-\alpha}}{\Gamma(1-\alpha)}g'(s)dsdt$$

(using Fubini theorem for change of order of integration.)

$$= -\int_0^T f(t)\frac{d}{dt}\int_t^T \frac{(s-t)^{1-\alpha}}{\Gamma(2-\alpha)}g'(s)dsdt$$

(using Leibnitz theorem for differentiation under integration.)

$$= \int_0^T f(t)\left(\frac{g(T)(T-t)^{-\alpha}}{\Gamma(1-\alpha)} - \frac{d}{dt}\int_t^T \frac{(s-t)^{1-\alpha}}{\Gamma(2-\alpha)}g'(s)ds\right)dt$$

$$(\because g(T)=0.)$$

$$= \int_0^T f(t)\left(\frac{d}{dt}\left(\frac{g(T)(T-t)^{1-\alpha}}{\Gamma(2-\alpha)}\right) - \frac{d}{dt}\int_t^T \frac{(s-t)^{1-\alpha}}{\Gamma(2-\alpha)}g'(s)ds\right)dt$$

$$= \int_0^T f(t)\left(\frac{d}{dt}\left(\frac{g(s)(s-t)^{1-\alpha}}{\Gamma(2-\alpha)}\right)\Big|_{s=t}^{s=T} - \frac{d}{dt}\int_t^T \frac{(s-t)^{1-\alpha}}{\Gamma(2-\alpha)}g'(s)ds\right)dt$$

$$= -\int_0^T f(t)\frac{d}{dt}\int_t^T \frac{(t-s)^{-\alpha}}{\Gamma(1-\alpha)}g(s)dsdt$$

(using integration by parts.)

$$= \int_0^T f(t)D_t^\alpha g(t)dt.$$

□

3. Solution of Primal System

3.1. Representation of the Solution

To derive the representation, we first focus on $t \in [0, t_1]$. We can rewrite (3) as

$$\partial_t u + (\beta * Au)_t = f(\cdot, t), u(0) = u_0, \qquad (9)$$

where $\beta(t) = \frac{t^{\alpha-1}}{\Gamma(\alpha)}$ and $Au = -\nabla^2 u$ is a symmetric, self-adjoint, uniformly elliptic operator with domain $D(A) = H^2(\Omega) \cap H_0^1(\Omega)$, the spectrum of A is entirely composed of a countable number of eigenvalues and we can set with finite multiplicities:

$$0 < \lambda_1 \leq \lambda_2 \leq \cdots \leq \lambda_n \leq \cdots.$$

By $\varphi_n \in H^2(\Omega) \cap H_0^1(\Omega)$, we denote the orthonormal eigenfunction corresponding to λ_n:

$$A\varphi_n = \lambda_n \varphi_n, \quad n = 1, 2, \cdots.$$

Then, the sequence $\{\varphi_n\}_{n \in \mathbb{N}}$ is an orthonormal basis in $L^2(\Omega)$. Since $u(t) \in L^2(\Omega)$, we have

$$u(t) = \sum_{j=1}^{\infty} u_j(t) \varphi_j,$$

where $u_j(t) = (u(t), \varphi_j)$ is the jth Fourier coefficient. Taking an inner product between (9) and φ_j, we have an infinite number of linear integro-differential equations:

$$\partial_t u_j(t) + \lambda_j (\beta * u_j)_t = f_j(\cdot, t), \tag{10}$$

where $f_j(\cdot, t) = (f(\cdot, t), \varphi_j)$ and $u_{j0} = (u_0, \varphi_j)$.

Taking Laplace Transform both sides of (10), we get

$$z\hat{u}_j(z) - u_{j0} + \lambda_j z^{1-\alpha} \hat{u}_j(z) = \hat{f}_j,$$

where $\hat{h}_j(z) = \int_0^{\infty} e^{-zt} u_j(t) dt$ is the Laplace Transform of u_j. Simplifying, we get

$$\hat{u}_j = \left(\frac{u_{j0} + \hat{f}_j(z)}{z + \lambda_j z^{1-\alpha}} \right).$$

By taking the inverse Laplace Transform, we get

$$\mathcal{L}^{-1} \left(\frac{1}{z + \lambda_j z^{1-\alpha}} \right) = E_\alpha(-\lambda_j t^\alpha). \tag{11}$$

Now, the representation for u_j of (10) is given by

$$u_j = E_\alpha(-\lambda_j t^\alpha) u_{j0} + \int_0^{\infty} E_\alpha(-\lambda_j (t-s)^\alpha) f_j(\cdot, s) ds. \tag{12}$$

Thus, a formal solution of (9) is given by

$$u(t) = \mathcal{E}(t) u_0 + \int_0^t \mathcal{E}(t-s) f(\cdot, s) ds, \tag{13}$$

where

$$\mathcal{E}(t) u_0 = \sum_{j=1}^{\infty} E_\alpha(-\lambda_j t^\alpha)(u_0, \varphi_j) \varphi_j, \tag{14}$$

$$u(\cdot, t) = \begin{cases} \mathcal{E}(t) u_0 + \int_0^t \mathcal{E}(t-s) f(\cdot, s) ds, & t \in [0, t_1], \\ \mathcal{E}(t) u_0 + \sum_{t_i < t} \mathcal{E}(t - t_i) I_i(u(\cdot, t_i)) + \int_0^t \mathcal{E}(t-s) f(\cdot, s) ds, & t \in (t_i, t_{i+1}], \quad i = 1, \cdots, P. \end{cases} \tag{15}$$

3.2. Weak Formulation

Rewriting the (3) in unified form, we get

$$\begin{cases} \partial_t u + \partial_t^{1-\alpha} A u = f(x,t) + \sum_{1 \leq i \leq P} I_i(u(\cdot, t_i))\delta(t - t_i) & \text{in } \Omega \times (0, T), \\ u = 0 & \text{on } \Gamma \times (0, T), \\ u(\cdot, 0) = u_0 & \text{in } \Omega. \end{cases} \quad (16)$$

A weak formulation of (16) is to find a $u \in PC(0, T; H_0^1(\Omega))$ such that

$$(\partial_t u, v) + (\partial_t^{1-\alpha} A u, v) = (f, v) + \sum_i (I_i(u(\cdot, t_i))\delta(t - t_i), v), v \in H_0^1(\Omega). \quad (17)$$

Thus, we have a variational form of (16) as follows:

$$(\partial_t u, v)dt + a(u, v) = l(v), \quad (18)$$

where,

$$a(u, v) = (\partial_t^{1-\alpha} A u, v) = \int_\Omega \partial_t^{1-\alpha} \nabla u \cdot \nabla v dx,$$

$$l(v) = (f, v)dt + \sum_{i=1}^P (I_i(u(t_i))\delta(t - t_i), v),$$

with the following conditions:

1. $a(\cdot, \cdot)$ is bounded or continuous i.e. $|a(u,v)|_{H_0^1(\Omega)} \leq C_1 \|u\|_{H_0^1(\Omega)} \|v\|_{H_0^1(\Omega)}$,
2. $a(\cdot, \cdot)$ is coercive i.e. $A(u, u) \geq C_2 \|u\|_{H_0^1(\Omega)}$,
3. l is continuous.

Definition 1. *A function $u : [0, T] \to H_0^1(\Omega)$ is called a weak solution of (3) if:*

(1) $u \in L^2(0, T; H_0^1(\Omega)) \cap PC(0, T; H_0^1(\Omega))$ *and* $\partial_t u \in L^2(0, T; H^{-1}(\Omega)) \cap PC(0, T; H^{-1}(\Omega))$,
(2) *For every $v \in H_0^1(\Omega)$, u satisfies (18),*
(3) $u(0) = u_0$.

Based on the above analysis, we can now formulate the following two theorems.

Theorem 1. *For every $f \in L^2(0, T; H^{-1}(\Omega))$ and $u_0 \in H_0^1(\Omega)$, there exists a unique weak solution $u \in L^2(0, T; H_0^1(\Omega)) \cap PC(0, T; H_0^1(\Omega))$ of (3).*

Proof. Existence and uniqueness of weak solution is followed by the Lax-Milgram theorem. □

Theorem 2. *For every $f \in L^2(0, T; H^{-1}(\Omega))$ and $u_0 \in H_0^1(\Omega)$, there exists a unique mild solution $u \in L^2(0, T; H_0^1(\Omega)) \cap PC(0, T; H_0^1(\Omega))$ of (3) and given by (15).*

4. Dual System

In order to establish approximate controllability, we also need to consider the *dual system* for (3), a similar strategy for partial differential equations of integer order (see Section 8 in [21] or Chapters 2 and 3 in [22] for example). The dual system for (3), which runs backward in time, is given by;

$$\begin{cases} -\partial_t v + D_t^{1-\alpha} A v = 0 & \text{in } \Omega \times (0,T), \\ \Delta v(\cdot, t_i) = I_i^*(v(\cdot, t_i)), & i = 1,2,3,\cdots,P, \\ v = 0 & \text{on } \Gamma \times (0,T), \\ v(\cdot, T) = v_0 & \text{in } \Omega. \end{cases} \quad (19)$$

4.1. Solution of Dual System

Proposition 1. *Let $v_0 \in L^2(\Omega)$. Then, there exists a unique solution of (19) and the solution is given by*

$$v(x,t) = \sum_{n=1}^{\infty} (T-t)^{\alpha-1} E_{\alpha,\alpha}(-\lambda_n (T-t)^{\alpha})(v_0, \varphi_n) \varphi_n(x)$$

$$+ \sum_{t<T-t_i} \sum_{n=1}^{\infty} (T-t-t_i)^{\alpha-1} E_{\alpha,\alpha}(-\lambda_n (T-t-t_i)^{\alpha})(I_i^*(v(\cdot,t_i)), \varphi_n) \varphi_n(x) \quad (20)$$

and has the following estimate:

$$\|v(\cdot,t)\|_{L^2(\Omega)} \leq C \left((T-t)^{\alpha-1} \|v_0\|_{L^2(\Omega)} + P \|I_{i_m}^*\|_{L^2(\Omega)} \left(\sum_{t<T-t_i} (T-t-t_i)^{2\alpha-2} \right)^{\frac{1}{2}} \right), \quad (21)$$

where $\|I_{i_m}\|_{L^2(\Omega)} = \sup_{1 \leq i \leq P} \{\|I_i\|_{L^2(\Omega)}\}$.
Moreover, the mapping $v : [0,T] \to L^2(\Omega)$ is analytically extended to $S_T := \{z \in \mathbb{C}; \operatorname{Re} z < T\}$.

Proof. Here, we establish existence and uniqueness of solution of (19) for $v_0 = 0$.
Multiplying (19) with φ_n and setting $v_n(t) = (v(\cdot,t), \varphi_n)$, we get

$$\partial_t v_n(t) + \lambda_n \partial_t^{1-\alpha} v_n(t) + \sum_{t<T-t_i} (I_i^*(v(t_i)), \varphi_n) = 0. \quad (22)$$

Since

$$|v_n(t)|^2 \leq \sum |v_n(t)|^2 = \|v(\cdot,t)\|_{L^2(\Omega)}^2 \to 0 \text{ as } t \to T,$$

we have

$$v_n(T) = 0. \quad (23)$$

From existence and uniqueness of the solution of the fractional differential equation (see [12]), we get

$$v_n(t) = 0, \quad n = 1,2,3,\cdots.$$

As $\{\varphi_n\}$ is a complete orthonormal system, we have

$$v = 0 \text{ in } \Omega \times (0,T).$$

Thus, Equation (19) has a unique solution.
Now, we show the estimate (21).

By (20), we have

$$\|v(\cdot,t)\|_{L^2(\Omega)}^2$$
$$\leq \left\| \sum_{n=1}^{\infty} (v_0, \varphi_n)(T-t)^{\alpha-1} E_{\alpha,\alpha}(-\lambda_n(T-t)^{\alpha}) \varphi_n \right\|_{L^2(\Omega)}^2$$
$$+ \left\| \sum_{t<T-t_i} \sum_{j=1}^{m_k} \left(\sum (I_i^*(v(\cdot,t_i)), \varphi_n) \varphi_n(x) \right) (T-t-t_i)^{\alpha-1} E_{\alpha,\alpha}(-\lambda_n(T-t-t_i)^{\alpha}) \right\|_{L^2(\Gamma)}^2$$
$$= \sum_{n=1}^{\infty} \left| (v_0, \varphi_n)(T-t)^{\alpha-1} E_{\alpha,\alpha}(-\mu_k(T-t)^{\alpha}) \right|^2$$
$$+ \sum_{n=1}^{\infty} \left| \sum_{t<T-t_i} (I_i^*(v(\cdot,t_i)), \varphi_n)(T-t-t_i)^{\alpha-1} E_{\alpha,\alpha}(-\mu_k(T-t-t_i)^{\alpha}) \right|^2$$
$$= C^2 \left(\sum_{n=1}^{\infty} |(v_0, \varphi_n)|^2 \right) (T-t)^{2\alpha-2} + C^2 \sum_{t<T-t_i} (T-t-t_i)^{2\alpha-2} \left(\sum_{n=1}^{\infty} |(I_i^*(v(\cdot,t_i)), \varphi_n)|^2 \right).$$

Therefore,

$$\|v(\cdot,t)\|_{L^2(\Omega)} \leq C \left((T-t)^{\alpha-1} \|v_0\|_{L^2(\Omega)} + P \|I_{i_m}^*\|_{L^2(\Omega)} \left(\sum_{t<T-t_i} (T-t-t_i)^{2\alpha-2} \right)^{\frac{1}{2}} \right).$$

Next, we show the analyticity of $v(\cdot,t)$ in $t \in S_T$.

We note that $E_{\alpha,\alpha}(-\lambda_n z)$ is an entire function (see [20] for example) and consequently each $(T-z)^{\alpha-1} E_{\alpha,\alpha}(-\lambda_n (T-z)^{\alpha})$ is analytic in $z \in S_T$. Therefore, $\sum_{n=1}^{N} (v_0, \varphi_n)(T-z)^{\alpha-1} E_{\alpha,\alpha}(-\lambda_n(T-z)^{\alpha}) \varphi_n$ in S_T.

If we fix $\delta > 0$ arbitrarily, then, for $z \in \mathbb{C}$ with $\operatorname{Re} z \leq T - \delta$, we have

$$\left\| \sum_{n=M}^{N} (v_0, \varphi_n)(T-z)^{\alpha-1} E_{\alpha,\alpha}(-\lambda_n(T-z)^{\alpha}) \varphi_n \right\|_{L^2(\Omega)}^2$$
$$= \sum_{n=M}^{N} \left| (v_0, \varphi_n)(T-z)^{\alpha-1} E_{\alpha,\alpha}(-\lambda_n(T-z)^{\alpha}) \right|^2$$
$$\leq C \sum_{n=M}^{N} |(v_0, \varphi_n)|^2 |T-z|^{2\alpha-2}$$
$$\leq C \delta^{2\alpha-2} \sum_{n=M}^{N} |(v_0, \varphi_n)|^2 \to 0 \quad \text{as } M, N \to \infty.$$

That is, (20) is uniformly convergent in $\{z \in \mathbb{C}; \operatorname{Re} z \leq T - \delta\}$. Hence, $v(\cdot,t)$ is also analytic in $t \in S_T$. □

4.2. Unique Continuation Property

Proposition 2. *Let ω be open in Ω and $v_0 \in L^2(\Omega)$. If a solution $v \in PC(0,T; H^2(\Omega) \cap H_0^1(\Omega))$ be the solution of (19) vanishing in $\omega \times (0,T)$, then $v = 0$ in $\Omega \times (0,T)$.*

Proof. Since $v(x,t) = 0$ in $\omega \times (0,T)$ and $v : [0,T) \to L^2(\Gamma)$ can be analytically extended to $S_T := \{z \in \mathbb{C}; \operatorname{Re} z < T\}$, we have

$$v(x,t) = \sum_{n=1}^{\infty}(v_0, \varphi_n)(T-t)^{\alpha-1} E_{\alpha,\alpha}(-\lambda_n(T-t)^\alpha)\varphi_n(x)$$
$$+ \sum_{t < T - t_i} \sum_{n=1}^{\infty} (I_i^*(v(\cdot,t_i)), \varphi_n)(T-t-t_i)^{\alpha-1} E_{\alpha,\alpha}(-\lambda_n(T-t-t_i)^\alpha)\varphi_n(x) \quad (24)$$
$$= 0, \quad x \in \omega, \ t \in (-\infty, T).$$

Let $\{\mu_k\}_{k\in\mathbb{N}}$ be all spectra of L without multiplicities and we denote by $\{\varphi_{kj}\}_{1\le j \le m_k}$ an orthonormal basis of $\operatorname{Ker}(\mu_k - L)$. By using these notations, we can rewrite (24) by

$$v(x,t) = \sum_{k=1}^{\infty} \left(\sum_{j=1}^{m_k} (v_0, \varphi_{kj}) \varphi_{kj}(x) \right) (T-t)^{\alpha-1} E_{\alpha,\alpha}(-\mu_k(T-t)^\alpha)$$
$$+ \sum_{t<T-t_i} \sum_{k=1}^{\infty} \left(\sum_{j=1}^{m_k} (I_i^*(v(\cdot,t_i)), \varphi_{kj}) \varphi_{kj}(x) \right) (T-t-t_i)^{\alpha-1} E_{\alpha,\alpha}(-\mu_k(T-t-t_i)^\alpha) \quad (25)$$
$$= 0, \quad x \in \omega, \ t \in (-\infty, T).$$

Then, for any $z \in \mathbb{C}$ with $\operatorname{Re} z = \xi > 0$ and $N \in \mathbb{N}$, we have

$$\left\| \sum_{k=1}^{N} \left(\sum_{j=1}^{m_k} (v_0, \varphi_{kj}) \varphi_{kj}(x) \right) e^{z(t-T)} (T-t)^{\alpha-1} E_{\alpha,\alpha}(-\mu_k(T-t)^\alpha) \right\|_{L^2(\Gamma)}^2$$
$$= \sum_{k=1}^{N} \left(\sum_{j=1}^{m_k} |(v_0, \varphi_{kj})|^2 \right) e^{2\xi(t-T)} \left| (T-t)^{\alpha-1} E_{\alpha,\alpha}(-\mu_k(T-t)^\alpha) \right|^2$$
$$\le C^2 e^{2\xi(t-T)} (T-t)^{2\alpha-2} \|v_0\|_{L^2(\Omega)}$$

and

$$\left\| \sum_{k=1}^{N} \sum_{t<T-t_i} \left(\sum_{j=1}^{m_k} (I_i^*(v(\cdot,t_i)), \varphi_{kj}) \varphi_{kj}(x) \right) (T-t-t_i)^{\alpha-1} E_{\alpha,\alpha}(-\mu_k(T-t-t_i)^\alpha) \right\|_{L^2(\Gamma)}^2$$
$$= \sum_{k=1}^{N} \sum_{t<t_i} \left(\sum_{j=1}^{m_k} |(I_i^*(v(\cdot,t_i)), \varphi_{kj})|^2 \right) e^{2\xi(t-T)} \left| (T-t-t_i)^{\alpha-1} E_{\alpha,\alpha}(-\mu_k(T-t-t_i)^\alpha) \right|^2$$
$$\le \sum_{k=1}^{N} \sum_{t<T-t_i} \left(\sum_{j=1}^{m_k} |(I_i^*(v(\cdot,t_i)), \varphi_{kj})|^2 \right) e^{2\xi(t-T)} \left| (T-t-t)^{\alpha-1} E_{\alpha,\alpha}(-\mu_k(T-t-t_i)^\alpha) \right|^2$$
$$\le P C^2 e^{2\xi(t-T)} \|I_{i_m}^*\|_{L^2(\Omega)}^2 \sum_{t<T-t_i} (T-t-t_i)^{2\alpha-2},$$

where $\|I_{i_m}\|_{L^2(\Omega)} = \sup_{1\le i \le P}\{\|I_i\|_{L^2(\Omega)}\}$.

Therefore,

$$\left\| \sum_{k=1}^{N} \left(\sum_{j=1}^{m_k} (v_0, \varphi_{kj}) \varphi_{kj}(x) \right) e^{z(t-T)} (T-t)^{\alpha-1} E_{\alpha,\alpha}(-\mu_k(T-t)^\alpha) \right\|_{L^2(\Gamma)}$$
$$\le C e^{\xi(t-T)} (T-t)^{\alpha-1} \|v_0\|_{L^2(\Omega)}$$

and

$$\left\|\sum_{k=1}^{N}\sum_{t<T-t_i}\left(\sum_{j=1}^{m_k}(I_i^*(v(\cdot,t_i)),\varphi_{kj})\varphi_{kj}(x)\right)(T-t-t_i)^{\alpha-1}E_{\alpha,\alpha}(-\mu_k(T-t-t_i)^\alpha)\right\|_{L^2(\Gamma)}$$

$$\leq PCe^{\xi(t-T)}\|I_{i_m}^*\|_{L^2(\Omega)}\left(\sum_{t<T-t_i}(T-t-t_i)^{2\alpha-2}\right)^{\frac{1}{2}}.$$

The right-hand sides of the two inequalities above are integrable on $(-\infty, T)$:

$$\int_{-\infty}^{T}e^{\xi(t-T)}(T-t)^{\alpha-1}dt = \frac{\Gamma(\alpha)}{\xi^\alpha}$$

and

$$\int_{-\infty}^{T-t_i}e^{\xi(t_i+t-T)}(T-t-t_i)^{\alpha-1}dt = \int_{0}^{\infty}e^{-\xi t}t^{\alpha-1}dt = \frac{\Gamma(\alpha)}{\xi^\alpha}.$$

Hence, the Lebesgue theorem yields that

$$\int_{-\infty}^{T}e^{z(t-T)}\left(\sum_{n=1}^{\infty}\left(\sum_{j=1}^{m_k}(v_0,\varphi_{kj})\varphi_{kj}(x)\right)(T-t)^{\alpha-1}E_{\alpha,\alpha}(-\mu_k(T-t)^\alpha)\right)dt + \int_{-\infty}^{T-t_i}e^{z(t_i+t-T)}$$

$$\times\left(\sum_{t<T-t_i}\sum_{n=1}^{\infty}\left(\sum_{j=1}^{m_k}(I_i^*(v(\cdot,t_i)),\varphi_{kj})\varphi_{kj}(x)\right)(T-t-t_i)^{\alpha-1}E_{\alpha,\alpha}(-\mu_k(T-t-t_i)^\alpha)\right)dt \quad (26)$$

$$=\sum_{k=1}^{\infty}\sum_{j=1}^{m_k}\frac{(v_0+\sum_{t<T-t_i}I_i^*(v(\cdot,t_i)),\varphi_{kj})}{z^\alpha+\mu_k}\varphi_{kj}(x),\quad \text{a.e. }x\in\Omega,\ \text{Re }z>0,$$

where we have used the Laplace transform formula;

$$\int_{0}^{\infty}e^{-zt}t^{\alpha-1}E_{\alpha,\alpha}(-\mu_k t^\alpha)dt = \frac{1}{z^\alpha+\mu_k},\quad \text{Re }z>0$$

(see (1.80) in p. 21 of [20]). By (25) and (26), we have

$$\sum_{k=1}^{\infty}\sum_{j=1}^{m_k}\frac{(v_0+\sum_{t<T-t_i}I_i^*(v(\cdot,t_i)),\varphi_{kj})}{z^\alpha+\mu_k}\varphi_{kj}(x)=0,\quad \text{a.e. }x\in\omega,\ \text{Re }z>0,$$

that is,

$$\sum_{k=1}^{\infty}\sum_{j=1}^{m_k}\frac{(v_0+\sum_{t<T-t_i}I_i^*(v(\cdot,t_i)),\varphi_{kj})}{\eta+\mu_k}\varphi_{kj}(x)=0,\quad \text{a.e. }x\in\omega,\ \text{Re }\eta>0.$$

By using analytic continuation in η, we have

$$\sum_{k=1}^{\infty}\sum_{j=1}^{m_k}\frac{(v_0+\sum_{t<T-t_i}I_i^*(v(\cdot,t_i)),\varphi_{kj})}{\eta+\mu_k}\varphi_{kj}(x)=0,\quad \text{a.e. }x\in\omega,\ \eta\in\mathbb{C}\setminus\{-\mu_k\}_{k\in\mathbb{N}}. \quad (27)$$

Then, we can take a suitable disk which includes $-\mu_\ell$ and does not include $\{-\mu_k\}_{k\neq\ell}$. By integrating (27) in the disk, we have

$$\sum_{j=1}^{m_\ell}(v_0+\sum_{t<T-t_i}I_i^*(v(\cdot,t_i)),\varphi_{\ell j})\varphi_{\ell j}(x)=0,\quad \text{a.e. }x\in\omega.$$

By setting $\widetilde{v}_\ell := \sum_{j=1}^{m_\ell}(v_0+\sum_{t<T-t_i}I_i^*(v(\cdot,t_i)),\varphi_{\ell j})\varphi_{\ell j}(x)$, we have

$$(A-\mu_\ell)\widetilde{v}_\ell=0\quad \text{in }\Omega\quad \text{and}\quad \widetilde{v}_\ell=0\quad \text{on }\omega.$$

Therefore, the unique continuation result for eigenvalue problem of elliptic operator (see [23,24]) implies

$$\tilde{v}_\ell(x) = \sum_{j=1}^{m_\ell}\left(v_0 + \sum_{t<T-t_i} I_i^*(v(\cdot,t_i)), \varphi_{\ell j}\right)\varphi_{\ell j}(x) = 0, \quad x \in \Omega$$

for each $\ell \in \mathbb{N}$. Since $\{\varphi_{\ell j}\}_{1\leq j\leq m_\ell}$ is linearly independent in Ω, we see that

$$\left(v_0 + \sum_{t<T-t_i} I_i^*(v(\cdot,t_i)), \varphi_{\ell j}\right) = 0, \quad 1 \leq j \leq m_\ell, \ \ell \in \mathbb{N}.$$

This implies $v = 0$ in $\Omega \times (0, T)$. □

5. Approximate Controllability

In this section, we complete the proof of our main theorems.

Theorem 3. *Let $0 < \alpha < 1$ and ω be an open set in Ω. Then, Equation (3) is approximately controllable for arbitrarily given $T > 0$. That is,*

$$\overline{\{u(\cdot, T); f \in C_0^\infty(\omega \times (0,T))\}} = L^2(\Omega), \tag{28}$$

where u is the solution to (3) and the closure on the left-hand side is taken in $L^2(\Omega)$.

We start the proof with a lemma.

Lemma 4. *If the conclusion of Theorem (3) is true for $u_0 \equiv 0$, then it is true for any $u_0 \in H_0^1(\Omega)$.*

Proof. Let $u_0 \in H_0^1(\Omega)$ and $u_T \in L^2(\Omega)$. Let $\epsilon > 0$. Let us introduce $\bar{u}$ the (mild) solution of

$$\begin{cases} \bar{u}_t + \partial_t^{1-\alpha} A\bar{u} = 0 & (x,t) \in \Omega \times (0,T), \\ \Delta u(\cdot,t_i) = I_i(u(\cdot,t_i)), & i = 1,2,3,\cdots,P, \\ \bar{u}(x,t) = 0, & t \in \Gamma \times (0,T), \\ \bar{u}(x,0) = u_0(x), & x \in \Omega. \end{cases}$$

Then, $\bar{u}(T) \in L^2(\Omega)$. Therefore, using the assumption of Lemma 4, there exists $f \in C_0^\infty(\omega \times (0,T))$ such that the solution w of

$$\begin{cases} \partial_t w + \partial_t^{1-\alpha} Aw = f(x,t) & (x,t) \in \Omega \times (0,T), \\ \Delta u(\cdot,t_i) = I_i(u(\cdot,t_i)), & i = 1,2,3,\cdots,P, \\ w(1,t) = 0, & t \in \Gamma \times (0,T), \\ w(x,0) = 0, & x \in \Omega, \end{cases}$$

satisfies

$$\|w(T) - (u_T - \bar{u}(T))\|_{L^2(\Omega)} \leq \epsilon.$$

One can easily see that $u(T) = w(T) + \bar{u}(T)$, so that the proof of Lemma 4 is achieved. □

We now assume that $u_0 \equiv 0$.

In order to complete the proof of Theorem 3, we will see that the unique continuation property for (19) is equivalent to the approximate controllability for (3) stated in Theorem 3.

Proof. Let u be a solution of (3) for $f \in C_0^\infty(\omega \times (0,T))$ and let v be a solution of (19) for $v_0 \in L^2(\Omega)$. Then, we see that

$$0 = \int_0^T \int_\Omega (\partial_t u + \partial_t^{1-\alpha} Au - f) v \, dxdt$$

$$= \int_0^T \int_\Omega (\partial_t u) v \, dxdt + \int_0^T \int_\Omega (\partial_t^{1-\alpha} Au) v \, dxdt$$

$$- \int_0^T \int_\Omega f v \, dxdt - \int_0^T \int_\Omega \sum_{1 \leq i \leq P} I_i(u(t_i)) \delta(t - t_i) v \, dxdt.$$

In the above equation, the first term is calculated as follows:

$$\int_0^{T-\delta} \int_\Omega (\partial_t u) v \, dxdt = \int_0^{T-\delta} \int_\Omega (\partial_t u) v \, dxdt$$

$$= \int_\Omega uv \Big|_{t=0}^{t=T-\delta} dx - \int_0^{T-\delta} \int_\Omega u(\partial_t v) \, dxdt$$

$$= \int_\Omega u(\cdot, T - \delta) v_0 \, dx + \int_0^{T-\delta} \int_\Omega u(\partial_t v) \, dxdt.$$

Here, we have used the integration in t by parts and the initial conditions in (3) and (19).

In terms of $u \in PC(0,T;H^2(\Omega))$ and $v \in PC(0,T;H^2(\Omega) \cap H_0^1(\Omega))$, we apply the Green formula to the second term, we have

$$\int_0^{T-\delta} \int_\Omega (\partial_t^{1-\alpha} Au) v \, dxdt = \int_0^{T-\delta} \int_\Omega (\partial_t^{1-\alpha} u)(Av) \, dxdt + \int_0^{T-\delta} \int_\Gamma \left(u \frac{\partial v}{\partial \nu_A} - \frac{\partial u}{\partial \nu_A} v \right) d\sigma_x dt$$

$$= - \int_0^{T-\delta} \int_\Omega u(D_t^{1-\alpha} Av) \, dxdt.$$

In the above calculation, we have used boundary conditions in (3) and (19).
Therefore, we have

$$0 = \int_0^{T-\delta} \int_\Omega (\partial_t u) v \, dxdt + \int_0^{T-\delta} \int_\Omega (\partial_t^{1-\alpha} Au) v \, dxdt - \int_0^{T-\delta} \int_\Omega f v \, dxdt$$

$$= \int_\Omega u(\cdot, T-\delta) v_0 \, dx + \int_0^{T-\delta} \int_\Omega u(D_t v) \, dxdt - \int_0^{T-\delta} \int_\Omega u(D_t^{1-\alpha} Av) \, dxdt$$

$$- \int_0^{T-\delta} \int_\Omega f v \, dxdt - \int_0^T \int_\Omega \sum_{1 \leq i \leq P} I_i(u(t_i)) \delta(t-t_i) v \, dxdt$$

$$= \int_\Omega u(\cdot, T-\delta) v_0 \, dx + \int_0^{T-\delta} \int_\Omega u(\partial_t v - D_t^{1-\alpha} Av) \, dxdt$$

$$- \int_0^{T-\delta} \int_\Omega f v \, dxdt - \int_\Omega \sum_{1 \leq i \leq P} I_i(u(t_i)) v \, dx$$

$$= \int_\Omega u(\cdot, T-\delta) v_0 \, dx - \int_0^{T-\delta} \int_\Omega f v \, dxdt - \sum_{1 \leq i \leq P} \int_\Omega I_i(u(t_i)) v \, dx.$$

Since $u \in PC([0,T], L^2(\Omega))$ and $v(\cdot, T) = v_0$ and taking $\delta \to 0$, we get

$$\int_\Omega u(\cdot, T) v_0 \, dx - \sum_{1 \leq i \leq P} \int_\Omega I_i(u(t_i)) v \, dx = \int_0^T \int_\Omega f v \, dxdt. \tag{29}$$

In order to prove density of $\overline{\{u(\cdot,T); f \in C_0^\infty(\omega \times (0,T))\}}$ in $L^2(\Omega)$, we have to show that, if $v_0 \in L^2(\Omega)$ satisfies

$$(u(\cdot,T), v_0) = \int_\Omega u(\cdot,T) v_0 dx = 0 \qquad (30)$$

for any $f \in C_0^\infty(\omega \times (0,T))$, then $v_0 \equiv 0$. This can be shown as follows: we have

$$\int_0^T \int_\Omega f v \, dx \, dt = 0$$

for any $f \in C_0^\infty(\omega \times (0,T))$. Then, by the fundamental theorem of the calculus of variations. we have

$$v(x,t) = 0, (x,t) \in \omega \times (0,T).$$

By proposition (2), we have

$$v(x,t) = 0, (x,t) \in \Omega \times (0,T).$$

By uniqueness of the solution of (1),

$$v_0(x) = 0, x \in \Omega,$$

which gives $\overline{\{u(\cdot,T); f \in C_0^\infty(\omega \times (0,T))\}}^\perp = \{0\}$. Hence, $\{u(\cdot,T); f \in C_0^\infty(\omega \times (0,T))\}$ is dense in $L^2(\Omega)$.

Thus, the proof of Theorem (3) is completed. □

6. Example

Example 1. *Consider the following relaxations' oscillation equation with fractional order given by*

$$\frac{\partial}{\partial t} u(x,t) = \frac{\partial^{1-\alpha}}{\partial t^{1-\alpha}} \frac{\partial^2}{\partial x^2} u(x,t) + f(x,t), \quad t \in I = (0,T), \; x \in \Omega = (0, \pi),$$
$$u(0,t) = u(\pi,t) = 0 \quad t \in (0,T),$$
$$u(x,0) = u_0, \quad x \in (0,\pi),$$
$$\Delta u(x,t_k) = -u(x,t_k) \quad k = 1,2,\cdots,N. \qquad (31)$$

Now, consider the corresponding system Let $u(t)x = u(x,t)$ and assume $f(x(t),t)$ to be a continuous function with respect to t that satisfies the Lipschitz condition in x. Define the operator $Au = \frac{\partial^2 u}{\partial x^2}$ with domain

$$D(A) = \{x \in L^2(0,\pi) : x, x' \text{ are absolutely continuous and } x, x', x'' \in L^2(0,\pi)\}.$$

It is well known that for $\alpha = 1$, sectorial operator, $A = \frac{\partial^2}{\partial x^2}$ generates an analytic semigroup and for $\alpha = 2$, sectorial operator, $A = \frac{\partial^2}{\partial x^2}$ generates a cosine family of operators.

Using the above notation, now consider the following system

$$\frac{\partial u}{\partial t} = \frac{\partial^{1-\alpha}}{\partial t^{1-\alpha}} Au, \quad t \in I = (0,T), \; x \in \Omega = (0,\pi),$$
$$u(0,t) = u(\pi,t) = 0 \quad t \in (0,T),$$
$$u(x,0) = 0, \quad x \in (0,\pi),$$
$$\Delta u(x,t_k) = -u(x,t_k) \quad k = 1,2,\cdots,N. \qquad (32)$$

The above problem can be posed as an abstract problem on $X = L^2(0,\pi) = U$, and hence it has a unique solution. Hence under the assumption of Theorem, the problem is approximately controllable.

Example 2. *By choosing the function $\cos(t^2)\exp(-t)$, we get the following relaxations oscillation equation with fractional order given by*

$$\partial_t^{1.8} u(t) + Au(t) = \cos(t^2)\exp(-t), u(0) = 1, u'(0) = 1, \tag{33}$$

where A is the operator mentioned above.

The graphical illustration of Example 2 is depicted in the Figure 1.

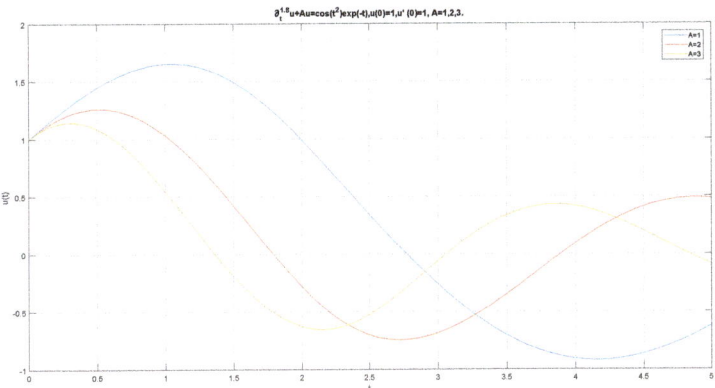

Figure 1. Comparison of solution of (33) with varied relaxation coefficients, $A = 1, 2$ and 3.

7. Discussion

This paper presents a fractional sub-diffusion equation of an impulsive system (3) and its dual (19). The unique continuation Property 2 of the dual system plays a crucial role in the proof of our main result, approximate controllability Theorem 3 of the primal system with an interior control acts on a sub-domain. As an example, the approximate controllability of a fractional relaxation-oscillation equation is discussed and simulated for different relaxation coefficients.

Author Contributions: Conceptualization, L.M. and S.A.; methodology, S.A. and M.H.; software, M.H.; validation, S.A. and H.M.S.; formal analysis, L.M. and H.M.S.; writing—original draft preparation, S.A. and M.H.; writing—review and editing, S.A. and H.M.S.; supervision, S.A. and H.M.S.

Funding: This research received no external funding.

Acknowledgments: The authors are thankful to the anonymous reviewers for their careful reading of the manuscript and constructive comments and suggestions. Lakshman Mahto would like to thank The Institute of Mathematical Sciences, Chennai, for support and hospitality during the postdoctoral work, where this work was initiated.

Conflicts of Interest: The authors declare no conflict of interest.

References

1. Metzler, R.; Klafter, J. The random walk's guide to anomalous diffusion: A fractional dynamics approach. *Phys. Rep.* **2000**, *339*, 1–77. [CrossRef]
2. Abada, N.; Benchohra, M.; Hammouche, H. Existence and controllability results for nondensely defined impulsive semilinear functional differential inclusions. *J. Differ. Equ.* **2009**, *246*, 3834–3863. [CrossRef]

3. Chu, J.; Nieto, J.J. Impulsive periodic solutions of first-order singular differential equations. *Bull. Lond. Math. Soc.* **2009**, *40*, 143–150. [CrossRef]
4. Fan, Z.; Li, G. Existence results for semilinear differential equations with nonlocal and impulsive conditions. *J. Funct. Anal.* **2010**, *258*, 1709–1727. [CrossRef]
5. Bainov, D.; Simeonov, P. *Impulsive Differential Equations: Periodic Solutions and Applications*; CRC Press: Boca Raton, FL, USA, 1993; Volume 66.
6. Sun, J.; Chen, H.; Nieto, J.J. Infinitely many solutions for second-order Hamiltonian system with impulsive effects. *Math. Comput. Model.* **2011**, *54*, 544–555. [CrossRef]
7. Lions, J.L. *Optimal Control of Systems Governed Partial Differential Equations*; Springer: New York, NY, USA, 1971.
8. Bergounioux, M.; Penalization, A. Method for Optimal Control of Elliptic Problems with State Constraints. *SIAM J. Control Optim.* **1992**, *30*, 305–323. [CrossRef]
9. Debbouche, A.; Torres, D.F.M. Approximate Controllability of Fractional Nonlocal Delay Semilinear Systems in Hilbert Spaces. *Int. J. Control* **2013**, *86*, 1577–1585. [CrossRef]
10. Debbouche, A.; Torres, D.F.M. Approximate Controllability of Fractional Delay Dynamic Inclusions with Nonlocal Control Conditions. *Appl. Math. Comput.* **2014**, *243*, 161–175. [CrossRef]
11. Khalida, A.; Benchohra, M.; Meghnafi, M. Controllability for impulsive fractional evolution equations with state-dependent delay. *Mem. Differ. Equ. Math. Phys.* **2018**, *73*, 1–20.
12. Mahto, L.; Abbas, S. Approximate controllability and existence of optimal control of impulsive fractional semilinear functional differential equations with non-local condition. *J. Abstr. Differ. Equ. Appl.* **2013**, *4*, 44–59.
13. Mahmudov, N.I. Partial-approximate controllability of nonlocal fractional evolution equations via approximating method. *Appl. Math. Comput.* **2018**, *334*, 227–238. [CrossRef]
14. Fujishiro, K.; Yamamoto, M. Approximate controllability for fractional diffusion equations by interior control. *Appl. Anal.* **2014**, *93*, 1793–1810. [CrossRef]
15. Love, E.R.; Young, M.L.C. On fractional integration by parts. *Proc. Lond. Math. Soc.* **1937**. [CrossRef]
16. Sakamoto, K.; Yamamoto, M. Initial value/boundary value problems for fractional diffusion-wave equations and applications to some inverse problems. *J. Math. Anal. Appl.* **2011**, *382*, 426–447. [CrossRef]
17. Samko, S.G.; Kilbas, A.A.; Marichev, O.I. *Fractional Integrals and Derivatives*; Gordon and Breach Science Publishers: Philadelphia, PA, USA, 1993.
18. Nigmatulin, R. The realization of the generalized transfer equation in a medium with fractal geometry. *Phys. Status Solidi B* **1986**, *133*, 425–430. [CrossRef]
19. Adams, R.A. *Sobolev Spaces*; Academic Press: New York, NY, USA, 1975.
20. Podlubny, I. *Fractional Differential Equations*; Academic Press: San Diego, CA, USA, 1999.
21. Russell, D.L. Controllability and stabilizability theory for linear partial differential equations: Recent progress and open questions. *SIAM Rev.* **1978**, *20*, 639–739. [CrossRef]
22. Micu, S.; Zuazua, E. An Introduction to the Controllability of Partial Differential Equations, Quelques Questions de Théorie du Contrôle. In *Collection Travaux en Cours*; Sari, T., Ed.; 2004; pp. 69–157. Available online: https://cel.archives-ouvertes.fr/cel-00392196/document (accessed on 7 January 2019).
23. Protter, M.H. Unique Continuation for Elliptic Equations. *Trans. Am. Math. Soc.* **1960**, *95*, 81–90. [CrossRef]
24. Kilbas, A.A.; Srivastava, H.M.; Trujillo, J.J. *Theory and Applications of Fractional Differential Equations*; North-Holland Mathematical Studies; Elsevier (North-Holland) Science Publishers: Amsterdam, The Netherlands; London, UK; New York, NY, USA, 2006; Volume 204.

© 2019 by the authors. Licensee MDPI, Basel, Switzerland. This article is an open access article distributed under the terms and conditions of the Creative Commons Attribution (CC BY) license (http://creativecommons.org/licenses/by/4.0/).

Article

The Extremal Solution To Conformable Fractional Differential Equations Involving Integral Boundary Condition

Shuman Meng [1] and Yujun Cui [2,*]

1. Department of Applied Mathematics, Shandong University of Science and Technology, Qingdao 266590, China; mx0609110@163.com
2. State Key Laboratory of Mining Disaster Prevention and Control Co-founded by Shandong Province and the Ministry of Science and Technology, Shandong University of Science and Technology, Qingdao 266590, China
* Correspondence: cyj720201@163.com

Received: 19 January 2019; Accepted: 14 February 2019; Published: 16 February 2019

Abstract: In this article, by using the monotone iterative technique coupled with the method of upper and lower solution, we obtain the existence of extremal iteration solutions to conformable fractional differential equations involving Riemann-Stieltjes integral boundary conditions. At the same time, the comparison principle of solving such problems is investigated. Finally, an example is given to illustrate our main results. It should be noted that the conformal fractional derivative is essentially a modified version of the first-order derivative. Our results show that such known results can be translated and stated in the setting of the so-called conformal fractional derivative.

Keywords: fractional differential equations; Riemann-stieltjes integral; monotone iterative method; upper and lower solutions

1. Introduction

Fractional calculus is an excellent tool for the description of the process of mathematical analysis in various areas of finance, physical systems, control systems and mechanics, and so forth [1–5]. Many methods are used to study various fractional differential equations, such as fixed point index theory [6], iterative method [7–9], theory of linear operator [10,11] sequential techniques, and regularization [12], fixed point theorems [13–17], the Mawhin continuation theorem for resonance [18–22], the variational method [23]. The definition of the fractional order derivative used in the aforementioned results is either the Caputo or the Riemann-Liouville fractional order derivative. Recently, Khalil et al. [24] gave a new simple fractional derivative called "the conformable fractional derivative" depending on the familiar limit definition of the derivative of a function and that break with other definitions. This new fractional derivative is called "the conformable fractional derivative" and this new theory is improved by Abdeljawad [25]. However, a conformal fractional derivative is not a fractional derivative, it is simply a first derivative multiplied by an additional simple factor. Therefore, this new definition seems to be a natural extension of the classical derivative. However, it has the advantage of being different from other fractional differentials. Firstly, it can integrate the standard properties of fractional derivatives. It is suitable for many extensions to the classical theorem of calculus, such as the derivative of the product and compound of two functions, the Rolle's and the mean value theorem, conformable integration by parts, fractional power series expansion and many more. Secondly, the conformal fractional derivative of the real function is zero, and the Riemann-Liouville fractional derivative does not satisfy this property. For the two iterative conformal differentials, the semigroup property is not satisfied, and the Caputo differential satisfies this. In particular, for functions that are not differentiable, in conformal sense; however, the function is differentiable. Some functions are not infinitely differentiable at some points; where there is no Taylor power series expansion, in conformal calculus theory, they do exist. This led many people wanting to explore it. Please see [26–34] for recent developments on conformable differentiation. For example, a mean value theorem of the conformable fractional calculus on arbitrary time scales is proved in [33], and whose results reconciled with familiar classical results when the operator T_α is of order $\alpha = 1$ and the time scale coincides with the set of real numbers. In [34], Asawasamrit and Ntoutas introduced a new definition of exponential notations and by employing the method of lower and upper solution combined with the monotone iterative technique, some new conditions for the existence of solutions are presented.

Motivated by the above works, we consider the existence of solutions for the following nonlinear conformable fractional differential equation involving integral boundary condition, using the method of upper and lower solutions and its associated monotone iterative technique

$$\begin{cases} D_\alpha x(t) = f(t, x(t)), \ t \in [0,1], \\ x(0) = \int_0^1 x(t) d\mu(t), \end{cases} \quad (1)$$

where $f \in C([0,1] \times \mathbb{R}, \mathbb{R})$, $\int_0^1 x(t) d\mu(t)$ denotes the Riemann-Stieltjes integral with positive Stieltjes measure of μ, and $D_\alpha f(t)$ stands for the conformable fractional derivative. Based on a comparison result, two monotone iterative sequences are obtained using the upper and lower solutions, and these two sequences approximate the extremal solutions of the given problem. For applications of the method of upper and lower solutions and monotone iterative technique to differential equations and differential systems such as ordinary differential equations [35–37], ordinary differential systems [38], fractional differential equations [39–42], fractional differential systems [43].

2. Preliminaries

In this section, we briefly show some necessary definitions and results which will be used in our main results.

Definition 1. *[24] Let $f : [0, +\infty) \to \mathbb{R}$ and $t > 0$. The conformable fractional derivative of order $0 < \alpha \leq 1$ is defined by*

$$D_\alpha f(t) = \lim_{\rho \to 0} \frac{f(t + \rho t^{1-\alpha}) - f(t)}{\rho}$$

for $t > 0$ and the conformable fractional derivative at 0 is defined as $D_\alpha f(0) = \lim_{t \to 0^+} (D_\alpha f)(t)$. If f is differentiable then $D_\alpha f(t) = t^{1-\alpha} f'(t)$.

Definition 2. *[24] Let $\alpha \in (0,1]$. The conformable fractional integral of a function $f : [0, +\infty) \to \mathbb{R}$ of order α is denoted by $I_\alpha f(t)$ and is defined as*

$$I_\alpha f(t) = \int_0^t s^{\alpha-1} f(s) ds.$$

Lemma 1. *[25] Let $f : (0, +\infty) \to \mathbb{R}$ be differentiable and $0 < \alpha \leq 1$. Then, for all $t > 0$ we have*

$$I_\alpha D_\alpha f(t) = f(t) - f(0).$$

Lemma 2. *[24] Let $\alpha \in (0,1]$, $l_1, l_2, q, k \in \mathbb{R}$, and the functions f, h be α-differentiable on $[0, +\infty)$. Then*

(i) $D_\alpha k = 0$ for all constant functions $f(t) = k$;
(ii) $D_\alpha(l_1 f + l_2 f) = l_1 D_\alpha f(t) + l_2 D_\alpha h(t)$;
(iii) $D_\alpha t^q = q t^{q-\alpha}$;

(iv) $D_\alpha(fh) = f(t)D_\alpha h(t) + h(t)D_\alpha f(t)$;

(v) $D_\alpha(\frac{f}{h}) = \frac{hD_\alpha f - fD_\alpha h}{h^2}$ when $h(t) \neq 0$.

Theorem 1. *[24] (Mean value theorem) Let $[a,b] \subset [0,+\infty)$, and let $f : [0,+\infty) \to \mathbb{R}$. Suppose that*

(1) *f is continuous on $[a,b]$;*

(2) *f is α-differentiable for some $\alpha \in (0,1]$ on $[a,b]$.*

Then there exists a constant $\xi \in (a,b)$, such that $D_\alpha f(\xi) = \frac{f(b)-f(a)}{\frac{1}{\alpha}b^\alpha - \frac{1}{\alpha}a^\alpha}$.

Definition 3. *A function $u \in C([0,1],\mathbb{R})$ is known as a lower solution of (1), if it satisfies*

$$D_\alpha u(t) \leq f(t, u(t)), t \in [0,1], \qquad (2)$$

$$u(0) \leq \int_0^1 u(t)d\mu(t). \qquad (3)$$

If inequalities (2), (3) are reversed, then u is an upper solution of problem (1).

Next, we present the following existence and uniqueness results for linear equations.

Lemma 3. *Let $0 < \alpha \leq 1$, $a \in \mathbb{R}$ and $M, N \in C([0,1],\mathbb{R})$. Then linear fractional differential equation involving integral boundary problem:*

$$\begin{cases} D_\alpha u(t) = -M(t)u(t) + N(t), \; t \in [0,1], \\ u(0) = \int_0^1 u(t)d\mu(t) + a \end{cases} \qquad (4)$$

has a unique solution provided $\triangle_\alpha = 1 - \int_0^1 e^{-\int_0^t s^{\alpha-1}M(s)ds} d\mu(t) \neq 0$.

Proof. Multiplying both sides of the first equation of the problem (4) by $e^{\int_0^t s^{\alpha-1}M(s)ds}$ and using Lemma 2, we can get

$$e^{\int_0^t s^{\alpha-1}M(s)ds} D_\alpha u(t) + M(t)u(t)e^{\int_0^t s^{\alpha-1}M(s)ds} = N(t)e^{\int_0^t s^{\alpha-1}M(s)ds}.$$

In other words, by means of the product rule (item (iv) of Lemma 2), the above equality turns to

$$D_\alpha[e^{\int_0^t s^{\alpha-1}M(s)ds} u(t)] = N(t)e^{\int_0^t s^{\alpha-1}M(s)ds}. \qquad (5)$$

Applying the conformable fractional integral of order α to both side of (5), we have

$$\begin{aligned} e^{\int_0^t s^{\alpha-1}M(s)ds} u(t) - u(0) &= I_\alpha[N(t)e^{\int_0^t s^{\alpha-1}M(s)ds}] \\ &= \int_0^t s^{\alpha-1}N(s)e^{\int_0^s \tau^{\alpha-1}M(\tau)d\tau} ds. \end{aligned}$$

Then

$$u(t) = e^{-\int_0^t s^{\alpha-1}M(s)ds}[u(0) + \int_0^t s^{\alpha-1}N(s)e^{\int_0^s \tau^{\alpha-1}M(\tau)d\tau} ds]. \qquad (6)$$

From the boundary condition of (4), we have

$$\begin{aligned} &\left(1 - \int_0^1 e^{-\int_0^t s^{\alpha-1}M(s)ds} d\mu(t)\right) u(0) \\ &= \int_0^1 e^{-\int_0^t s^{\alpha-1}M(s)ds} \int_0^t s^{\alpha-1}N(s)e^{\int_0^s \tau^{\alpha-1}M(\tau)d\tau} ds d\mu(t) + a. \end{aligned}$$

On account of condition $\triangle_\alpha \neq 0$, then

$$u(0) = \frac{\int_0^1 e^{-\int_0^t s^{\alpha-1}M(s)ds} \int_0^t s^{\alpha-1}N(s)e^{\int_0^s \tau^{\alpha-1}M(\tau)d\tau} ds d\mu(t) + a}{1 - \int_0^1 e^{-\int_0^t s^{\alpha-1}M(s)ds} d\mu(t)},$$

thus problem (4) has a unique solution. The proof is finished. □

In the next Lemma, we discuss comparison results for the linear problem which play a key role in the proof of the main result.

Lemma 4. *Let $0 < \alpha \leq 1$. Suppose that $M, u \in C([0,1], \mathbb{R})$ satisfies*

$$\begin{cases} D_\alpha u(t) \leq -M(t)u(t), & t \in [0,1], \\ u(0) \leq \int_0^1 u(t) d\mu(t). \end{cases}$$

Then $u(t) \leq 0$ on $[0,1]$ provided $\triangle_\alpha > 0$.

Proof. Let $N(t) = D_\alpha u(t) + M(t)u(t)$ and $a = u(0) - \int_0^1 u(t) d\mu(t)$, we know that $N(t) \leq 0, a \leq 0$ and

$$\begin{cases} D_\alpha u(t) = -M(t)u(t) + N(t), & t \in [0,1], \\ u(0) = \int_0^1 u(t) d\mu(t) + a. \end{cases}$$

Using $\triangle_\alpha > 0$, we have

$$u(0) = \frac{\int_0^1 e^{-\int_0^t s^{\alpha-1}M(s)ds} \int_0^t s^{\alpha-1}N(s)e^{\int_0^s \tau^{\alpha-1}M(\tau)d\tau} ds d\mu(t) + a}{1 - \int_0^1 e^{-\int_0^t s^{\alpha-1}M(s)ds} d\mu(t)} \leq 0.$$

Then by (6), we can conclude that

$$u(t) \leq e^{-\int_0^t s^{\alpha-1}M(s)ds} u(0) \leq 0.$$

The proof is complete. □

3. Main Results

In this section, we prove the existence of extremal solutions for conformable fractional differential equations involving integral boundary condition. For convenience, we list some assumptions.

(H_1): $f : [0,1] \times \mathbb{R} \to \mathbb{R}$ is continuous.

(H_2): Assume that $v_0, w_0 \in E = C[0,1]$ is lower and upper solution of problem (1), and $v_0(t) \leq w_0(t)$.

(H_3): There exists a function $M \in E$ with $\triangle_\alpha > 0$ which satisfies

$$f(t,x) - f(t,\bar{x}) \leq M(t)(\bar{x} - x),$$

for $v_0(t) \leq x \leq \bar{x} \leq w_0(t)$.

Theorem 2. *Assume that (H_1), (H_2), (H_3) hold. Then there exist monotone iterative sequences $\{v_n\}_{n=0}^\infty, \{w_n\}_{n=0}^\infty \subset E$ such that*

$$\lim_{n \to \infty} v_n = v, \quad \lim_{n \to \infty} w_n = w$$

uniformly on $[0,1]$, and v, w are the extremal solutions of problem (1) in the sector $[v_0, w_0] = \{g \in E : v_0(t) \leq g(t) \leq w_0(t), 0 \leq t \leq 1\}$.

Proof. For all $v_n, w_n \in E$, let

$$\begin{cases} D_\alpha v_{n+1}(t) = f(t, v_n(t)) - M(t)(v_{n+1}(t) - v_n(t)), & t \in [0,1], \\ D_\alpha w_{n+1}(t) = f(t, w_n(t)) - M(t)(w_{n+1}(t) - w_n(t)), & t \in [0,1], \\ v_{n+1}(0) = \int_0^1 v_{n+1}(t)d\mu(t), \quad w_{n+1}(0) = \int_0^1 w_{n+1}(t)d\mu(t). \end{cases} \quad (7)$$

Thus, the iterative sequences $\{v_n\}$ and $\{w_n\}$ can be constructed by Lemma 3.

Firstly, we shall prove that

$$v_n \leq v_{n+1} \leq w_{n+1} \leq w_n, \ n = 0, 1, 2, \ldots.$$

Let $p = v_0 - v_1$. According to (7) and Definition 3, we have

$$\begin{cases} D_\alpha p(t) = D_\alpha v_0(t) - D_\alpha v_1(t) \leq f(t, v_0(t)) - f(t, v_0(t)) + M(t)(v_1(t) - v_0(t)), & t \in [0,1], \\ p(0) \leq \int_0^1 v_0(t)d\mu(t) - \int_0^1 v_1(t)d\mu(t), \end{cases}$$

i.e.,

$$\begin{cases} D_\alpha p(t) \leq -M(t)p(t), & t \in [0,1], \\ p(0) \leq \int_0^1 p(t)d\mu(t). \end{cases}$$

Therefore, by Lemma 4, we have $v_0(t) \leq v_1(t)$. Similarly, we can prove that $w_1(t) \leq w_0(t), t \in [0,1]$. Now, let $r = v_1 - w_1$, according to (7) and (H_3), we have

$$\begin{cases} D_\alpha r(t) &= f(t, v_0(t)) - f(t, w_0(t)) - M(t)(v_1(t) - v_0(t) - w_1(t) + w_0(t)) \\ &\leq M(t)(w_0(t) - v_0(t)) - M(t)(v_1(t) - v_0(t) - w_1(t) + w_0(t)) \\ &= -M(t)r(t), \\ r(0) &= \int_0^1 r(t)d\mu(t). \end{cases}$$

By Lemma 4, we have $v_1(t) \leq w_1(t), t \in [0,1]$.

Secondly, we show that v_1, w_1 are lower and upper solutions of (1), respectively.

$$\begin{cases} D_\alpha v_1(t) &= f(t, v_0(t)) - M(t)(v_1(t) - v_0(t)) - f(t, v_1(t)) + f(t, v_1(t)) \\ &\leq M(t)(v_1(t) - v_0(t)) - M(t)(v_1(t) - v_0(t)) + f(t, v_1(t)) \\ &= f(t, v_1(t)), \\ v_1(0) &= \int_0^1 v_1(t)d\mu(t). \end{cases}$$

According to (H_3) and Definition 3, we deduce that v_1 is a lower solution of (1). Similarly, w_1 is a upper solutions of (1). By the above arguments and mathematical induction, it is clear that

$$v_0 \leq \cdots \leq v_n \leq v_{n+1} \leq w_{n+1} \leq w_n \leq \cdots \leq w_0, \ n = 0, 1, 2, \ldots. \quad (8)$$

Thirdly, we show that $\lim_{n \to \infty} v_n = v$, $\lim_{n \to \infty} w_n = w$. Hence, we need to conclude that v_n, w_n are uniformly bounded and equicontinuous on $[0,1]$. Obviously, the uniform boundedness of sequences v_n, w_n follows from (8). Thus, there exists $L > 0$ such that

$$|f(t, v_n(t)) - M(t)(v_{n+1}(t) - v_n(t))| \leq L$$

and

$$|f(t, w_n(t)) - M(t)(w_{n+1}(t) - w_n(t))| \leq L.$$

Using Theorem 1, we get

$$
\begin{aligned}
|v_n(t_1) - v_n(t_2)| &= \frac{1}{\alpha}|D_\alpha v_n(\xi)||t_1^\alpha - t_2^\alpha| \\
&= \frac{1}{\alpha}|f(\xi, v_{n-1}(\xi)) - M(\xi)(v_n(\xi) - v_{n-1}(\xi))||t_1^\alpha - t_2^\alpha|.
\end{aligned}
$$

Therefore, $\{v_n\}$ are equicontinuous. Similarly, we obtain that $\{w_n\}$ are equicontinuous too. By Arzela-Ascoli Theorems, we conclude that $\{v_n\}$, $\{w_n\}$ have subsequences $\{v_{n_k}\}$, $\{w_{n_k}\}$ such that $\{v_{n_k}\} \to v$, and $\{w_{n_k}\} \to w$ when $k \to \infty$. This together with the monotonicity of sequences $\{v_n\}$ and $\{w_n\}$ implies

$$\lim_{n\to\infty} v_n(t) = v(t), \quad \lim_{n\to\infty} w_n(t) = w(t)$$

uniformly on $[0, 1]$. Please note that the sequence $\{v_n\}$ satisfies

$$
\begin{cases}
v_n(t) = e^{-\int_0^t s^{\alpha-1} M(s) ds}[v_{n-1}(0) + Rv_{n-1}(t)], & t \in [0,1], \\
v_n(0) = \int_0^1 v_n(t) d\mu(t), & n = 1, 2, \ldots,
\end{cases}
\quad (9)
$$

where

$$Rv_{n-1}(t) = \int_0^t s^{\alpha-1}[f(t, v_{n-1}(s)) + M(s)v_{n-1}(s)] e^{\int_0^s \tau^{\alpha-1} M(\tau) d\tau} ds.$$

Let $n \to \infty$ in (9). We have

$$
\begin{cases}
v(t) = e^{-\int_0^t s^{\alpha-1} M(s) ds}[v(0) + Rv(t)], & t \in [0,1], \\
v(0) = \int_0^1 v(t) d\mu(t).
\end{cases}
$$

This shows that v is a solution of the nonlinear problem (1). Similarly, we obtain w is a solution of the nonlinear problem (1) too. And

$$v_0(t) \leq v(t) \leq w(t) \leq w_0(t), \quad t \in [0,1].$$

Finally, we are going to prove that v, w are minimal and maximal solutions of (1) in the sector $[v_0, w_0]$. In the following, we show this using induction arguments. Suppose that $g(t)$ is any solution of (1) in the $[v_0, w_0]$ that is

$$v_0(t) \leq g(t) \leq w_0(t), \quad t \in [0,1].$$

Assume that $v_n(t) \leq g(t) \leq w_n(t)$ hold. Let $p(t) = v_{n+1}(t) - g(t)$, we have

$$
\begin{cases}
D_\alpha p(t) &= D_\alpha v_{n+1}(t) - D_\alpha g(t) \\
&= f(t, v_n(t)) - M(t)(v_{n+1}(t) - v_n(t)) - f(t, g(t)) \\
&\leq M(t)(g(t) - v_n(t)) - M(t)(v_{n+1}(t) - v_n(t)) \\
&= -M(t)p(t), \\
p(0) &= \int_0^1 p(t) d\mu(t).
\end{cases}
$$

Then, by Lemma 4, we have $v_{n+1}(t) \leq g(t), t \in [0,1]$. By similar method, we can show that $g(t) \leq w_{n+1}(t), t \in [0,1]$. Therefore,

$$v_n \leq g \leq w_n, \quad n = 1, 2, \ldots.$$

By taking $n \to \infty$ in the above inequalities, we get that $v \leq g \leq w$. That is v, w are extremal solutions of problem (1) in $[v_0, w_0]$. Thus, the proof is finished. □

Example 1. *Consider the following nonlinear problem:*

$$\begin{cases} D_{\frac{1}{2}}x(t) = -\frac{2}{9}(1+x(t))^3 + 9\sin\frac{x^2(t)}{4}, & t \in [0,1], \\ x(0) = \frac{1}{3}x(\frac{1}{4}) + \frac{1}{6}x(\frac{1}{2}). \end{cases} \quad (10)$$

Let

$$\mu(t) = \begin{cases} 0, & t \in [0, \frac{1}{4}), \\ \frac{1}{3}, & t \in [\frac{1}{4}, \frac{1}{2}), \\ \frac{1}{2}, & t \in [\frac{1}{2}, 1]. \end{cases}$$

Obviously, $\alpha = \frac{1}{2}$, $f(t,x) = -\frac{2}{9}(1+x)^3 + 9\sin\frac{x^2}{4}$ and

$$\int_0^1 x(t)d\mu(t) = \frac{1}{3}x(\frac{1}{4}) + \frac{1}{6}x(\frac{1}{2}).$$

We can get

$$\int_0^1 d\mu(t) = \frac{1}{2}.$$

Take

$$v_0(t) = -2, \quad w_0(t) = 0,$$

then,

$$\begin{cases} D_{\frac{1}{2}}v_0(t) = 0 < \frac{2}{9} + 9\sin 1 = f(t, v_0(t)), \\ v_0(0) = -2 < -1 = \int_0^1 v_0(t)d\mu(t), \end{cases}$$

and

$$\begin{cases} D_{\frac{1}{2}}w_0(t) = 0 > -\frac{2}{9} = f(t, w_0(t)), \\ w_0(0) = 0 = \int_0^1 w_0(t)d\mu(t). \end{cases}$$

Then v_0, w_0 are lower and upper solutions of (10). When $M(t) = 1$, it is easy to verify that assumption (H_3) holds. In addition,

$$\int_0^1 e^{-\int_0^t s^{\alpha-1}M(s)ds} d\mu(t) = \int_0^1 e^{-t^{\frac{1}{2}}} d\mu(t) < \int_0^1 d\mu(t) = \frac{1}{2} < 1.$$

By Theorem 2, problem (10) has an extremal iterative solution in $[v_0, w_0]$.

4. Conclusions

In this article, on the integral boundary value problem for conformable fractional differential equations, we use the monotone iterative technique to investigate the existence results for extremal solutions for Equation (1). At the same time, two sequences are obtained using the upper and lower solutions, and these two sequences approximate the extremal solutions of nonlinear differential equations. It is clear that the method of using the upper and lower solutions is a very effective method for studying the solvability of conformable fractional differential equations. However, almost all the results derived in the paper are more-or-less straightforward extensions of well-known results from the theory of the first-order differential equations, since the conformal fractional derivative is essentially a modified version of the first-order derivative.

Author Contributions: The authors have made the same contribution.

Funding: This work was partially supported by the Natural Science Foundation of China (11371221, 11571207), the Shandong Natural Science Foundation (ZR2018MA011), and the Tai'shan Scholar Engineering Construction Fund of Shandong Province of China.

Conflicts of Interest: The authors declare that they have no competing interests.

References

1. Agarwal, R.P.; Benchohra, M.; Hamani, S. A survey on existence results for boundary value problems of nonlinear fractional differential equations and inclusions. *Acta. Appl. Math.* **2010**, *109*, 973–1033. [CrossRef]
2. Kilbas, A.A.; Srivastava, H.M.; Trujillo, J.J. *Theory and Applications of Fractional Differential Equations*; Elsevier: Amsterdam, The Netherlands, 2006.
3. Bucur, C.; Valdinoci, E. *Nonlocal diffusion and applications. (Lecture Notes of the Unione Matematica Italiana)*; Springer: Cham, Switzerland; Unione Matematica Italiana: Bologna, Italy, 2016; Volume 20, p. xii+155.
4. Kochubei, A.N. Distributed order calculus and equations of ultraslow diffusion. *J. Math. Anal. Appl.* **2008**, *340*, 252–281. [CrossRef]
5. Lu, C.; Fu, C.; Yang, H. Time-fractional generalized Boussinesq equation for Rossby solitary waves with dissipation effect in stratified fluid and conservation laws as well as exact solutions. *Appl. Math. Comput.* **2018**, *327*, 104–116. [CrossRef]
6. Jiang, J.; Liu, W.; Wang, H. Positive solutions to singular Dirichlet-type boundary value problems of nonlinear fractional differential equations. *Adv. Differ. Equ.* **2018**, *2018*, 169. [CrossRef]
7. Song, Q.; Dong, X.; Bai, Z.; Chen, B. Existence for fractional Dirichlet boundary value problem under barrier strip conditions. *J. Nonlinear Sci. Appl.* **2017**, *10*, 3592–3598. [CrossRef]
8. Wu, J.; Zhang, X.; Liu, L.; Wu, Y.; Cui, Y. The convergence analysis and error estimation for unique solution of a p-Laplacian fractional differential equation with singular decreasing nonlinearity. *Bound. Value Probl.* **2018**, *2018*, 82. [CrossRef]
9. He, J.; Zhang, X.; Liu, L.; Wu, Y.; Cui, Y. Existence and asymptotic analysis of positive solutions for a singular fractional differential equation with nonlocal boundary conditions. *Bound. Value Probl.* **2018**, *2018*, 189. [CrossRef]
10. Cui, Y.; Ma, W.; Wang, X.; Su, X. Uniqueness theorem of differential system with coupled integral boundary conditions. *Electron. J. Qual. Theory Differ. Equ.* **2018**, *9*, 1–10. [CrossRef]
11. Yue, Z.; Zou, Y. New uniqueness results for fractional differential equation with dependence on the first order derivative. *Adv. Differ. Equ.* **2019**, *2019*, 38. [CrossRef]
12. Guo, L.; Liu, L.; Wu, Y. Existence of positive solutions for singular fractional differential equations with infinite-point boundary conditions. *Nonlinear Anal. Model. Control* **2016**, *21*, 635–650. [CrossRef]
13. Hao, X.; Wang, H.; Liu, L.; Cui, Y. Positive solutions for a system of nonlinear fractional nonlocal boundary value problems with parameters and p-Laplacian operator. *Bound. Value Probl.* **2017**, *2017*, 182. [CrossRef]
14. Jiang, J.; Liu, L. Existence of solutions for a sequential fractional differential system with coupled boundary conditions. *Bound. Value Probl.* **2016**, *2016*, 159. [CrossRef]
15. Sun, Q.; Ji, H.; Cui, Y. Positive solutions for boundary value problems of fractional differential equation with integral boundary conditions. *J. Funct. Spaces* **2018**, *2018*, 6461930. [CrossRef]
16. Zhai, C.; Wang, W.; Li, H. A uniqueness method to a new Hadamard fractional differential system with four-point boundary conditions. *J. Inequal. Appl.* **2018**, *2018*, 207. [CrossRef]
17. Zuo, M.; Hao, X.; Liu, L.; Cui, Y. Existence results for impulsive fractional integro-differential equation of mixed type with constant coefficient and antiperiodic boundary conditions. *Bound. Value Probl.* **2017**, *2017*, 161. [CrossRef]
18. Jiang, W.; Kosmatov, N. Existence results for a functional boundary value problem of fractional differential equations. *Bound. Value Probl.* **2018**, *2018*, 72. [CrossRef]
19. Kosmatov, N.; Jiang, W. Resonant functional problems of fractional order. *Chaos Solitons Fractals* **2016**, *91*, 573–579. [CrossRef]
20. Qi, T.; Liu, Y.; Zou, Y. Existence result for a class of coupled fractional differential systems with integral boundary value conditions. *J. Nonlinear Sci. Appl.* **2017**, *10*, 4034–4045. [CrossRef]

21. Sun, Q.; Meng, S.; Cui, Y. Existence results for fractional order differential equation with nonlocal Erdélyi-Kober and generalized Riemann-Liouville type integral boundary conditions at resonance. *Adv. Differ. Equ.* **2018**, *2018*, 243. [CrossRef]
22. Tian, Y.; Bai, Z. Existence results for the three-point impulsive boundary value problem involving fractional differential equations. *Comput. Math. Appl.* **2010**, *59*, 2601–2609. [CrossRef]
23. Wang, Y.; Liu, Y.; Cui, Y. Infinitely many solutions for impulsive fractional boundary value problem with p-Laplacian. *Bound. Value Probl.* **2018**, *2018*, 94. [CrossRef]
24. Khalil, R.; al Horani, M.; Yousef, A.; Sababheh, M. A new definition of fractional derivatuive. *J. Comput. Appl. Math.* **2014**, *264*, 65–70. [CrossRef]
25. Abdeljawad, T. On conformable fractional calculus. *J. Comput. Appl. Math.* **2015**, *279*, 57–66. [CrossRef]
26. Ammi, M.R.S.; Torres, D.F.M. Existence of solution to a nonlocal conformable fractional thermistor problem. *Commun. Fac. Sci. Univ. Ank. Ser. A1 Math. Stat.* **2019**, *68*, 1061–1072.
27. Anderson, D.R.; Avery, R.I. Fractional-order boundary value problem with Sturm-Liouville boundary conditions. *Electron. J. Differ. Equ.* **2015**, *29*, 1–10.
28. Anderson, D.R.; Ulness, D.J. Results for conformable differential equations. *preprint* **2016**, in progress.
29. Benkhettou, N.; Hassani, S.; Torres, D.F.M. A conformable fractional calculus on arbitrary time scales. *J. King Saud Univ. Sci.* **2016**, *28*, 93–98. [CrossRef]
30. Eslami, M.; Rezazadeh, H. The first integral method for Wu-Zhang system with conformable time-fractional derivative. *Calcolo* **2016**, *53*, 475–485. [CrossRef]
31. Sitho, S.; Ntouyas, S.K.; Agarwal, P.; Tariboon, J. Noninstantaneous impulsive inequalities via conformable fractional calculus. *J. Inequal. Appl.* **2018**, *2018*, 261. [CrossRef] [PubMed]
32. Ünal, E.; Gökdogan, A.; Çelik, E. Solutions of sequential conformable fractional differential equations around an ordinary point and conformable fractional Hermite differential equation. *Br. J. Appl. Sci. Technol.* **2015**, *10*, 1–11.
33. Nwaeze, E.R. A mean value theorem for the conformable fractional calculus on arbitrary time scales. *Progr. Fract. Differ. Appl.* **2016**, *4*, 287–291. [CrossRef]
34. Asawasamrit, S.; Ntoutas, S.K.; Thiramanus, P.; Tariboon, J. Periodic boundary value problems for impulsive conformable fractional integro-differential equations. *Bound. Value Probl.* **2016**, *2016*, 122. [CrossRef]
35. Cui, Y.; Zou, Y. Existence of solutions for second-order integral boundary value problems. *Nonlinear Anal. Model. Control* **2016**, *6*, 828–838. [CrossRef]
36. Cui, Y.; Zou, Y. Monotone iterative technique for $(k, n-k)$ conjugate boundary value problems. *Electron. J. Qual. Theory Differ. Equ.* **2015**, *69*, 1–11. [CrossRef]
37. Jankowski, T. Monotone iterative method for first-differential equations at resonance. *Appl. Math. Comput.* **2014**, *233*, 20–28. [CrossRef]
38. Cui, Y.; Zou, Y. Monotone iterative method for differential systems with coupled integral boundary value problems. *Bound. Value Probl.* **2013**, *2013*, 245. [CrossRef]
39. Liu, S.; Wang, H.; Li, X.; Li, H. The extremal iteration solution to a coupled system of nonlinear conformable fractional differential equations. *J. Nonlinear Sci. Appl.* **2017**, *10*, 5082–5089. [CrossRef]
40. Wang, G. Monotone iterative technique for boundary value problems of a nonlinear fractional differential equations with deviating arguments. *J. Comput. Appl. Math.* **2012**, *236*, 2425–2430. [CrossRef]
41. Bai, Z.; Zhang, S.; Sun, S.; Yin, C. Monotone iterative method for a class of fractional differential equations. *Electron. J. Differ. Equ.* **2016**, *2016*, 6.
42. Cui, Y.; Sun, Q.; Su, X. Monotone iterative technique for nonlinear boundary value problems of fractional order $p \in (2,3]$. *Adv. Differ. Equ.* **2017**, *2017*, 248. [CrossRef]
43. Cui, Y.; Zou, Y. Existence results and the monotone iterative technique for nonlinear fractional differential systems with coupled four-point boundary value problems. *Abstr. Appl. Anal.* **2014**, *2014*, 242591. [CrossRef]

© 2019 by the authors. Licensee MDPI, Basel, Switzerland. This article is an open access article distributed under the terms and conditions of the Creative Commons Attribution (CC BY) license (http://creativecommons.org/licenses/by/4.0/).

Article

Hankel and Toeplitz Determinants for a Subclass of q-Starlike Functions Associated with a General Conic Domain

Hari M. Srivastava [1,2,*], Qazi Zahoor Ahmad [3], Nasir Khan [4], Nazar Khan [3] and Bilal Khan [3]

1. Department of Mathematics and Statistics, University of Victoria, Victoria, BC V8W 3R4, Canada
2. Department of Medical Research, China Medical University Hospital, China Medical University, Taichung 40402, Taiwan
3. Department of Mathematics, Abbottabad University of Science and Technology, Abbottabad 22010, Pakistan; zahoorqazi5@gmail.com (Q.Z.A.); nazarmaths@gmail.com (N.K.); bilalmaths789@gmail.com (B.K.)
4. Department of Mathematics, FATA University, Akhorwal (Darra Adam Khel), FR Kohat 26000, Pakistan; dr.nasirkhan@fu.edu.pk
* Correspondence: harimsri@math.uvic.ca

Received: 29 January 2019; Accepted: 12 February 2019; Published: 15 February 2019

Abstract: By using a certain general conic domain as well as the quantum (or q-) calculus, here we define and investigate a new subclass of normalized analytic and starlike functions in the open unit disk $\mathbb{U}$. In particular, we find the Hankel determinant and the Toeplitz matrices for this newly-defined class of analytic q-starlike functions. We also highlight some known consequences of our main results.

Keywords: analytic functions; starlike and q-starlike functions; q-derivative operator; q-hypergeometric functions; conic and generalized conic domains; Hankel determinant; Toeplitz matrices

MSC: Primary 05A30, 30C45; Secondary 11B65, 47B38

1. Introduction and Definitions

Let the class of functions, which are analytic in the open unit disk

$$\mathbb{U} = \{z : z \in \mathbb{C} \text{ and } |z| < 1\},$$

be denoted by $\mathcal{L}(\mathbb{U})$. Also let $\mathcal{A}$ denote the class of all functions f, which are analytic in the open unit disk $\mathbb{U}$ and normalized by

$$f(0) = 0 \quad \text{and} \quad f'(0) = 1.$$

Then, clearly, each $f \in \mathcal{A}$ has a Taylor–Maclaurin series representation as follows:

$$f(z) = z + \sum_{n=2}^{\infty} a_n z^n \qquad (z \in \mathbb{U}). \tag{1}$$

Suppose that $\mathcal{S}$ is the subclass of the analytic function class $\mathcal{A}$, which consists of all functions which are also univalent in $\mathbb{U}$.

A function $f \in \mathcal{A}$ is said to be starlike in $\mathbb{U}$ if it satisfies the following inequality:

$$\Re\left(\frac{zf'(z)}{f(z)}\right) > 0 \qquad (z \in \mathbb{U}).$$

We denote by $\mathcal{S}^*$ the class of all such starlike functions in $\mathbb{U}$.

For two functions f and g, analytic in $\mathbb{U}$, we say that the function f is subordinate to the function g and write this subordination as follows:

$$f \prec g \quad \text{or} \quad f(z) \prec g(z),$$

if there exists a Schwarz function w which is analytic in $\mathbb{U}$, with

$$w(0) = 0 \quad \text{and} \quad |w(z)| < 1,$$

such that

$$f(z) = g(w(z)).$$

In the case when the function g is univalent in $\mathbb{U}$, then we have the following equivalence (see, for example, [1]; see also [2]):

$$f(z) \prec g(z) \quad (z \in \mathbb{U}) \iff f(0) = g(0) \text{ and } f(\mathbb{U}) \subset g(\mathbb{U}).$$

Next, for a function $f \in \mathcal{A}$ given by (1) and another function $g \in \mathcal{A}$ given by

$$g(z) = z + \sum_{n=2}^{\infty} b_n z^n \quad (z \in \mathbb{U}),$$

the convolution (or the Hadamard product) of f and g is defined here by

$$(f * g)(z) := z + \sum_{n=2}^{\infty} a_n b_n z^n =: (g * f)(z). \tag{2}$$

Let $\mathcal{P}$ denote the well-known Carathéodory class of functions p, analytic in the open unit disk $\mathbb{U}$, which are normalized by

$$p(z) = 1 + \sum_{n=1}^{\infty} c_n z^n, \tag{3}$$

such that

$$\Re(p(z)) > 0 \quad (z \in \mathbb{U}).$$

Following the works of Kanas et al. (see [3,4]; see also [5]), we introduce the conic domain Ω_k ($k \geqq 0$) as follows:

$$\Omega_k = \left\{ u + iv : u > k\sqrt{(u-1)^2 + v^2} \right\}. \tag{4}$$

In fact, subjected to the conic domain Ω_k ($k \geqq 0$), Kanas and Wiśniowska (see [3,4]; see also [6]) studied the corresponding class k-$\mathcal{ST}$ of k-starlike functions in $\mathbb{U}$ (see Definition 1 below). For fixed k, Ω_k represents the conic region bounded successively by the imaginary axis ($k = 0$), by a parabola ($k = 1$), by the right branch of a hyperbola ($0 < k < 1$), and by an ellipse ($k > 1$).

For these conic regions, the following functions play the role of extremal functions.

$$p_k(z) = \begin{cases} \dfrac{1+z}{1-z} = 1 + 2z + 2z^2 + \cdots & (k = 0) \\[1em] 1 + \dfrac{2}{\pi^2} \left[\log\left(\dfrac{1+\sqrt{z}}{1-\sqrt{z}}\right)\right]^2 & (k = 1) \\[1em] 1 + \dfrac{2}{1-k^2} \sinh^2\left[\left(\dfrac{2}{\pi}\arccos k\right)\arctan\left(h\sqrt{z}\right)\right] & (0 \leqq k < 1) \\[1em] 1 + \dfrac{1}{k^2 - 1}\left[1 + \sin\left(\dfrac{\pi}{2K(\kappa)} \int_0^{\frac{u(z)}{\sqrt{\kappa}}} \dfrac{dt}{\sqrt{(1-t^2)(1-\kappa^2 t^2)}}\right)\right] & (k > 1), \end{cases} \tag{5}$$

where

$$u(z) = \frac{z - \sqrt{\kappa}}{1 - \sqrt{\kappa} z} \qquad (z \in \mathbb{U}),$$

and $\kappa \in (0,1)$ is so chosen that

$$k = \cosh\left(\frac{\pi K'(\kappa)}{4K(\kappa)}\right).$$

Here $K(\kappa)$ is Legendre's complete elliptic integral of first kind and

$$K'(\kappa) = K\left(\sqrt{1-\kappa^2}\right),$$

that is, $K'(\kappa)$ is the complementary integral of $K(\kappa)$ (see, for example, ([7], p. 326, Equation 9.4 (209))). Indeed, from (5), we have

$$p_k(z) = 1 + p_1 z + p_2 z^2 + p_3 z^3 + \cdots. \tag{6}$$

The class k-$\mathcal{ST}$ is defined as follows.

Definition 1. *A function $f \in \mathcal{A}$ is said to be in the class k-$\mathcal{ST}$ if and only if*

$$\frac{zf'(z)}{f(z)} \prec p_k(z) \quad (\forall z \in \mathbb{U}; \ k \geqq 0).$$

We now recall some basic definitions and concept details of the q-calculus which will be used in this paper (see, for example, ([7], p. 346 et seq.)). Throughout the paper, unless otherwise mentioned, we suppose that $0 < q < 1$ and

$$\mathbb{N} = \{1, 2, 3 \cdots\} = \mathbb{N}_0 \setminus \{0\} \qquad (\mathbb{N}_0 := \{0, 1, 2, \cdots\}).$$

Definition 2. *Let $q \in (0,1)$ and define the q-number $[\lambda]_q$ by*

$$[\lambda]_q = \begin{cases} \dfrac{1-q^\lambda}{1-q} & (\lambda \in \mathbb{C}) \\[1em] \sum_{k=0}^{n-1} q^k = 1 + q + q^2 + \cdots + q^{n-1} & (\lambda = n \in \mathbb{N}). \end{cases}$$

Definition 3. *Let $q \in (0,1)$ and define the q-factorial $[n]_q!$ by*

$$[n]_q! = \begin{cases} 1 & (n = 0) \\[0.5em] \prod_{k=1}^{n} [k]_q & (n \in \mathbb{N}). \end{cases}$$

Definition 4 (see [8,9]). *The q-derivative (or q-difference) operator D_q of a function f defined, in a given subset of $\mathbb{C}$, by*

$$(D_q f)(z) = \begin{cases} \dfrac{f(z) - f(qz)}{(1-q)z} & (z \neq 0) \\ f'(0) & (z = 0), \end{cases} \qquad (7)$$

provided that $f'(0)$ exists.

From Definition 4, we can observe that

$$\lim_{q \to 1^-} (D_q f)(z) = \lim_{q \to 1^-} \frac{f(z) - f(qz)}{(1-q)z} = f'(z)$$

for a differentiable function f in a given subset of $\mathbb{C}$. It is also known from (1) and (7) that

$$(D_q f)(z) = 1 + \sum_{n=2}^{\infty} [n]_q a_n z^{n-1}. \qquad (8)$$

Definition 5. *The q-Pochhammer symbol $[\xi]_{n,q}$ ($\xi \in \mathbb{C}$; $n \in \mathbb{N}_0$) is defined as follows:*

$$[\xi]_{n,q} = \frac{(q^\xi; q)_n}{(1-q)^n} = \begin{cases} 1 & (n=0) \\ [\xi]_q [\xi+1]_q [\xi+2]_q \cdots [\xi+n-1]_q & (n \in \mathbb{N}). \end{cases}$$

Moreover, the q-gamma function is defined by the following recurrence relation:

$$\Gamma_q(z+1) = [z]_q \Gamma_q(z) \quad \text{and} \quad \Gamma_q(1) = 1.$$

Definition 6 (see [10]). *For $f \in \mathcal{A}$, let the q-Ruscheweyh derivative operator $\mathcal{R}_q^\lambda$ be defined, in terms of the Hadamard product (or convolution) given by (2), as follows:*

$$\mathcal{R}_q^\lambda f(z) = f(z) * \mathcal{F}_{q,\lambda+1}(z) \qquad (z \in \mathbb{U}; \lambda > -1),$$

where

$$\mathcal{F}_{q,\lambda+1}(z) = z + \sum_{n=2}^{\infty} \frac{\Gamma_q(\lambda+n)}{[n-1]_q! \Gamma_q(\lambda+1)} z^n = z + \sum_{n=2}^{\infty} \frac{[\lambda+1]_{q,n-1}}{[n-1]_q!} z^n.$$

We next define a certain q-integral operator by using the same technique as that used by Noor [11].

Definition 7. *For $f \in \mathcal{A}$, let the q-integral operator $\mathcal{F}_{q,\lambda}$ be defined by*

$$\mathcal{F}_{q,\lambda+1}^{-1}(z) * \mathcal{F}_{q,\lambda+1}(z) = z(D_q f)(z).$$

Then

$$\mathcal{I}_q^\lambda f(z) = f(z) * \mathcal{F}_{q,\lambda+1}^{-1}(z)$$

$$= z + \sum_{n=2}^{\infty} \psi_{n-1} a_n z^n \qquad (z \in \mathbb{U}; \lambda > -1), \qquad (9)$$

where

$$\mathcal{F}_{q,\lambda+1}^{-1}(z) = z + \sum_{n=2}^{\infty} \psi_{n-1} z^n$$

and
$$\psi_{n-1} = \frac{[n]_q! \Gamma_q(\lambda+1)}{\Gamma_q(\lambda+n)} = \frac{[n]_q!}{[\lambda+1]_{q,n-1}}.$$

Clearly, we have
$$\mathcal{I}_q^0 f(z) = z(D_q f)(z) \quad \text{and} \quad \mathcal{I}_q^1 f(z) = f(z).$$

We note also that, in the limit case when $q \to 1-$, the q-integral operator $\mathcal{F}_{q,\lambda}$ given by Definition 7 would reduce to the integral operator which was studied by Noor [11].

The following identity can be easily verified:

$$zD_q\left(\mathcal{I}_q^{\lambda+1} f(z)\right) = \left(1 + \frac{[\lambda]_q}{q^\lambda}\right) \mathcal{I}_q^\lambda f(z) - \frac{[\lambda]_q}{q^\lambda} \mathcal{I}_q^{\lambda+1} f(z). \tag{10}$$

When $q \to 1-$, this last identity in (10) implies that

$$z\left(\mathcal{I}^{\lambda+1} f(z)\right)' = (1+\lambda) \mathcal{I}^\lambda f(z) - \lambda \mathcal{I}^{\lambda+1} f(z),$$

which is the well-known recurrence relation for the above-mentioned integral operator which was studied by Noor [11].

In geometric function theory, several subclasses belonging to the class of normalized analytic functions class $\mathcal{A}$ have already been investigated in different aspects. The above-defined q-calculus gives valuable tools that have been extensively used in order to investigate several subclasses of $\mathcal{A}$. Ismail et al. [12] were the first who used the q-derivative operator D_q to study the q-calculus analogous of the class $\mathcal{S}^*$ of starlike functions in $\mathbb{U}$ (see Definition 8 below). However, a firm footing of the q-calculus in the context of geometric function theory was presented mainly and basic (or q-) hypergeometric functions were first used in geometric function theory in a book chapter by Srivastava (see, for details, ([13], p. 347 et seq.); see also [14]).

Definition 8 (see [12]). *A function $f \in \mathcal{A}$ is said to belong to the class $\mathcal{S}_q^*$ if*

$$f(0) = f'(0) - 1 = 0 \tag{11}$$

and
$$\left| \frac{z}{f(z)} (D_q f) z - \frac{1}{1-q} \right| \leq \frac{1}{1-q}. \tag{12}$$

It is readily observed that, as $q \to 1-$, the closed disk:

$$\left| w - \frac{1}{1-q} \right| \leq \frac{1}{1-q}$$

becomes the right-half plane and the class $\mathcal{S}_q^*$ of q-starlike functions reduces to the familiar class $\mathcal{S}^*$ of normalized starlike functions in $\mathbb{U}$ with respect to the origin ($z = 0$). Equivalently, by using the principle of subordination between analytic functions, we can rewrite the conditions in (11) and (12) as follows (see [15]):

$$\frac{z}{f(z)} (D_q f)(z) \prec \widehat{p}(z) \qquad \left(\widehat{p}(z) = \frac{1+z}{1-qz}\right). \tag{13}$$

The notation $\mathcal{S}_q^*$ was used by Sahoo and Sharma [16].

Now, making use of the principle of subordination between analytic functions and the above-mentioned q-calculus, we present the following definition.

Definition 9. A function p is said to be in the class $k\text{-}\mathcal{P}_q$ if and only if

$$p(z) \prec \frac{2p_k(z)}{(1+q)+(1-q)\,p_k(z)},$$

where $p_k(z)$ is defined by (5).

Geometrically, the function $p(z) \in k\text{-}\mathcal{P}_q$ takes on all values from the domain $\Omega_{k,q}$ $(k \geqq 0)$ which is defined as follows:

$$\Omega_{k,q} = \left\{ w : \Re\left(\frac{(1+q)\,w}{(q-1)\,w+2} \right) > k \left| \frac{(1+q)\,w}{(q-1)\,w+2} - 1 \right| \right\}.$$

The domain $\Omega_{k,q}$ represents a generalized conic region.
It can be seen that

$$\lim_{q \to 1-} \Omega_{k,q} = \Omega_k,$$

where Ω_k is the conic domain considered by Kanas and Wiśniowska [3]. Below, we give some basic facts about the class $k\text{-}\mathcal{P}_q$.

Remark 1. First of all, we see that

$$k\text{-}\mathcal{P}_q \subseteq \mathcal{P}\left[\frac{2k}{2k+1+q} \right],$$

where $\mathcal{P}\left[\frac{2k}{2k+1+q}\right]$ is the well-known class of functions with real part greater than $\frac{2k}{2k+1+q}$. Secondly, we have

$$\lim_{q \to 1-} k\text{-}\mathcal{P}_q = \mathcal{P}(p_k),$$

where $\mathcal{P}(p_k)$ is the well-known function class introduced by Kanas and Wiśniowska [3]. Thirdly, we have

$$\lim_{q \to 1-} 0\text{-}\mathcal{P}_q = \mathcal{P},$$

where $\mathcal{P}$ is the well-known class of analytic functions with positive real part.

Definition 10. A function f is said to be in the class $\mathcal{ST}(k, \lambda, q)$ if and only if

$$\frac{z\left(D_q \mathcal{I}_q^\lambda f\right)(z)}{f(z)} \in k\text{-}\mathcal{P}_q \qquad (k \geqq 0;\ \lambda \geqq 0),$$

or, equivalently,

$$\Re\left(\frac{(1+q)\frac{z(D_q \mathcal{I}_q^\lambda f)(z)}{f(z)}}{(q-1)\frac{z(D_q \mathcal{I}_q^\lambda f)(z)}{f(z)}+2} \right) > k \left| \frac{(1+q)\frac{z(D_q \mathcal{I}_q^\lambda f)(z)}{f(z)}}{(q-1)\frac{z(D_q \mathcal{I}_q^\lambda f)(z)}{f(z)}+2} - 1 \right|.$$

Remark 2. First of all, it is easily seen that

$$\mathcal{ST}(0,1,q) = \mathcal{S}_q^*,$$

where $\mathcal{S}_q^*$ is the function class introduced and studied by Ismail et al. [12]. Secondly, we have

$$\lim_{q \to 1-} \mathcal{ST}(k,1,q) = k\text{-}\mathcal{ST},$$

where k-$\mathcal{ST}$ is a function class introduced and studied by Kanas and Wiśniowska [4]. Finally, we have

$$\lim_{q \to 1-} \mathcal{ST}(0,1,q) = \mathcal{S}^*,$$

where $\mathcal{S}^*$ is the well-known class of starlike functions in $\mathbb{U}$ with respect to the origin $(z = 0)$.

Remark 3. *Further studies of the new q-starlike function class $\mathcal{ST}(k, \lambda, q)$, as well as of its more consequences, can next be determined and investigated in future papers.*

Let $n \in \mathbb{N}_0$ and $j \in \mathbb{N}$. The following jth Hankel determinant was considered by Noonan and Thomas [17]:

$$\mathcal{H}_j(n) = \begin{vmatrix} a_n & a_{n+1} & \cdots & a_{n+j-1} \\ a_{n+1} & \cdot & & \cdot \\ \cdot & \cdot & & \cdot \\ \cdot & \cdot & & \cdot \\ a_{n+j-1} & \cdot & \cdots & a_{n+2(j-1)} \end{vmatrix},$$

where $a_1 = 1$. In fact, this determinant has been studied by several authors, and sharp upper bounds on $\mathcal{H}_2(2)$ were obtained by several authors (see [18–20]) for various classes of functions. It is well-known that the Fekete–Szegö functional $|a_3 - a_2^2|$ can be represented in terms of the Hankel determinant as $\mathcal{H}_2(1)$. This functional has been further generalized as $|a_3 - \mu a_2^2|$ for some real or complex μ. Fekete and Szegö gave sharp estimates of $|a_3 - \mu a_2^2|$ for μ real and $f \in \mathcal{S}$, the class of normalized univalent functions in $\mathbb{U}$. It is also known that the functional $|a_2 a_4 - a_3^2|$ is equivalent to $\mathcal{H}_2(2)$ (see [18]). Babalola [21] studied the Hankel determinant $\mathcal{H}_3(1)$ for some subclasses of normalized analytic functions in $\mathbb{U}$. The symmetric Toeplitz determinant $\mathcal{T}_j(n)$ is defined by

$$\mathcal{T}_j(n) = \begin{vmatrix} a_n & a_{n+1} & \cdots & a_{n+j-1} \\ a_{n+1} & \cdot & & \cdot \\ \cdot & \cdot & & \cdot \\ \cdot & \cdot & & \cdot \\ a_{n+j-1} & \cdot & \cdots & a_n \end{vmatrix},$$

so that

$$\mathcal{T}_2(2) = \begin{vmatrix} a_2 & a_3 \\ a_3 & a_2 \end{vmatrix}, \quad \mathcal{T}_2(3) = \begin{vmatrix} a_3 & a_4 \\ a_4 & a_3 \end{vmatrix}, \quad \mathcal{T}_3(2) = \begin{vmatrix} a_2 & a_3 & a_4 \\ a_3 & a_2 & a_3 \\ a_4 & a_3 & a_2 \end{vmatrix},$$

and so on.

For $f \in \mathcal{S}$, the problem of finding the best possible bounds for $||a_{n+1}| - |a_n||$ has a long history (see, for details, [22]). It is a known fact from [22] that

$$||a_{n+1}| - |a_n|| < c$$

for a constant c. However, the problem of finding exact values of the constant c for $\mathcal{S}$ and its various subclasses has proved to be difficult. In a very recent investigation, Thomas and Abdul-Halim [23] succeeded in obtaining some sharp estimates for $\mathcal{T}_j(n)$ for the first few values of n and j involving symmetric Toeplitz determinants whose entries are the coefficients a_n of starlike and close-to-convex functions.

In the present investigation, our focus is on the Hankel determinant and the Toeplitz matrices for the function class $\mathcal{ST}(k, \lambda, q)$ given by Definition 10.

2. A Set of Lemmas

In order to prove our main results in this paper, we need each of the following lemmas.

Lemma 1 (see [20]). *If the function $p(z)$ given by (3) is in the Carathéodory class $\mathcal{P}$ of analytic functions with positive real part in $\mathbb{U}$, then*

$$2c_2 = c_1^2 + x\left(4 - c_1^2\right)$$

and

$$4c_3 = c_1^3 + 2\left(4 - c_1^2\right)c_1 x - c_1\left(4 - c_1^2\right)x^2 + 2\left(4 - c_1^2\right)\left(1 - |x|^2\right)z$$

for some $x, z \in \mathbb{C}$ with $|x| \leq 1$ and $|z| \leq 1$.

Lemma 2 (see [24]). *Let the function $p(z)$ given by (3) be in the Carathéodory class $\mathcal{P}$ of analytic functions with positive real part in $\mathbb{U}$. Also let $\mu \in \mathbb{C}$. Then*

$$|c_n - \mu c_k c_{n-k}| \leq 2\max(1, |2\mu - 1|) \qquad (1 \leq k \leq n-1).$$

Lemma 3 (see [22]). *Let the function $p(z)$ given by (3) be in the Carathéodory class $\mathcal{P}$ of analytic functions with positive real part in $\mathbb{U}$. Then*

$$|c_n| \leq 2 \qquad (n \in \mathbb{N}).$$

This last inequality is sharp.

3. Main Results

Throughout this section, unless otherwise mentioned, we suppose that

$$q \in (0,1), \quad \lambda > -1 \quad \text{and} \quad k \in [0,1].$$

Theorem 1. *If the function $f(z)$ given by (1) belongs to the class $\mathcal{ST}(k, \lambda, q)$, where $k \in [0,1]$, then*

$$|a_2| \leq \frac{(1+q) p_1}{2q\psi_1},$$

$$a_3 \leq \frac{1}{2q\psi_2}\left(p_1 + \left|p_2 - p_1 + \frac{(q^2+1) p_1^2}{2q}\right|\right)$$

and

$$a_4 \leq \frac{(1+q)}{4(q+q^2+q^3) \psi_3}\left(2p_1 + 4\left|p_2 - p_1 + \frac{(2+q^2) p_1^2}{4q}\right|\right.$$
$$+ \left|2p_3 + 2p_1 - 4p_2 - \frac{(2(1+q^2) - q) p_1^2}{q} + \frac{(4q^2 - 3q + 2)}{q} p_1 p_2\right.$$
$$+ \left.\left.\frac{(q^2 + 2q - 1)}{2q^2} p_1^3\right|\right), \tag{14}$$

where p_j ($j = 1, 2, 3$) are positive and are the coefficients of the functions $p_k(z)$ defined by (6). Each of the above results is sharp for the function $g(z)$ given by

$$g(z) = \frac{2p_k(z)}{(1+q) + (1-q) p_k(z)}.$$

Proof. Let $f(z) \in \mathcal{ST}(k, \lambda, q)$. Then, we have

$$\frac{z(D_q f)(z)}{f(z)} = \mathfrak{q}(z) \prec S_k(z), \tag{15}$$

where

$$S_k(z) = \frac{2p_k(z)}{(1+q) + (1-q)p_k(z)},$$

and the functions $p_k(z)$ are defined by (6).

We now define the function $p(z)$ with $p(0) = 1$ and with a positive real part in $\mathbb{U}$ as follows:

$$p(z) = \frac{1 + S_k^{-1}(\mathfrak{q}(z))}{1 - S_k^{-1}(\mathfrak{q}(z))} = 1 + c_1 z + c_2 z^2 + \cdots. \tag{16}$$

After some simple computation involving (16), we get

$$\mathfrak{q}(z) = S_k\left(\frac{p(z) + 1}{p(z) - 1}\right).$$

We thus find that

$$S_k\left(\frac{p(z)+1}{p(z)-1}\right)$$
$$= 1 + \left(\frac{q+1}{2}\right)\left[\frac{p_1 c_1}{2}z + \left\{\frac{p_1 c_2}{2} + \left(\frac{p_2}{4} - \frac{p_1}{4} + \frac{(q-1)p_1^2}{8}\right)c_1^2\right\}z^2\right.$$
$$+ \left\{\frac{p_1 c_3}{2} + \left(\frac{p_2}{2} - \frac{p_1}{2} + \frac{(q-1)p_1^2}{4}\right)c_1 c_2\right.$$
$$\left.\left.+ \left(\frac{p_1}{8} - \frac{p_2}{4} - \frac{(q-1)p_1^2}{8} + \frac{p_3}{8} - \frac{(q-1)p_1 p_2}{8} + \frac{(q-1)^2 p_1^3}{32}\right)c_1^3\right\}z^3\right] + \cdots. \tag{17}$$

Now, upon expanding the left-hand side of (15), we have

$$\frac{z(D_q \mathcal{I}_q^\lambda f)(z)}{f(z)} = 1 + q\psi_1 a_2 z + \left\{(q+q^2)\psi_2 a_3 - q\psi_1^2 a_2^2\right\}z^2$$
$$+ \left\{(q+q^2+q^3)\psi_3 a_4 - (2q+q^2)\psi_1 \psi_2 a_2 a_3 + q\psi_1^3 a_2^3\right\}z^3 + \cdots. \tag{18}$$

Finally, by comparing the corresponding coefficients in (17) and (18) along with Lemma 3, we obtain the result asserted by Theorem 1. □

Theorem 2. *If the function $f(z)$ given by (1) belongs to the class $\mathcal{ST}(k, \lambda, q)$, then*

$$T_3(2) \leq \left[\left(\frac{1+q}{2q\psi_1}\right)p_1^2 + \left(\frac{1+q}{4(q+q^2+q^3)\psi_3}\right)[\Omega_1 + \Omega_2]\right]$$
$$\cdot \left[4\left(\frac{(1+q)^2}{16q^2\psi_1^2}\right)p_1^2 + 16|\Omega_3| + \frac{p_1^2}{4q^2\psi_2^2} + 2\Omega_5 p_1^2 \left|2 - \frac{\Omega_4}{\Omega_5 p_1^2}\right|\right],$$

where

$$\Omega_1 = 2p_1 + 4\left|p_2 - p_1 + \frac{(2+q^2)}{4q}p_1^2\right|,$$

$$\Omega_2 = \left|2p_3 + 2p_1 - 4p_2 - \left(2\left(1+q^2\right) - q\right)p_1^2\right.$$

$$\left. + \left(\frac{4q^2 - 3q + 2}{q}\right)p_1p_2 + \left(\frac{q^2+q+1}{2q^2}p_1^3\right)\right|,$$

$$\Omega_3 = \frac{1}{2q^2\psi_2^2}\left(\frac{p_2}{4} - \frac{p_1}{4} + \frac{(q^2+1)p_1^2}{8q}\right)^2 - \Omega_5 \cdot \left[\frac{p_3}{4} + \frac{p_1}{4} - \frac{p_2}{2}\right.$$

$$\left. - \frac{[2(1+q^2) - q]p_1^2}{8q} + \frac{4q^2 - 3q + 2}{8q}p_1p_2 + \left(\frac{q^2+2q-1}{16q^2}\right)p_1^3\right],$$

$$\Omega_4 = \frac{p_1}{2q^2\psi_2^2}\left(\frac{p_2}{4} - \frac{p_1}{4} + \frac{(q^2+1)p_1^2}{8q}\right) - \Omega_5 p_1\left(p_2 - p_1 + \frac{(2+q^2)p_1^2}{4q}\right),$$

$$\Omega_5 = \frac{(1+q)^2}{16q^2(1+q+q^2)\psi_1\psi_3}$$

and p_j ($j = 1, 2$) are positive and are the coefficients of the functions $p_k(z)$ defined by (6).

Proof. Upon comparing the corresponding coefficients in (17) and (18), we find that

$$a_2 = \frac{(1+q)p_1c_1}{4q\psi_1}, \tag{19}$$

$$a_3 = \frac{1}{2q\psi_2}\left[\frac{p_1c_2}{2} + \left(\frac{p_2}{4} - \frac{p_1}{4} + \frac{(q^2+1)p_1^2}{8q}\right)c_1^2\right], \tag{20}$$

$$a_4 = \frac{(1+q)}{4(q+q^2+q^3)\psi_3}\left[p_1c_3 + \left(p_2 - p_1 + \frac{(2+q^2)p_1^2}{4q}\right)c_1c_2\right.$$

$$\left. + \left(\frac{p_3}{4} + \frac{p_1}{4} - \frac{p_2}{2} - \frac{(2(1+q^2) - q)p_1^2}{8q} + \frac{(4q^2 - 3q + 2)}{8q}p_1p_2\right.\right.$$

$$\left.\left. + \frac{(q^2+2q-1)}{16q^2}p_1^3\right)c_1^3\right]. \tag{21}$$

By a simple computation, $\mathcal{T}_3(2)$ can be written as follows:

$$\mathcal{T}_3(2) = (a_2 - a_4)\left(a_2^2 - 2a_3^2 + a_2a_4\right).$$

Now, if $f \in \mathcal{ST}(k, \lambda, q)$, then it is clearly seen that

$$|a_2 - a_4| \leq |a_2| + |a_4|$$

$$\leq \left(\frac{1+q}{2q\psi_1}\right)p_1^2 + \left(\frac{1+q}{4(q+q^2+q^3)\psi_3}\right)(\Omega_1 + \Omega_2).$$

We need to maximize $\left|a_2^2 - 2a_3^2 + a_2 a_4\right|$ for a function $f \in \mathcal{ST}(k,\lambda,q)$. So, by writing a_2, a_3, and a_4 in terms of c_1, c_2, and c_3, with the help of (19)–(21), we get

$$\left|a_2^2 - 2a_3^2 + a_2 a_4\right|$$
$$= \left|\left(\frac{(1+q)^2}{16q^2\psi_1^2}\right) p_1^2 c_1^2 - \Omega_3 c_1^4 - \Omega_4 c_1^2 c_2 - \frac{p_1^2}{8q^2\psi_2^2} c_2^2 + \Omega_5 p_1^2 c_1 c_3\right|. \tag{22}$$

Finally, by applying the trigonometric inequalities, Lemmas 2 and 3 along with (22), we obtain the result asserted by Theorem 2. □

As an application of Theorem 2, we first set $\psi_{n-1} = 1$ and $k = 0$ and then let $q \to 1-$. We thus arrive at the following known result.

Corollary 1 (see [25]). *If the function $f(z)$ given by (1) belongs to the class $\mathcal{S}^*$, then*

$$T_3(2) \leqq 84.$$

Theorem 3. *If the function $f(z)$ given by (1) belongs to the class $\mathcal{ST}(k,\lambda,q)$, then*

$$\left|a_2 a_4 - a_3^2\right| \leqq \frac{1}{4q^2 \psi_2^2} p_1^2, \tag{23}$$

where $k \in [0,1]$ and p_j ($j = 1,2,3$) are positive and are the coefficients of the functions $p_k(z)$ defined by (6).

Proof. Making use of (19)–(21), we find that

$$a_2 a_4 - a_3^2 = \frac{A(q)}{16q^2\psi_1\psi_3} p_1^2 c_1 c_3 + \left(\frac{A(q)\psi_2^2 - \psi_1\psi_3}{16q^2\psi_1\psi_2^2\psi_3}\right) p_1 p_2 - \frac{A(q)\psi_2^2 - \psi_1\psi_3}{16q^2\psi_1\psi_2^2\psi_3} p_1^2$$
$$+ \frac{A(q)(2+q^2)\psi_2^2 - 2(1+q^2)\psi_1\psi_3}{64q^2\psi_1\psi_3} p_1^3\right) c_1^2 c_2 + \frac{1}{16q^2\psi_2^2} p_1^2 c_2^2$$
$$+ \left[\frac{A(q)}{64q^2\psi_1\psi_3} p_1 p_3 + \left(\frac{A(q)\psi_2^2 - \psi_1\psi_3}{64q^2\psi_1\psi_2^2\psi_3}\right) p_1^2 + \left(\frac{\psi_1\psi_3 - A(q)\psi_2^2}{32q^2\psi_1\psi_2^2\psi_3}\right) p_1 p_2\right.$$
$$+ \left(\frac{2(1+q^2)\psi_1\psi_3 - (2(1+q^2)-q)A(q)\psi_2^2}{128q^3\psi_1\psi_2^2\psi_3}\right) p_1^3$$
$$+ \left(\frac{A(q)(4q^2 - 3q + 2)\psi_2^2 - 2(1+q^2)\psi_1\psi_3}{128q^3\psi_1\psi_2^2\psi_3}\right) p_1^2 p_2$$
$$+ \left.\left(\frac{A(q)(q^2 + 2q - 1)\psi_2^2 - (1+q^2)^2\psi_1\psi_3}{256q^4\psi_1\psi_2^2\psi_3}\right) p_1^4 - \frac{1}{64q^2\psi_2^2} p_2^2\right] c_1^4, \tag{24}$$

where

$$A(q) = \frac{(1+q)^2}{1+q+q^2}.$$

We substitute the values of c_2 and c_3 from the above Lemma and, for simplicity, take $Y = 4 - c_1^2$ and $Z = (1 - |x|^2)z$. Without loss of generality, we assume that $c = c_1$ ($0 \leq c \leq 2$), so that

$$a_2 a_4 - a_3^2 = \left[\frac{q(1-q) A(q) \psi_2^2}{128 q^2 \psi_1 \psi_3} p_1^3 + \frac{A(q)}{64 q^2 \psi_1 \psi_3} p_1 p_3 \right.$$
$$+ \left(\frac{A(q)(4q^2 - 3q + 2)\psi_2^2 - 2(1+q^2)\psi_1\psi_3}{128 q^3 \psi_1 \psi_2^2 \psi_3} \right) p_1^2 p_2$$
$$+ \left(\frac{A(q)(q^2 + 2q - 1)\psi_2^2 - (1+q^2)^2 \psi_1 \psi_3}{256 q^4 \psi_1 \psi_2^2 \psi_3} \right) p_1^4 - \left. \frac{1}{64 q^2 \psi_2^2} p_2^2 \right] c^4$$
$$+ \left[\frac{A(q)\psi_2^2 - \psi_1 \psi_3}{32 q^2 \psi_1 \psi_2^2 \psi_3} p_1 p_2 + \frac{A(q)(2 + q^2)\psi_2^2 - 2(1+q^2)\psi_1 \psi_3}{128 q^2 \psi_1 \psi_3} p_1^3 \right] c^2 xY$$
$$\cdot \left[-\frac{A(q)}{64 q^2 \psi_1 \psi_3} p_1^2 c^2 Y x^2 - \frac{1}{64 q^2 \psi_2^2} p_1^2 x^2 Y^2 + \frac{A(q)}{32 q^2 \psi_1 \psi_3} p_1^2 c Y Z \right]. \tag{25}$$

Upon setting $Z = (1 - |x|^2)z$ and taking the moduli in (25) and using trigonometric inequality, we find that

$$\left| a_2 a_4 - a_3^2 \right| \leq |\lambda_1| c^4 + |\lambda_2| |x| Y c^2 + \frac{A(q)}{64 q^2 \psi_1 \psi_3} p_1^2 Y |x|^2 c^2$$
$$+ \frac{1}{64 q^2 \psi_2^2} p_1^2 |x|^2 Y^2 + \frac{A(q)}{32 q^2 \psi_1 \psi_3} p_1^2 c^2 Y (1 - |x|^2)$$
$$= \Lambda(c, |x|), \tag{26}$$

where

$$\lambda_1 = \frac{q(1-q) A(q) \psi_2^2}{128 q^2 \psi_1 \psi_3} p_1^3 + \frac{A(q)}{64 q^2 \psi_1 \psi_3} p_1 p_3$$
$$+ \left(\frac{A(q)(4q^2 - 3q + 2)\psi_2^2 - 2(1+q^2)\psi_1\psi_3}{128 q^3 \psi_1 \psi_2^2 \psi_3} \right) p_1^2 p_2$$
$$+ \left(\frac{A(q)(q^2 + 2q - 1)\psi_2^2 - (1+q^2)^2 \psi_1 \psi_3}{256 q^4 \psi_1 \psi_2^2 \psi_3} \right) p_1^4 - \frac{1}{64 q^2 \psi_2^2} p_2^2$$
$$\lambda_2 = \frac{A(q)\psi_2^2 - \psi_1 \psi_3}{32 q^2 \psi_1 \psi_2^2 \psi_3}; p_1 p_2 + \frac{A(q)(2 + q^2)\psi_2^2 - 2(1+q^2)\psi_1 \psi_3}{128 q^2 \psi_1 \psi_3} p_1^3.$$

Now, trivially, we have

$$\Lambda'(|x|) > 0$$

on $[0, 1]$, and so

$$\Lambda(|x|) \leqq \Lambda(1).$$

Hence, by puting $Y = 4 - c_1^2$ and after some simplification, we have

$$\left| a_2 a_4 - a_3^2 \right| = \left(|\lambda_1| - |\lambda_2| + \frac{\psi_1 \psi_3 - A(q) \psi_2^2}{64 q^2 \psi_1 \psi_3} p_1^2 \right) c^4$$
$$+ \left(4|\lambda_2| + \left(\frac{A(q) \psi_2^2 - \psi_1 \psi_3}{16 q^2 \psi_1 \psi_3} p_1^2 \right) \right) c^2 + \frac{1}{4 q^2 \psi_2^2} p_1^2$$
$$= G(c). \tag{27}$$

For optimum value of $G(c)$, we consider $G'(c) = 0$, which implies that $c = 0$. So $G(c)$ has a maximum value at $c = 0$. We therefore conclude that the maximum value of $G(c)$ is given by

$$\frac{1}{4q^2\psi_2^2}p_1^2,$$

which occurs at $c = 0$ or

$$c^2 = -\frac{128|\lambda_2|q^2\psi_1\psi_3 + 4A(q)\psi_2^2 - 2\psi_1\psi_3 p_1^2}{(64q^2(|\lambda_1| - |\lambda_2|)\psi_1\psi_3 + \psi_1\psi_3 - A(q)\psi_2^2 p_1^2)}.$$

This completes the proof of Theorem 3. □

If we put $\psi_{n-1} = 1$ and let $q \to 1-$ in Theorem 3, we have the following known result.

Corollary 2 (see [26]). *If the function $f(z)$ given by (1) belongs to the class k-$\mathcal{ST}$, where $k \in [0,1]$, then*

$$\left|a_2 a_4 - a_3^2\right| \leqq \frac{p_1^2}{4}.$$

If we put

$$p_1 = 2 \quad \text{and} \quad \psi_{n-1} = 1,$$

by letting $q \to 1-$ in Theorem 3, we have the following known result.

Corollary 3 (see [18]). *If $f \in \mathcal{S}^*$, then*

$$\left|a_2 a_4 - a_3^2\right| \leqq 1.$$

By letting $k = 1$, $\psi_{n-1} = 1$, $q \to 1-$ and

$$p_1 = \frac{8}{\pi^2}, \quad p_2 = \frac{16}{3\pi^2} \quad \text{and} \quad p_3 = \frac{184}{45\pi^2}$$

in Theorem 3, we have the following known result.

Corollary 4 (see [27]). *If the function $f(z)$ given by (1) belong to the class $\mathcal{SP}$, then*

$$\left|a_2 a_4 - a_3^2\right| \leqq \frac{16}{\pi^4}.$$

4. Concluding Remarks and Observations

Motivated significantly by a number of recent works, we have made use of a certain general conic domain and the quantum (or q-) calculus in order to define and investigate a new subclass of normalized analytic functions in the open unit disk $\mathbb{U}$, which we have referred to as q-starlike functions. For this q-starlike function class, we have successfully derived several properties and characteristics. In particular, we have found the Hankel determinant and the Toeplitz matrices for this newly-defined class of q-starlike functions. We also highlight some known consequences of our main results which are stated and proved as theorems and corollaries.

Author Contributions: conceptualization, Q.Z.A. and N.K. (Nazar Khan); methodology, N.K. (Nasir Khan); software, B.K.; validation, H.M.S.; formal analysis, H.M.S.; writing—original draft preparation, H.M.S.; writing—review and editing, H.M.S.; supervision, H.M.S.

Funding: This research received no external funding.

Conflicts of Interest: The authors declare no conflict of interest.

References

1. Miller, S.S.; Mocanu, P.T. Differential subordination and univalent functions. *Mich. Math. J.* **1981**, *28*, 157–171. [CrossRef]
2. Miller, S.S.; Mocanu, P.T. *Differential Subordination: Theory and Applications*; Series on Monographs and Textbooks in Pure and Applied Mathematics, No. 225; Marcel Dekker Incorporated: New York, NY, USA; Basel, Switzerland, 2000.
3. Kanas, S.; Wiśniowska, A. Conic regions and k-uniform Convexity. *J. Comput. Appl. Math.* **1999**, *105*, 327–336. [CrossRef]
4. Kanas, S.; Wiśniowska, A. Conic domains and starlike Functions. *Rev. Roum. Math. Pures Appl.* **2000**, *45*, 647–657.
5. Kanas, S.; Srivastava, H.M. Linear operators associated with k-uniformly convex functions. *Integral Transform. Spec. Funct.* **2000**, *9*, 121–132. [CrossRef]
6. Mahmood, S.; Jabeen, M.; Malik, S.N.; Srivastava, H.M.; Manzoor, R.; Riaz, S.M.J. Some coefficient inequalities of q-starlike functions associated with conic domain defined by q-derivative. *J. Funct. Spaces* **2018**, *2018*, 8492072. [CrossRef]
7. Srivastava, H.M.; Karlsson, P.W. *Multiple Gaussian Hypergeometric Series*; Ellis Horwood Limited: Chichester, UK, 1985.
8. Jackson, F.H. On q-definite integrals. *Quart. J. Pure Appl. Math.* **1910**, *41*, 193–203.
9. Jackson, F.H. q-difference equations. *Am. J. Math.* **1910**, *32*, 305–314. [CrossRef]
10. Kanas, S.; Răducanu, D. Some class of analytic functions related to conic domains. *Math. Slov.* **2014**, *64*, 1183–1196. [CrossRef]
11. Noor, K.I. On new classes of integral operators. *J. Nat. Geom.* **1999**, *16*, 71–80.
12. Ismail, M.E.H.; Merkes, E.; Styer, D. A generalization of starlike functions. *Complex Var. Theory Appl.* **1990**, *14*, 77–84. [CrossRef]
13. Srivastava, H.M. Univalent functions, fractional calculus, and associated generalized hypergeometric functions. In *Univalent Functions, Fractional Calculus and Their Applications*; Srivastava, H.M., Owa, S., Eds.; Ellis Horwood Limited: Chichester, UK, 1989; pp. 329–354.
14. Srivastava, H.M.; Bansal, D. Close-to-convexity of a certain family of q-Mittag-Leffer functions. *J. Nonlinear Var. Anal.* **2017**, *1*, 61–69.
15. Uçar, H.E.Ö. Coefficient inequality for q-starlike Functions. *Appl. Math. Comput.* **2016**, *76*, 122–126.
16. Sahoo, S.K.; Sharma, N.L. On a generalization of close-to-convex functions. *Ann. Polon. Math.* **2015**, *113*, 93–108. [CrossRef]
17. Noonan, J.W.; Thomas, D.K. On the second Hankel derminant of areally mean p-valent functions. *Trans. Am. Math. Soc.* **1976**, *223*, 337–346.
18. Janteng, A.; Abdul-Halim, S.; Darus, M. Hankel determinant for starlike and convex functions. *Int. J. Math. Anal.* **2007**, *1*, 619–625.
19. Mishra, A.K.; Gochhayat, P. Second Hankel determinant for a class of analytic functions defined by fractional derivative. *Internat. J. Math. Math. Sci.* **2008**, *2008*, 153280. [CrossRef]
20. Singh, G.; Singh, G. On the second Hankel determinant for a new subclass of analytic functions. *J. Math. Sci. Appl.* **2014**, *2*, 1–3.
21. Babalola, K.O. On $H_3(1)$ Hankel determinant for some classes of univalent functions. *Inequal. Theory Appl.* **2007**, *6*, 1–7.
22. Duren, P.L. *Univalent Functions (Grundlehren der Mathematischen Wissenschaften 259)*; Springer: New York, NY, USA; Berlin/Heidelberg, Germany; Tokyo, Japan, 1983.
23. Thomas, D.K.; Abdul-Halim, S. Toeplitz matrices whose elements are the coefficients of starlike and close-to-convex functions. *Bull. Malays. Math. Sci. Soc.* **2017**, *40*, 1781–1790. [CrossRef]
24. Efraimidis, I. A generalization of Livingston's coefficient inequalities for functions with positive real part. *J. Math. Anal. Appl.* **2016**, *435*, 369–379. [CrossRef]
25. Ali, M.F.; Thomas, D.K.; Vasudevarao, A. Toeplitz determinants whose element are the coefficients of univalent functions. *Bull. Aust. Math. Soc.* **2018**, *97*, 253–264. [CrossRef]

26. Ramachandran, C.; Annamalai, S. On Hankel and Toeplitz determinants for some special class of analytic functions involving conical domains defined by subordination. *Internat. J. Engrg. Res. Technol.* **2016**, *5*, 553–561.
27. Lee, S.K.; Ravichandran, V.; Supramaniam, S. Bounds for the second Hankel determinant of certain univalent functions. *J. Inequal. Appl.* **2013**, *2013*, 281–297. [CrossRef]

 © 2019 by the authors. Licensee MDPI, Basel, Switzerland. This article is an open access article distributed under the terms and conditions of the Creative Commons Attribution (CC BY) license (http://creativecommons.org/licenses/by/4.0/).

Article

A Hermite Polynomial Approach for Solving the SIR Model of Epidemics

Aydin Secer [1,*], Neslihan Ozdemir [1] and Mustafa Bayram [2]

1. Department of Mathematical Engineering, Yildiz Technical University, Istanbul 34200, Turkey; ozdemirn@yildiz.edu.tr
2. Department of Computer Engineering, Gelisim University, Istanbul 34315, Turkey; mbayram@gelisim.edu.tr
* Correspondence: asecer@yildiz.edu.tr; Tel.: +90-532-062-1353

Received: 2 October 2018; Accepted: 27 November 2018; Published: 5 December 2018

Abstract: In this paper, the problem of the spread of a non-fatal disease in a population is solved by using the Hermite collocation method. Mathematical modeling of the problem corresponds to a three-dimensional system of nonlinear ODEs. The presented scheme reduces the problem to a nonlinear algebraic equation system by expanding the approximate solutions by using Hermite polynomials with unknown coefficients. These coefficients of the Hermite polynomials are computed by using the matrix operations of derivatives together with the collocation method. Maple software is used to carry out the computations. In addition, comparison of our method with the Homotopy perturbation method (HPM) and Laplece-Adomian decomposition method (LADM) proves accuracy of solution.

Keywords: SIR model; Hermite collocation method; approximate solution; Hermite polynomials and series; collocation points

1. Introduction

Systems of ordinary differential equations are useful in representing some real life problems in terms of the mathematical expressions, which abound in the fields of biological, physical, engineering, financial or sociological fields. It is well known that many nonlinear problems in these fields can be well modeled by systems of ordinary differential equations. However, finding exact solutions of systems of ordinary differential equations involving nonlinear terms can be extremely difficult in most of the situations. In addition, we know that exact solutions of most realistic systems of ordinary differential equations cannot be found, so we need numerical and approximate methods for finding approximate solutions.There are a lot of methods that have been studied by many researchers to solve the systems of ordinary differential equations. Some of these methods are the multi-step method proposed by Hojjati et al. [1], the collocation method presented by Mastorakis [2], the Adomian decomposition method improves [3], the exponential Galerkin method introduced by Yüzbaşı and Karaçayır [4], the exponential collocation method proposed by Yüzbaşı [5], the Galerkin finite element method given by Al-Omari et al. [6].

In this study, we are interested in the SIR model, a model of an epidemic of an infectious disease in a population. This model comprises three types of individuals: those who might be susceptible to the disease, those who might be infected with the disease , and those who might have recovered or be immune from the disease. The model thus has three classes or states.

The following system determines the progress of the disease [7]:

$$\frac{dS}{dt} = -\beta S(t)I(t)$$
$$\frac{dI}{dt} = \beta S(t)I(t) - \gamma I(t) \qquad (1)$$
$$\frac{dR}{dt} = \gamma I(t)$$

with initial conditions

$$S(0) = N_S, I(0) = N_I, R(0) = N_R. \qquad (2)$$

N_S = the number of susceptible individuals in the population at time t.
N_I = the number of infected individuals in the population at time t.
N_R = the number of recovered individuals in the population at time t.
N = the population size.
β = the transmissivity rate.
γ = is the recovery rate. Note that, at any given time, an individual can only be in one of the three groups. Thus, $N_S + N_I + N_R = N$.

Finding exact solutions of SIR models is important because biologists could use it to design and run experiments to observe the spread of infectious diseases by introducing natural initial conditions. Through these experiments, as well as through mathematical modelling, one can learn the ways on how to control the spread of epidemics. It is extremely difficulty to obtain the exact solutions for such problems that actually represented such phenomena. It is a big task for scientific community to search for appropriate methods. Within two decades, to obtain approximate solutions of Equation (1), some authors have studied this model using different methods. For Equation (1), Argub and El-Ajou used Homotopy Analysis Method for different parameter values [7], Awawdeh et al. used Homotopy Analysis Method [8], Biazar used the Adomian decomposition method [9], Rafei et al. applied homotopy perturbation method [10], Ibrahim et al. applied Differential Transformation Approach [11]. In [12], this system was solved using Laplace-Adomian decomposition method. In [13], Harman and Johnston solved the epidemic model using stochastic Galerkin method. Equation (1) was solved using 4th order Runge-Kutta method by Kousar et al. [14] and using Euler, Runge Kutta-2 and Runge-Kutta-4 methods by Hussain et al. [15].

The collocation method has become progressively favourite to solve differential equations. This method can reduce the complexity of solving the systems of ordinary differential equations for epidemic models with high dimensions and it is very useful in contributing highly accurate solutions to differential equations. In this study, Hermite polynomials, a class of the orthogonal polynomials $\{H_0(t), H_1(t), \ldots, H_L(t)\}$ that are orthogonol on $(-\infty, \infty)$, are used. Hermite polynomials have advantages over other orthogonal polynomials. Hermite collocation method (HCM) has been used to solve systems of nonlinear ordinary differential equations with special initial conditions. The most important advantage of the presented method is that it transforms this system (1) into a nonlinear system of algebraic equations which can be easily solved. Until recently, HCM has been used to obtain solutions to a higher-order linear Fredholm integro differential equations in [16], to linear fractional order Systems of differential equations in [17], to differential difference equations in [18], to fractional order differential equations in [19] and to the neutral functional-differential equations with proportional delays in [20].

2. The Hermite Collocation Method (HCM)

In this section, we present the Hermite collocation method to obtain approximate solutions to Equation (1) in the truncated Hermite series form

$$S(t) = \sum_{l=0}^{L} c_{1,l} H_l(t), \ I(t) = \sum_{l=0}^{L} c_{2,l} H_l(t) \text{ and } R(t) = \sum_{l=0}^{L} c_{3,l} H_l(t). \tag{3}$$

Here, $c_{1,l}, c_{2,l}$, and $c_{3,l}$ ($l = 0, 1, 2, \ldots, L$) are the unknown Hermite coefficients, L is any positive number where $L \geq m$ (m is the number of equations in the system), and $H_l(t), l = 0, 1, 2, \ldots, L$ are the Hermite polynomials. The Hermite polynomials are identified by

$$H_l(t) = l! \sum_{m=0}^{L} \frac{(-1)^m}{m!(l-2m)!} (2t)^{l-2m}, \ l \in \mathbb{N}, \ 0 \leq t \leq \infty \tag{4}$$

where $L = l/2$ if l is even and $L = (l-1)/2$ if l is odd.

We represent Equation (1) in the form of matrices. Firstly, we write the approximate solutions of Equation (1):

$$\begin{aligned} S(t) &= H(t) C_1 \\ I(t) &= H(t) C_2 \\ R(t) &= H(t) C_3 \end{aligned} \tag{5}$$

where

$$H(t) = \begin{bmatrix} H_0(t) & H_1(t) & \cdots & H_{L-1}(t) & H_L(t) \end{bmatrix}, C_1 = \begin{bmatrix} c_{1,0} & c_{1,1} & \cdots & c_{1,L} \end{bmatrix}^T$$

$$C_2 = \begin{bmatrix} c_{2,0} & c_{2,1} & \cdots & c_{2,L} \end{bmatrix}^T, C_3 = \begin{bmatrix} c_{3,0} & c_{3,1} & \cdots & c_{3,L} \end{bmatrix}^T$$

If L is an odd number,

$$\underbrace{\begin{bmatrix} H_0(t) \\ H_1(t) \\ \vdots \\ H_{L-1}(t) \\ H_L(t) \end{bmatrix}}_{H^T(t)} = \underbrace{\begin{bmatrix} 2^0 & 0 & \cdots & 0 & 0 \\ 0 & 2^1 & \cdots & 0 & 0 \\ \vdots & \vdots & \ddots & \vdots & \vdots \\ (-1)^{(\frac{L-5}{2})} \frac{2^0}{0!} \frac{(L-1)}{(\frac{L-1}{2})!} & \cdots & & 2^{L-1} & 0 \\ 0 & (-1)^{(\frac{L-1}{2})} \frac{2^1}{1!} \frac{(L)}{(\frac{L-1}{2})!} & 0 & \cdots & 0 & 2^L \end{bmatrix}}_{F} \underbrace{\begin{bmatrix} 1 \\ t \\ \vdots \\ t^{L-1} \\ t^L \end{bmatrix}}_{X^T(t)}. \tag{6}$$

If L is an even number,

$$\underbrace{\begin{bmatrix} H_0(t) \\ H_1(t) \\ \vdots \\ H_{L-1}(t) \\ H_L(t) \end{bmatrix}}_{H^T(t)} = \underbrace{\begin{bmatrix} 2^0 & 0 & \cdots & 0 & 0 \\ 0 & 2^1 & \cdots & 0 & 0 \\ \vdots & \vdots & \ddots & \vdots & \vdots \\ 0 & (-1)^{(\frac{L-2}{2})} \frac{2^1}{1!} \frac{(L-1)}{(\frac{L-2}{2})!} & \cdots & 2^{L-1} & 0 \\ (-1)^{(\frac{L-4}{2})} \frac{2^0}{0!} \frac{(L)}{(\frac{L}{2})!} & 0 & \cdots & 0 & 2^L \end{bmatrix}}_{F} \underbrace{\begin{bmatrix} 1 \\ t \\ \vdots \\ t^{L-1} \\ t^L \end{bmatrix}}_{X^T(t)} \tag{7}$$

where $X(t) = \begin{bmatrix} 1 & t & t^2 & \cdots & t^L \end{bmatrix}$.

Therefore, we can write the following equations:

$$S(t) = X(t)F^T C_1$$
$$I(t) = X(t)F^T C_3 \qquad (8)$$
$$R(t) = X(t)F^T C_3.$$

The relation between the matrix $X(t)$ and its derivative $X^{(1)}(t)$ is

$$X^{(1)}(t) = X(t)B^T \qquad (9)$$

where

$$B^T = \begin{bmatrix} 0 & 1 & 0 & \cdots & 0 \\ 0 & 0 & 2 & \cdots & 0 \\ \vdots & \vdots & \vdots & \ddots & \vdots \\ 0 & 0 & 0 & \cdots & L \\ 0 & 0 & 0 & \cdots & 0 \end{bmatrix}.$$

From Equations (8) and (9), we obtain the following equations:

$$S^{(1)}(t) = X(t)B^T F^T C_1, \ I^{(1)}(t) = X(t)B^T F^T C_2, \ R^{(1)}(t) = X(t)B^T F^T C_3. \qquad (10)$$

Thus, we can construct the matrices $v(t)$ and $v^{(1)}(t)$ as follows:

$$v(t) = \overline{X}\,\overline{F}C \text{ and } v^{(1)}(t) = \overline{X}\,\overline{B}\,\overline{F}C \qquad (11)$$

where

$$v(t) = \begin{bmatrix} S(t) \\ I(t) \\ R(t) \end{bmatrix}, v^{(1)}(t) = \begin{bmatrix} S^{(1)}(t) \\ I^{(1)}(t) \\ R^{(1)}(t) \end{bmatrix}, \overline{X}(t) = \begin{bmatrix} X(t) & 0 & 0 \\ 0 & X(t) & 0 \\ 0 & 0 & X(t) \end{bmatrix}, \overline{F} = \begin{bmatrix} F^T & 0 & 0 \\ 0 & F^T & 0 \\ 0 & 0 & F^T \end{bmatrix}$$

and

$$\overline{B} = \begin{bmatrix} B^T & 0 & 0 \\ 0 & B^T & 0 \\ 0 & 0 & B^T \end{bmatrix}, C = \begin{bmatrix} C_1 \\ C_2 \\ C_3 \end{bmatrix}.$$

We can express Equation (1) in the matrix form

$$v^{(1)}(t) - Kv(t) - Mv_{1,2}(t) = g \qquad (12)$$

where

$$g = \begin{bmatrix} 0 \\ 0 \\ 0 \end{bmatrix}, K = \begin{bmatrix} 0 & 0 & 0 \\ 0 & -\gamma & 0 \\ 0 & \gamma & 0 \end{bmatrix}, M = \begin{bmatrix} -\beta \\ \beta \\ 0 \end{bmatrix}, v_{1,2} = \begin{bmatrix} S(t)I(t) \end{bmatrix}.$$

Now let us determine the unknown coefficients $c_{1,l}, c_{2,l}$, and $c_{3,l}$. We can use the collocation points defined by

$$t_i = a + \frac{b-a}{L}i, \ i = 0, 1, \ldots, L \qquad (13)$$

for an interval $a \leq t \leq b$.

By using the collocation points in Equation (12), we obtain the following system of matrix equations:

$$v^{(1)}(t_i) - Kv(t_i) - Mv_{1,2}(t_i) = g. \tag{14}$$

$$V^{(1)} = \begin{bmatrix} v^{(1)}(t_0) \\ v^{(1)}(t_1) \\ \vdots \\ v^{(1)}(t_L) \end{bmatrix}, \overline{K} = \begin{bmatrix} K & 0 & \cdots & 0 \\ 0 & K & \cdots & 0 \\ \vdots & \vdots & \ddots & \vdots \\ 0 & 0 & \cdots & K \end{bmatrix}_{(L+1)\times(L+1)}, v = \begin{bmatrix} v(t_0) \\ v(t_1) \\ \vdots \\ v(t_L) \end{bmatrix}, G = \begin{bmatrix} g \\ g \\ \vdots \\ g \end{bmatrix}_{(L+1)\times 1}$$

$$\widetilde{V} = \begin{bmatrix} v_{1,2}(t_0) \\ v_{1,2}(t_1) \\ \vdots \\ v_{1,2}(t_L) \end{bmatrix}, \overline{M} = \begin{bmatrix} M & 0 & \cdots & 0 \\ 0 & M & \cdots & 0 \\ \vdots & \vdots & \ddots & \vdots \\ 0 & 0 & \cdots & M \end{bmatrix}_{(L+1)\times(L+1)}.$$

By aid of the upper matrices, Equation (1) can be written in the following matrix form:

$$V^{(1)} - \overline{K}V - \overline{M}\,\widetilde{\overline{V}} = G. \tag{15}$$

By putting the collocation points of Equation (13) in Equation (11), because we can write recurrence relations

$$v(t_i) = \overline{X}(t_i)\overline{F}C \text{ and } v^{(1)}(t_i) = \overline{X}(t_i)\,\overline{B}\,\overline{F}C,$$

we can write

$$V = X\overline{F}C \text{ and } V^{(1)} = X\overline{B}\overline{F}C \tag{16}$$

so that

$$X = \begin{bmatrix} \overline{X}(t_0) & \overline{X}(t_1) & \cdots & \overline{X}(t_L) \end{bmatrix}^T, \overline{X}(t_i) = \begin{bmatrix} X(t_i) & 0 & 0 \\ 0 & X(t_i) & 0 \\ 0 & 0 & X(t_i) \end{bmatrix}.$$

Let us put the collocation points into the $v_{1,2}(t)$. We then obtain the matrix form

$$\widetilde{V} = \begin{bmatrix} v_{1,2}(t_0) \\ v_{1,2}(t_1) \\ \vdots \\ v_{1,2}(t_L) \end{bmatrix} = \begin{bmatrix} I(t_0) & 0 & \cdots & 0 \\ 0 & I(t_1) & \cdots & 0 \\ \vdots & \vdots & \ddots & \vdots \\ 0 & 0 & \cdots & I(t_L) \end{bmatrix} \begin{bmatrix} S(t_0) \\ S(t_1) \\ \vdots \\ S(t_L) \end{bmatrix} = \overline{I}\,\overline{S} \tag{17}$$

where

$$\overline{I} = \widetilde{X}\,\widetilde{F}\,\overline{C_2}, \text{ and } \overline{S} = \widetilde{\widetilde{T}}\widetilde{F}C \tag{18}$$

so that

$$\tilde{X} = \begin{bmatrix} X(t_0) & 0 & \cdots & 0 \\ 0 & X(t_1) & \cdots & 0 \\ \vdots & \vdots & \ddots & \vdots \\ 0 & 0 & \cdots & X(t_L) \end{bmatrix}, \overline{C}_2 = \begin{bmatrix} C_2 & 0 & \cdots & 0 \\ 0 & C_2 & \cdots & 0 \\ \vdots & \vdots & \ddots & \vdots \\ 0 & 0 & \cdots & C_2 \end{bmatrix}_{(L+1)\times(L+1)}$$

$$\overline{F} = \begin{bmatrix} F^T & 0 & \cdots & 0 \\ 0 & F^T & \cdots & 0 \\ \vdots & \vdots & \ddots & \vdots \\ 0 & 0 & \cdots & F^T \end{bmatrix}_{(L+1)\times(L+1)}$$

$$\widetilde{\widetilde{X}} = \begin{bmatrix} X(t_0) \\ X(t_1) \\ \vdots \\ X(t_L) \end{bmatrix}, \widetilde{\widetilde{F}} = \begin{bmatrix} F^T & S & S \end{bmatrix}, S = \begin{bmatrix} 0 & 0 & \cdots & 0 \\ 0 & 0 & \cdots & 0 \\ \vdots & \vdots & \ddots & \vdots \\ 0 & 0 & \cdots & 0 \end{bmatrix}_{(L+1)\times(L+1)}.$$

From Equations (16)–(18), we obtain the fundamental matrix equation

$$\{X\overline{B}\,\overline{F} - \overline{K}X\overline{F} - \overline{M}\tilde{X}\,\overline{FC}_2\widetilde{\widetilde{X}}\,\widetilde{\widetilde{F}}\}C = G. \tag{19}$$

Shortly, Equation (19) can be written as

$$WC = G \text{ or } [W; G] \tag{20}$$

$$W = X\overline{B}\,\overline{F} - \overline{K}X\overline{F} - \overline{M}\tilde{X}\,\overline{FC}_2\widetilde{\widetilde{X}}\,\widetilde{\widetilde{F}}. \tag{21}$$

Equation (21) subtends a system of $3(L+1)$ nonlinear algebraic equations with the unknown Hermite coefficients $c_{1,l}, c_{2,l}$, and $c_{3,l}$. By placing $t \to 0$ in Equation (5), the matrix forms of the initial conditions can be expressed by

$$\begin{aligned} S(t) &= H(0)C_1 = [N_S] \\ I(t) &= H(0)C_2 = [N_I] \\ R(t) &= H(0)C_3 = [N_R]. \end{aligned} \tag{22}$$

That is, these matrix forms can be expressed by

$$\begin{aligned} U_1 &= S(0) = \begin{bmatrix} c_{1,0} & c_{1,1} & \cdots & c_{1,L} \end{bmatrix} \\ U_2 &= I(0) = \begin{bmatrix} c_{2,0} & c_{2,1} & \cdots & c_{2,L} \end{bmatrix} \\ U_3 &= R(0) = \begin{bmatrix} c_{3,0} & c_{3,1} & \cdots & c_{3,L} \end{bmatrix}. \end{aligned} \tag{23}$$

When the rows in the matrices in Equation (23) are replaced with any three rows of the matrix in Equation (20), we obtain the solution to Equation (1) under initial conditions. Thereby, we get the augmented matrix

$$\widetilde{W}C = \widetilde{G}, \tag{24}$$

which is an algebraic system. To determine the coefficients, this system must be solved. The determined coefficients $c_{i,0}, c_{i,1}, \cdots, c_{i,L}$, $(i = 1, 2, 3)$ are substituted into Equation (3), and we can then obtain approximate solutions.

3. Error Estimate for the Solution

We can here check the accuracy of the proposed method. Since the $S_L(t), I_L(t), R_L(t)$ is an approximate solution to Equation (1), once these functions and their first derivative are substituted into Equation (1), the obtained equations should satisfied approximately, in short, for $t = t_r \in [0, R]$, $r = 0, 1, \ldots$

$$\begin{aligned} E_{1,L}(t_r) &= |S'(t_r) + \beta S(t_r) I(t_r)| \cong 0 \\ E_{2,L}(t_r) &= |I'(t_r) - \beta S(t_r) I(t_r) + \gamma I(t_r)| \cong 0 \\ E_{3,L}(t_r) &= |R'(t_r) - \gamma I(t_r)| \cong 0, \end{aligned} \tag{25}$$

and $E_{i,L} \leq 10^{-k_r}$, $i = 1, 2, 3$ (k_r any positive contant). If $max 10^{-k_r} = 10^{-k}$ is prescribed, the truncation limit L is increased until the difference $E_{i,L}(t_r)$, $(i = 1, 2, 3)$ at each of the points becomes smaller than the prescribed 10^{-k} [21,22].

4. Illustrative Example

In this section, to show the accuracy and efficiency of the presented method, the SIR model of epidemics, given in Equation (1), is solved with it. For the SIR model, the following parameter values that given in [9] are used. Numerical calculations were performed using Maple software.

$$N_S = 20, N_I = 15, N_R = 10, \beta = 0.01, \gamma = 0.02.$$

In interval $0 \leq t \leq 1$, we obtain approximate solutions for $L = 5$ using the presented method; in turn, approximate solutions with five terms:

$$\begin{aligned} S(t) =\ &20.00000000 - 2.999999999t - 0.04499957685t^2 + 0.02804671200t^3 \\ &+ 0.0008058035642t^4 - 0.0003329149155t^5 \\ I(t) =\ &15.00000000 + 2.699999999t + 0.01799970470t^2 - 0.02816759777t^3 - 0.0006626352597t^4 \\ &+ 0.0003329022387t^5 \\ R(t) =\ &10.00000000 + 0.3000000000t + 0.2699987216t^2 + 0.0001208857706t^3 - 0.0001431683045t^4 \\ &+ 0.126768711610^{-7} t^5. \end{aligned}$$

The approximate solutions of this system were presented by Rafei et al. using the homotopy perturbation method (HPM) [10]. The homotopy perturbation method is a efficient method for finding solutions of ordinary/partial differential equations without the need for a linearization process. The obtained approximate solutions with five terms:

$$\begin{aligned} S(t) &= 20 - 3t - 0.045t^2 + 0.02805t^3 + 0.0007953750t^4 - 0.0003165502t^5 \\ I(t) &= 15 + 2.7t + 0.018t^2 - 0.02817t^3 - 0.0006545250t^4 + 0.0003191683t^5 \\ R(t) &= 10 + 0.3t + 0.027t^2 + 0.00012t^3 - 0.0001408500t^4 - 0.0000021681t^5. \end{aligned}$$

The approximate solutions of this system were also presented by Dogan and Akin using Laplace-Adomian decomposition method (LADM) [12]. The LADM provides us with an approximate solution in the form of infinite series. The obtained approximate solutions with five terms:

$$S(t) = 20 - 3t - 0.045t^2 + 0.02805t^3 + 0.000795375t^4 - 0.00031655t^5$$
$$I(t) = 15 + 2.7t + 0.018t^2 - 0.02817t^3 - 0.000654525t^4 + 0.000319168t^5$$
$$R(t) = 10 + 0.3t + 0.027t^2 + 0.00012t^3 - 0.00014085t^4 - 0.0000021868t^5.$$

We know that Equation (1) has no exact solution. So, we compared the obtained results using Hermite collocation method with the obtained results using HPM presented [10] and the obtained results using LADM presented [12].

From Figures 1–3, it is clear that the results obtained using HCM is very efficient.

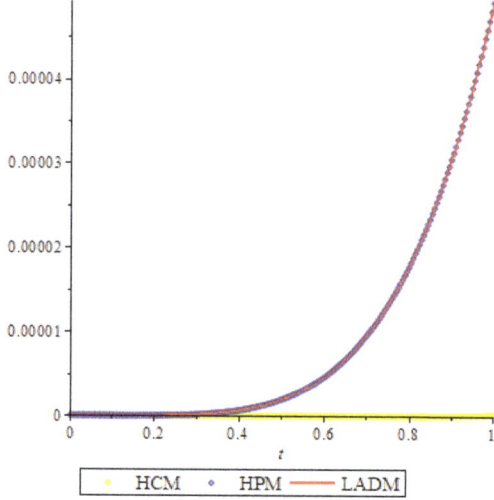

Figure 1. Comparison of the error function $E_{1,5}(t)$ for $S(t)$.

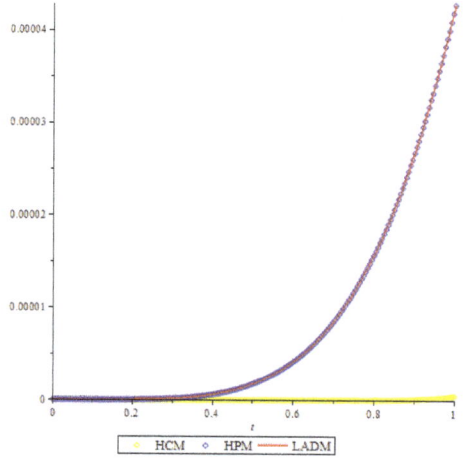

Figure 2. Comparison of the error function $E_{2,5}(t)$ for $I(t)$.

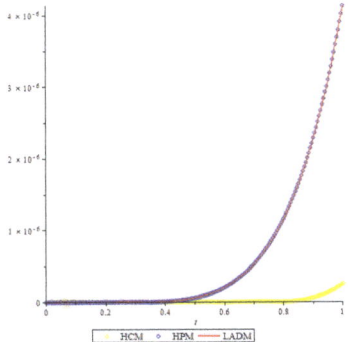

Figure 3. Comparison of the error function $E_{3,5}(t)$ for $R(t)$.

5. Conclusions

In this study, the Hermite Collocation Method was applied to obtain the approximate solutions of SIR model. We showed the accuracy and efficiency of the presented method with an example. To show the correctness of the obtained approximate solutions, we put the obtained approximate solutions back into Equation (1) with the aid of Maple software. Thus, it gives extra measure for confidence of the obtained approximate solutions. The obtained approximate results and the error values are compared with the error values and the approximate solutions obtained with homotopy perturbation method (HPM) [10] and Laplace-Adomian decomposition method [12]. These comparisons reveal that our method is more efficient and useful to find approximate solution the SIR model of epidemics. From Tables 1–3, it is seen that the numerical solutions of the HPM [10] and the LADM [12] are almost same. Therefore, it is observed that the presented method is an alternative way for the solution of nonlinear ODEs system that have no analytic solution. The greatest advantage of the presented method is that all of above computations can be computed easily in very shorter time by using the computer code written in Maple software.

Table 1. The values of $S(t)$, and the residual errors ER_S for HPM, HCM and LADM.

t	$S(t)$ (HPM)	ER_S (HPM)	$S(t)$ (HCM for L = 5)	ER_S (HCM for L = 5)	$S(t)$ (LADM)	ER_S (LADM)
0.2	19.39842557	$2.241556564 \times 10^{-8}$	19.39842556	$1.007967528 \times 10^{-9}$	19.39842557	$2.241715696 \times 10^{-8}$
0.3	19.09671302	$1.639339936 \times 10^{-7}$	19.09671301	$5.224692456 \times 10^{-9}$	19.09671302	$1.639420311 \times 10^{-7}$
0.4	18.79461232	$6.642448998 \times 10^{-7}$	18.79461227	$1.020070225 \times 10^{-9}$	18.79461232	$6.642702518 \times 10^{-7}$
0.5	18.49229607	$1.945779417 \times 10^{-6}$	18.49229590	$4.808788220 \times 10^{-9}$	18.49229607	$1.945841205 \times 10^{-6}$
0.6	18.18993727	$4.638804798 \times 10^{-6}$	18.18993677	$1.036285111 \times 10^{-9}$	18.18993727	$4.638932739 \times 10^{-6}$
0.7	17.88770892	$9.586736659 \times 10^{-6}$	17.88770774	$2.046575227 \times 10^{-9}$	17.88770892	$9.586973403 \times 10^{-6}$
0.8	17.58578366	$1.783241468 \times 10^{-5}$	17.58578115	$1.056548098 \times 10^{-9}$	17.58578366	$1.783281825 \times 10^{-5}$
0.9	17.28433338	$3.058543511 \times 10^{-5}$	17.28432849	$2.255631029 \times 10^{-8}$	17.28433338	$3.058608116 \times 10^{-5}$
1.0	16.98352883	$4.917070631 \times 10^{-5}$	16.98352001	$1.991812037 \times 10^{-7}$	16.98352883	$4.917169073 \times 10^{-5}$

Table 2. The values of $I(t)$, and the residual errors ER_I for HPM, HCM and LADM.

t	$I(t)$ (HPM)	ER_I (HPM)	$I(t)$ (HCM for L = 5)	ER_I (HCM for L = 5)	$I(t)$ (LADM)	ER_I (LADM)
0.2	15.54049369	$2.037288852 \times 10^{-8}$	15.54049370	$1.007967528 \times 10^{-9}$	15.54049369	$2.037528176 \times 10^{-8}$
0.3	15.81085489	$1.484224142 \times 10^{-7}$	15.81085488	$5.224692456 \times 10^{-9}$	15.81085489	$1.484345163 \times 10^{-7}$
0.4	16.08106363	$5.988792320 \times 10^{-7}$	16.08106367	$1.020070225 \times 10^{-9}$	16.08106363	$5.989174454 \times 10^{-7}$
0.5	16.35094781	$1.746299230 \times 10^{-6}$	16.35094797	$4.808788220 \times 10^{-9}$	16.35094781	$1.746392455 \times 10^{-6}$
0.6	16.62033527	$4.142434258 \times 10^{-6}$	16.62033570	$1.036285111 \times 10^{-9}$	16.62033527	$4.142627465 \times 10^{-6}$
0.7	16.88905418	$8.513884329 \times 10^{-6}$	16.88905522	$2.046575227 \times 10^{-9}$	16.88905418	$8.514242143 \times 10^{-6}$
0.8	17.15693346	$1.574071331 \times 10^{-5}$	17.15693567	$1.056548098 \times 10^{-9}$	17.15693345	$1.574132364 \times 10^{-5}$
0.9	17.42380311	$2.681612132 \times 10^{-5}$	17.42380741	$2.255631029 \times 10^{-8}$	17.42380310	$2.681709896 \times 10^{-5}$
1.0	17.68949465	$4.278734031 \times 10^{-5}$	17.68950236	$1.991812037 \times 10^{-7}$	17.68949464	$4.278883073 \times 10^{-5}$

Table 3. The values of $R(t)$, and the residual errors ER_R for HPM, HCM and LADM.

t	$R(t)$ (HPM)	ER_R (HPM)	$R(t)$ (HCM for L = 5)	ER_R (HCM for L = 5)	$R(t)$ (LADM)	ER_R (LADM)
0.2	10.06108073	$1.55732288 \times 10^{-9}$	10.06108073	1.780152×10^{-12}	10.06108073	1.5573248×10^{-9}
0.3	10.09243209	$2.71342062 \times 10^{-9}$	10.09243209	$2.99878840 \times 10^{-9}$	10.09243209	2.7134352×10^{-9}
0.4	10.12432405	$7.76566784 \times 10^{-9}$	10.12432405	3.55561×10^{-12}	10.12432405	7.7656064×10^{-9}
0.5	10.15675613	$5.88551875 \times 10^{-8}$	10.15675613	2.9916773×10^{-9}	10.15675613	$5.88550000 \times 10^{-8}$
0.6	10.18972750	$2.047705402 \times 10^{-7}$	10.18972751	5.3304×10^{-12}	10.18972750	$2.047700736 \times 10^{-7}$
0.7	10.22323698	$5.326273240 \times 10^{-7}$	10.22323703	6.997166×10^{-9}	10.22323698	$5.32626315 \times 10^{-7}$
0.8	10.25728304	$1.170101371 \times 10^{-6}$	10.25728317	7.109×10^{-12}	10.25728304	$1.170099405 \times 10^{-6}$
0.9	10.29186379	$2.293088789 \times 10^{-6}$	10.29186411	6.2910522×10^{-8}	10.29186379	$2.293085246 \times 10^{-6}$
1.0	10.32697698	4.133366×10^{-6}	10.32697760	$4.548599737 \times 10^{-7}$	10.32697698	4.13336×10^{-6}

Author Contributions: All authors contributed equally to the writing of this paper. All authors read and approved the final manuscript.

Funding: This research received no external funding.

Conflicts of Interest: The authors declare no conflict of interest.

References

1. Hojjati, G.; Ardabili, M.R.; Hosseini, S.M. A-EBDF: An adaptive method for numerical solution of stiff systems of ODEs. *Math. Comput. Simul.* **2004**, *66*, 33–41. [CrossRef] [CrossRef]
2. Mastorakis, N.E. Numerical solution of non-linear ordinary differential equations via collocation method (finite elements) and genetic algorithms. *WSEAS Trans. Inf. Sci. Appl.* **2005**, *2*, 467–473. [CrossRef]
3. Shawagfeh, N.; Kaya, D. Comparing numerical methods for the solutions of systems of ordinary differential equations. *Appl. Math. Lett.* **2004**, *17*, 323–328. [CrossRef] [CrossRef]
4. Yüzbaşı, Ş.; Karaçayır, M. An exponential Galerkin method for solutions of HIV infection model of CD4$^+$ T-cells. *Comput. Biol. Chem.* **2017**, *67*, 205–212. [CrossRef] [CrossRef] [PubMed]
5. Yüzbaşı, Ş. An exponential collocation method for the solutions of the HIV infection model of CD4$^+$ T cells. *Int. J. Biomath.* **2016**, *9*, 885–893. [CrossRef] [CrossRef]
6. Al-Omari, A.; Schüttler, H.B.; Arnold, J.; Taha, T. Solving nonlinear systems of first order ordinary differential equations using a Galerkin finite element method. *IEEE Access* **2013**, *1*, 408–417. [CrossRef] [CrossRef]
7. Arqub, O.A.; El-Ajou, A. Solution of the fractional epidemic model by homotopy analysis method. *J. King Saud Univ.-Sci.* **2013**, *25*, 73–81. [CrossRef] [CrossRef]
8. Fadi, A.; Adawi, A.; Mustafa, Z. Solutions of the SIR models of epidemics using HAM. *Chaos Soliton Fract.* **2009**, *42*, 3047–3052. [CrossRef]
9. Biazar, J. Solution of the epidemic model by Adomian decomposition method. *Appl. Math. Comput.* **2006**, *173*, 1101–1106. [CrossRef] [CrossRef]
10. Rafei, M.; Ganji, D.D.; Daniali, H. Solution of the epidemic model by homotopy perturbation method. *Appl. Math. Comput.* **2007**, *187*, 1056–1062. [CrossRef] [CrossRef]
11. Ibrahim, S.F.M.; Ismail, S.M. Differential Transformation Approach to a SIR Epidemic Model with Constant Vaccination. *J. Am. Sci.* **2017**, *8*, 764–769. [CrossRef]
12. Dogan, N.; Akin, Ö. Series solution of epidemic model. *TWMS J. Appl. Eng. Math.* **2012**, *2*, 238–244. [CrossRef]
13. Harman, D.B.; Johnston, P.R. Applying the stochastic galerkin method to epidemic models with uncertainty in the parameters. *Math. Biosci.* **2016**, *277*, 25–37. [CrossRef] [CrossRef] [PubMed]
14. Kousar, N.; Mahmood, R.; Ghalib, M.A. Numerical Study of SIR Epidemic Model. *Int. J. Sci. Basic Appl. Res.* **2012**, *25*, 354–363. [CrossRef]
15. Hussain, R.; Ali, A.; Chaudary, A.; Jarral, F.; Yasmeen, T.; Chaudary, F. Numerical solution for mathematical model of Ebola Virus. *Int. J. Adv. Res.* **2017**, *5*, 1532–1538. [CrossRef] [CrossRef]
16. Pirim, A.N.; Şahin, N.; Sezer, M. A Hermite collocation method for the approximate solutions of high order linear Fredholm integro-differential equations. *Numer. Methods Part. Differ. Equat.* **2011**, *6*, 1707–1721. [CrossRef]
17. Pirim, A.N.; Ayaz, F. A New Technique for Solving Fractional Order Systems: Hermite Collocation Method. *Springer Phys. Math. Stat.* **2016**, *7*, 2307. [CrossRef]

18. Gülsu, M.; Yalman, H.; Öztürk, Y.; Sezer, M. A New Hermite Collocation Method for Solving Differential Difference Equations. *Appl. Appl. Math.* **2011**, *6*, 1856–1869. [CrossRef]
19. Pirim, N.A.; Ayaz, F. Hermite collocation method for fractional order differential equations. *IJOCTA* **2018**, *8*, 228–236.[CrossRef]
20. Ibis, B.; Bayram, M. Numerical solution of the neutral functional-differential equations with proportional delays via collocation method based on Hermite polynomials. *Commun. Math. Model. Appl.* **2016**, *1*, 22–30. [CrossRef]
21. Gülsu, M.; Sezer, M. On the solution of the Riccati Equation by the Taylor Matrix Method. *Appl. Math. Comput.* **2006**, *176*, 414–421. [CrossRef] [CrossRef]
22. Erdem, K.; Yalçınbaş, S. Bernoulli Polynomial Approach to HighOrder Linear Differential-Difference Equations. *AIP Conf. Proc.* **2012**, *1479*, 360–364. [CrossRef]

© 2018 by the authors. Licensee MDPI, Basel, Switzerland. This article is an open access article distributed under the terms and conditions of the Creative Commons Attribution (CC BY) license (http://creativecommons.org/licenses/by/4.0/).

Article

On a Length Problem for Univalent Functions

Mamoru Nunokawa [1], Janusz Sokół [2] and Nak Eun Cho [3],*

1. Department of Mathematics, Gunma University, Hoshikuki-cho 798-8, Chuou-Ward, Chiba 260-0808, Japan; mamoru_nuno@doctor.nifty.jp
2. Faculty of Mathematics and Natural Sciences, University of Rzeszów, 1 Prof. St. Pigoń Street, 35-310 Rzeszów, Poland; jsokol@prz.edu.pl or jsokol@ur.edu.pl
3. Department of Applied Mathematics, College of Natural Sciences, Pukyong National University, Pusan 608-737, Korea
* Correspondence: necho@pknu.ac.kr

Received: 11 October 2018; Accepted: 16 November 2018; Published: 19 November 2018

Abstract: Let g be an analytic function with the normalization in the open unit disk. Let $L(r)$ be the length of $g(\{z : |z| = r\})$. In this paper we present a correspondence between g and $L(r)$ for the case when g is not necessary univalent. Furthermore, some other results related to the length of analytic functions are also discussed.

Keywords: analytic functions; starlike functions; univalent functions; length problems

MSC: 30C45; 30C80

---

1. Introduction

Let $\mathcal{A}$ be the family of functions of the form

$$g(z) = z + \sum_{n=2}^{\infty} a_n z^n \tag{1}$$

which are analytic in the open unit disk $\mathbb{D} = \{z \in \mathbb{C} : |z| < 1\}$. Let $\mathcal{S}$ denote the subfamily of $\mathcal{A}$ consisting of all univalent functions in $\mathbb{D}$.

Let $C(r)$ denote the image curve of the $|z| = r < 1$ under the function $g \in \mathcal{A}$ which bound the area $A(r)$. Furthermore, let $L(r)$ be the length of $C(r)$ and $M(r) = \max_{|z|=r<1} |g(z)|$.

If $g \in \mathcal{A}$ satisfies

$$\mathfrak{Re}\left\{\frac{zg'(z)}{g(z)}\right\} > 0, \ z \in \mathbb{D},$$

then g is said to be starlike with respect to the origin in $\mathbb{D}$ and we write $g \in \mathcal{S}^*$. It is known (for details, see [1,2]) that $\mathcal{S}^* \subset \mathcal{S}$.

The aim of the present paper is to prove, using a modified methodology, that in the following implication

$$g \in \mathcal{S}^* \quad \Rightarrow \quad L(r) = \mathcal{O}\left(M(r) \log \frac{1}{1-r}\right) \quad \text{as} \quad r \to 1, \tag{2}$$

where $\mathcal{O}$ denotes the Landau's symbol, the assumption that g is starlike univalent can be changed by a weaker one. Result (2) was proved by Keogh [3]. Moreover, some other length problems for analytic functions are investigated. Several interesting developments related to length problems for univalent functions were considered in [4–15].

2. Main Results

Theorem 1. *Let g be of the form (1) and suppose that*

$$\left|\frac{zg'(z)}{g(z)}\right| \leq \left|\frac{1+z}{1-z}\right|, \quad z \in \mathbb{D}. \tag{3}$$

Then

$$L(r) = \mathcal{O}\left(M(r)\log\frac{1}{1-r}\right) \quad \text{as} \quad r \to 1,$$

where

$$M(r) = \max_{|z|=r<1} |g(z)|$$

and $\mathcal{O}$ means Landau's symbol.

Proof. Let $z = re^{i\nu}$. We have $g \neq 0$ in $\mathbb{D} \setminus \{0\}$. In fact, if $g = 0$ in $\mathbb{D}$, it contradicts hypothesis (3). Applying [3] (Theorem 1) and the hypothesis of Theorem 1, we have

$$L(r) = \int_0^{2\pi} |zg'(z)|d\nu = \int_0^{2\pi} \left|\frac{zg'(z)}{g(z)}\right| |g(z)|d\nu$$

$$\leq M(r)\int_0^{2\pi} \left|\frac{zg'(z)}{g(z)}\right| d\nu \leq M(r)\int_0^{2\pi} \left|\frac{1+re^{i\nu}}{1-re^{i\nu}}\right| d\nu$$

$$\leq M(r)\left(2\pi + 4\log\frac{1+r}{1-r}\right) \quad \text{as} \quad r \to 1.$$

□

Remark 1. *If g satisfies the condition of Theorem 1, then g is not necessary univalent in $\mathbb{D}$. It is well known that if $g \in \mathcal{S}$, then it follows that*

$$\frac{1-|z|}{1+|z|} \leq \left|\frac{zg'(z)}{g(z)}\right| \leq \frac{1+|z|}{1-|z|}, \quad z \in \mathbb{D}$$

(for details, see [1] (Vol. 1, p. 69)).
    If $g \in \mathcal{A}$ satisfies

$$\mathfrak{Re}\left\{\frac{zg'(z)}{g^{1-\gamma}(z)h^\gamma(z)}\right\} > 0, \quad z \in \mathbb{D}$$

for some $h \in \mathcal{S}^*$ and some $\gamma \in (0,\infty)$, then g is said to be a Bazilevič function of type γ [13]. The class of Bazilevič functions of type γ is denoted by $g \in \mathcal{B}(\gamma)$. We note that Theorem 1 improves the implication (2) by Keogh [3] and it is also related to Theorem 3 given by Thomas [13].

We will need the following Tsuji's result.

Lemma 1 ([16] (p. 226)). *(Theorem 3) If $0 \leq r < R$ and $z = e^{i\nu}$, then*

$$\frac{R-r}{R+r} \leq \mathfrak{Re}\left\{\frac{Re^{i\phi}+z}{Re^{i\phi}-z}\right\} = \frac{R^2-r^2}{R^2-2Rr\cos(\phi-\nu)+r^2} \leq \frac{R+r}{R-r}. \tag{4}$$

Moreover,

$$\frac{1}{2\pi}\int_0^{2\pi}\frac{R^2-r^2}{R^2-2Rr\cos(\phi-\nu)+r^2}d\nu = 1. \tag{5}$$

Theorem 2. Let g be of the form (1) and suppose that

$$\left|\frac{zg'(z)}{g(z)}\right| \leq \left|\frac{1+z}{1-z}\right|, \quad z \in \mathbb{D} \tag{6}$$

and

$$M(r,\beta) = \max_{|z|=r<1} |g(z)| \leq \left|\frac{1+z}{1-z}\right|^\beta, \tag{7}$$

where $1 < \beta$. Then

$$L(r) = \mathcal{O}\left(\frac{1}{(1-r)^\beta}\right) \quad \text{as} \quad r \to 1,$$

where $\mathcal{O}$ means Landau's symbol.

Proof. From the hypotheses (6) and (7), it follows that

$$L(r) = \int_0^{2\pi} |zg'(z)| d\nu = \int_0^{2\pi} \left|\frac{zg'(z)}{g(z)}\right| |g(z)| d\nu$$

$$\leq \int_0^{2\pi} \left|\frac{1+z}{1-z}\right| \left|\frac{1+z}{1-z}\right|^\beta d\nu \leq 2^{1+\beta} \int_0^{2\pi} \frac{1}{|1-z|^{1+\beta}} d\nu$$

$$= \frac{2^{1+\beta}}{(1-r)^{\beta-1}} \int_0^{2\pi} \frac{1}{1-2r\cos\nu+r^2} d\nu.$$

From (5), we have

$$\int_0^{2\pi} \frac{1}{1-2r\cos\nu+r^2} d\nu = \frac{2\pi}{1-r^2}.$$

Hence, we obtain

$$L(r) \leq \frac{2^{1+\beta}}{(1-r)^{\beta-1}} \frac{2\pi}{1-r^2}$$

$$= \mathcal{O}\left(\frac{1}{(1-r)^\beta}\right) \quad \text{as} \quad r \to 1.$$

Therefore, we complete the proof of Theorem 2. □

Let us recall the following Fejér-Riesz's result.

Lemma 2 ([16]). Let h be analytic in $\mathbb{D}$ and continuous on $\overline{\mathbb{D}}$. Then

$$\int_{-1}^1 |h(z)|^p |dz| \leq \frac{1}{2} \int_{|z|=1} |h(z)|^p |dz|,$$

where $p > 0$.

Theorem 3. Let g be of the form (1) and suppose that

$$\frac{1-|z|}{1+|z|} \leq \left|\frac{zg'(z)}{g(z)}\right| \leq \frac{1+|z|}{1-|z|}, \quad z \in \mathbb{D}. \tag{8}$$

Then

$$\mathcal{O}\left(m(r) \log \frac{1}{1-r}\right) \leq L(r) \leq \mathcal{O}\left(\frac{M(r)}{1-r}\right) \quad \text{as} \quad r \to 1,$$

where

$$m(r) = \min_{|z|=r<1} |g(z)|, \quad M(r) = \max_{|z|=r<1} |g(z)| \tag{9}$$

and $\mathcal{O}$ means Landau's symbol.

Proof. From the assumption, we have

$$L(r) = \int_0^{2\pi} |zg'(z)|\, d\nu = \int_0^{2\pi} \left|\frac{zg'(z)}{g(z)}\right| |g(z)|\, d\nu$$

$$\geq m(r) \int_0^{2\pi} \left|\frac{zg'(z)}{g(z)}\right| d\nu$$

because $g(z) \neq 0$ in $\mathbb{D} \setminus \{0\}$. In fact, if $g(z) = 0$ in $\mathbb{D}$, it contradicts hypothesis (8). Applying Fejér-Riesz's Lemma 2, we have

$$L(r) \geq m(r) \int_0^{2\pi} \left|\frac{zg'(z)}{g(z)}\right| d\nu \geq 2m(r) \int_{-r}^{r} \frac{1-\rho}{1+\rho}\, d\rho$$

$$\geq 2m(r) \log \frac{1+r}{1-r} - 2r$$

$$= \mathcal{O}\left(m(r) \log \frac{1}{(1-r)}\right) \quad \text{as} \quad r \to 1.$$

While, we obtain

$$L(r) = \int_0^{2\pi} |zg'(z)|\, d\nu = \int_0^{2\pi} \left|\frac{zg'(z)}{g(z)}\right| |g(z)|\, d\nu$$

$$= M(r) \int_0^{2\pi} \frac{1+|z|}{1-|z|}\, d\nu = 2\pi M(r) \frac{1+r}{1-r}$$

$$= \mathcal{O}\left(\frac{M(r)}{1-r}\right) \quad \text{as} \quad r \to 1.$$

Therefore, we complete the proof of Theorem 3. □

From Theorem 3, we have the following result.

Corollary 1. *Let g be of the form (1) and suppose that g is univalent in $\mathbb{D}$. Then we have*

$$\mathcal{O}\left(m(r) \log \frac{1}{1-r}\right) \leq L(r) \leq \mathcal{O}\left(\frac{M(r)}{1-r}\right) \quad \text{as} \quad r \to 1,$$

where $m(r)$ and $M(r)$ are given by (9), respectively.

Proof. From the hypothesis, we have

$$\frac{1-|z|}{1+|z|} \leq \left|\frac{zg'(z)}{g(z)}\right| \leq \frac{1+|z|}{1-|z|}, \quad z \in \mathbb{D},$$

which completes the proof. □

Lemma 3 ([17] (p. 280) and [18] (p. 491)).

$$\int_0^{2\pi} \frac{d\nu}{|1-re^{i\nu}|^\beta} = \begin{cases} \mathcal{O}\left((1-r)^{1-\beta}\right) & \text{for the case } 1 < \beta, \\ \mathcal{O}\left(\log \frac{1}{1-r}\right) & \text{for the case } \beta = 1, \\ \mathcal{O}(1) & \text{for the case } 0 \leq \beta < 1, \end{cases}$$

where $0 < r < 1$, $0 \leq v \leq 2\pi$, $0 \leq \beta$ and $\mathcal{O}$ means Landau's symbol.

Theorem 4. *Let g be of the form (1) and suppose that*

$$\left|\frac{zg'(z)}{g(z)}\right| \leq \frac{1}{1-|z|}, \quad z \in \mathbb{D} \tag{10}$$

and

$$|g(z)| \leq \frac{1}{|1-z|^\beta}, \quad z \in \mathbb{D}. \tag{11}$$

Then

$$L(r) \leq \begin{cases} \mathcal{O}\left((1-r)^{-3/2}\right) & \text{for } 1 < \beta \leq 3/2, \\ \mathcal{O}\left((1-r)^{-3/2} \log \frac{1}{1-r}\right) & \text{for the case } \beta = 3/2, \\ \mathcal{O}\left((1-r)^{-\beta}\right) & \text{for the case } 3/2 < \beta, \end{cases}$$

where $0 < |z| = r < 1$ and $\mathcal{O}$ means Landau's symbol.

Proof. From the hypothesis (10), it follows that $g(z) \neq 0$ in $\mathbb{D} \setminus \{0\}$. Then we have

$$L(r) = \int_0^{2\pi} \left|re^{iv}g'(re^{iv})\right| dv = \int_0^{2\pi} \left|\frac{zg'(z)}{g(z)}\right| |g(z)| dv$$

$$< \int_0^{2\pi} \left(\frac{1}{1-|z|}\right)\left(\frac{1}{|1-z|^\beta}\right) dv$$

$$= \int_0^{2\pi} \left(\frac{1}{|1-z|}\right)\left(\frac{1}{|1-z|^{\beta-1}}\right)\left(\frac{1}{1-|z|}\right) dv$$

$$\leq \left(\int_0^{2\pi} \frac{1}{|1-z|^2} dv\right)^{1/2} \left(\int_0^{2\pi} \left(\frac{1}{|1-z|^{2\beta-2}}\right) \frac{1}{(1-|z|)^2} dv\right)^{1/2}.$$

Applying Hayman's Lemma 3, we have

$$L(r) \leq \left(\frac{1}{1-r^2}\right)^{1/2} \left(\frac{1}{1-r}\right) \mathcal{O}(1)$$

$$= \mathcal{O}\left(\frac{1}{(1-r)^{3/2}}\right) \quad \text{as} \quad r \to 1$$

for the case $1 < \beta < 3/2$,

$$L(r) \leq \left(\frac{1}{1-r^2}\right)^{1/2} \left(\frac{1}{1-r}\right) \mathcal{O}\left(\log \frac{1}{1-r}\right)$$

$$= \mathcal{O}\left(\frac{1}{(1-r)^{3/2}} \log \frac{1}{1-r}\right) \quad \text{as} \quad r \to 1$$

for the case $\beta = 3/2$ and

$$L(r) = \left(\frac{1}{1-r^2}\right)^{1/2} \left(\frac{1}{1-r}\right) \left(\frac{1}{1-r}\right)^{(2\beta-3)/2} \quad \text{as} \quad r \to 1$$

for the case $3/2 < \beta$. □

Lemma 4 ([16] (p. 227)). *If $g(z) = u(z) + iv(z)$ is analytic in $|z| \leq R$, then*

$$g(z) = \frac{1}{2\pi} \int_0^{2\pi} u(Re^{i\phi}) \frac{Re^{i\phi} + z}{Re^{i\phi} - z} d\phi + iv(0). \tag{12}$$

Moreover, if $|z| < R$ and $v(0) = 0$, then

$$|g(z)| = \frac{1}{2\pi}\int_0^{2\pi} |u(Re^{i\phi})|\left|\frac{Re^{i\phi}+z}{Re^{i\phi}-z}\right| d\phi.$$

Theorem 5. *Let g be of the form* (1). *Then*

$$M(r) = \mathcal{O}\left(A(r)\log\frac{1}{1-r}\right) \text{ as } r \to 1, \qquad (13)$$

where $0 < |z| = r < 1$ and $\mathcal{O}$ means Landau's symbol.

Proof. It follows that

$$M(r) = \max_{|z|=r<1}\left|\int_0^z g'(s)ds\right| = \max_{|z|=r<1}\left|\int_0^r g'(\rho e^{iv})d\rho\right|.$$

Applying (12), we have

$$\begin{aligned} M(r) &= \max_{|z|=r<1}\left|\frac{1}{2\pi}\int_0^r\int_0^{2\pi}\mathfrak{Re}g'(te^{iv})\frac{te^{i\phi}+\rho e^{iv}}{te^{i\phi}-\rho e^{iv}}d\phi d\rho\right| \\ &\leq \max_{|z|=r<1}\frac{1}{2\pi}\int_0^r\int_0^{2\pi}|g'(te^{iv})|\left|\frac{te^{i\phi}+\rho e^{iv}}{te^{i\phi}-\rho e^{iv}}\right|d\phi d\rho, \end{aligned}$$

where $0 \leq \rho \leq r < t < 1$. Then, applying Schwarz's lemma, we have

$$\begin{aligned} M(r) &\leq \max_{|z|=r<1}\left(\frac{1}{2\pi}\int_0^r\int_0^{2\pi}|g'(te^{iv})|^2 d\phi d\rho\right)^{1/2}\left(\int_0^r\int_0^{2\pi}\left|\frac{te^{i\phi}+\rho e^{iv}}{te^{i\phi}-\rho e^{iv}}\right|^2 d\phi d\rho\right)^{1/2} \\ &\leq \max_{|z|=r<1}(I_1)^{1/2}(I_2)^{1/2}, \text{ say.} \end{aligned}$$

Putting $0 < r_1 < r$ and $t = \sqrt{(1+\rho^2)/2}$, we have

$$\rho d\rho = 2\sqrt{\frac{1+\rho^2}{2}}dt < 2dt.$$

Then we have

$$\begin{aligned} I_1 &= \frac{1}{2\pi}\int_0^{r_1}\int_0^{2\pi}|g'(te^{i\phi})|^2 d\phi d\rho + \frac{1}{2\pi r_1^2}\int_{\sqrt{(1+r_1^2)/2}}^{\sqrt{(1+r^2)/2}}\int_0^{2\pi} t|g'(te^{i\phi})|^2 d\phi dt \\ &\leq C + \frac{1}{2\pi r_1^2}A\left(\sqrt{\frac{1+r^2}{2}}\right) \\ &= C + \frac{1}{2\pi r_1^2}A\left(\sqrt{\frac{1+r^2}{2r^2}}r\right) \\ &= \mathcal{O}(A(r)) \text{ as } r \to 1, \end{aligned}$$

where C is a bounded positive constant. On the other hand, putting $t \to 1^-$, we have

$$\begin{aligned} I_2 &= \int_0^r \int_0^{2\pi} \left| \frac{te^{i\phi} + \rho e^{iv}}{te^{i\phi} - \rho e^{iv}} \right|^2 d\phi d\rho \\ &\leq \int_0^r \int_0^{2\pi} \frac{4}{|te^{i\phi} - \rho e^{iv}|^2} d\phi d\rho \\ &= \int_0^r \int_0^{2\pi} \frac{4}{t^2 - 2\rho t \cos(\phi - v) + \rho^2} d\phi d\rho. \end{aligned}$$

Using (5), we have

$$\begin{aligned} I_2 &\leq 8\pi \int_0^r \frac{1}{t^2 - \rho^2} d\rho \\ &= \frac{4\pi}{t} \int_0^r \left(\frac{1}{t+\rho} + \frac{1}{t-\rho} \right) d\rho \\ &= \frac{4\pi}{t} \log \frac{t+r}{t-r} \to \mathcal{O}\left(\log \frac{1}{1-r} \right) \text{ as } r \to 1. \end{aligned}$$

Therefore we complete the proof of (13). □

Remark 2. *In Theorem 5, we do not suppose that g is univalent in $|z| < 1$ and therefore, it improves the result by Pommerenke [2].*

Author Contributions: All authors contributed equally.

Funding: The authors would like to express their gratitude to the referees for many valuable suggestions regarding the previous version of this paper. This research was supported by the Basic Science Research Program through the National Research Foundation of Korea (NRF) funded by the Ministry of Education, Science and Technology (No. 2016R1D1A1A09916450).

Conflicts of Interest: The authors declare no conflict of interest.

References

1. Goodman, A.W. *Univalent Functions*; Mariner Publishing Co.: Tampa, FL, USA, 1983.
2. Pommerenke, C. Über nahezu konvexe analytische Functionen. *Arch. Math.* **1965**, *16*, 344–347. [CrossRef]
3. Keogh, F.R. Some theorems on conformal mapping of bounded star-shaped domain. *Proc. Lond. Math. Soc.* **1959**, *9*, 481–491. [CrossRef]
4. Guariglia, E. Fractional Derivative of the Riemann zeta function. In *Fractional Dynamics*; Cattani, C., Srivastava, H.M., Yang, X.J., Eds.; De Gruyter: Berlin, Germany, 2015; Volume 21, pp. 357–368.
5. Guariglia, E.; Silvestrov, S. A functional equation for the Riemann zeta fractional derivative. *AIP Conf. Proc.* **2017**, *1798*, 020063.
6. Lian, H.H.; Ge, W. Calculus of variations for a boundary value problem of differential system on the half line. *Math. Comput. Appl.* **2009**, *58*, 58–64. [CrossRef]
7. Nunokawa, M. On Bazilevič and convex functions. *Trans. Am. Math. Soc.* **1969**, *143*, 337–341.
8. Nunokawa, M. A note on convex and Bazilevič functions. *Proc. Am. Math. Soc.* **1970**, *24*, 332–335.
9. Nunokawa, M.; Owa, S.; Hayami, T.; Kuroki, K. Some properties of univalent functions. *Int. J. Pure Appl. Math.* **2009**, *52*, 603–609.
10. Nunokawa, M.; Sokół, J. On some lenght problems for analytic functions. *Osaka J. Math.* **2014**, *51*, 695–707.
11. Nunokawa, M.; Sokół, J. On some lenght problems for univalent functions. *Math. Meth. Appl. Sci.* **2016**, *39*, 1662–1666. [CrossRef]
12. Nunokawa, M.; Sokół, J. On some lenght problems for close-to-convex functions. *Stud. Sci. Math. Hung.* **2018**, in press.
13. Thomas, D.K. On Bazilevič functions. *Trans. Am. Math. Soc.* **1968**, *132*, 353–361.

14. Thomas, D.K. On starlike and close-to-convex univalent functions. *J. Lond. Math. Soc.* **1967**, *42*, 427–435. [CrossRef]
15. Thomas, D.K. A note on starlike functions. *J. Lond. Math. Soc.* **1968**, *43*, 703–706. [CrossRef]
16. Tsuji, M. *Complex Functions Theory*; Maki Book Company: Tokyo, Japan, 1968. (In Japanese)
17. Hayman, W.F. The asymptotic behaviour of *p*-valent functions. *Proc. Lond. Math. Soc.* **1955**, *3*, 257–284. [CrossRef]
18. Eenigenburg, P. On the radius of curvature for convex analytic functions. *Can. J. Math.* **1970**, *22*, 486–491. [CrossRef]

© 2018 by the authors. Licensee MDPI, Basel, Switzerland. This article is an open access article distributed under the terms and conditions of the Creative Commons Attribution (CC BY) license (http://creativecommons.org/licenses/by/4.0/).

Article

A New Operational Matrix of Fractional Derivatives to Solve Systems of Fractional Differential Equations via Legendre Wavelets

Aydin Secer [1],* and Selvi Altun [2]

1. Department of Mathematical Engineering, Yildiz Technical University, Istanbul 34200, Turkey
2. Yildiz Technical University, Istanbul 34200, Turkey; selvialtun89@gmail.com
* Correspondence: asecer@yildiz.edu.tr; Tel.: +90-532-062-1353

Received: 2 October 2018; Accepted: 31 October 2018; Published: 5 November 2018

Abstract: This paper introduces a new numerical approach to solving a system of fractional differential equations (FDEs) using the Legendre wavelet operational matrix method (LWOMM). We first formulated the operational matrix of fractional derivatives in some special conditions using some notable characteristics of Legendre wavelets and shifted Legendre polynomials. Then, the system of fractional differential equations was transformed into a system of algebraic equations by using these operational matrices. At the end of this paper, several examples are presented to illustrate the effectivity and correctness of the proposed approach. Comparing the methodology with several recognized methods demonstrates that the advantages of the Legendre wavelet operational matrix method are its accuracy and the understandability of the calculations.

Keywords: Legendre wavelet; operational matrix; systems of fractional order differential equations; Liouville_Caputo sense

1. Introduction

Differential and integral operators are the basis of mathematical models, and they are also used as a means of understanding the working principles of natural and artificial systems. Therefore, differential and integral equations are of great importance both theoretically and practically. Such equations have a wide range of applications, including in the physical sciences (such as in physics and engineering) as well as in social science. Systems of differential equations, as differential equations, are often used in issues such as theories of elasticity, dynamics, fluid mechanics, oscillation, and quantum dynamics.

Interest in differential and integral operators has led to the exploration of fractional differential and integral operators by examining these issues further in depth. Owing to a question, the origin of fractional calculus arose in a message from Leibniz to L'Hôpital in 1695. Fractional calculus has received attention in recent years due to its ability to simplify numerous physical, engineering, and economics phenomena such as the fluid dynamic traffic model, damping laws, continuum and statistical mechanics, diffusion processes, solid mechanics, control theory, colored noise, viscoelasticity, electrochemistry, and electromagnetism, among others.

Because a variety of solutions of fractional differential equations (FDEs) cannot be found analytically, numerical and approximate methods are needed. There are a lot of techniques that have been studied by many researchers in solving FDEs and the system of such equations numerically. Some of these techniques are the Adomian decomposition method presented by Song and Wang [1], the collocation method, the improved operational matrix method [2–4], the perturbation iteration method introduced by Şenol and Dolapçı [5], the computational matrix method illustrated by Khader et al. [6], the differential transform method demonstrated by Ertürk and Momani [7], the variational iteration method, the Laplace transform method given by Gupta et al. [8], and the fractional complex transform method studied by

Ghazanfari and Ghazanfari [9], among others. Kilbas et al. [10] inclusively examined fractional differential and fractional integro-differential equations. In addition, numerical solutions of FDEs and the system of such equations have been presented using the Legendre polynomial operational matrix method [11], Bernstein operational matrix method [12], Genocchi operational matrix method [13], Jacobi operational matrix method [14], Chebyshev wavelet operational matrix method [15], polynomial least squares method (PLSM) [16], Legendre wavelet-like operational matrix method (LWPT) [17], and the Genocchi wavelet-like operational matrix method [18].

This paper focuses on the numerical analysis of a system of fractional order differential equations using the Legendre wavelet operational matrix method. The most important advantage of the proposed method is that it presents an understandable procedure to reduce FDEs and the system of such equations to a system of algebraic equations. First, we begin by presenting some basic definitions and fundamental relations in Sections 2 and 3, respectively. Then, in Section 4, the operational matrix of the fractional derivate is natively formulated to linear and nonlinear systems of fractional differential equations. Section 5 presents five illustrative examples that were tested with the introduced method. Finally, the last section includes the conclusions.

2. Basic Definitions

The Liouville_Caputo fractional_order derivative, shifted Legendre polynomials, and Legendre wavelets are defined below [19,20].

Definition 1. *The Liouville_Caputo fractional derivative of u is defined as [19]*

$$D^\alpha u(t) = \frac{1}{\Gamma(n-\alpha)} \int_0^t \frac{u^{(n)}(\xi)}{(t-\xi)^{\alpha+1-n}} d\xi, \quad n-1 < \alpha \leq n, \; n \in \mathbb{N}. \tag{1}$$

Some characteristics of the Liouville_Caputo fractional derivative are as follows:

$$D^\alpha C = 0, \tag{2}$$

where C is a constant. In addition, there is

$$D^\alpha t^\beta = \begin{cases} 0, & \beta \in \mathbb{N}_0 \text{ and } \beta < \lceil \alpha \rceil \\ \frac{\Gamma(\beta+1)}{\Gamma(\beta+1-\alpha)} x^{\beta-\alpha}, & \beta \in \mathbb{N}_0 \text{ and } \beta \geq \lceil \alpha \rceil \text{ or } \beta \notin \mathbb{N} \text{ and } \beta > \lfloor \alpha \rfloor \end{cases} \tag{3}$$

in which $\lfloor \alpha \rfloor$ and $\lceil \alpha \rceil$ respectively imply that the largest integer is less than or equal to α, and the smallest integer is greater than or equal to α.

The Liouville_Caputo fractional order derivative is a linear operation of the integer order derivative

$$D^\alpha(\eta u(t) + \zeta v(t)) = \eta D^\alpha u(t) + \zeta D^\alpha v(t), \tag{4}$$

where η and ζ are constant.

Definition 2. *Let a and b respectively be the parameters of dilation and translation of a single function called the mother wavelet. If a and b change continuously, then we obtain the following family of continuous wavelets [21,22]:*

$$\psi_{ab}(t) = |a|^{-1/2} \psi\left(\frac{t-b}{a}\right), \quad a, b \in \mathbb{R}, \; a \neq 0. \tag{5}$$

Definition 3. Let $P_m(t)$ imply the shifted Legendre polynomials of order m. Then $P_m(t)$ can be formulated as [21]

$$P_m(t) = \sum_{k=0}^{m} (-1)^{m+k} \frac{(m+k)!}{(m-k)!} \frac{t^k}{(k!)^2}, \tag{6}$$

and the orthogonality condition is

$$\int_0^1 P_m(t) P_n(t) dt = \begin{cases} \frac{1}{2m+1}, & \text{for } m = n \\ 0, & \text{for } m \neq n \end{cases}. \tag{7}$$

Definition 4. Let n and k be any positive integer, m be the order of shifted Legendre polynomials, and t be the normalized time. Then the Legendre wavelets $\psi_{nm}(t) = \psi(k, n, m, t)$ are defined on the interval $[0, 1]$ by [21,22].

$$\psi_{nm}(t) = \begin{cases} 2^{\frac{k+1}{2}} \sqrt{m + \frac{1}{2}} P_m(2^k t - n), & \frac{n}{2^k} \leq t \leq \frac{n+1}{2^k} \\ 0, & \text{otherwise} \end{cases}, \tag{8}$$

where $m = 0, 1, \ldots, M$; $n = 0, 1, \ldots, (2^k - 1)$. The coefficient $\sqrt{\frac{m+1}{2}}$ is for orthonormality.

Definition 5. Let $u(t)$ and $v(t)$ be functions defined over $[0, 1]$ and then expanded in the terms of the Legendre wavelet as [21,22]

$$u(t) = \sum_{n=0}^{\infty} \sum_{m=0}^{\infty} c_{nm} \psi_{nm}(t), \tag{9}$$

where $c_{nm} = (u(t), \psi_{nm}(t))$, in which $(.,.)$ implies the inner product. If the infinite series in Equation (9) is truncated, then it can be expressed as

$$u(t) \cong \sum_{n=0}^{2^k-1} \sum_{m=0}^{M} c_{nm} \psi_{nm}(t) = C^T \psi(t), \tag{10}$$

where C and $\psi(t)$ are matrices, as presented by

$$\begin{aligned} C &= \left[c_{0,0}, c_{0,1}, \ldots, c_{0,M}, \ldots, c_{2,M}, \ldots, c_{2^k-1,0}, c_{2^k-1,1}, \ldots, c_{2^k-1,M} \right]^T \\ \psi &= \left[\psi_{0,0}, \psi_{0,1}, \ldots, \psi_{0,M}, \ldots, \psi_{2,M}, \ldots, \psi_{2^k-1,0}, \psi_{2^k-1,1}, \ldots, \psi_{2^k-1,M} \right]^T \end{aligned}. \tag{11}$$

3. Fundamental Relations

Saadatmandi and Dehghan [11] derived the operational matrix of a fractional derivative by using shifted Legendre polynomials. In this section, we show how we derived the Legendre wavelet operational matrix of fractional derivatives in some special conditions by drawing from Saadatmandi and Dehghan [11]. Additionally, the theorem and corollary related to the Legendre wavelet operational matrix of derivatives illustrated by Mohammadi [21] are cited here as follows.

Theorem 1. Let $\psi(t)$ be the Legendre wavelet vector introduced in Equation (8). Then $\psi(t)$ is expressed as [21,22]

$$\frac{d\psi(t)}{dt} = D\psi(t), \tag{12}$$

where D is the $2^k(M+1)$ operational matrix of the derivative, which can be stated as

$$D = \begin{bmatrix} U & O & \cdots & O \\ O & U & \cdots & O \\ \vdots & \vdots & \ddots & \vdots \\ O & O & \cdots & U \end{bmatrix}, \quad (13)$$

where U is an $(M+1)(M+1)$ matrix and its (r,s)th element is written as

$$U_{r,s} = \begin{cases} 2^{k+1}\sqrt{(2r-1)(2s-1)}, & r=2,\ldots,(M+1),\ s=1,\ldots,r-1 \text{ and } (r+s) \text{ odd} \\ 0, & \text{otherwise} \end{cases}. \quad (14)$$

Corollary 1. *Using Equation (12), the operational matrix for the nth derivative can be stated as [21]*

$$\frac{d^n \psi(t)}{dt^n} = D^n \psi(t), \quad (15)$$

where D^n is the nth power of matrix D.

Lemma 1. *Let $\psi(t)$ be the Legendre wavelets vector introduced in Equation (8). Assuming that $k=0$, then*

$$D^\alpha \psi_r(t) = 0, \quad r = 0,1,\ldots,\lceil \alpha \rceil - 1,\ \alpha > 0. \quad (16)$$

Proof. The desired result can be obtained by using Equations (2) and (4) in Equation (8). □

Theorem 2. *Let $\psi(t)$ be the Legendre wavelets vector introduced in Equation (8). Supposing that $k=0$ and $\alpha > 0$, then*

$$D^\alpha \psi(t) \cong D^{(\alpha)} \psi(t), \quad (17)$$

where $D^{(\alpha)}$ is the $(M+1) \times (M+1)$ operational matrix of the fractional derivative of the order $\alpha > 0$, $N-1 < \alpha \leq N$ in the Liouville_Caputo sense and can be stated as

$$D^{(\alpha)} = \begin{pmatrix} 0 & 0 & \cdots & 0 \\ \vdots & \vdots & \cdots & \vdots \\ 0 & 0 & \cdots & 0 \\ \sum_{h=\lceil \alpha \rceil}^{\lceil \alpha \rceil} \zeta_{\lceil \alpha \rceil,0,h} & \sum_{h=\lceil \alpha \rceil}^{\lceil \alpha \rceil} \zeta_{\lceil \alpha \rceil,1,h} & \cdots & \sum_{h=\lceil \alpha \rceil}^{\lceil \alpha \rceil} \zeta_{\lceil \alpha \rceil,m,h} \\ \vdots & \vdots & \cdots & \vdots \\ \sum_{h=\lceil \alpha \rceil}^{r} \zeta_{r,0,h} & \sum_{h=\lceil \alpha \rceil}^{r} \zeta_{r,1,h} & \cdots & \sum_{h=\lceil \alpha \rceil}^{r} \zeta_{r,m,h} \\ \vdots & \vdots & \cdots & \vdots \\ \sum_{h=\lceil \alpha \rceil}^{m} \zeta_{m,0,h} & \sum_{h=\lceil \alpha \rceil}^{m} \zeta_{m,1,h} & \cdots & \sum_{h=\lceil \alpha \rceil}^{m} \zeta_{m,m,h} \end{pmatrix}, \quad (18)$$

where $\zeta_{r,s,h}$ is written as

$$\zeta_{r,s,h} = \sqrt{2r+1}\sqrt{2s+1}\sum_{l=0}^{s} \frac{(-1)^{r+s+h+l}(r+h)!(s+l)!}{(r-h)!h!\Gamma(h-\alpha+1)(s-l)!(l!)^2(h+l-\alpha+1)}. \quad (19)$$

Consider in $D^{(\alpha)}$ that the first $\lceil \alpha \rceil$ rows are all zero.

Proof. Presume that $\psi_r(t)$ is the rth element of the vector $\psi(t)$ introduced in Equation (11), where $r = nM + (m+1)$, $m = 0, 1, \ldots, M$, $n = 0, 1, \ldots, (2^k - 1)$. Then $\psi_r(t)$ can be stated as

$$\psi_r(t) = 2^{\frac{k+1}{2}}\sqrt{r+\frac{1}{2}}P_r(2^k t - n)\chi_{[\frac{n}{2^k}, \frac{n+1}{2^k}]}. \tag{20}$$

Accepting that $k = 0$, and by using the shifted Legendre polynomial, we obtain

$$\psi_r(t) = \sqrt{2}\sqrt{r+\frac{1}{2}}\sum_{h=0}^{r} \frac{(-1)^{r+h}(r+h)!}{(r-h)!(h!)^2} t^h \chi_{[0,1]}. \tag{21}$$

If we use Equations (3), (4), and (21), then we have

$$\begin{aligned} D^{(\alpha)}\psi_r(t) &= \sqrt{2}\sqrt{r+\frac{1}{2}}\sum_{h=0}^{r} \frac{(-1)^{r+h}(r+h)!}{(r-h)!(h!)^2} D^{\alpha}(t^h)\chi_{[0,1]} \\ &= \sqrt{2r+1}\sum_{h=\lceil\alpha\rceil}^{r} \frac{(-1)^{r+h}(r+h)!}{(r-h)!(h!)\Gamma(h-\alpha+1)} t^{h-\alpha}\chi_{[0,1]}, \quad r = \lceil\alpha\rceil, \ldots, m. \end{aligned} \tag{22}$$

Approximating $t^{h-\alpha}$ by $(m+1)$ terms of the Legendre wavelets, then we obtain

$$t^{h-\alpha} \cong \sum_{s}^{m} b_{h,s}\psi_s(t), \tag{23}$$

where

$$\begin{aligned} b_{h,s} &= \int_0^1 t^{h-\alpha}\psi_s(t)dt = \sqrt{2}\sqrt{s+\frac{1}{2}}\sum_{l=0}^{s} \frac{(-1)^{s+l}(s+l)!}{(s-l)!(l!)^2}\int_0^1 t^{h+l-\alpha}dt \\ &= \sqrt{2s+1}\sum_{l=0}^{s} \frac{(-1)^{s+l}(s+l)!}{(s-l)!(l!)^2(h+l-\alpha+1)} \end{aligned}. \tag{24}$$

Utilizing Equations (22) and (24), we get

$$\begin{aligned} D^{\alpha}\psi_r(t) &\cong \sqrt{2r+1}\sum_{h=\lceil\alpha\rceil}^{r}\sum_{s=0}^{m} \frac{(-1)^{r+h}(r+h)!}{(r-h)!(h!)\Gamma(h-\alpha+1)} b_{h,s}\psi_s(t)\chi_{[0,1]} \\ &= \sum_{s=0}^{m}\left(\sum_{h=\lceil\alpha\rceil}^{r} \zeta_{r,s,h}\right)\psi_s(t)\chi_{[0,1]}, \quad r = \lceil\alpha\rceil, \ldots, m \end{aligned} \tag{25}$$

in which $\zeta_{r,s,h}$ is presented in Equation (19). In addition, if we use Lemma 1, then we can write

$$D^{\alpha}\psi_r(t) = 0, \quad r = 0, 1, \ldots, \lceil\alpha\rceil - 1, \quad \alpha > 0. \tag{26}$$

Combining Equations (25) and (27), the result can be obtained. □

4. Solving Systems of Fractional Order Differential Equations

In this section, the Legendre wavelet operational matrix method was implemented to obtain the numerical solution of the system of fractional order differential equations. Consider a system of fractional differential equations as follows:

$$\begin{aligned} D^{\eta_1}u_1(t) &= U_1(t, u_1, u_2, \ldots, u_m), \\ D^{\eta_2}u_2(t) &= U_2(t, u_1, u_2, \ldots, u_m), \\ &\vdots \\ D^{\eta_n}u_m(t) &= U_m(t, u_1, u_2, \ldots, u_m), \end{aligned} \tag{27}$$

where U_i is a linear/nonlinear function of $t, u_1, u_2, \ldots, u_m$, D^{η_i} is the derivative of u_i with the order of η_i in the Liouville–Caputo sense and $N - 1 \leq \eta_i < N$, and they are subjected to the initial conditions

$$
\begin{array}{lllll}
u_1(t_0) = u_{10}, & \frac{du_1}{dt}(t_0) = u_{11}, & \frac{d^2u_1}{dt^2}(t_0) = u_{12}, & \ldots, & \frac{d^{n-1}u_1}{dt^{n-1}}(t_0) = u_{1(n-1)} \\
u_2(t_0) = u_{20}, & \frac{du_2}{dt}(t_0) = u_{21}, & \frac{d^2u_2}{dt^2}(t_0) = u_{22}, & \ldots, & \frac{d^{n-1}u_2}{dt^{n-1}}(t_0) = u_{2(n-1)} \\
\vdots & \vdots & \vdots & & \vdots \\
u_m(t_0) = u_{m0}, & \frac{du_m}{dt}(t_0) = u_{m1}, & \frac{d^2u_m}{dt^2}(t_0) = u_{m2}, & \ldots, & \frac{d^{n-1}u_m}{dt^{n-1}}(t_0) = u_{m(n-1)}
\end{array}
\tag{28}
$$

First of all, approximating $u_1(t), u_2(t), \ldots, u_m(t)$ and $D^{\eta_1}u_1(t), D^{\eta_2}u_2(t), \ldots, D^{\eta_n}u_m(t)$, we obtain

$$
\begin{aligned}
u_1(t) &\approx \sum_{n=0}^{2^k-1} \sum_{m=0}^{M} c_{1n,m}\psi_{n,m} = C_1^T \psi(t) \\
u_2(t) &\approx \sum_{n=0}^{2^k-1} \sum_{m=0}^{M} c_{2n,m}\psi_{n,m} = C_2^T \psi(t) \\
&\vdots \\
u_m(t) &\approx \sum_{n=0}^{2^k-1} \sum_{m=0}^{M} c_{nn,m}\psi_{n,m} = C_m^T \psi(t)
\end{aligned}
\tag{29}
$$

where C_i, $i = 1, 2, \ldots, m$ is an unknown vector and $\psi(t)$ is the vector introduced in Equation (8). If we utilize Equation (17), then we have

$$
\begin{aligned}
D^{\eta_1}u_1(t) &\approx C_1^T D^{(\eta_1)} \psi(t) \\
D^{\eta_2}u_2(t) &\approx C_2^T D^{(\eta_2)} \psi(t) \\
&\vdots \\
D^{\eta_n}u_m(t) &\approx C_m^T D^{(\eta_n)} \psi(t)
\end{aligned}
\tag{30}
$$

Substituting Equations (29) and (30) into Equation (27), we obtain

$$
\begin{aligned}
R_1(t) &= C_1^T D^{(\eta_1)} \psi(t) - U_1(t, C_1^T \psi(t), C_2^T \psi(t), \ldots, C_m^T \psi(t)) \\
R_2(t) &= C_2^T D^{(\eta_2)} \psi(t) - U_2(t, C_1^T \psi(t), C_2^T \psi(t), \ldots, C_m^T \psi(t)) \\
&\vdots \\
R_m(t) &= C_m^T D^{(\eta_n)} \psi(t) - U_m(t, C_1^T \psi(t), C_2^T \psi(t), \ldots, C_m^T \psi(t))
\end{aligned}
\tag{31}
$$

If U_i is a linear function of $t, u_1, u_2, \ldots, u_m$, then we produce $2^k(M+1) - mn$ linear equations by implementing

$$
\int_0^1 \psi_j(t) R_i(t) dt = 0, \quad j = 1, \ldots, 2^k(M+1) - mn, \quad i = 1, 2, \ldots, m.
\tag{32}
$$

Also, by substituting the initial conditions in Equation (28) into Equation (30), then we obtain

$$
\begin{array}{lllll}
u_1(t_0) \approx C_1^T \psi(t_0) = u_{10}, & \frac{du_1}{dt}(t_0) \approx C_1^T D\psi(t_0) = u_{11}, & \ldots, & \frac{d^{n-1}u_1}{dt^{n-1}}(t_0) \approx C_1^T D^{n-1}\psi(t_0) = u_{1(n-1)} \\
u_2(t_0) \approx C_2^T \psi(t_0) = u_{20}, & \frac{du_2}{dt}(t_0) \approx C_2^T D\psi(t_0) = u_{21}, & \ldots, & \frac{d^{n-1}u_2}{dt^{n-1}}(t_0) \approx C_2^T D^{n-1}\psi(t_0) = u_{2(n-1)} \\
\vdots & \vdots & & \vdots \\
u_m(t_0) \approx C_m^T \psi(t_0) = u_{m0}, & \frac{du_m}{dt}(t_0) \approx C_m^T D\psi(t_0) = u_{m1}, & \ldots, & \frac{d^{n-1}u_m}{dt^{n-1}}(t_0) \approx C_m^T D^{n-1}\psi(t_0) = u_{m(n-1)}
\end{array}
\tag{33}
$$

A $2^k(M+1)$ set of linear equations is generated by combining Equations (32) and (33). The solution of these linear equations can be obtained for unknown coefficients of the vector C. Consequently, $u_1(t), u_2(t), \ldots, u_m(t)$, introduced in Equation (27), can be computed.

If U_i is a nonlinear function of $t, u_1, u_2, \ldots, u_m$, then we first compute $R_1(t), R_2(t), \ldots, R_m(t)$ at $2^k(M+1) - mn$ points and, for a better result, use the first $2^k(M+1) - mn$ roots of shifted Legendre $P_{2^k(M+1)}(t)$. Then these equations, collectively with Equation (33), produce $2^k(M+1)$ nonlinear equations. The solution of these nonlinear equations can be obtained by employing Newton's iterative method. Consequently, $u_1(t), u_2(t), \ldots, u_m(t)$, introduced in Equation (27), can be computed.

5. Illustrative Examples

In this section, to show the applicability and powerfulness of the introduced method, we present the solutions to five linear and nonlinear systems of fractional order differential equations.

Example 1. *We first considered the following linear system of fractional differential equations [7,8]:*

$$\begin{aligned} D^\alpha u(t) &= u(t) + v(t) \\ D^\alpha v(t) &= -u(t) + v(t) \end{aligned},$$

subject to

$$u(0) = 0, \ v(0) = 1.$$

The exact solution of this system when $\alpha = 1$ is known to be

$$u(t) = e^t \sin t, \ v(t) = e^t \cos t.$$

This example was examined for $M = 2$, $k = 0$, and $\alpha = 0.9, 0.7, 0.5$. When the obtained results were matched against the exact solution when $\alpha = 1$, as demonstrated in Figure 1, we can clearly observe that when α approached 1, our results approached the exact solution. We also solved this problem by using Legendre polynomial operational matrix method (LPOMM), and we compared the results with the LWOMM. The numerical computations for $u(t)$ and $v(t)$ when $\alpha = 0.9$ are revealed in Tables 1 and 2.

Table 1. Numerical solutions of $u(t)$ when $\alpha = 0.9$ attained by the introduced method and the LPOMM for Example 1.

t	u_{LWOMM}	u_{LPOMM}	Absolute Error
0.0	0.3×10^{-9}	0.0000000000	0.3×10^{-9}
0.1	0.1483784330	0.1483784325	0.12×10^{-9}
0.2	0.3217645283	0.3217645277	0.633×10^{-9}
0.3	0.5201582862	0.5201582855	0.65×10^{-9}
0.4	0.7435597067	0.7435597059	0.83×10^{-9}
0.5	0.9919687898	0.9919687890	0.8×10^{-9}
0.6	1.265385536	1.265385535	0.53×10^{-9}
0.7	1.563809944	1.563809943	0.95×10^{-9}
0.8	1.887242014	1.887242014	0.733×10^{-9}
0.9	2.235681748	2.235681748	0.62×10^{-9}
1.0	2.609129144	2.609129144	0.3×10^{-9}

Table 2. Numerical solutions of $v(t)$ when $\alpha = 0.9$ attained by the introduced method and the LPOMM for Example 1.

t	v_{LWOMM}	v_{LPOMM}	Absolute Error
0.0	1.000000000	1.000000000	0.2×10^{-9}
0.1	1.152270899	1.152270900	-0.5×10^{-9}
0.2	1.274801858	1.274801858	0.599×10^{-9}
0.3	1.367592877	1.367592878	-0.67×10^{-9}
0.4	1.430643956	1.430643957	-0.13×10^{-8}
0.5	1.463955094	1.463955094	-0.1×10^{-8}
0.6	1.467526291	1.467526293	-0.13×10^{-8}
0.7	1.441357549	1.441357551	-0.171×10^{-8}
0.8	1.385448866	1.385448868	-0.1461×10^{-8}
0.9	1.299800244	1.299800245	-0.15×10^{-8}
1.0	1.184411680	1.184411682	-0.18×10^{-8}

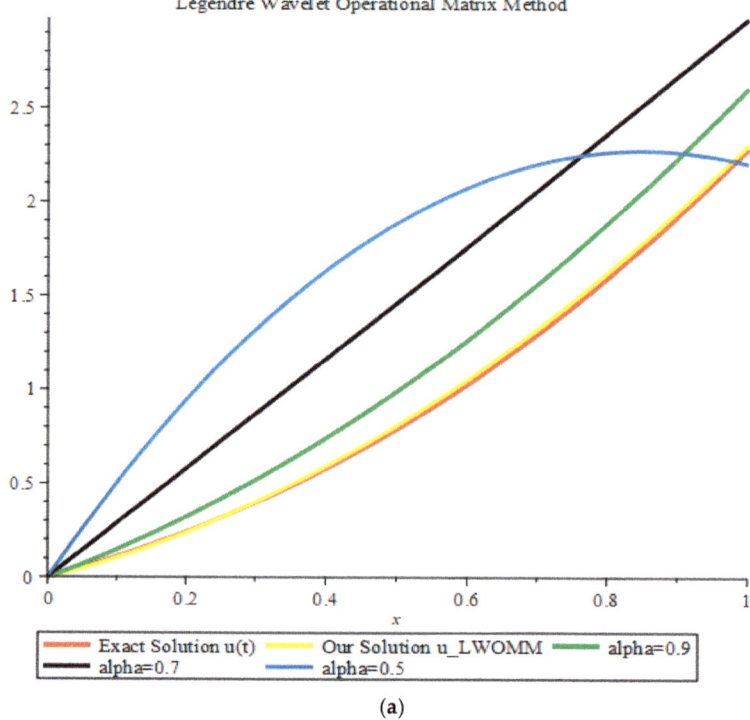

(a)

Figure 1. Cont.

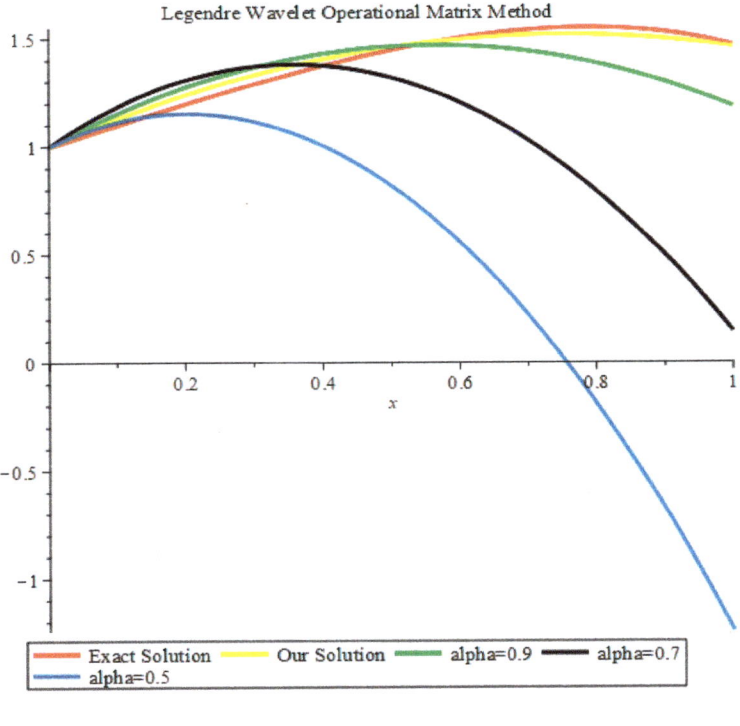

Figure 1. Comparison of our solutions and the exact solution when $\alpha = 0.9, 0.7, 0.5$ in Example 1: (a) Our solution $u(t)$; and (b) Our solution $v(t)$.

Example 2. *We considered the following nonlinear system of fractional differential equations [13]:*

$$D^{\frac{3}{2}}u(t) = -8u(t) + v^2(t) - 4t^6 + 4t^3 + \frac{8t^{\frac{3}{2}}}{\sqrt{\pi}} - 1$$
$$D^{\frac{1}{2}}v(t) = t^2 Du(t) + v(t) - 3t^4 - 2t^3 + \frac{32t^{\frac{5}{2}}}{5\sqrt{\pi}} - 1$$

$$u(0) = 0, \ v(0) = 1, \ u(1) = 1, \ v(1) = 3, \ u'(0) = 0, \ u'(1) = 3.$$

The exact solution of this system is known to be

$$u(t) = t^3, \ v(t) = 2t^3 + 1$$

Using the parameters $M = 3$ and $k = 0$, we applied both the proposed method and the LPOMM to solve this problem and show that our approach is more efficient and useful. Our numerical results supported the idea that our solution approaches the exact solution more than the approximate solution LPOMM. Comparisons of the approximate and exact solutions are presented in Tables 3 and 4.

Table 3. The numerical results attained by using the introduced method in comparison to the approximate solution LPOMM and the exact solution $u(t)$ in Example 2.

t	Exact Solution	u_{LWOMM}	u_{LPOMM}
0.0	0.000	-0.12×10^{-9}	0.000000000000
0.1	0.001	0.01000000005	0.001000000000
0.2	0.008	0.02000000016	0.008000000000
0.3	0.027	0.03750000021	0.027000000000
0.4	0.064	0.07000000020	0.064000000000
0.5	0.125	0.12500000001	0.125000000000
0.6	0.216	0.21000000000	0.216000000000
0.7	0.343	0.33249999998	0.343000000000
0.8	0.512	0.49999999996	0.512000000000
0.9	0.729	0.71999999993	0.729000000000
1.0	1.000	0.99999999989	1.000000000000

Table 4. The numerical results attained by using the introduced method in comparison with the approximate solution LPOMM and exact solution $v(t)$ in Example 2.

t	Exact Solution	v_{LWOMM}	v_{LPOMM}
0.0	1.000	1.000000000	0.9999999998
0.1	1.002	1.034841367	1.165803114
0.2	1.016	1.057587628	1.283854751
0.3	1.054	1.086537668	1.374911456
0.4	1.128	1.139990370	1.459729770
0.5	1.250	1.236244618	1.559066242
0.6	1.432	1.393599296	1.693677414
0.7	1.686	1.630353290	1.884319831
0.8	2.024	1.964805482	2.151750038
0.9	2.458	2.415254758	2.516724579
1.0	3.000	3.000000000	3.000000000

Example 3. We considered the following nonlinear system of fractional differential equations with the initial conditions [8]

$$D^\alpha u(t) = \frac{u(t)}{2}$$
$$D^\alpha v(t) = u^2(t) + v(t) ,$$
$$u(0) = 1, \ v(0) = 0.$$

The exact solution of this system when $\alpha = 1$ is known to be

$$u(t) = e^{(\frac{t}{2})}, \ v(t) = te^t.$$

The parameters $M = 2$, $k = 0$, and $\alpha = 0.5, 0.7, 0.9$ were utilized. A comparison of our results and the exact solution when $\alpha = 1$ is displayed in Figure 2. The figures support that when α approximated 1, our results approximated the exact solution. We also solved this problem by using the LPOMM, and compared the results to the LWOMM. Finally, we present the numerical computations for $u(t)$ and $v(t)$ when $\alpha = 0.9$ in Tables 5 and 6.

6. Conclusions

In this paper, a system of fractional_order differential equations was examined by drawing from a new operational matrix of the fractional derivative in some special conditions. We also systematized a very operational algorithm in order to attain the solution of the linear and nonlinear systems of fractional differential equations in Maple. All numerical results and graphical presentations generated by Maple affirmed that the Legendre wavelet operational matrix method is very effective and applicable. As the next step, the method introduced in this paper can be applied to fractional partial differential equations and the system of such equations, fractional integral equations and the system of such equations, and fractional integro-differential equations. These equations are at least as important as fractional differential equations and they are very significant in science, engineering, and technology.

Author Contributions: All authors contributed equally to the writing of this paper. All authors read and approved the final manuscript.

Funding: This research received no external funding.

Conflicts of Interest: The authors declare no conflict of interest.

References

1. Song, L.; Wang, W. A new improved Adomian decomposition method and its application to fractional differential equations. *Appl. Math. Model.* **2013**, *37*, 1590–1598. [CrossRef]
2. Balaji, S. Legendre wavelet operational matrix method for solution of fractional order Riccati differential equation. *J. Eqypt. Math. Soc.* **2015**, *23*, 263–270. [CrossRef]
3. Rehman, M.; Khan, R.A. The Legendre wavelet method for solving fractional differential equations. *Commun. Nonlinear Sci. Numer. Simul.* **2011**, *16*, 4163–4173. [CrossRef]
4. Mohammadi, F.; Hosseini, M.M.; Mohyud-Din, S.T. A new operational matrix for Legendre wavelets and its applications for solving fractional order boundary value problems. *Int. J. Syst. Sci.* **2011**, *32*, 7371–7378. [CrossRef]
5. Şenol, M.; Dolapçı, İ.T. On the perturbation-iteration algorithm for fractional differential equations. *J. King Saud Univ.-Sci.* **2016**, *28*, 69–74. [CrossRef]
6. Khader, M.M.; Danaf, T.S.; Hendy, A.S. A computational matrix method for solving systems of high order fractional differential equations. *Appl. Math. Model.* **2013**, *37*, 4035–4050. [CrossRef]
7. Ertürk, V.S.; Momani, S. Solving systems of fractional differential equations using differential transform method. *J. Comput. Appl. Math.* **2008**, *215*, 142–151. [CrossRef]
8. Gupta, S.; Kumar, D.; Singh, J. Numerical study for systems of fractional differential equations via Laplace transform. *J. Egypt. Math. Soc.* **2015**, *23*, 256–262. [CrossRef]
9. Ghazanfari, B.; Ghazanfari, A.G. Solving system of fractional differential eqautions by fractional complex transform method. *Asian J. Appl. Sci.* **2012**, *5*, 438–444. [CrossRef]
10. Kilbas, A.A.; Srivastava, H.M.; Trujillo, J.J. *Theory and Applications of Fractional Differential Equations*, 1st ed.; Elsevier Science: Amsterdam, The Netherlands, 2006; p. 540.
11. Saadatmandi, A.; Dehghan, M. A new operational matrix for solving fractional-order differential equations. *Comput. Math. Appl.* **2010**, *59*, 1326–1336. [CrossRef]
12. Alshbool, M.H.T.; Bataineh, A.S.; Hashim, I.; Işık, O.R. Solution of fractional-order differential equations based on the operational matrices of new fractional Bernstein functions. *J. King Saud Univ.-Sci.* **2017**, *29*, 1–18. [CrossRef]
13. Isah, A.; Phang, C. New operational matrix of derivative for solving non-linear fractional differential equations via Genocchi polynomials. *J. King Saud Univ.-Sci.* **2017**. [CrossRef]
14. Doha, E.H.; Bhrawy, A.H.; Ezz-Eldien, S.S. A new Jacobi operational matrix: An application for solving fractional differential equations. *Appl. Math. Model.* **2012**, *36*, 4931–4943. [CrossRef]
15. Mohammadi, F. Numerical solution of Bagley-Torvik equation using Chebyshev wavelet operational matrix of fractional derivative. *Int. J. Adv. Appl. Math. Mech.* **2014**, *2*, 83–91.

16. Bota, C.; Caruntu, B. Approximate analytical solutions of the fractional-order brusselator system using the polynomial least squares method. *Corp. Adv. Differ. Equ.* **2015**. [CrossRef]
17. Chang, P.; Isah, A. Legendre wavelet operational matrix of fractional derivative through wavelet-polynomial transformation and its applications in solving fractional order brusselator system. *J. Phys. Conf. Ser.* **2016**, *693*, 012001. [CrossRef]
18. Isah, A.; Phang, C. Genocchi wavelet-like operational matrix and its application for solving non-linear fractional differential equations. *Open Phys.* **2016**, *14*, 463–472. [CrossRef]
19. Miller, K.S.; Ross, B. *An Introduction to the Fractional Calculus and Fractional Differential Equations*, 1st ed.; Wiley: Hoboken, NJ, USA, 1993; p. 384.
20. Oldham, K.B.; Spanier, J. *The Fractional Calculus*, 1st ed.; Academic Press: New York, NY, USA, 1974; p. 234.
21. Mohammadi, F.; Hosseini, M.M. A new Legendre wavelet operational matrix of derivative and its applications in solving the singular ordinary differential equations. *J. Frankl. Inst.* **2011**, *348*, 1787–1796. [CrossRef]
22. Mohammadi, F.; Hosseini, M.M.; Mohyud-Din, S.T. Legendre wavelet Galerkin method for solving ordinary differential equations with non-analytic solution. *Int. J. Syst. Sci.* **2011**, *42*, 579–585. [CrossRef]

© 2018 by the authors. Licensee MDPI, Basel, Switzerland. This article is an open access article distributed under the terms and conditions of the Creative Commons Attribution (CC BY) license (http://creativecommons.org/licenses/by/4.0/).

MDPI
St. Alban-Anlage 66
4052 Basel
Switzerland
Tel. +41 61 683 77 34
Fax +41 61 302 89 18
www.mdpi.com

Mathematics Editorial Office
E-mail: mathematics@mdpi.com
www.mdpi.com/journal/mathematics

www.ingramcontent.com/pod-product-compliance
Lightning Source LLC
LaVergne TN
LVHW070223100526
838202LV00015B/2081

Table 5. Our solutions $u(t)$ when $\alpha = 0.9$ attained by the presented method and the LPOMM for Example 3.

t	u_{LWOMM}	u_{LPOMM}	Absolute Error
0.0	1.000000000	1.000000000	-0.37×10^{-10}
0.1	1.064816320	1.064816320	-0.132×10^{-9}
0.2	1.130248588	1.130248589	-0.1375×10^{-9}
0.3	1.196296807	1.196296807	0.44×10^{-10}
0.4	1.262960974	1.262960975	-0.808×10^{-9}
0.5	1.330241090	1.330241091	-0.532×10^{-9}
0.6	1.398137156	1.398137156	-0.168×10^{-9}
0.7	1.466649171	1.466649171	-0.756×10^{-9}
0.8	1.535777134	1.535777134	-0.1375×10^{-9}
0.9	1.605521046	1.605521046	0.468×10^{-9}
1.0	1.675880908	1.675880908	0.163×10^{-9}

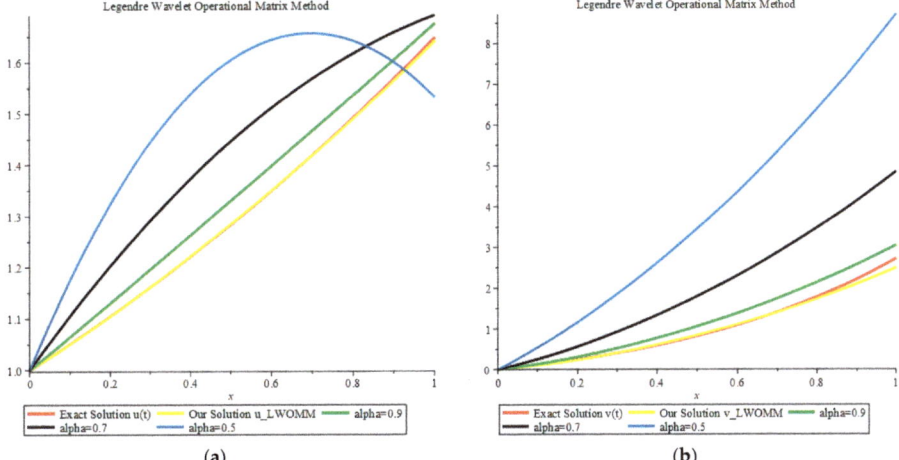

Figure 2. Comparison of our solutions to the exact solution when $\alpha = 0.9, 0.7, 0.5$ for Example 3: (**a**) Our solution $u(t)$; and (**b**) Our solution $v(t)$.

Table 6. Our solutions $v(t)$ when $\alpha = 0.9$ attained by the presented method and the LPOMM for Example 3.

t	v_{LWOMM}	v_{LPOMM}	Absolute Error
0.0	-0.1×10^{-9}	0.0000000000	-0.1×10^{-9}
0.1	0.1381762603	0.1381762609	-0.7×10^{-9}
0.2	0.3133603361	0.3133603363	-0.2×10^{-9}
0.3	0.5255522259	0.5255522263	-0.37×10^{-9}
0.4	0.7747519304	0.7747519309	-0.5×10^{-9}
0.5	1.060959450	1.060959450	-0.5×10^{-9}
0.6	1.384174783	1.384174784	-0.8×10^{-9}
0.7	1.744397932	1.744397932	-0.47×10^{-9}
0.8	2.141628894	2.141628895	-0.7×10^{-9}
0.9	2.575867672	2.575867672	0.3×10^{-9}
1.0	3.047114264	3.047114264	-0.1×10^{-9}

Example 4. *We considered the following nonlinear system of FDEs with initial conditions [13]*

$$D^\alpha u(t) = -1002u(t) + 1000v^2(t)$$
$$D^\alpha v(t) = u(t) - v(t) - v^2(t)$$

$$u(0) = 1, \quad v(0) = 1$$

The exact solution of this system when $\alpha = 1$ is known to be

$$u(t) = e^{-2t}, \quad v(t) = e^{-t}$$

This example was analyzed for $M = 3, k = 0$, and $\alpha = 0.9, 0.7, 0.5$. When the obtained results were matched against the exact solution when $\alpha = 1$, as demonstrated in Figure 3, we can clearly observe that when α approached 1, our results approached the exact solution. We also solved this problem by using the LPOMM, and compared the results with the LWOMM. The numerical computations for $u(t)$ and $v(t)$ when $\alpha = 0.99$ are also revealed in Tables 7 and 8.

Table 7. Numerical solutions of $u(t)$ when $\alpha = 0.99$ for Example 4.

t	u_{LWOMM}	u_{LPOMM}	Absolute Error
0.0	1.000000000	1.000000000	-0.4×10^{-10}
0.1	0.8144351529	0.8144351528	-0.21×10^{-10}
0.2	0.6639425233	0.6639425233	-0.15×10^{-9}
0.3	0.5429947229	0.5429947230	-0.34×10^{-9}
0.4	0.4460643636	0.4460643636	-0.42×10^{-9}
0.5	0.3676240568	0.3676240568	0
0.6	0.3021464142	0.3021464147	-0.68×10^{-9}
0.7	0.2441040487	0.2441040481	-0.24×10^{-9}
0.8	0.1879695699	0.1879695690	0.53×10^{-9}
0.9	0.1282155920	0.1282155905	0.461×10^{-9}
1.0	0.0593147248	0.0593147222	0.144×10^{-8}

Table 8. Numerical solutions of $v(t)$ when $\alpha = 0.99$ for Example 4.

t	v_{LWOMM}	v_{LPOMM}	Absolute Error
0.0	1.000000000	0.9999999999	0.79×10^{-10}
0.1	0.9025601837	0.9025601837	0.1498×10^{-9}
0.2	0.8152646487	0.8152646488	-0.119×10^{-9}
0.3	0.7371116577	0.7371116578	-0.8×10^{-10}
0.4	0.6670994737	0.6670994739	-0.11×10^{-9}
0.5	0.6042263597	0.6042263600	-0.24×10^{-9}
0.6	0.5474905785	0.5474905788	-0.19×10^{-9}
0.7	0.4958903932	0.4958903935	-0.26×10^{-9}
0.8	0.4484240664	0.4484240670	-0.443×10^{-9}
0.9	0.4040898614	0.4040898620	-0.5898×10^{-9}
1.0	0.3618860408	0.3618860417	-0.719×10^{-9}

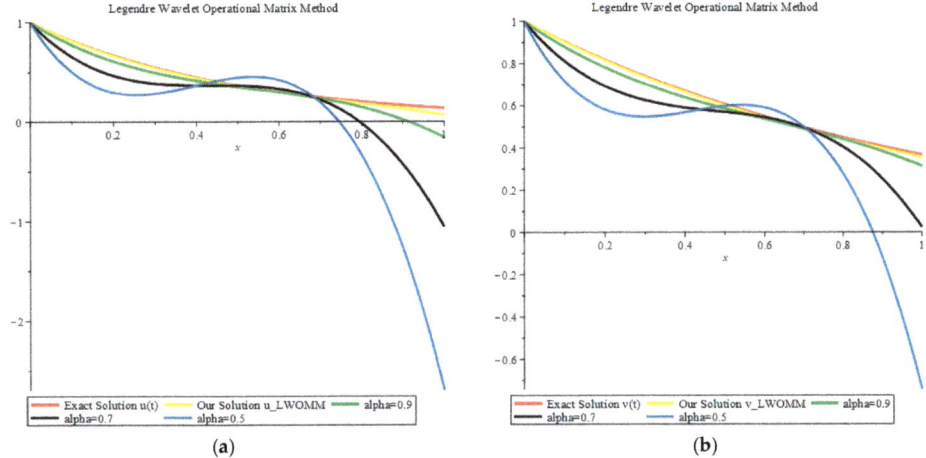

Figure 3. Comparison of our solutions to the exact solution when $\alpha = 0.9, 0.7, 0.5$ for Example 4: (**a**) Our solution $u(t)$; and (**b**) Our solution $v(t)$.

Example 5. *We considered the following fractional order Brusselator system [16,17]:*

$$D^\alpha u(t) = -2u(t) + u^2(t)v(t)$$
$$D^\alpha v(t) = u(t) - u^2(t)v(t)$$

$$u(0) = 1, \quad v(0) = 1.$$

The approximate solutions of this system when $\alpha = 1$ and $\alpha = 0.98$ were presented by Chang and Isah using the LWPT [17] and by Bota and Caruntu using the PLSM [16]. These solutions when $\alpha = 98$ are given by

$$u_{LWPT}(t) = 1 - 1.0791t + 0.2711t^2 - 0.0638t^3, \quad v_{LWPT}(t) = 1 + 0.0151t + 0.4185t^2 - 0.2624t^3$$
$$u_{PLSM}(t) = 1 - 1.08655t + 0.311138t^2 + 0.0243682t^3, \quad v_{PLSM}(t) = 1 + 0.0349127t + 0.333424t^2 - 0.184414t^3$$

The parameters $M = 2$, $k = 0$, and $\alpha = 0.98$ were used. A comparison of our results to the approximate solutions introduced by Bota and Caruntu [16] and Chang and Isah [17] when $\alpha = 0.98$ is displayed in Figure 4. Finally, we also present the numerical computations for $u(t)$ and $v(t)$ when $\alpha = 0.98$ in Tables 9 and 10.

Table 9. Numerical solutions of $u(t)$ when $\alpha = 0.98$ obtained by the introduced method, the LWPT, and the PLSM for Example 5.

t	u_{LWOMM}	u_{LWPT}	u_{PLSM}
0.0	1.000000000	1.0000000	1.0000000000
0.1	0.8942024826	0.8947372	0.8944807482
0.2	0.7950696916	0.7945136	0.7953304656
0.3	0.7026016268	0.6989464	0.7026953614
0.4	0.6167982883	0.6076528	0.6167216448
0.5	0.5376596761	0.5202500	0.5375555250
0.6	0.4651857902	0.4363552	0.4653432112
0.7	0.3993766306	0.3555856	0.4002309126
0.8	0.3402321973	0.2775584	0.3423648384
0.9	0.2877524902	0.2018908	0.2918911978
1.0	0.2419375095	0.1282000	0.2489562000

Table 10. Numerical solutions of $v(t)$ when $\alpha = 0.98$ obtained by the introduced method, the LWPT, and the PLSM for Example 5.

t	v_{LWOMM}	v_{LWPT}	v_{PLSM}
0.0	1.000000000	1.0000000	1.000000000
0.1	1.008069307	1.0054326	1.006641096
0.2	1.019479961	1.0176608	1.018844188
0.3	1.034231959	1.0351102	1.035502792
0.4	1.052325304	1.0562064	1.055510424
0.5	1.073759995	1.0793750	1.077760600
0.6	1.098536032	1.1030416	1.101146836
0.7	1.126653415	1.1256318	1.124562648
0.8	1.158112143	1.1455712	1.146901552
0.9	1.192912217	1.1612854	1.167057064
1.0	1.231053638	1.1712000	1.183922700

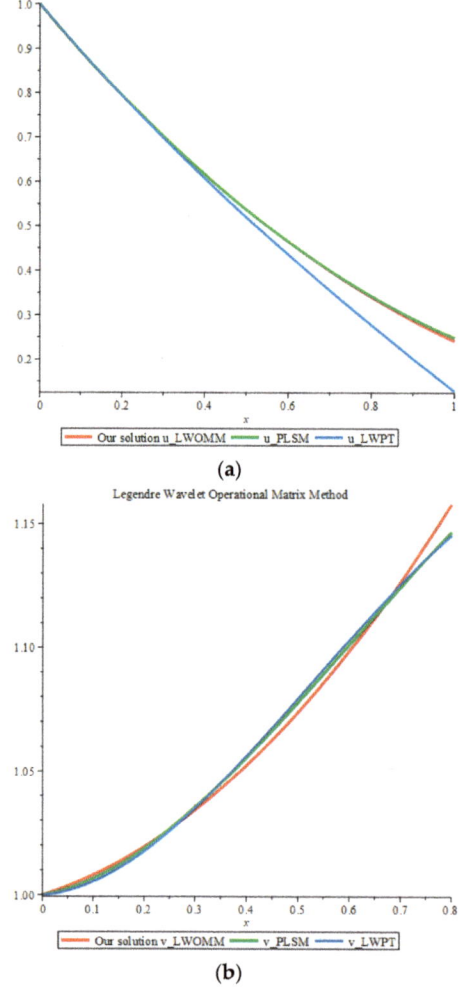

Figure 4. Comparison of our solutions to the approximate solution LWPT and the approximate solution PLSM when $\alpha = 0.98$ for Example 5: (**a**) Our solution $u(t)$; and (**b**) Our solution $v(t)$.